Fungal Biology

For further volumes:
http://www.springer.com/series/11224

Juan-Francisco Martín • Carlos García-Estrada
Susanne Zeilinger
Editors

Biosynthesis and Molecular Genetics of Fungal Secondary Metabolites

Springer

Editors
Juan-Francisco Martín, Ph.D.
Department of Molecular Biology
University of León
León, Spain

Carlos García-Estrada, D.V.M., Ph.D.
Parque Científico de León
Instituto de Biotecnología de León
(INBIOTEC)
León, Spain

Susanne Zeilinger
Institute of Chemical Engineering
Vienna University of Technology
Vienna, Austria

ISSN 2198-7777 ISSN 2198-7785 (electronic)
ISBN 978-1-4939-1190-5 ISBN 978-1-4939-1191-2 (eBook)
DOI 10.1007/978-1-4939-1191-2
Springer New York Heidelberg Dordrecht London

Library of Congress Control Number: 2014946216

Printed on acid-free paper

Springer is part of Springer Science+Business Media (www.springer.com)

Preface

The Wonderful World of Fungal Secondary Metabolites

There are thousands of fungal species in nature but only a handful of them, most of them ascomycetes, have been studied in detail. Studies on the model fungi *Neurospora crassa*, *Aspergillus nidulans*, *Aspergillus niger*, *Penicillium chrysogenum*, and others, in comparison with the yeast *Saccharomyces cerevisiae*, have provided the basic core of scientific knowledge on the vegetative metabolism and morphological differentiation of filamentous fungi. However, the biochemistry and molecular genetics of fungal secondary metabolites are less known due to their large diversity.

Some fungal products are extremely beneficial to combat tumors or bacterial and fungal infections, and others contribute to control cholesterol metabolism to improve human health. A large number of fungal metabolites, the mycotoxins, are highly toxic for humans and for the livestock. They also affect soil-dwelling worms or other organisms and, therefore, have a profound ecological interest. Finally other fungal metabolites provide the vivid colors (e.g., β[beta]-carotene, astaxanthin) of some fungi.

During the last decades, there has been an intense effort to elucidate the biosynthesis pathways of fungal secondary metabolites to characterize the genes that encode the biosynthetic enzymes and the regulatory mechanisms that control their expression. One interesting finding is that genes encoding fungal secondary metabolites are clustered together, as occurs also with the bacterial genes for secondary metabolites. This is in contrast to fungal primary metabolism genes, which are frequently scattered in the genome. However, in contrast to the bacterial gene clusters, most of the fungal secondary metabolite genes are expressed as monocistronic transcripts from individual promoters. This raises the question of possible unbalanced levels of the different mRNAs of the genes in a pathway and the need of temporal and spatial coordination of their expression. Furthermore, expression of the secondary metabolites in fungi is correlated with differentiation and with the formation of either sexual or asexual spores, including cleistothecia and other types of differentiated cells.

Fungal secondary metabolites are complex chemical molecules that are formed by a few basic mechanisms with multiple late modifications of their chemical structures. The basic mechanisms include enzymes such as non-ribosomal peptide synthetases (NRPSs), polyketide synthases (PKSs), terpene synthases and cyclases, and less known "condensing" enzymes that use as substrates a variety of activated precursors.

In this book we bring together 15 review articles by expert scientists on the best known secondary metabolites that serve as model of the different biosynthetic types of fungal secondary metabolites. Each chapter presents an updated review of the medical, agricultural, food and feed applications, and the ecological relevance of each compound.

Furthermore, we provide descriptions of the present status of knowledge on the molecular genetics and biosynthesis of each of these compounds. All together the expertise of the authors of those chapters provides an impressive overview of the actual knowledge of the world of fungal secondary metabolites.

León, Spain Juan-Francisco Martín

Contents

Contributors

Ikuro Abe, Ph.D. Graduate School of Pharmaceutical Sciences, The University of Tokyo, Tokyo, Japan

Jennifer Alcaino, Sc.D. Departamento de Ciencias Ecológicas, Facultad de Ciencias, Universidad de Chile, Santaigo, Chile

Javier Ávalos, Ph.D. Department of Genetics, Faculty of Biology, University of Seville, Sevilla, Spain

Katharina Walburga von Bargen, Dr. rer. nat. University Münster, Institute of Food Chemistry, Münster, Germany

Sandra Bloemendal, Ph.D. Christian Doppler Laboratory for Fungal Biotechnology, Lehrstuhl für Allgemeine und Molekulare Botanik, Ruhr-Universität Bochum, Bochum, Germany

Ditlev Egeskov Brodersen, Ph.D. Department of Molecular Biology and Genetics, Aarhus University, Aarhus C, Denmark

Marcelo Baeza, Ph.D. Departamento de Ciencias Ecológicas, Facultad de Ciencias, Universidad de Chile, Santiago, Chile

Marta Castrillo Department of Genetics, Faculty of Biology, University of Seville, Sevilla, Spain

Victor Cifuentes, Sc.D. Departamento de Ciencias Ecológicas, Facultad de Ciencias, Universidad de Chile, Santiago, Chile

Arnold L. Demain, Ph.D., M.S., B.S. Research Institute for Scientists Emeriti (R.I.S.E.), Drew University, Madison, NJ, USA

Violeta Díaz-Sánchez, Ph.D. Department of Genetics, Faculty of Biology, University of Seville, Sevilla, Spain

David Dietrich, B.Sc., Ph.D. Department of Chemistry, University of Alberta, Edmonton, AB, Canada

Anna Adele Fabbri, Ph.D. Department of Environmental Biology, Università Sapienza, Roma, Italy

Corrado Fanelli, Ph.D. Department of Environmental Biology, Università Sapienza, Roma, Italy

Patricia Romans Fuertes, M.Sc. Department of Biotechnology, Chemistry and Environmental Engineering, Aalborg University, Aalborg, Denmark

Carlos García-Estrada, D.V.M., Ph.D. INBIOTEC (Institute of Biotechnology of León), Parque Científico de León, León, Spain

Jorge García-Martínez Department of Genetics, Faculty of Biology, University of Seville, Sevilla, Spain

Henriette Giese, Ph.D. Department of Biotechnology, Chemistry and Environmental Engineering, Aalborg University, Aalborg, Denmark

Frederik Teilfeldt Hansen, Ph.D. Department of Molecular Biology and Genetics, Aarhus University, Aarhus C, Denmark

Hans-Ulrich Humpf, Dr. rer. nat. University of Münster, Institute of Food Chemistry, Münster, Germany

Karin Kleigrewe, Dr. rer. nat. University Münster, Institute of Food Chemistry, Münster, Germany

Michael Knudsen, Ph.D. Bioinformatics Research Center, Aarhus University, Aarhus, Denmark

Ulrich Kück, Ph.D. Christian Doppler Laboratory for Fungal Biotechnology, Lehrstuhl für Allgemeine und Molekulare Botanik, Ruhr-Universität Bochum, Bochum, Germany

Alfons Lawen, Dipl.-Chem., Dr. rer. nat. Department of Biochemistry and Molecular Biology, School of Biomedical Sciences, Monash University, Melbourne, VIC, Australia

T. Verne Lee, Ph.D. AgResearch Structural Biology Laboratory, School of Biological Sciences, University of Auckland, Auckland, New Zealand

M. Carmen Limón, Ph.D. Department of Genetics, Faculty of Biology, University of Seville, Sevilla, Spain

John E. Linz, M.S., Ph.D. Department of Food Science and Human Nutrition, Microbiology and Molecular Genetics, Michigan State University, East Lansing, MI, USA

Paloma Liras, Ph.D. Department of Molecular Biology, Microbiology Section, University of León, León, Spain

Juan-Francisco Martín, Ph.D. Department of Molecular Biology, Microbiology Section, University of León, León, Spain

Yudai Matsuda, M.Sc. Graduate School of Pharmaceutical Sciences, The University of Tokyo, Tokyo, Japan

Lisa Neubauer Institut für Biologie und Biotechnologie der Pflanzen, Westfälische Wilhelms Universität Münster, Münster, Germany

Eva-Maria Niehaus University of Münster, Institute of Biology und Biotechnology of Plants, Münster, Germany

Claus Olesen, M.Sc. Department of Molecular Biology and Genetics, Aarhus University, Aarhus C, Denmark

Christian Nørgaard Storm Pedersen, Ph.D. Bioinformatics Research Center, Aarhus University, Aarhus C, Denmark

Massimo Reverberi, Ph.D. Department of Environmental Biology, Università Sapienza, Roma, Italy

Ludmila V. Roze, Ph.D. Department of Plant Biology, Michigan State University, East Lansing, MI, USA

Macarena Ruger-Herreros, Pharm.D. Department of Genetics, Faculty of Biology, University of Seville, Sevilla, Spain

Teis Esben Sondergaard, Ph.D. Department of Biotechnology, Chemistry and Environmental Engineering, Aalborg University, Aalborg, Denmark

Jens Laurids Sørensen, Ph.D. Department of Biotechnology, Chemistry and Environmental Engineering, Aalborg University, Aalborg, Denmark

Lena Studt, Ph.D., Dr. rer. nat. Institute for Biology and Biotechnology of Plants, University of Münster, Münster, Germany

Bettina Tudzynski, Ph.D., Dr. habil. rer. nat. University of Münster, Institute of Biology and Biotechnology of Plants, Münster, NRW, Germany

Paul Tudzynski, Dr. rer. nat. Institut für Biologie und Biotechnologie der Pflanzen, Westfälische Wilhelms Universität Münster, Münster, Germany

John C. Vederas, B.Sc., Ph.D. Department of Chemistry, University of Alberta, Edmonton, AB, Canada

Tony Velkov, Ph.D. Department of Pharmaceutics, Monash University, Parkville, VIC, Australia

Josephine M. Wee, B.Sc. Department of Food Science and Human Nutrition, Center for Integrative Toxicology, Michigan State University, East Lansing, MI, USA

Chapter 1
Valuable Secondary Metabolites from Fungi

Arnold L. Demain

Introduction

A major contribution of microbes to the health and well-being of people began back in 1928, when Alexander Fleming discovered in a Petri dish seeded with *Staphylococcus aureus* that a compound produced by a mold killed the bacterium. The mold, *Penicillium notatum*, produced an active agent, which was named penicillin. Fleming's discovery began the microbial drug era. By using the same method, other naturally occurring substances, like chloramphenicol and streptomycin, were later isolated from bacterial fermentations. Naturally occurring antibiotics are produced by fermentation, an old technique that can be traced back almost 8,000 years, initially for beer and wine production, and recorded in the written history of ancient Egypt and Mesopotamia. During the last 4,000 years, *Penicillium roqueforti* has been utilized for cheese production and for the past 3,000 years, soy sauce in Asia and bread in Egypt represented examples of traditional fermentations [1].

Natural products (NPs) with high commercial value can be produced via primary or secondary metabolism. The present review deals with secondary metabolites. Due to technical improvements in screening programs and separation and isolation techniques, the number of natural compounds discovered exceeds one million [2]. Among them, 50–60 % are produced by plants (alkaloids, flavonoids, terpenoids, steroids, carbohydrates, etc.) and 5 % of these plant products have a microbial origin. From all the reported natural products, about 20–25 % show biological activity and of these, approximately 10 % have been obtained from microbes. Microorganisms produce many compounds with biological activity. From the 22,500 biologically active compounds so far obtained from microbes, about 40 % are produced by fungi [2, 3]. The role of fungi in the production of antibiotics and other drugs for treatment of noninfective diseases has been dramatic [4].

A.L. Demain, Ph.D., M.S., B.S. (✉)
Research Institute for Scientists Emeriti (R.I.S.E.), Drew University, Madison, NJ, USA
e-mail: ademain@drew.edu

J.-F. Martín et al. (eds.), *Biosynthesis and Molecular Genetics of Fungal Secondary Metabolites*, Fungal Biology, DOI 10.1007/978-1-4939-1191-2_1,

Biosynthetic genes are present in clusters coding for large, multidomain, and multi-modular enzymes such as polyketide synthases, prenyltransferases, nonribosomal peptide synthases, and terpene cyclases. Genes adjacent to the biosynthetic gene clusters encode regulatory proteins, oxidases, hydroxylases, and transporters. Aspergilli usually contain 30–40 secondary metabolite gene clusters. Strategies to activate silent genes have been reviewed by Brakhage and Schroekh [3].

Currently, with less than 1 % of the microbial world having been cultured, there have been significant advances in microbial techniques for growth of uncultured organisms as a potential source of new chemicals [5]. Furthermore, metagenomics—i.e., the extraction of DNA from soil, plants, and marine habitats and its incorporation into known organisms—is allowing access to a vast untapped reservoir of genetic and metabolic diversity [6, 7]. The potential for discovery of new secondary metabolites with beneficial use for humans is great. A method to predict secondary metabolite gene clusters in filamentous fungi has recently been devised [8].

Microbes normally produce secondary metabolites in only tiny amounts due to the evolution of regulatory mechanisms that limit production to a low level. Such a level is probably enough to allow the organism to compete with other organisms and/or coexist with other living species in nature. The industrial microbiologist, however, desires a strain that will overproduce the molecule of interest. Development of higher-producing strains involves mutagenesis and, more recently, recombinant DNA technologies [9]. Although some metabolites of interest can be made by plants or animals, or by chemical synthesis, the recombinant microbe is usually the "creature of choice." Thousandfold increases in production of small molecules have been obtained by mutagenesis and/or genetic engineering. Other important parts of industrial production include creating a proper nutritional environment for the organism to grow and produce its product, and the avoidance of negative effects such as inhibition and/or repression by carbon sources, nitrogen sources, phosphorus sources, metals, and the final product itself. Avoidance of enzyme decay is also desired [4, 10].

Applications of Microbial Natural Products

Over the years, the pharmaceutical industry extended their antibiotic screening programs to other areas [11, 12]. Since microorganisms are such a prolific source of structurally diverse bioactive metabolites, the industry extended their screening programs in order to look for microbes with activity in other disease areas. As a result of this move, some of the most important products of the pharmaceutical industry were obtained. For example, the immunosuppressants have revolutionized medicine by facilitating organ transplantation [13]. Other products include antitumor drugs, hypocholesterolemic drugs, enzyme inhibitors, gastrointestinal motor stimulator agents, ruminant growth stimulants, insecticides, herbicides, antiparasitics versus coccidia and helminths, and other pharmacological activities. Catalyzed by the use of simple enzyme assays for screening prior to testing in intact animals or in the field, further applications are emerging in various areas of pharmacology and agriculture.

Antibiotics

Of the 12,000 antibiotics known in 1955, filamentous fungi produced 22 % [14, 15]. The beta-lactams are the most important class of antibiotics in terms of use. They constitute a major part of the antibiotic market. Included are the penicillins, cephalosporins, clavulanic acid, and the carbapenems. Of these, fungi are responsible for production of penicillins and cephalosporins. The natural penicillin G and the biosynthetic penicillin V had a market of $4.4 billion by the late 1990s. Major markets also included semisynthetic penicillins and cephalosporins with a market of $11 billion. In 2006, the market for cephalosporins amounted to $9.4 billion and that for penicillins was $6.7 billion. By 2003, production of all beta-lactams had reached over 60,000 t. The titer of penicillin is over 100 g L^{-1} and that for cephalosporin C is about 35 g L^{-1} [16, 17]. Recovery yields are more than 90 %. There have been more than 15,000 molecules based on penicillin that have been made by semisynthesis or by total synthesis. By the mid 1990s, 160 antibiotics and their derivatives were already on the market [15, 18]. The market in 2000 was $35 billion. Despite these impressive figures, more antibiotics are needed to combat evolving pathogens, naturally resistant microbes, and bacteria and fungi that have developed resistance to current antibiotics. A new and approved cephalosporin is ceftobiprole, which is active against methicillin-resistant *S. aureus* (MRSA) and is not hydrolyzed by a number of beta-lactamases from Gram-positive bacteria [19]. Another antibiotic of note is cerulenin, an antifungal agent produced by *Acremonium caerelens*. It was the first inhibitor of fatty acid biosynthesis discovered [20]. It alkylates and inactivates the active site nucleophylic cysteine of the ketosynthase enzyme of fatty acid synthetase by epoxide ring opening. Other properties that are desired in new antibiotics are improved pharmacological properties, ability to combat viruses and parasites, and improved potency and safety.

Pharmacological Agents

Years ago, noninfectious diseases were mainly treated with synthetic compounds. Despite testing thousands of synthetic chemicals, only a handful of promising structures was obtained. As new synthetic lead compounds became extremely difficult to find, microbial products came into play. Poor or toxic antibiotics produced by fungi such as cyclosporin A or mycotoxins such as ergot alkaloids, gibberellins, zearelanone were then successfully applied in medicine and agriculture. This led to the use of fungal products as immunosuppressive agents, hypocholesterolemic drugs, antitumor agents, and for other applications.

Hypocholesterolemic Agents

Only about 30 % of cholesterol in humans comes from the diet. The rest is synthesized by the body, predominantly in the liver. Many people cannot control their level of cholesterol at a healthy level by diet alone and require hypocholesterolemic

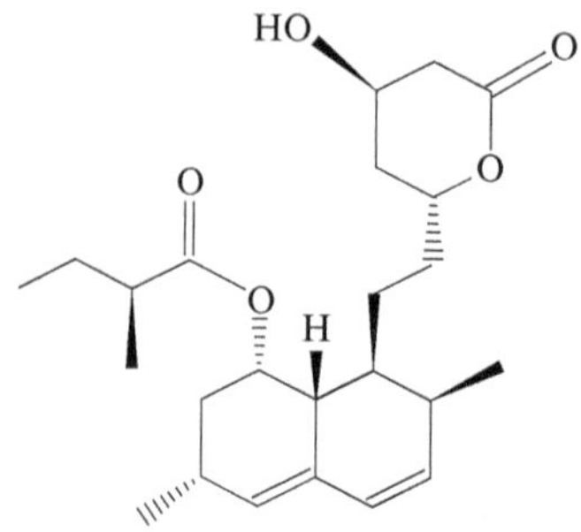

Fig. 1.1 Chemical structure of lovastatin

agents. High blood cholesterol leads to atherosclerosis, which is a chronic, progressive disease characterized by continuous accumulation of atheromatous plaque within the arterial wall, causing stenosis and ischemia. Atherosclerosis is a leading cause of human death. The last two decades have witnessed the introduction of a variety of anti-atherosclerotic therapies. The statins form a class of hypolipidemic drugs, formed as secondary metabolites by fungi, and used to lower cholesterol by inhibiting the rate-limiting enzyme of the mevalonate pathway of cholesterol biosynthesis; i.e., 3-hydroxymethyl glutaryl-CoA (HMG-CoA) reductase. Inhibition of this enzyme in the liver stimulates low-density lipoprotein (LDL) receptors, resulting in an increased clearance of LDL from the bloodstream and a decrease in blood cholesterol levels. They can reduce total plasma cholesterol by 20–40 %. Through their cholesterol-lowering effect, they reduce risk of cardiovascular disease, prevent stroke, and reduce development of peripheral vascular disease [21].

Currently, there are a number of statins in clinical use. They reached an annual market of nearly $30 billion before one became a generic pharmaceutical. The history of the statins has been described by Akira Endo, the discoverer of the first statin, compactin (mevastatin; ML-236B) [22]. This first member of the group was isolated as an antibiotic product of *Penicillium brevicompactum* [23]. At about the same time, it was found by Endo and coworkers as a cholesterolemic product of *Penicillium citrinum* [24]. Although compactin was not of commercial importance, its derivatives achieved strong medical and commercial success. Lovastatin (monacolin K; mevinolin; Mevacor™), was isolated in broths of *Monascus rubra* and *Aspergillus terreus* [25, 26]. Lovastatin, developed by Merck & Co. and approved by the US Food and Drug Administration (FDA) in 1987, was the first commercially marketed statin. In its chemical structure, lovastatin has a hexahydronaphthalene skeleton substituted with a *p*-hydroxy-lactone moiety (Fig. 1.1).

A semisynthetic derivative of lovastatin is Zocor® (simvastatin), one of the main hypocholesterolemic drugs, selling for $7 billion per year before becoming generic. An unexpected effect of simvastatin is its beneficial activity on pulmonary artery hypertension [27]. Another surprising effect is its antiviral activity [28]. Simvastatin is active against RNA viruses and acts as monotherapy against chronic hepatitis C virus in humans. It has been shown to act in vitro against hepatitis B virus (HBV). This virus infects 400 million people and is the most common infectious disease agent in the world. The virus causes hepatocellular cancer, which is the leading cause of cancer death. Nucleotide analogs (lamivudine, adefovir, tenofovir, entecavir, telbuvidine) were approved for HBV infections but they only work on 11–17 % of patients. Simvastatin is synergistic with these nucleotide analogs.

Statins also have antithrombotic, anti-inflammatory, and antioxidant effects [29]. They have shown activity against multiple sclerosis, artherosclerosis, Alzheimer's Disease, and ischemic stroke [30, 31]. However, these applications have not yet been approved since more clinical studies are required. The neuroprotective effect of statins has been demonstrated in an in vitro model of Alzheimer's disease using primary cultures of cortical neurons [32]. The effect did not appear to be due to cholesterol lowering but rather to reduction in formation of isoprenyl intermediates of the cholesterol biosynthetic process. Lovastatin has shown antitumor activity against embryonal carcinoma and neuroblastoma cells [33].

Although simvastatin is usually made from lovastatin chemically in a multistep process, an enzymatic/bioconversion process using recombinant *Escherichia coli* has been developed [34]. Another statin, pravastatin ($3.6 billion in sales per year), is made via different biotransformation processes from compactin by *Streptomyces carbophilus* [35] and *Actinomadura* sp. [36]. Other genera involved in production of statins are *Doratomyces, Eupenicillium, Gymnoascus, Hypomyces, Paecilomyces, Phoma, Trichoderma,* and *Pleurotus* [37]. A synthetic compound, modeled from the structure of the natural statins, is Lipitor®, which was the leading drug of the entire pharmaceutical industry in terms of market (about $14 billion per year) for many years.

Anticancer Drugs

More than 12 million new cases of cancer were diagnosed in the world in 2008; 6.6 million cases were in men and 6.0 million in women, resulting in 7.6 million cancer-related deaths. The tumor types with the highest incidence were lung (12.7 %), breast (10.9 %), and colorectal (9.8 %). Some of the anticancer drugs in clinical use are secondary metabolites derived from plants and fungi. Among the approved products are taxol and camptothecin.

Taxol (paclitaxel) was first isolated from the Pacific yew tree, *Taxus brevifolia* [38] and later found to be a fungal secondary metabolite [39]. It is a steroidal alkaloid diterpene alkaloid that has a characteristic *N*-benzoylphenyl isoserine side chain and a tetracycline ring (Fig. 1.2). It inhibits rapidly dividing mammalian cancer cells by promoting tubulin polymerization and interfering with normal microtubule breakdown during cell division. The benzoyl group of the molecule is particularly crucial for maintaining the strong bioactivity of taxol. The drug also inhibits several fungi (species of *Pythium, Phytophthora, Aphanomyces*) by the same mechanism. In 1992, taxol was approved for refractory ovarian cancer and today is used against breast cancer and advanced forms of Kaposi's sarcoma [40]. A formulation in which paclitaxel is bound to albumin is sold under the trademark Abraxane®. Taxol sales amounted to $1.6 billion in 2006 for Bristol Myers-Squibb, representing 10 % of the company's pharmaceutical sales and its third largest selling product. It has reached $3.7 billion annual sales in international markets.

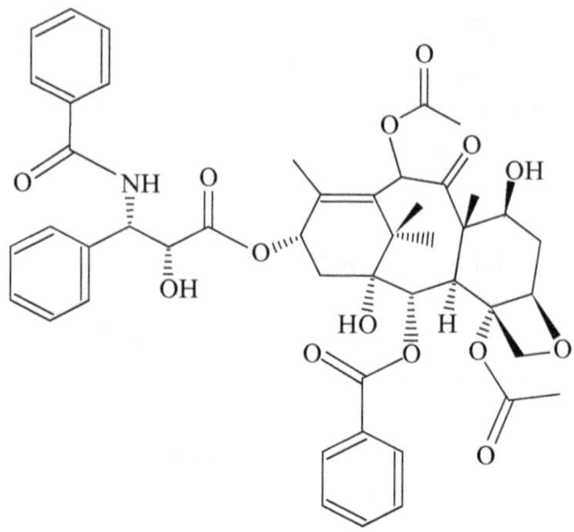

Fig. 1.2 Chemical structure of taxol. The benzoyl group is located in the left side of the structure

Although synthetic methods for taxol production have been tried, the chemical molecular structure is so complex that commercial synthetic production is unfeasible. Currently, Italy, the UK, the Netherlands, and other Western countries are engaged in the production of taxol by plant cell fermentation technology. Taxol production by plant cell culture of *Taxus* sp. was reported to be at 67 mg L^{-1} [41]. However, addition of methyl jasmonate, a plant signal transducer, increased production to 110 mg L^{-1}.

As stated previously, taxol has also been found to be a fungal metabolite [39, 42]. Fungi such as *Taxomyces andreanae, Pestalotiopsis microspora, Tubercularia* sp., *Phyllosticta citricarpa, Nodulisporium sylviforme, Colletotrichum gloeosporoides, Colletotrichum annutum, Fusarium maire*, and *Pestalotiopsis versicolor* produce it [39, 43–49]. The endophyte *F. maire* produces 225 μg L^{-1}. Production by *P. citricarpa* amounted to 265 μg L^{-1} [50]. Production was reported at 417 μg L^{-1} by submerged fermentation with an engineered strain of the endophytic fungus *Ozonium* sp. (EFY-21). The transformed strain overproduced the rate-limiting enzyme of taxol biosynthesis, taxadiene synthase [51]. Another endophytic fungus, *Phoma betae*, isolated from the medicinal tree *Ginkgo biloba*, produced taxol at 795 μg L^{-1} [52]. *Cladosporium cladosporoides*, an endophyte of the *Taxus media* tree, produced 800 μg L^{-1} of taxol [53]. *Metarhizium anisopiliae* H-27, isolated from the tree *Taxus chinensis*, yielded 846 μg L^{-1} [54]. Although a review of taxol production by endophytic fungi indicated that strain improvement had resulted in levels of only 0.4–1.0 mg L^{-1} [55], it was reported that another fungus, *Alternaria alternate* var. *monosporus*, from the bark of *Taxus yunanensis*, after ultraviolet and nitrosoguanidine mutagenesis, could produce taxol at 227 mg L^{-1} [56]. The endophytic fungus *P. versicolor*, from the plant *Taxus cuspidata*, produced 478 μg L^{-1} [44] and *C. annutum* from *Capsicum annuum* made 687 μg L^{-1} [45].

Another important antitumor agent is camptothecin, a modified monoterpene indole alkaloid produced by certain plants (angiosperms) and by the endophytic fungus, *Entrophospora infrequens*. The fungus was isolated from the plant

Nathapodytes foetida [38]. In view of the low concentration of camptothecin in tree roots and poor yield from chemical synthesis, the fungal fermentation is very promising for industrial production of camptothecin. It is used for recurrent colon cancer and has unusual activity against lung, ovarian, and uterine cancer [57]. Colon cancer is the second-leading cause of cancer fatalities in the USA and the third most common cancer among US citizens. Camptothecin is known commercially as Camptosar and Campto and achieved sales of $1 billion in 2003 [58]. Camptothecin's water-soluble derivatives irinotecan and topotecan have been approved and are used clinically. Metastatic colorectal cancer is treated by irinotecan whereas topotecan has use for ovarian cancer, cervical cancer, and small-cell lung cancer. A review of the activities of camptothecin and its many small and macromolecular derivatives has been published by Venditto and Simanek [59].

The cellular target of camptothecin is type I DNA topoisomerase. When patients become resistant to irinotecan, its use can be prolonged by combining it with the monoclonal antibody Erbitux (Cetuximab). Erbitux blocks a protein that stimulates tumor growth and the combination helps metastatic colorectal cancer patients expressing epidermal growth factor receptor (EGFR). This protein is expressed in 80 % of advanced metastatic colorectal cancers. The drug combination reduces invasion of normal tissues by tumor cells and the spread of tumors to new areas.

Angiogenesis, the recruitment of new blood vessels, is necessary for tumors to obtain oxygen and nutrients. Tumors actively secrete growth factors that trigger angiogenesis. Anti-angiogenesis therapy is now known as one of four cancer treatments; the other three are surgery, radiotherapy, and chemotherapy. By the end of 2007, 23 anti-angiogenesis drugs were in Phase III clinical trials and more than 30 were in Phase II. Fumagillin, a secondary metabolite of *Aspergillus fumigatus*, was one of the first agents found to act as an anti-angiogenesis compound. Next to come along were its oxidation product ovalacin and the fumagillin analog TNP-470 (=AGM-1470). TNP-470 binds to and inhibits type 2 methionine aminopeptidase. This interferes with amino-terminal processing of methionine, which may lead to inactivation of enzymes essential for growth of endothelial cells. In animal models, TNP-470 effectively treated many types of tumors and metastases.

Inhibitors of farnesyltransferase (FTIs) have anticancer activity because farnesylation is required for activation of Ras, a necessary step in cancer progression. They also induce apoptosis in cancer cells. The fungus *Phoma* sp. FL-415 produces an FTI known as TAN-1813 [60].

Immunosuppressant Drugs

An individual's immune system is capable of distinguishing between native and foreign antigens and to mount a response only against the latter. Suppressor cells are critical in the regulation of the normal immune response. The suppression of the immune response, either by drugs or radiation, in order to prevent the rejection of grafts or transplants or to control autoimmune diseases, is called immunosuppression.

Microbial compounds capable of suppressing the immune response have been discovered as fungal secondary metabolites. Cyclosporin A was originally discovered in the 1970s as a narrow-spectrum antifungal peptide produced by the mold, *Tolypocladium nivenum* (previously *Tolupocladium inflatum*) in an aerobic fermentation [61]. Cyclosporins are a family of neutral, highly lipophilic, cyclic undecapeptides containing some unusual amino acids, synthesized by a nonribosomal peptide synthetase, cyclosporin synthetase. Discovery of the immunosuppressive activity of this secondary metabolite led to use in heart, liver, and kidney transplants and to the overwhelming success of the organ transplant field [62]. Cyclosporin was approved for use in 1983. It is thought to bind to the cytosolic protein cyclophilin (immunophilin) of immunocompetent lymphocytes, especially *T*-lymphocytes. This complex of cyclosporin and cyclophilin inhibits calcineurin, which under normal circumstances is responsible for activating the transcription of interleukin-2. It also inhibits lymphokine production and interleukin release and therefore leads to a reduced function of effector *T*-cells. Annual world sales of cyclosporin A are approximately $2 billion. Cyclosporin A also has activity against corona viruses [63].

Studies on the mode of action of cyclosporin, and the later-developed immunosuppressants from actinomycetes, such as sirolimus (a rapamycin) and FK-506 (tacrolimus), have markedly expanded current knowledge of *T*-cell activation and proliferation. These agents act by interacting with an intracellular protein (an immunophilin), thus forming a novel complex that selectively disrupts the signal transduction events of lymphocyte activation.

Their targets are inhibitors of signal transduction cascades in microbes and humans. In humans, the signal transduction pathway is required for activation of *T* cells.

A very old broad-spectrum antibiotic, actually the first antibiotic ever discovered, is mycophenolic acid, which has an interesting history. Bartolomeo Gosio (1863–1944), an Italian physician, discovered the compound in 1893 [64]. Gosio isolated a fungus from spoiled corn, which he named *Penicillium glaucum*, which was later reclassified as *P. brevicompactum*. He isolated crystals of the compound from culture filtrates in 1896 and found it to inhibit growth of *Bacillus anthracis*. This was the first time an antibiotic had been crystallized and the first time that a pure compound had ever been shown to have antibiotic activity. The work was forgotten but fortunately the compound was rediscovered by Alsberg and Black [65] and given the name mycophenolic acid. They used a strain originally isolated from spoiled corn in Italy called *Penicillium stoloniferum*, a synonym of *P. brevicompactum*. The chemical structure was elucidated many years later (1952) by Birkinshaw and coworkers [66] in England. Mycophenolic acid has antibacterial, antifungal, antiviral, antitumor, antipsoriasis, and immunosuppressive activities. Its antiviral activity is exerted against yellow fever, dengue virus, and Japanese encephalitis virus [67]. It was never commercialized as an antibiotic because of its toxicity, but its 2-morpholinoethylester was approved as a new immunosuppressant for kidney transplantation in 1995 and for heart transplants in 1998 [68]. The ester is called mycophenolate mofetil (CellCept) and is a prodrug that is hydrolyzed to mycophenolic acid in the body. It is sometimes used along with cyclosporin in kidney, liver, and heart transplants. Mycophenolic acid also appears to have anti-angiogenic activity [69].

Applications of Mycotoxins

Fungi produce poisons called mycotoxins, which, strangely enough, have been harnessed as medically useful agents. These agents (e.g., ergot alkaloids) caused fatal poisoning of humans and animals (ergotism) for centuries by consumption of bread made from grain contaminated with species of the fungus *Claviceps*. However, mycotoxins later were found useful for angina pectoris, hypertonia, serotonin-related disturbances, inhibition of protein release in agalactorrhea, reduction in bleeding after childbirth, and prevention of implantation in early pregnancy [70, 71]. Their physiological activities include inhibition of action of adrenalin, noradrenalin, and serotonin, as well as the contraction of smooth muscles of the uterus. Antibiotic activity is also possessed by some ergot alkaloids.

Members of the genus *Gibberella* produce zearelanone and gibberellins. Zearelanone is an estrogen made by *Gibbberella zeae* (syn. *Fusarium graminearum*) [72]. Its reduced derivative zeranol is used as an anabolic agent in sheep and cattle, which increases growth and feed efficiency. Giberellic acid, a member of the mycotoxin group known as gibberellins, is a product of *Gibberella fujikori* and causes "foolish rice seedling" disease in rice [73]. Gibberellins are employed to speed up the malting of barley, improve the quality of malt, increase the yield of vegetables, and cut the time in half for obtaining lettuce and sugar beet seed crops. They are isoprenoid growth regulators, controlling flowering, seed germination, and stem elongation [74]. More than 25 t are produced annually with a market of over $100 billion.

Inhibitors of Enzyme Activity

Enzyme inhibitors have received increased attention as useful tools, not only for the study of enzyme structures and reaction mechanisms, but also for potential utilization in medicine and agriculture. Several enzyme inhibitors with various industrial uses have been isolated from microbes [75]. Among the most important are the statins and hypocholesterolemic drugs discussed previously. Fungal products are also used as enzyme inhibitors against cancer, diabetes, poisoning, and Alzheimer's disease. The enzymes inhibited include acetylcholinesterase, protein kinase, tyrosine kinase, glycosidases, and others [76].

Pigments

Since 800 AD, *Monascus purpurea* has been grown on rice to prepare koji or Angkak (red rice), which is used as a traditional Chinese food and medicine [77]. Monascorubramine and rubropunctatin are water-soluble red pigments formed upon reaction of the orange pigments monascorubrin and rubropunctatin with amino acids in fermentation media [78]. The fungus is used to prepare red rice, wine, soybean cheese, meat, and fish. It is authorized in Japan and China for food use. There are 54 known *Monascus* pigments. They have an amazing number of activities:

antimicrobial, anticancer, anti-mutagenesis, antidiabetes, anti-obesity, anti-inflammatory, cholesterol-lowering, immunosuppressive, and hypotensive [79, 80]. Nutritional control of the formation of the red pigments has been described in a series of publications by Lin and Demain [81–84].

Phaffia rhodozyma (*Xanthophyllomyces dendrorhous*) is a heterobasidiomycetous yeast that has become the most important microbial source for preparation of the carotenoid astaxanthin [85, 86]. This oxygenated carotenoid pigment is used in the feed, food, and cosmetic industries. It is responsible for the orange to pink color of salmonid flesh and the reddish color of boiled crustacean shells. Feeding of pen-reared salmonids with a diet containing this yeast induces pigmentation of the white muscle [87]. It is a very good antioxidant, 10 times more active than beta-carotene and 100 times more than alpha-tocopherol. It is the second most important carotenoid. Astaxanthin enhances the immune system, and protects skin from radiation injury and cancer. It can be produced synthetically as hydroxyl-astaxanthin from petrochemicals with a selling price of $2,500 per kg. However, the natural product is favored because the synthetic product is a mixture of stereoisomers. Natural astaxanthin is more stable than the synthetic version and more bioavailable. The natural product is present in algae and fish as mono- and di-esters of fatty acids. However, it is difficult to hydrolyze the esters from algae, which limits its usage to trout and salmon. The yeast product is better since it is the 97 % free, non-esterified (3R, 3'R) stereoisomer. The natural product is more expensive ($7,000 per kg) than synthetic astaxanthin ($2,500 per kg). The astaxanthin market was $219 million in 2007 with 97 % being synthetic. Most of the production processes with the yeast yield levels of astaxanthin lower than 100 mg L^{-1}. However, white light improved production to 420 mg L^{-1} [88] and mutant strain UBv-AX2 can make 580 mg L^{-1} [89].

Sweeteners

Thaumatin, a protein produced by the plant *Thaumatococcus danielli*, can also be produced by *P. roqueforti and Aspergillus niger* var *awamori* [90]. Thaumatin is intensely sweet (i.e., 3,000 times sweeter than sucrose) and is approved as a food-grade ingredient. Production by *A. niger* var *awamori* was improved from 2 mg L^{-1} up to 14 mg L^{-1} by increasing gene dosage and use of a strong promoter [91]. The sweetener xylitol, normally produced by *Pichia stipitis*, can be produced by recombinant *Saccharomyces cerevisiae* in higher concentrations by transforming the *XYL1* gene of *P. stipitis* into *S. cerevisiae*. The gene encodes a xylose reductase [92].

Conclusion

Microorganisms have greatly contributed for about 85 years to the development of medicine and agriculture. However, due to different situations, pathogenic microbes have become resistant to many antibiotics creating a dangerous situation and therefore the need for new antibiotics is imperative. Unfortunately, most of the large pharmaceutical companies have abandoned the search for new antimicrobial compounds. Due to economics, they have concluded that drugs directed against chronic diseases offer a better revenue stream than do antimicrobial agents, for which the length of treatment is short and government restriction is likely. Some small pharmaceutical and biotechnology companies are still developing antibiotics but most depend on venture capital rather than sales income, and with the present regulations, face huge barriers to enter into the market. These barriers were raised with the best intentions of ensuring public safety but they are having the opposite effect; i.e., termination of antibiotic development while resistance continues to increase [93]. However, there are some new bright possibilities. One of the more promising is the utilization of uncultivated microorganisms. Considering that 99 % of bacteria and 95 % of fungi have not yet been cultivated in the laboratory, efforts to find means to grow such uncultured microorganisms is proceeding and succeeding [5]. Furthermore, researchers are now extracting bacterial DNA from soil samples, cloning large fragments into, for example, bacterial artificial chromosomes, expressing them in a host bacterium and screening the library for new antibiotics. This metagenomic effort could open up the exciting possibility of a large untapped pool from which new natural products could be discovered [94]. Another exciting possibility is that of genome mining [95]. In addition to these relatively new techniques, chemical and biological modification of old antibiotics could still supply new and powerful drugs. These comments also apply to non-antibiotics such as antitumor agents and other microbial products. In addition, natural products must continue to be tested for desirable therapeutic activities. I believe that significant progress in identifying new antibiotics, oncology therapeutics, and other useful medicines will be made, probably not by the big pharmaceutical companies, but by biotechnology companies and small research groups from institutes and universities.

References

1. Hölker U, Höfer M, Lenz J. Biotechnological advantages of laboratory-scale solid-state fermentation with fungi. Appl Microbiol Biotechnol. 2004;64:175–86.
2. Berdy J. Bioactive microbial metabolites. A personal view. J Antibiot. 2005;58:1–26.
3. Brakhage AA, Schroekh V. Fungal secondary metabolites. Strategies to activate silent gene clusters. Fungal Genet Biol. 2011;48:15–22.
4. Demain AL, Velasco J, Adrio JL. Industrial mycology: past, present, and future. In: An Z, editor. Handbook of industrial mycology. New York: Marcel Dekker; 2004. p. 1–25.
5. Kaeberlein T, Lewis K, Epstein SS. Isolating "uncultivable" microorganisms in pure culture in a simulated natural environment. Science. 2002;296:1127–9.
6. Colwell RR. Fulfilling the promise of biotechnology. Biotechnol Adv. 2002;20:215–28.
7. Gaudilliere B, Bernardelli P, Berna P. To market, to market-2000. In: Doherty AM, editor. Annual reports in medicinal chemistry, vol. 36. Amsterdam: Academic; 2001. p. 293–318. Chapter 28.
8. Anderson MR, Nielsen JB, Klitgaard A, Petersen LM, Zachariasen M, Hansen TJ, et al. Accurate prediction of secondary metabolite gene clusters in filamentous fungi. Proc Natl Acad Sci U S A. 2013;110:24–5.

9. Adrio JL, Demain AL. Fungal biotechnology. Int Microbiol. 2003;6:191–9.
10. Brakhage A. Regulation of fungal secondary metabolism. Nat Rev Microbiol. 2013; 11:21–32.
11. Cardenas ME, Sanfridson A, Cutler NS, Heitman J. Signal-transduction cascades as targets for therapeutic intervention by natural products. Trends Biotechnol. 1998;16:427–33.
12. Kremer L, Douglas JD, Baulard AR, Morehouse C, Guy MR. Thiolactomycin and related analogues as novel anti-mycobacterial agents targeting KasA and KasB condensing enzymes in *Mycobacterium tuberculosis*. J Biol Chem. 2000;275:16857–64.
13. Verdine GL. The combinatorial chemistry of nature. Nature. 1996;384:11–3.
14. Berdy J. Are actinomycetes exhausted as a source of secondary metabolites? In: Proceedings of 9th international symposium on the biology of actinomycetes, Part 1. New York: Allerton; 1995. pp. 3–23.
15. Strohl WR. Industrial antibiotics: today and the future. In: Strohl WR, editor. Biotechnology of antibiotics. New York: Marcel Dekker; 1997. p. 1–47.
16. Masurekar P. Nutritional and engineering aspects of microbial process development. Prog Drug Res. 2008;65:292–328.
17. Yang Y, Xia J, Li J, Chu J, Li L, Wang Y. A novel impeller configuration to improve fungal physiology performance and energy conservation for cephalosporin C production. J Biotechnol. 2012;161:250–6.
18. Brown KS. Pharmaceutical and biotech firms taking on drug-resistant microbes. The Scientist. 1996;10(1):8–9.
19. Shang S, Shanley CA, Caraway ML, Orme EA, Henao-Tamayo M, Hascall-Dove L, et al. Activities of TMC207, rifampin, and pyrazinamide against *Mycobacterium tuberculosis* infection in guinea pigs. Antimicrob Agents Chemother. 2010;54:956–9.
20. Vance D, Goldberg I, Mitsuhashi O, Bloch K, Omura S, Nomura S. Inhibition of fatty acid synthetases by the antibiotic cerulenin. Biochem Biophys Res Commun. 1972;48:649–56.
21. Nicholls SJ, Tuzcu EM, Sipahi I, Grasso AW, Schoenhagen P, Hu T, et al. Statins, high-density lipoprotein cholesterol, and regression of coronary atherosclerosis. JAMA. 2007;297: 499–508.
22. Endo A. A historical perspective on the discovery of statins. Proc Jpn Acad Ser B. 2010;86:484–92.
23. Brown AG, Smale TC, King TJ, Hasenkamp R, Thompson RH. Crystal and molecular structure of compactin: A new antifungal metabolite from *Penicillium brevicompactum*. J Chem Soc Perkin Trans. 1976;1:1165–70.
24. Endo A, Kuroda M, Tsujita Y. ML-236B and ML-236C, new inhibitors of cholesterolgenesis produced by *Penicillium citrinin*. J Antibiot. 1976;29:1346–8.
25. Alberts AW, Chen J, Kuron G, Hunt V, Huff J, Hoffman C, Rothrock J, Lopez M, Joshua H, Harris E, et al. Mevinolin, a highly potent competitive inhibitor of hydroxylmethylglutaryl-coenzyme A reductase and a cholesterol-lowering agent. Proc Natl Acad Sci U S A. 1980;77:3957–61.
26. Endo A, Monacolin K. A new hypocholesterolemic agent produced by Monascus species. J Antibiot. 1979;32:852–4.
27. Liu Z-Q, Liu B, Yu L, Wang X-Q, Wang J, Liu H-M. Simvastatin has beneficial effect on pulmonary artery hypertension by inhibiting NF-kB expression. Mol Cell Biochem. 2011;354:77–82.
28. Bader T, Fazili J, Madhoun M, Aston C, Hughes D, Rizvi S, et al. Fluvastatin inhibits hepatitis C replication in humans. Am J Gastroenterol. 2008;103:1383–9.
29. Makris GC, Geroulakos G, Makris MC, Mikhailidis D, Falagas ME. The pleiotropic effects of statins and omega-3 fatty acids against sepsis: a new perspective. Expert Opin Investig Drugs. 2010;19:809–14.
30. Menge T, Hartung H-P, Stueve O. Statins—a cure-all for the brain? Nat Rev Neurosci. 2005;6:325–31.
31. Puttananjaiah M-KH, Dhale MA, Gaonkar V, Keni S. Statins: 3-Hydroxy-3-methylglutaryl-CoA (HMG-CoA) reductase inhibitors demonstrate anti-atherosclerotic character due to their antioxidant capacity. Appl Biochem Biotechnol. 2011;163:215–22.

32. Fonseca ACRG, Proenca T, Resende R, Oliviera CR, Pereira CMF. Neuroprotective effect of statins in an in vitro model of Alzheimer's disease. J Alzheimers Dis. 2009;17:503–17.
33. Arnold DE, Gagne C, Niknejad N, McBurney MW, Dimitroulakos J. Lovastatin induces neuronal differentiation and apoptosis of embryonal carcinoma and neuroblastoma cells: enhanced differentiation and apoptosis in combination with dbcAMP. Mol Cell Biochem. 2010;345: 1–11.
34. Xie X, Tang Y. Efficient synthesis of simvastatin by use of whole-cell biocatalysts. Appl Environ Microbiol. 2007;73:2054–60.
35. Serizawa N, Matsuoka T. A two-component-type cytochrome P-450 monooxygenase system in a prokaryote that catalyzes hydroxylation of ML-236B to pravastatin, a tissue-selective inhibitor of 3-hydroxy-3-methylglutaryl coenzyme A reductase. Biochim Biophys Acta. 1991;1084:35–40.
36. Peng Y, Demain AL. A new hydroxylase system in *Actinomadura* sp. cells converting compactin to pravastatin. J Ind Microbiol Biotechnol. 1998;20:373–5.
37. Alarcon J, Aguila S, Arancibia-Avila P, Fuentes O, Zamorano-Ponce E, Hernandez M. Production and purification of statins from *Pleurotus ostreatus* (Basidiomycetes) strains. Z Naturforsch C. 2003;58:62–4.
38. Wall ME, Wani MC. Camptothecin and taxol: from discovery to clinic. J Ethnopharmacol. 1996;51:239–54.
39. Stierle A, Strobel G, Stierle D. Taxol and taxane production by *Taxomyces andreanae*, an endophytic fungus of Pacific yew. Science. 1993;260:214–6.
40. Newman DJ, Cragg GM. Natural products as sources of new drugs over the last 25 years. J Nat Prod. 2007;70:461–77.
41. Sabater-Jara AB, Tudela LR, Lopez-Perez AJ. In vitro culture of *Taxus* sp.: strategies to increase cell growth and taxoid production. Phytochem Rev. 2010;9:343–56.
42. Flores-Bustamante ZR, Rivera-Orduna FN, Martinez-Cardenas A, Flores-Cotera LB. Microbial paclitaxel: advances and perspectives. J Antibiot. 2010;63:460–7.
43. Gangadevi V, Muthumary J. Isolation of *Colletotrichum gloeosporiodes*, a novel endophytic taxol-producing fungus from the leaves of a medicinal plant. Mycol Balc. 2008;5:1–4.
44. Kumaran RS, Kim HJ, Hur B-K. Taxol promising fungal endophyte, *Pestalotiopsis* species isolated from *Taxus cuspidata*. J Biosci Bioeng. 2010;110:541–6.
45. Kumaran RS, Jung H, Kim HJ. In vitro screening of taxol, an anticancer drug produced by the fungus *Colletotricum capsici*. Eng Life Sci. 2011;3:264–71.
46. Li J-Y, Strobel G, Sidhu R, Hess WM, Ford EJ. Endophytic taxol-producing fungi from bald cypress, *Taxodium distichum*. Microbiology. 1996;142:2223–6.
47. Wang JF, Li GL, Lu HY, Zhang ZH, Huang YJ, Su WJ. Taxol from *Tubercularia* sp. strain TF5, an endophytic fungus of *Taxus mairei*. FEMS Microbiol Lett. 2000;193:249–53.
48. Xu F, Tao W, Cheng L, Guo L. Strain improvement and optimization of the media of taxol-producing fungus *Fusarium maire*. Biochem Eng J. 2006;31:67–73.
49. Zhao K, Zhou D, Ping W, Ge J. Study on the preparation and regeneration of protoplast from taxol-producing fungus *Nodulisporium sylviforme*. Nat Sci. 2004;2:52–9.
50. Kumaran RS, Muthumary JP, Hur B-K. Taxol from *Phyllosticta citricarpa*, a leaf spot fungus of the angiosperm *Citrus medica*. J Biosci Bioeng. 2008;106:103–6.
51. Wei Y, Liu L, Zhou X, Lin J, Sun X, Tang K. Engineering taxol biosynthetic pathway for improving taxol yield in taxol-producing endophytic fungus EFY-21 (*Ozonium* sp.). Afr J Biotechnol. 2012;11:9094–101.
52. Kumaran RS, Choi Y-K, Lee S, Jeon HJ, Jung H, Kim HJ. Isolation of taxol, an anticancer drug produced by the endophytic fungus, *Phoma betae*. Afr J Biotechnol. 2012;11:950–60.
53. Zhang P, Zhou P-P, Yu L-J. An endophytic taxol-producing fungus from *Taxus media, Cladosporium cladosporoides* MD2. Curr Microbiol. 2009;59:227–32.
54. Liu K, Ding X, Deng B, Chen W. Isolation and characterization of endophytic taxol-producing fungi from *Taxus chinensis*. J Ind Microbiol Biotechnol. 2009;36:1171–7.
55. Zhou X, Zhu H, Liu L, Lin J, Tang K. A review: recent advances and future prospects of taxol-producing endophytic fungi. Appl Microbiol Biotechnol. 2010;86:1707–17.

56. Duan L-L, Chen H-R, Chen J-P, Li W-P, Hong L. Screening the high-yield paclitaxel producing strain *Alternaria alternate* var *monosporus*. Chin J Antibiot. 2008;33:650–2.
57. Amna T, Puri SC, Verma V, Sharma JP, Khajuria RK, Spiteller M, et al. Bioreactor studies on the endophytic fungus *Entrophospora* for the production of an anticancer alkaloid camptothecin. Can J Microbiol. 2006;52:189–96.
58. Lorence A, Nessler CL. Camptothecin, over four decades of surprising findings. Phytochemistry. 2004;65:2735–49.
59. Venditto VJ, Simanek EE. Cancer therapies utilizing the camptothecins: a review of the in vivo literature. Mol Pharm. 2010;7:307–49.
60. Bernardes N, Seruca R, Chakrabarty AM, Fialho AM. Microbial-based therapy of cancer. Current progress and future prospects. Bioeng Bugs. 2010;1:178–90.
61. Borel JF, Feurer C, Gabler HU, Stahelin H. Biological effects of cyclosporine A: a new anti-lymphocytic agent. Agents Action. 1976;6:468–75.
62. Borel JF. History of the discovery of cyclosporin and of its early pharmacological development. Wien Klin Wochenschr. 2002;114:433–7.
63. de Wilde AH, Zevenhoven-Dobbe JC, van der Meer Y, Theil V, Narayanan K, Makino S, et al. Cyclosporin A inhibits the replication of diverse coronaviruses. J Gen Virol. 2011;92:2542–8.
64. Bentley R. Bartolomeo Gosio, 1863–1944: an appreciation. Adv Appl Microbiol. 2001;48: 229–50.
65. Alsberg CL, Black OF. USDA Bur Plant Ind, Bull No. 270, Washington: Government Printing Office; 1913.
66. Birkinshaw JH, Raistrick H, Ross DJ. Studies in the biochemistry of micro-organisms. 86. The molecular constitution of mycophenolic acid, a metabolic product of *Penicillium brevicompactum* Dierckx. Part 3. Further observations on the structural formula for mycophenolic acid. Biochem J. 1952;50:630–4.
67. Sebastian L, Madhusudana SN, Ravi V, Desai A. Mycophenolic acid inhibits replication of Japanese Encephalitis Virus. Chemotherapy. 2011;57:56–61.
68. Lee WA, Gu L, Kikszal AR, Chu N, Leung K, Nelson PH. Bioavailability improvement of mycophenolic acid through amino ester derivatization. Pharm Res. 1990;7:161–6.
69. Chong CR, Quian DZ, Pan F, Wei Y, Pili R, Sullivan Jr DJ, et al. Identification of type 1 inosine monophosphate dehydrogenase as an antiangiogenic drug target. J Med Chem. 2006;49: 2677–80.
70. Bentley R. Microbial secondary metabolites play important roles in medicine: prospects to discovery of new drugs. Perspect Biol Med. 1997;40:364–94.
71. Vining LC, Taber WA. Ergot alkaloids. In: Rose AH, editor. Secondary products of metabolism, vol. 3. London: Academic; 1979. p. 389–420.
72. Hidy PH, Baldwin RS, Greasham RL, Keith CL, McMullen JR. Zearelanone and some derivatives: production and biological activities. Adv Appl Microbiol. 1977;22:59–82.
73. Jefferys EG. The gibberellin fermentation. Adv Appl Microbiol. 1970;13:283–316.
74. Tudzinski B. Biosynthesis of gibberellins in *Gibberella fujikuroi*: biomolecular aspects. Appl Microbiol Biotechnol. 1999;52:298–310.
75. Umezawa H. Enzyme inhibitors of microbial origin. Tokyo: University of Tokyo; 1972.
76. Paterson RRM. Fungal enzyme inhibitors as pharmaceuticals, toxins and scourge of PCR. Curr Enzyme Inhib. 2008;4:46–59.
77. Ma J, Li Y, Ye Q, Li J, Hua Y, Ju D, et al. Constituents of red yeast rice, a traditional Chinese food and medicine. J Agric Food Chem. 2000;48:5220–5.
78. Juzlova P, Martinkova L, Kren V. Secondary metabolites of the fungus *Monascus*: a review. J Industr Microbiol. 1996;16:163–70.
79. Feng Y, Shao Y, Chen F. *Monascus* pigments. Appl Microbiol Biotechnol. 2012;96:1421–40.
80. Lee B-H, Pan T-M. Benefit of *Monascus*-fermented products for hypertension prevention: a review. Appl Microbiol Biotechnol. 2012;94:1151–61.
81. Lin TF, Demain AL. Effect of nutrition of *Monascus* on formation of red pigments. Appl Microbiol Biotechnol. 1991;36:70–5.

82. Lin TF, Demain AL. Resting cell studies on formation of water-soluble red pigments by *Monascus* sp. J Ind Microbiol Biotechnol. 1993;12:361–7.
83. Lin TF, Demain AL. Leucine interference in the production of water-soluble red *Monascus* pigments. Arch Microbiol. 1994;162:114–9.
84. Lin TF, Demain AL. Negative effect of ammonium nitrate as nitrogen source on the production of water-soluble red pigments by *Monascus* sp. Appl Microbiol Biotechnol. 1995;43:701–5.
85. Andrewes AG, Phaff HJ, Starr MP. Carotenoids of *Phaffia rhodozyma*, a red-pigmented fermenting yeast. Phytochemistry. 1976;15:1003–7.
86. Rodriguez-Saiz M, de la Fuente JL, Barredo JL. *Xanthophyllomyces dendrorhous* for the industrial production of astaxanthin. Appl Microbiol Biotechnol. 2010;88:645–58.
87. Johnson EA, Villa TG, Lewis MJ. *Phaffia rhodozyma* as an astaxanthin source in animal diets. Aquaculture. 1980;20:123–34.
88. de la Fuente JL, Rodriguez-Saiz M, Schleissner C, Diez B, Peiro E, et al. High-titer production of astaxanthin by the semi-industrial fermentation of *Xanthophyllomyces dendrorhous*. J Biotechnol. 2010;145:144–6.
89. Jacobson GK, Jolly SO, Sedmak JJ, Skatrud TJ, Wasileski JM. Astaxanthin over-producing strains of *Phaffia rhodozyma*. Method for their cultivation and their use in animal feeds. US Patent 6015684; 1999.
90. Faus I. Recent developments in the characterization and biotechnological production of sweet-tasting proteins. Appl Microbiol Biotechnol. 2000;53:145–51.
91. Moralejo FJ, Cardoza RE, Gutiierez S, Martin JF. Thaumatin production in *Aspergillus awamori* by use of expression cassettes with strong fungal promoters and high gene dosage. Appl Environ Microbiol. 1999;65:1168–74.
92. Hallborn J, Walfridsson M, Airaksinen U, Ojamo H, Hahn-Hagerdal B, Penttila M, et al. Xylitol production by recombinant *Saccharomyces cerevisiae*. Biotechnology. 1991;9: 1090–5.
93. Livermore DM. The need for new antibiotics. Clin Microbiol Infect. 2004;10:1–9.
94. Clardy J, Fischbach MA, Walsh CT. New antibiotics from bacterial natural products. Nat Biotechnol. 2006;24:1541–50.
95. Scheffler R, Colmer S, Tynan H, Demain AL, Gullo VP. Antimicrobials, drug discovery, and genome mining. Appl Microbiol Biotechnol. 2013;97:969–78.

Chapter 2
Penicillins

Carlos García-Estrada and Juan-Francisco Martín

Introduction

The Discovery of Penicillin and the Improvement of the Penicillin Production Process

The discovery of antibiotics represents one of the most important events in the history of medicine and entailed a revolution in modern chemotherapy, since these compounds have contributed to drastically reduce the mortality rate of human and animal bacterial diseases. Antibiotics are commonly classified according to their chemical structure or mechanism of action into different families, including beta-lactam antibiotics, macrolides, quinolones, tetracyclines, aminoglycosides, etc. Beta-lactam antibiotics (mainly penicillins and cephalosporins), with millions of prescriptions worldwide, stand out from the rest of families due to their high activity and low toxicity.

The year that is considered as the starting point for the history of antibiotics is that of the discovery of penicillin. In September 1928, Sir Alexander Fleming discovered the antimicrobial activity generated by a fungus contaminating a Petri dish culture of *Staphylococcus* sp. Fleming, who worked at the St. Mary's Hospital in London, initially identified the mold responsible for the observed antibacterial effect as *Penicillium rubrum* [1]. Although Fleming tried to concentrate and purify the antibiotic, he was not successful due to the instability of penicillin. It was not until 1932 when the Fleming's isolate was identified as *Penicillium notatum* and the active compound inhibiting the bacterial growth was dubbed penicillin [2].

C. García-Estrada, D.V.M., Ph.D. (✉)
INBIOTEC (Institute of Biotechnology of León), Parque Científico de León, León, Spain
e-mail: c.gestrada@unileon.es

J.-F. Martín, Ph.D.
Department of Molecular Biology, Microbiology Section, University of León, León, Spain

J.-F. Martín et al. (eds.), *Biosynthesis and Molecular Genetics of Fungal Secondary Metabolites*, Fungal Biology, DOI 10.1007/978-1-4939-1191-2_2,

The instability of penicillin purified from culture broths did not allow Fleming to extend his work to clinical studies and the fate of this compound seemed to be not very promising. In fact, during the following years the work with penicillin was set aside, but the interest in penicillin's therapeutic properties by a multidisciplinary group from the Sir William Dunn School of Pathology at Oxford University (led by Howard Walter Florey, Ernst Boris Chain and Norman Heatley) resumed the work with this antibiotic in 1938. One of the members of the team, Edward Abraham, used the newly discovered technique of alumina column chromatography to remove impurities from the penicillin prior to clinical trials. In August 1940, the methodology for penicillin production and its use to treat infections in mice were characterized [3]. These initial successful experiments highlighted the medical relevance of penicillin and encouraged scientists to carry out the first clinical trials in human beings.

Massive production of penicillin could not be accomplished in England due to World War II. Therefore, Florey and Heatley traveled to the United States in June 1941 to try to convince the American pharmaceutical industry to produce penicillin on a large scale. They contacted the Northern Regional Research Laboratory (NRRL) in Peoria, Illinois and various pharmaceutical companies. In 1942, the first patients were treated with the penicillin produced by Merck & Co., Inc., under the auspices of the US Office of Scientific Research and Development (OSRD).

The improvement of the penicillin production process was making progress in parallel at the NRRL. Significant increases in penicillin yields were achieved by Andrew Moyer, who used sucrose and corn-steep liquor in the fermentation medium, replaced the surface growth by submerged cultures [4, 5] and fed the fermentation process with penicillin side chain precursors, such as phenylacetic acid [6]. At the same time, various *Penicillium* strains were screened for productivity under the direction of Kenneth Raper at the NRRL. Soil samples from all around the world were sent to the NRRL, but the most productive one came from a moldy cantaloupe obtained from a local market in Peoria [7]. This strain was *Penicillium chrysogenum* NRRL 1951, which was X-ray mutated by Milislav Demerec at the Carnegie Institution of Washington (Cold Spring Harbor, NY, USA) and later exposed to UV radiation at the University of Wisconsin (origin of the Wisconsin line of strains) with a subsequent increase in productivity.

Large-scale production of penicillin was achieved due to the development of the industrial technology, such as the fermentation process, recovery, and purification. The tremendous cooperative effort among universities and industrial laboratories in England and the United States during World War II led to multiple large-scale clinical trials to treat those wounded in battle in England and in the United States. Penicillin production began to increase dramatically by early 1944; and on March 1, 1944, Pfizer opened the first commercial plant for large-scale production of penicillin by submerged culture in Brooklyn, New York. In March 1945, penicillin was available to the American consumer on the open market.

In Oxford, Chain and Abraham, aided by the X-ray crystallography work of Dorothy Hodgkin, were able to establish in 1945 the four-membered beta-lactam ring fused to a thiazolidine ring structure of the penicillin molecule [8]. On December 10, 1945, Ernst B. Chain, Howard W. Florey, and Alexander Fleming received the

Nobel Prize in Physiology or Medicine "for the discovery of penicillin and its curative effect in various infectious diseases." Finally, penicillin was on sale to the general public in the United Kingdom as a prescription-only drug on June 1, 1946.

Another important milestone that opened a new era of chemotherapy was the detection and isolation of 6-aminopenicillanic acid (6-APA) in fermented broths in the 1950s [9, 10]. This penicillin precursor is the basis of semisynthetic penicillins, which are synthesized through the addition of different side chains to 6-APA by a chemical process. Pharmaceutical companies implemented their own *P. chrysogenum* improvement programs, which together with medium modifications and the optimization of the process have provided a production of 100,000 times more penicillin than Fleming's original isolate. Current overproducing mutants reach penicillin titers of more than 50 mg/mL (83,300 u.i./mL) in industrial fermentations [11].

The modern history of penicillin has been related to the characterization of the biosynthetic pathway, including biosynthetic and ancillary enzymes and genes, regulation of the biosynthetic process, and, more recently after the publication of the *P. chrysogenum* genome in 2008 [12], to the global analysis of the molecular mechanisms underlying the increased productivity of industrial strains using the modern genomics, transcriptomics, proteomics, and metabolomics techniques [13–15].

Structure and Mechanism of Action of Penicillin

Antibiotics from the beta-lactam group are included into the peptidic class of antibiotics. Beta-lactam antibiotics (penicillins, cephalosporins, monolactams, clavulanic acid, and carbapenems) have a common structure, which consists of a four-membered beta-lactam ring [16]. With the exception of monolactams, which have only the beta-lactam ring, the rest of beta-lactam antibiotics possess a bicyclic system, the second ring being the structure that allows their classification [16, 17]. Penicillins have a central core of 6-APA, which is constituted by the beta-lactam ring fused to a five-membered thiazolidinic ring (Fig. 2.1). The side chain fused to the core structure confers hydrophobic or hydrophilic properties. Hence, some penicillins are hydrophobic, such as penicillin G or benzylpenicillin (phenylacetic acid as side chain) and

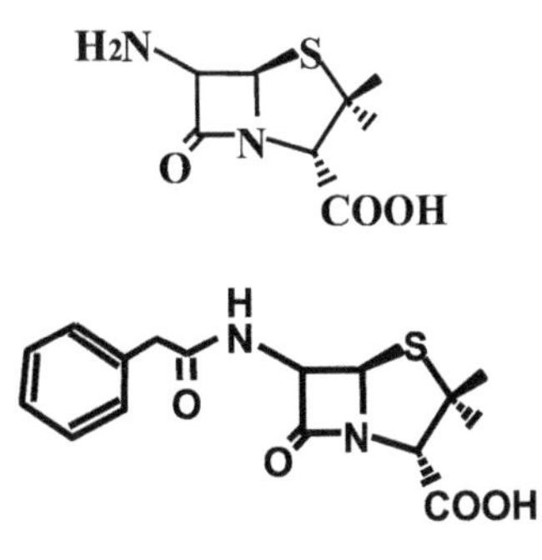

Fig. 2.1 Central core of 6-APA (*top*) and chemical structure of benzylpenicillin (*bottom*)

penicillin V or phenoxymethylpenicillin (phenoxyacetic acid as side chain), whereas other penicillins have a hydrophilic side chain of L-α(alpha)-aminoadipic acid, as in the case of isopenicillin N (IPN).

Hydrophobic penicillins are exclusively synthesized by filamentous fungi (*Penicillium allii-sativi*, *Penicillium chrysogenum*, *Penicillium dipodomyis*, *Penicillium flavigenum*, *Penicillium griseofulvum*, *Penicillium nalgiovense*, *Penicillium rubens*, *Penicillium tardochrysogenum*, *Penicillium vanluykii*, and *Aspergillus nidulans*) [18–20]. However, hydrophilic penicillins are produced by filamentous fungi (*P. chrysogenum, Acremonium chrysogenum, A. nidulans*), several Actinomycetes (*Streptomyces* sp.) and some Gram-negative bacteria [16]. Naturally occurring penicillins are the basis for semisynthetic penicillins, which are produced after the chemical addition of different side chains to 6-APA and have improved pharmacological properties.

The structure of beta-lactam antibiotics is highly related to the mechanism of action of these compounds. Beta-lactam antibiotics are bactericidal agents that block the last step of the bacterial cell wall biosynthesis, which consists of the crosslinking of peptidoglycan chains. Penicillin inhibits the activity of PBPs (Penicillin Binding Proteins), namely DD-transpeptidase and DD-carboxypeptidase, which are responsible for the peptidoglycan crosslinking process. The structure of penicillins is closely related to acyl-D-alanyl-D-alanine (the last two amino acids of the pentapeptide that links the peptidoglycan molecule), which is the natural substrate of PBPs. Therefore, penicillin enters the active site and forms a stable penicillin-enzyme complex, which produces the irreversible inhibition of PBPs. As a consequence of enzyme inactivation, the bacterial cell wall becomes osmotically unstable, which leads to the bacterial cell autolysis [21, 22]. In addition, penicillins trigger the activation of cell wall hydrolases and autolysins, which further destroy the bacteria [23]. Obviously, the activity of these antibiotics is higher in Gram-positive bacteria, since the peptidoglycan is the major constituent of the cell wall in this group of microorganisms.

Penicillin Biosynthesis

As indicated previously, penicillins have a peptidic nature, although they are synthesized in a non-ribosomal manner. The penicillin biosynthetic pathway has been studied in detail from the biochemical and molecular points of view [24] (Fig. 2.2a). The characterization of the penicillin biosynthetic enzymes and genes began during the decade of 1980s. The three penicillin biosynthetic genes are clustered in *P. chrysogenum* [25] and in *A. nidulans* [26, 27], showing a divergent orientation of the *pcbAB* and *pcbC* genes and the localization of the *penDE* gene immediately downstream of the *pcbC* gene (Fig. 2.2b). Mapping of the cluster inside the genome of these filamentous fungi indicated that the biosynthetic genes are located on chromosome I in *P. chrysogenum* [28] and on chromosome VI in *A. nidulans* [27]. More recently, the availability of genome sequences from different filamentous fungi (including the penicillin producer ascomycetes *P. chrysogenum* and *A. nidulans*) has provided more information on novel biosynthetic and ancillary genes.

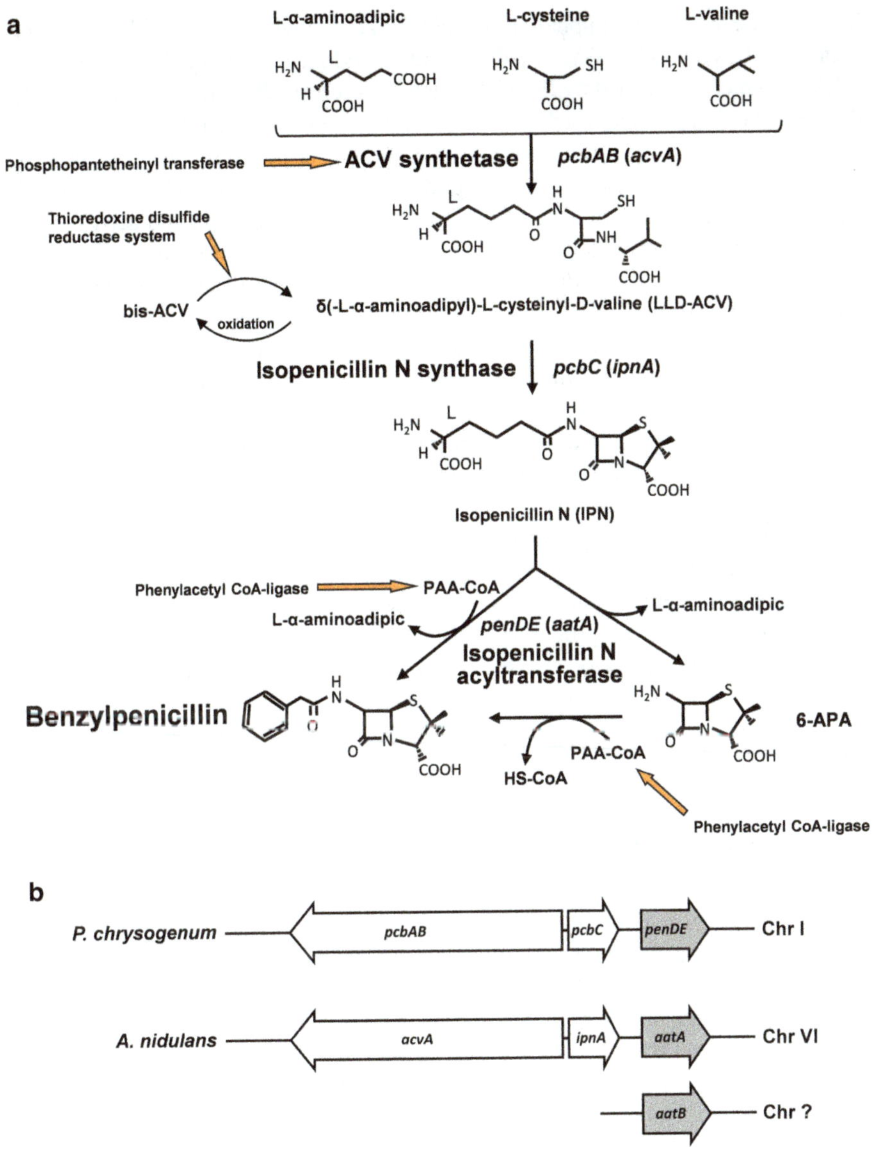

Fig. 2.2 (**a**) Penicillin biosynthetic pathway. Structural and ancillary proteins are represented. (**b**) Organization of the penicillin biosynthetic genes

Tripeptide Biosynthesis

Penicillin biosynthesis begins with the non-ribosomal condensation of three amino acids: L-α(alpha)-aminoadipic acid (non-proteinogenic amino acid formed by a specific pathway related to lysine biosynthesis), L-cysteine, and L-valine, which constitute the tripeptide δ(delta)-L(α[alpha]-aminoadipyl)-L-cysteinyl-D-valine (ACV). This step is catalyzed by the 426-kDa trimodular and multidomain enzyme ACV synthetase (ACVS), which is encoded by the 11-kbp intron-free *pcbAB* (or *acvA*) gene [25, 29]. This enzyme was purified from *P. chrysogenum* in the late 1990s [30].

The ACVS contains different catalytic domains necessary for the biosynthesis, epimerization, and release of the tripeptide. Each of the three modules of ACVS include: (1) adenylate-forming or activation domain (recognition and activation of the three precursor amino acids); (2) thiolation or amioacyl carrier domain (including a conserved serine residue that binds a thiol-containing phosphopantetheine cofactor derived from CoA); and (3) condensation domain (peptide bond formation between two activated amino acids on adjacent modules, thus catalyzing the elongation reaction). At the end of the third module there is an epimerase domain, which catalyzes the conversion of the precursor amino acid L-valine into D-valine [31]. Adjacent to this domain and integrated into the C-terminal region of the ACVS, there is a thioesterase domain, which is involved in the release of the tripeptide from the enzyme [32–35].

Initial studies on *P. chrysogenum* ACVS associated this protein to membrane structures that were identified as Golgi-like organelles [36]. Additional cell fractionation experiments located ACVS attached to or inside vacuoles [37, 38]. However, the biochemical properties of this enzyme, such as in vitro optimal pH 8.4 (which resulted to be above that of the vacuolar pH), cofactor requirement, and protease sensitivity indicated that this enzyme was a cytosolic enzyme. This fact was confirmed by immunoelectron microscopy [39].

Tripeptide Cyclization

During the second step in the penicillin biosynthetic pathway, four hydrogen atoms from the ACV tripeptide are removed, with the subsequent formation of the bicyclic structure (penam nucleus) of IPN. IPN is the second penicillin precursor molecule in the pathway and, unlike ACV, has antimicrobial properties.

This cyclization reaction is catalyzed by the 38-kDa IPN synthase or cyclase, which is an intermolecular dioxygenase that requires Fe^{2+}, molecular oxygen and ascorbate [40–43] and only accepts ACV in its reduced state (the oxidized bis-ACV has to be previously reduced before it can be converted). The IPN synthase co-localizes with ACVS in the cytosol [37, 39] and it behaves as a soluble enzyme [40, 44], although its activity in cell-free extracts seems to be stimulated by the addition of Triton X-100 or sonication [45].

The 1.1-kbp gene encoding IPN synthase (*pcbC* or *ipnA*) has no introns and was cloned from *P. chrysogenum* during the 1980s [46, 47]. The *pcbC*-encoded IPN synthase was purified from *P. chrysogenum* [40, 46] and its crystallization revealed a bread roll-like structure [48, 49].

Side Chain Replacement: Formation of Hydrophobic Penicillins

Filamentous fungi producing hydrophobic penicillins are able to hydrolyze the α(alpha)-aminoadipic side chain of IPN [50] and replace it with hydrophobic acyl molecules, which have to be activated by aryl-CoA ligases before they become substrates for this reaction. Side chain replacement is catalyzed by the acyl-CoA: IPN acyltransferase (IAT), which accepts a wide range of side chains as substrate. Thus, natural penicillins, such as penicillin F (D3-hexenoic as side chain) and K (octanoic acid as side chain) are synthesized under natural conditions. However, feeding the culture media with phenylacetic or phenoxyacetic acids directs the biosynthesis mainly towards benzylpenicillin (penicillin G) or phenoxymethyl-penicillin (penicillin V), respectively.

A two-step enzymatic process has been proposed for this reaction [51]. During the first step, the amidohydrolase activity removes the L-α(alpha)-aminoadipate side chain of IPN, thus forming 6-APA. Next, the acyl-CoA: 6-APA acyltransferase activity introduces the new activated acyl side chain. These two activities fall upon the IAT, which is synthesized as a preprotein of 40 kDa termed proacyltransferase or proIAT. The proacyltransferase undergoes an autocatalytic self-processing between residues Gly102-Cys103 in *P. chrysogenum*. The processed protein constitutes a heterodimer with subunits α(alpha) (11 kDa, corresponding to the N-terminal fragment) and β(beta) (29 kDa, corresponding to the C-terminal region) [52–55]. The self-processing of IAT is an important differential factor between *P. chrysogenum* and *A. nidulans*. The *A. nidulans* proIAT remains predominantly unprocessed as a 40-kDa protein through several purification steps, whereas the *P. chrysogenum* enzyme is efficiently self-processed, rapidly dissociating into the 29-kDa and 11-kDa subunits [56]. It has also been reported that the *P. chrysogenum* IAT is posttranslationally regulated by its preprotein, which interferes with the self-processing [57].

The IAT is located inside the microbodies (peroxisomes), since it bears a consensus peroxisomal targeting sequence of type 1 (PTS1) at the C-terminus [37, 58, 59]. It is now well established that peroxisomes are required for efficient penicillin biosynthesis in *P. chrysogenum* [60] and that the penicillin biosynthetic pathway is compartmentalized between the cytosol and peroxisomes [24]. The distinct subcellular organization of penicillin biosynthesis implies transport of enzymes, precursors, intermediates, and products through these compartments [61].

The 40-kDa IAT is encoded by the *penDE* (*aatA*) gene, which unlike the other two genes in the pathway, contains three introns. It was cloned from *P. chrysogenum* [52, 54] and *A. nidulans* [27, 54]. The presence of introns within this gene initially suggested a fungal origin [16], although this issue was not so clear because no known close eukaryotic or prokaryotic analogous genes had been identified. However, in silico analysis of the genomes of several ascomycetes (including *A. nidulans* and *P. chrysogenum*) has recently allowed the identification of a putative gene paralogue of the *penDE* (*aatA*), which has been designated *aatB* in *A. nidulans* [62] or *ial* in *P. chrysogenum* [63]. These genes are not clustered with the rest of the penicillin biosynthetic genes (Fig. 2.2b).

The *aatB* gene of *A. nidulans* shows an expression profile similar to that of the *aatA* gene and contains three introns, but unlike the IAT encoded by the *aatA* gene, the protein encoded by the *aatB* gene lacks a PTS1 and is located on the cytosol. Despite this subcellular localization, it is claimed to be involved in penicillin biosynthesis [62]. Regarding the *ial* gene of *P. chrysogenum*, it contains two introns and is expressed very poorly or not expressed at all in several *P. chrysogenum* strains. This gene encodes a protein that lacks the PTS1 signal and has no IAT activity [63]. Therefore, *aatB* and *ial* genes differ in function and appear to have had a different evolutionary origin.

Ancillary Proteins of the Penicillin Biosynthetic Process

In addition to the three penicillin biosynthetic proteins, the activity of some enzymes from primary metabolism is required for the biosynthesis of penicillin. These activities are phosphopantetheinyl transferase (PPTase), disulfide reductase, and acyl-CoA ligase (ACL).

The ACVS protein is synthesized as an inactive apoprotein, whose activation is achieved by means of the addition of a 4′-phosphopantetheine arm, derived from CoA, to the thiolation domain of ACVS. This reaction is catalyzed by the cytosolic PPTase, which belongs to a class of enzymes responsible for the posttranslational modification of enzymes involved in fatty acid, polyketides, and non-ribosomal peptides biosynthesis. It has been reported that PPTase is necessary for penicillin biosynthesis both in *A. nidulans* [64–66] and in *P. chrysogenum* [34].

Once the ACV tripeptide is synthesized, the aeration conditions that are present inside the culture medium rapidly oxidize the monomer to the disulfide form, thus forming the bis-ACV dimer. This dimer cannot be substrate for IPN synthase, since this enzyme only accepts ACV as the reduced monomer. In addition, bis-ACV inhibits the activity of ACVS [30]. Therefore, an enzymatic activity is necessary to reduce the bis-ACV and allow the reincorporation of this molecule to the penicillin biosynthetic pathway. This is achieved by means of the cytosolic NADPH-dependent thioredoxin disulfide reductase system (TrxAB), which has been fully characterized in *P. chrysogenum* [67].

The side chain replacement of IPN that is catalyzed by IAT requires that precursor acyl molecules (i.e., phenylacetic acid or phenoxyacetic acid) are activated as CoA thioesters. This activation is achieved by means of the ACL activity. In 1997, an enzyme with phenylacetyl-CoA ligase (PCL) activity and a PTS1 signal was identified in peroxisomes [68]. Some years later the PCL-encoding *phl* gene of *P. chrysogenum* was cloned and its direct involvement in the penicillin biosynthesis process was demonstrated [69]. It was also seen that this gene was not the only one encoding a PCL activity. In fact, a second gene (*phlB*) was cloned later, which was proposed to encode a PCL enzyme with a PTS1 signal involved in the activation of phenylacetic acid [70]. However, studies from Koetsier et al. [71, 72] revealed that the *phlB* gene (also named *aclA*) is not involved in the activation of phenylacetic acid, since it encodes a broad spectrum ACL protein that activates adipic acid.

A recent study has identified a third gene (*phlC*) in *P. chrysogenum* that encodes a protein with a PTS1 signal and exhibits PCL activity [73]. The peroxisomal location of PCL enzymes represents a clear benefit for the coordination of the last step of penicillin biosynthesis, since IAT is also targeted to these organelles.

Unlike the penicillin biosynthetic genes *pcbAB*, *pcbC*, and *penDE*, which constitute a functional cluster, those genes encoding the aforementioned ancillary proteins are not clustered together and are located on different genomic regions from the penicillin gene cluster.

Transport Processes Involved in Penicillin Biosynthesis and Secretion

The compartmentalization of the penicillin biosynthetic pathway implies that precursors, intermediates, enzymes, and the final product have to be transported through different subcellular organelles. This organization facilitates the supply of cofactors and precursors from primary metabolism in an optimal way. Not only peroxisomes are important organelles in the penicillin biosynthetic process (side chain activation and further addition to IPN/6-APA), but mitochondria and vacuoles as well.

Mitochondria are essential for the biosynthesis of the precursor amino acid α(alpha)-aminoadipic acid [74, 75], whereas vacuoles control the concentration of free amino acids in the cytosol, namely L-α(alpha)-aminoadipic acid and L-cysteine, which may be toxic at moderate concentrations [76, 77].

Transport of Precursors: Amino Acids and Phenylacetic Acid

Although filamentous fungi are capable of synthesizing all amino acids, the latter can be taken up from the culture medium to the inner cell to serve as carbon or nitrogen sources or as building blocks [78, 79]. Three active amino acid permeases have been characterized so far in *P. chrysogenum* [80] and up to nine amino acid transport systems have been identified in this filamentous fungus:

1. System I for L-methionine [81]
2. System II for L-cysteine [82]
3. System III for all amino acids [83, 84]
4. System IV for acidic amino acids
5. System V for L-proline
6. System VI for L-lysine and L-arginine
7. System VII for L-arginine
8. System VIII for L-lysine
9. System IX for L-cysteine [84]

L-α(alpha)-aminoadipic acid is mainly taken up through the general amino acid permease [85].

PenV, a transporter of the MFS class that is related to the supply of amino acids from the vacuolar lumen to the vacuole-anchored ACVS, has been recently characterized [86], providing evidence of a compartmentalized storage of precursor amino acids for non-ribosomal peptides.

Side chain precursors phenylacetic and phenoxyacetic acids are weak acids that rapidly enter *P. chrysogenum* cells by passive diffusion and distribute along the membrane according to the pH transmembrane gradient [87, 88]. An active transport system for the uptake of phenylacetic acid was also postulated by other authors [89–91]. The main differences between both studies were the phenylacetic acid concentrations and the type of strain (low versus high penicillin production). It was concluded after the detailed analysis of both conditions that during penicillin biosynthesis, there is a high utilization of side chain precursors, which enter the cell mainly through passive diffusion [92].

Since the activation of the side chain precursor occurs inside microbodies (see above), this compound has to be transported inside these organelles. A gene encoding a phenylacetic acid transporter (*paaT*) has been recently characterized. It encodes a drug/H^+ antiporter of 12 TMSs located in the peroxisomal membrane [93] that is involved in phenylacetic acid resistance and penicillin production. In parallel, another group reported that this transporter (also named PenT) stimulates penicillin production probably through enhancing the translocation of penicillin precursors across fungal cellular membrane [94].

Transport of Intermediates

In addition to benzylpenicillin, the three intermediates of the benzylpenicillin biosynthetic pathway (ACV, IPN and 6-APA) are secreted to the culture medium by *P. chrysogenum* [95, 96]. It has been proposed that ACV may be secreted to the culture medium through the glutathione export system [97, 98]. Once the tripeptide is outside the cell, this compound is oxidized to the dimer form and not taken up from the culture medium as either monomer or dimer [99].

IPN is synthesized in the cytosol and therefore it has to be incorporated inside the microbody for its conversion to benzylpenicillin. Due to the hydrophilic nature of this compound, its diffusion across the microbody membrane is unlikely. The peroxisomal transporter of IPN has not yet been characterized in *P. chrysogenum*. However, in the cephalosporin producer fungus *A. chrysogenum*, a peroxisomal membrane protein (CefP) likely involved in the import of IPN into the peroxisomes has been identified [100], which suggests the presence of an orthologous IPN import peroxisomal protein in *P. chrysogenum*. As it was indicated before, a fraction of the IPN is secreted to the extracellular culture. The hydrophilic nature of this intermediate suggests an active transport instead of passive diffusion. In fact, it has been reported that transport of IPN from the culture medium back to the cytosol is very inefficient [99].

Therefore, a transporter seems to be required to release IPN to the culture medium. Although such a transporter has not yet been identified in *P. chrysogenum*, the expression of the *A. chrysogenum cefT* gene (encoding a plasma membrane protein of the MFS class of transporters) in *P. chrysogenum* led to an increase in the secretion of hydrophilic penicillins (including IPN) and adipoyl-7-amino-3-carbamoyloxymethyl-3-cephem-4-carboxylic acid (ad7-ACCCA), a carbamoylated derivate of adipoyl-7-aminodeacetoxy-cephalosporanic acid [101, 102]. This indicates that an orthologous transporter may be present in *P. chrysogenum*. In fact, it was observed that *P. chrysogenum* contains an endogenous gene similar to *cefT*, which may be involved in the well-known secretion of IPN to the culture medium [101].

Unlike IPN, 6-APA (the last intermediate in the biosynthetic process that is formed inside microbodies by the IAT) is a more hydrophobic compound. This is because 6-APA lacks a carboxyl group compared to IPN, which facilitates passive diffusion process. This was confirmed by the fact that 6-APA uptake from the culture medium is achieved by a concentration-dependent diffusion process [99].

Transport of Hydrophobic Penicillins

Aromatic penicillins synthesized by *P. chrysogenum* are amphipathic, moderately hydrophobic, and negatively charged compounds [103]. These characteristics promote secretion by diffusion mechanisms through the peroxisomal and plasma membranes.

Several reasons point to an active transport process. Firstly, some physical properties of the plasma membrane, such as the high packaging degree of the lipidic bilayer, different superficial and electrochemistry factors, may drastically inhibit passive diffusion of penicillins [104]. On the other hand, passive diffusion itself cannot explain the outside–inside distribution of penicillins, whose concentrations are much higher in the culture medium than in the cytosol [105]. However, authentic benzylpenicillin transporters in the peroxisomal or in the fungal membrane have not been scientifically confirmed in *P. chrysogenum*. The search for transporters in *A. nidulans* led to the identification of an ABC transporter (encoded by the *atrD* gene), which is related in some way to the secretion of penicillin [106].

Another secretion mechanism mediated by vesicles, which would be a mechanism between passive diffusion and active transport, cannot be ruled out. Vesicular transport is involved in many processes in eukaryotes, such as aflatoxins biosynthesis, cell wall synthesis, pheromones secretion, and bile acids transport. The fusion of peroxisomes with secretory vesicles that may unload the penicillin formed by an exocytosis mechanism has been proposed [24, 107], although there is no evidence supporting this mechanism of penicillin release. This vesicle-mediated mechanism may be different from the known pexophagy (autophagic degradation of peroxisomes) because a *P. chrysogenum* mutant defective in pexophagy produced normal or even higher levels of penicillin [108].

Regulation of Penicillin Biosynthesis

Penicillin biosynthesis is probably one of the processes that have been more deeply studied over the last three decades. Penicillin production takes place, preferentially, under stress nutrient and low growth rate conditions. Limitations in the carbon, nitrogen, or phosphorous availability, together with other factors (pH, aeration, certain amino acids, or media composition) strongly influence the production process. This process is subjected to complex regulatory processes controlled by different transcription factors [24, 109–112], although no penicillin pathway-specific regulatory genes have been found in the amplified region containing the three biosynthetic genes [113, 114]. This indicates that penicillin biosynthesis is controlled directly by global regulators rather than by pathway-specific ones.

Carbon Catabolite Regulation

Penicillin biosynthesis in *P. chrysogenum* is strongly regulated by glucose and sucrose and to lower extent by maltose, fructose, and galactose, but not by lactose. The repressing effect of glucose on penicillin biosynthesis is greatly enhanced by high phosphate concentrations [110]. Unlike lactose, glucose reduces the L-α(alpha)-aminoadipic acid pool, likely reducing the synthesis of ACV. The formation of ACV and IPN are also repressed by high glucose concentrations [115] in *P. chrysogenum* AS-P-78, a penicillin high-producing strain. The IAT specific activity in the wild-type strain *P. chrysogenum* NRRL 1951 was reduced in cultures grown in the presence of glucose [116]. In *A. nidulans*, glucose regulation of *ipnA* (*pcbC*) takes place, at least in part, at the transcriptional level, with a reduction in the IPN synthase activity when the cultures were grown in the presence of glucose. The effect of glucose on the *aatA* gene was posttranscriptionally mediated and the IAT specific activity in *A. nidulans* was also reduced in cultures grown in the presence of glucose [116].

Carbon catabolite regulation of primary metabolism is mediated in *A. nidulans* by the zinc finger transcription factor CreA. However, *A. nidulans creA* mutants still exhibit glucose-mediated repression of *ipnA* transcript levels [117], which suggest that in *A. nidulans* a second CreA-independent mechanism of carbon repression is involved in the control of penicillin biosynthesis. It is not yet clear whether *P. chrysogenum* has a similar mechanism of carbon regulation mediated by CreA.

pH Regulation

Beta-lactam antibiotic production is strongly influenced by the external pH, which exerts its regulation in filamentous fungi by the Cys2-His2 transcriptional activator PacC [118, 119]. PacC activates the transcription of genes expressed at alkaline pH and represses the transcription of genes expressed at acidic pH. In *P. chrysogenum*, seven and eight putative PacC binding sites are present in the *pcbAB-pcbC* intergenic region and in the promoter region of *penDE*, respectively [120]. Penicillins are produced at high levels at alkaline pH, since under these conditions a proteolytic processing activates PacC [121].

Nitrogen Source Regulation

Nitrogen regulation in fungi is mediated by AreA in *A. nidulans* and its homologue in *P. chrysogenum* NRE. Regulation is exerted by binding to the consensus sequence GATA [122]. In *P. chrysogenum*, the bidirectional promoter region *pcbAB-pcbC* contains a total of six GATA motifs. Only NRE strongly interacts in vitro with a site that contains two of these GATA motifs. In this binding site, the two GATA motifs, which are separated by 27 bp, are arranged in a head-to-head fashion [123].

In *A. nidulans*, there is only one GATA motif in the bidirectional *pcbAB-pcbC* promoter region and no evidence on a possible nitrogen-dependent regulation of penicillin biosynthesis in this fungus has been reported so far. However, when the *pcbC* promoter from *P. chrysogenum* was introduced in *A. nidulans*, expression was sensitive to nitrogen regulation [124], which suggests similar mechanisms of nitrogen regulation in both penicillin producers.

Phosphate Source Regulation

An excess in phosphate enhances glucose repression during penicillin biosynthesis. In a phosphate-limited complex medium, the glucose repression of penicillin levels is about 13 % when sugar is added at inoculation time, whereas it increases to 59 % when the medium is supplemented with 100 mM inorganic phosphate [110]. Inorganic phosphate has little or no effect per se on penicillin production under non-repressing conditions.

Amino Acids as Mediators of Regulation

It is well-known that the addition of L-lysine to the fermentation medium in *A. nidulans* and *P. chrysogenum* decreases penicillin production [116, 125]. L-threonine, L-aspartate, L-glutamate, and L-cysteine led to increased *acvA* gene fusion expression, but had no effect on *ipnA* gene fusion expression. L-methionine (at concentrations above 10 mM), L-leucine, L-isoleucine, L-phenylalanine, L-valine, L-histidine, and L-lysine led to the repression of both *acvA* and *ipnA* gene fusion expression, which was dependent on the amino acid concentration. L-tyrosine, L-tryptophan, L-proline, and L-α(alpha)-aminoadipic acid had no major effects on *acvA* gene fusion expression, but led to the repression of *ipnA* gene fusion expression. L-serine and L-arginine did not show any effect on the expression of either of these gene fusions at any concentration. The negative effects of L-histidine and L-valine were due to reduced activation by PacC under the acidic conditions caused by these amino acids. However, the repressive effects of L-lysine and L-methionine acted independently of PacC by unknown mechanisms.

Regulation by Aeration Conditions

A good aeration of mycelia is a prerequisite for higher beta-lactam titers [126, 127]. Some enzymes, such as IPN synthase, require oxygen for their activity. However, Renno et al. (1992) showed that the expression of *pcbAB* and *pcbC* in *P. chrysogenum* can also be induced in response to stress by the reduction of O_2 levels [128].

Regulation by Polyamines

Some polyamines act as inducers of the penicillin biosynthetic process in *P. chrysogenum*. Biosynthesis of penicillin G is stimulated by 1,3-diaminopropane (1,3-DAP) and spermidine, which induce the transcription of *pcbAB*, *pcbC*, *penDE* [129], and *laeA* [130]. These polyamines promoted a deep reorganization of the proteome and increased the intracellular content of vesicles that derived to vacuoles in late stages [131].

Regulation by Corn-Steep Liquor

Corn-steep liquor (CSL) is a by-product of the cornstarch manufacturing process and has been used as a regular component of the microbiological culture media.

The addition of CSL to a production medium greatly increases penicillin yields in *P. chrysogenum* [132] and *A. nidulans* [133].

The most important result obtained after the addition of CSL in *A. nidulans* is a great increase in the expression of the penicillin biosynthetic gene cluster. The stimulatory effect still occurs even in the presence of repressing carbon sources of penicillin biosynthesis [26, 134, 135]. However, the molecular mechanisms connecting the presence of CSL and the increased gene expression are still unknown, although the composition of CSL (high content of amino acids, polypeptides, polyamines, minerals, etc.) likely favors the biosynthesis of penicillin.

Global Regulators

As it was indicated before, the lack of a penicillin pathway-specific regulator indicates that penicillin biosynthesis is controlled directly by global regulators. In fact, several global regulators, such as LaeA [136, 137] and some other proteins from the velvet complex: VeA [138–140], PcVelB, PcVelC, and PcVosA [141], have been reported to control the penicillin biosynthetic process.

PcRFX1, which is a homologue of the *A. chrysogenum* cephalosporin regulator CPCR1, has also been confirmed to control the whole penicillin biosynthetic process [142].

Penicillin biosynthetic genes are also regulated by other proteins, such as the protein AnCF (*A. nidulans* CCAAT-binding factor constituted by subunits HapB, HapC, and HapE) formerly known as PENR1 (Penicillin regulator 1) [143, 144], AnBH1 (PENR2) [145], PTA1 [146], and Pga1 [147].

Molecular Bases for the Improvement of Penicillin Production in *P. chrysogenum*

As a result of the industrial strain improvement programs, *P. chrysogenum* was tamed to produce impressive penicillin titers. Some of the modifications that were introduced in this filamentous fungus after chemical and physical mutagenesis have been fully characterized and are described as follows.

The Amplification of the Penicillin Gene Cluster

It is well known that the genomic region containing the penicillin gene cluster undergoes tandem amplification in penicillin high-producing strains of *P. chrysogenum*. Unlike the wild-type strain (NRRL 1951) and the laboratory reference strain (Wisconsin 54-1255), which contain one copy of this region, penicillin high-producing strains bear several copies of the amplified region, whose length varies (56.8 or 106.5 kb) depending on the strain [148, 149]. The mechanism responsible for this phenomenon of amplification is not fully characterized, but it has been suggested that the conserved hexanucleotides flanking the borders of the amplified region may act as "hot spots" for recombination [148].

Interestingly, in addition to the three penicillin biosynthetic genes *pcbAB*, *pcbC*, and *penDE* the amplified region contains several ORFs, which also undergo amplification [113, 114]. However, these ORFs may play an ancillary nonessential role in the biosynthesis, regulation, or secretion of penicillin, since the only presence of the three penicillin biosynthetic genes was sufficient to restore the antibiotic production in a mutant lacking the whole penicillin gene cluster [99, 114].

Although penicillin high-producing strains contain several copies of the amplified region, there is no linear correlation between penicillin titers and the penicillin gene copy number, transcript and protein levels [149–152], which suggests that the strain improvement programs introduced other modifications that play an important role in the improvement of penicillin production [14].

Decrease in the Catabolism of Phenylacetic Acid

The benzylpenicillin side chain precursor phenylacetic acid is a weak acid that is toxic according to the concentration and the pH of the culture medium. This compound can be metabolized at least by two different ways in *P. chrysogenum*; incorporation to IPN/6-APA to give rise to benzylpenicillin or catabolic degradation via the homogentisate pathway [153–156].

The first step in the homogentisate pathway consists of the 2-hydroxylation of phenylacetic acid and the formation of 2-hydroxyphenylacetate, which is sequentially catabolized to fumarate and acetoacetate. The 2-hydroxylation step is catalyzed by phenylacetate 2-hydroxylase (EC: 1.14.13), which is a microsomal cytochrome P450 monooxygenase encoded by the *pahA* gene. Sequencing of the *pahA* gene revealed that, whereas the wild-type strain of *P. chrysogenum* (NRRL 1951) contains a C at position 598, the 49-133 strain (and also its derived strain Wisconsin 54-1255) underwent a point mutation in that position (C was replaced by T). This mutation modifies the protein (L181F mutation) and leads to a reduction in the hydroxylase activity. Therefore, in the Wisconsin 54-1255 strain (and presumably also in derived strains) the catabolism of phenylacetic acid is diminished, which is responsible for the accumulation of the side chain precursor and the increase in the production of benzylpenicillin [157].

The importance of the phenylacetate 2-hydroxylase in the production of penicillin was highlighted when the sequence of the *pahA* genes from *P. notatum* (the Fleming's original isolate) and *P. chrysogenum* NRRL 1951 (the wild-type strain isolated from a cantaloupe in Peoria, IL) were compared to each other. The *pahA* gene of the wild-type NRRL 1951 strain shows a point mutation (C1357T) that is translated in a modification in the protein (A394V). This modification is also present in the Wisconsin 54-1255 strain (and presumably also in derived strains), with the consequent decrease in the phenylacetate 2-hydroxylase activity and the increase in penicillin production [158]. Therefore, the consecutive accumulation of point mutations in the first enzyme of the phenylacetic acid catabolism confirms the historical selection of *P. chrysogenum* as industrial producer of penicillin and represents another reason why this filamentous fungus became such a good penicillin producer.

Increase in the Number of Microbodies (Peroxisomes)

Microbodies are organelles involved in the last step of the penicillin biosynthetic pathway, namely the substitution of the side chain catalyzed by IAT and the activation of the side chain precursor by a PCL activity. It has been reported that there is a positive correlation between the microbody abundance and the increase in penicillin titers [159]. In fact, it has been confirmed that the number of microbodies is increased in penicillin high-producing strains [12].

The fact that the penicillin biosynthetic pathway is compartmentalized between the cytosol and microbodies has been critical for productivity and it seems that *P. chrysogenum* has lost the ability to synthesize penicillins in the cytosol [58], since peroxisomes are essential for an efficient biosynthesis of penicillin [60].

Global Metabolic Reorganizations

The modifications indicated above were individually characterized along the past decades and represented the tip of the iceberg. It was not until the recent "omics" era when a global vision of the modifications introduced by the strain improvement programs was provided.

The publication of the complete *P. chrysogenum* Wisconsin 54-1255 genome sequence [12] represented the ultimate springboard for the development of several studies aimed to decipher the secrets underlying the production of penicillin.

Transcriptomics studies indicated that the expression of the genes involved in the biosynthesis of penicillin amino acid precursors (cysteine, valine, and α[alpha]-aminoadipic acid) is increased in high-producing strains. The same trend is followed by some genes encoding microbody proteins. Interestingly, it was also observed that the transcription of the penicillin biosynthetic genes *pcbAB*, *pcbC*, and *penDE* was only twofold higher in high-producing strains [12], which indicates that the phenomenon of tandem amplification of the penicillin gene cluster is not the main reason for the increased productivity in industrial strains.

The recent implementation of the proteomics techniques to *P. chrysogenum* [14], allowed the comparative analysis of three strains (very low-producing wild-type NRRL 1951, low-producing Wisconsin 54-1255, and high-producing AS-P-78), which were representative of different steps of the industrial strain improvement program. A global vision of the differentially represented proteins revealed that virulence and plant cell wall degradative proteins were predominant in the wild-type strain. It is also interesting to note that in the high-producing strain some secondary metabolism pathways were lost (synthesis of terpenoids, pigments, etc.), which would have redirected the metabolic fluxes towards the biosynthesis of beta-lactam antibiotics due to the availability of more precursors and energy. Cysteine biosynthesis (a penicillin amino acid precursor) is also favored in the high-producing strain, together with the formation of NADPH (there is a positive correlation between NADPH and penicillin biosynthesis) through the pentose phosphate pathway. Other proteins that are overrepresented in the high-producing strain belong to the oxidative stress response network [14], which indicates that this is an adaptative mechanism aimed to penicillin overproduction.

Conclusion

The current "omics" era has allowed scientists, 85 years after Fleming's discovery of the penicillin-producing fungus, to start to understand the molecular mechanisms of the increased productivity resulting from six decades of classical industrial strain improvement programs. The recent advances in microbiology, biochemistry, genetics, and molecular biology have contributed to the knowledge of the structural, ancillary, and regulatory genes and enzymes, together with the characterization of the subcellular compartments involved in the biosynthesis of beta-lactam antibiotics.

Data provided by genomics, transcriptomics, proteomics, and metabolomics have revealed the main modifications on primary and secondary metabolism that took place during the taming process of *P. chrysogenum*. These modifications resulted in a careful rebalancing of metabolic and cellular processes, which are responsible for the impressive penicillin titers reached by current industrial strains. These findings will be useful for the scientific community from industry and academia, not only to continue the exploitation of the successful correlation between *P. chrysogenum* and penicillin production, but also to explore the production of other interesting compounds by this filamentous fungus, which has long proven its quality as a versatile cell factory [15].

Despite the old age of penicillin, this wonder drug still continues to provide health to animals and humans, mainly in the form of semisynthetic penicillins and in the combination of amoxicillin with clavulanic acid. The properties exhibited by this beta-lactam antibiotic assure it a long life in medicine as a therapeutic drug and make it to deserve the title of the twentieth century magic bullet.

References

1. Fleming A. On the antibacterial action of a *Penicillium* with special reference to their use in the isolation of *B. influenzae*. Br J Exp Pathol. 1929;10:226–36.
2. Clutterbuck PW, Lovell R, Raistrick H. Studies in the biochemistry of the microorganisms. XXVI. The formation from glucose by members of the *Penicillium chrysogenum* species of a pigment, an alkali soluble protein and penicillin. The antibacterial substance of Fleming. Biochem J. 1932;26:1907–18.
3. Chain E, Florey HW, Gardner AD, Heatley NG, Jennnings MA, Orr-Ewing J, Sanders AG. Penicillin as a chemotherapeutic agent. Lancet. 1940;II:226–8.
4. Moyer AJ, Coghill RD. Penicillin VIII. Production of penicillin in surface cultures. J Bacteriol. 1946;51:57–78.
5. Moyer AJ, Coghill RD. Penicillin IX. The laboratory scale production of penicillin in submerged cultures by *Penicillium notatum* Westling (NRRL 832). J Bacteriol. 1946;51:79–93.
6. Moyer AJ, Coghill RD. Penicillin X. The effect of phenylacetic acid on penicillin production. J Bacteriol. 1947;53:329–41.
7. Raper KB. The development of improved penicillin-producing molds. Ann N Y Acad Sci. 1946;58:41–56.
8. Crowfoot D, Bunn CW, Rogers-Low BW, Turner-Jones A. X-ray crystallographic investigation of the structure of penicillin. In: Clarke HT, Johnson JR, Robinson R, editors. Chemistry of penicillin. Princeton: Princeton University; 1949. p. 310–67.

9. Kato K. Further notes on penicillin-nucleus. J Antibiot (Tokyo). 1953;6:184–5.
10. Batchelor FR, Doyle PD, Nayler JHC, Rolinson GN. Synthesis of penicillin: 6-amino-penicillanic acid in penicillin fermentations. Nature. 1959;183:257–8.
11. Elander RP. Industrial production of beta-lactam antibiotics. Appl Microbiol Biotechnol. 2003;61:385–92.
12. van den Berg MA, Albang R, Albermann K, Badger JH, Daran JM, Driessen AJ, Garcia-Estrada C, Fedorova ND, Harris DM, Heijne WH, Joardar V, Kiel JA, Kovalchuk A, Martín JF, Nierman WC, Nijland JG, Pronk JT, Roubos JA, van der Klei IJ, van Peij NN, Veenhuis M, von Döhren H, Wagner C, Wortman J, Bovenberg RAL. Genome sequencing and analysis of the filamentous fungus *Penicillium chrysogenum*. Nat Biotechnol. 2008;26:1161–68.
13. Harris DM, van der Krogt ZA, Klaassen P, Raamsdonk LM, Hage S, van den Berg MA, Bovenberg RA, Pronk JT, Daran JM. Exploring and dissecting genome-wide gene expression responses of *Penicillium chrysogenum* to phenylacetic acid consumption and penicillin G production. BMC Genomics. 2009;10:75.
14. Jami MS, Barreiro C, García-Estrada C, Martín JF. Proteome analysis of the penicillin producer *Penicillium chrysogenum*: Characterization of protein changes during the industrial strain improvement. Mol Cell Proteomics. 2010;9:1182–98.
15. Jami MS, García-Estrada C, Barreiro C, Cuadrado AA, Salehi-Najafabadi Z, Martín JF. The *Penicillium chrysogenum* extracellular proteome. Conversion from a food-rotting strain to a versatile cell factory for white biotechnology. Mol Cell Proteomics. 2010;9:2729–44.
16. Aharonowitz Y, Cohen G, Martin JF. Penicillin and cephalosporin biosynthetic genes: structure, organization, regulation, and evolution. Annu Rev Microbiol. 1992;46:461–95.
17. O'Sullivan J, Sykes RB. β-Lactam antibiotics. In: Pape H, Rehm H-J, editors. Biotechnology, a comprehensive treatise in 8 volumes, vol. 4. Weinheim: VCH Verlagsgesellschaft; 1986. p. 247–81.
18. Laich F, Fierro F, Cardoza RE, Martin JF. Organization of the gene cluster for biosynthesis of penicillin in *Penicillium nalgiovense* and antibiotic production in cured dry sausages. Appl Environ Microbiol. 1999;65:1236–40.
19. Laich F, Fierro F, Martín JF. Production of penicillin by fungi growing on food products: identification of a complete penicillin gene cluster in *Penicillium griseofulvum* and a truncated cluster in *Penicillium verrucosum*. Appl Environ Microbiol. 2002;68:1211–9.
20. Houbraken J, Frisvad JC, Seifert KA, Overy DP, Tuthill DM, Valdez JG, Samson RA. New penicillin-producing Penicillium species and an overview of section Chrysogena. Persoonia. 2012;29:78–100.
21. Giesbrecht P, Franz M, Krüeger D, Labischinski H, Wecke J. Bacteriolysis of Staphylococci is only a side-effect of penicillin induced death. In: Kleinkauf H, von Döhren H, editors. 50 years of penicillin application, history and trends. Czech Republic: PUBLIC Ltd; 1991. p. 353–63.
22. Frère JM, Nguyen-Distéche M, Coyette J, Joris B. Mode of action: interaction with the penicillin binding proteins. In: Page MI, editor. The chemistry of β-lactams. London: Blackie Academic and Professional; 1993. p. 148–97.
23. Kong KF, Schneper L, Mathee K. Beta-lactam antibiotics: From antibiosis to resistance and bacteriology. Acta Pathol Microbiol Immunol Scandinavica. 2010;118:1–36.
24. Martín JF, Ullán RV, García-Estrada C. Regulation and compartmentalization of beta-lactam biosynthesis. Microb Biotechnol. 2010;3:285–99.
25. Díez B, Gutiérrez S, Barredo JL, van Solingen P, van der Voort LHM, Martín JF. The cluster of penicillin biosynthetic genes. Identification and characterization of the pcbAB gene encoding the α-aminoadipyl-cysteinylvaline synthetase and linkage to the pcbC and penDE genes. J Biol Chem. 1990;265:16358–65.
26. MacCabe AP, Riach MB, Unkles SE, Kinghorn JR. The *Aspergillus nidulans npeA* locus consists of three contiguous genes required for penicillin biosynthesis. EMBO J. 1990;9:279–87.
27. Montenegro E, Barredo JL, Gutiérrez S, Díez B, Alvarez E, Martín JF. Cloning, characterization of the acyl-CoA: 6 amino penicillanic acid acyltransferase gene of *Aspergillus nidulans* and linkage to the isopenicillin N synthase gene. Mol Gen Genet. 1990;221:322–30.

28. Fierro F, Gutiérrez S, Díez B, Martín JF. Resolution of four large chromosomes in penicillin-producing filamentous fungi: the penicillin gene cluster is located on chromosome II (9.6 Mb) in *Penicillium notatum* and chromosome I (10.4 Mb) in *Penicillium chrysogenum*. Mol Gen Genet. 1993;241:573–8.
29. Smith DJ, Burnham MK, Edwards J, Earl AJ, Turner G. Cloning and heterologous expression of the penicillin biosynthetic gene cluster from *Penicillum chrysogenum*. Biotechnology (NY). 1990;8:39–41.
30. Theilgaard HBA, Kristiansen KN, Enriksen CM, Nielsen J. Purification and characterization of δ-(L-α-aminoadipyl)-L-cysteinyl-D-valine synthetase from *Penicillium chrysogenum*. Biochem J. 1997;327:185–91.
31. Martín JF. α-Aminoadipyl-cysteinyl-valine synthetases in β-lactam producing organisms. From Abraham's discoveries to novel concepts of non-ribosomal peptide synthesis. J Antibiot. 2000;53:1008–21.
32. Baldwin JE, Bird JW, Field RA, O'Callaghan NM, Schofield CJ, Willis AC. Isolation y partial characterisation of ACV synthetase from *Cephalosporium acremonium* and *Streptomyces clavuligerus*: evidence for the presence of phosphopantothenate in ACV synthetase. J Antibiot. 1991;44:241–8.
33. Stachelhaus T, Marahiel MA. Modular structure of genes encoding multifunctional peptide synthetases required for non-ribosomal peptide synthesis. FEMS Microbiol Lett. 1995;125:3–14.
34. García-Estrada C, Ullán RV, Velasco-Conde T, Godio RP, Teijeira F, Vaca I, Feltrer R, Kosalková K, Mauriz E, Martín JF. Post-translational enzyme modification by the phosphopantethenyl transferase is required for lysine and penicillin biosynthesis but not for roquefortine or fatty acid formation in *Penicillium chrysogenum*. Biochem J. 2008;415:317–24.
35. Wu X, García-Estrada C, Vaca I, Martín JF. Motifs in the C-terminal region of the *Penicillium chrysogenum* ACV synthetase are essential for valine epimerization and processivity of tripeptide formation. Biochimie. 2012;94:354–64.
36. Kurylowicz W, Kurzatkowski W, Kurzatkowski J. Biosynthesis of benzylpenicillin by *Penicillium chrysogenum* and its Golgi apparatus. Arch Immunol Ther Exp (Warsz). 1987;35:699–724.
37. Müller WH, van der Krift TP, Krouwer AJ, Wösten HA, van der Voort LH, Smaal EB, Verkleij AJ. Localization of the pathway of the penicillin biosynthesis in *Penicillium chrysogenum*. EMBO J. 1991;10:489–95.
38. Lendenfeld T, Ghali D, Wolschek M, Kubicek-Pranz EM, Kubicek CP. Subcellular compartmentation of penicillin biosynthesis in *Penicillium chrysogenum*. The amino acid precursors are derived from the vacuole. J Biol Chem. 1993;268:665–71.
39. van der Lende TR, van de Kamp M, Berg M, Sjollema K, Bovenberg RA, Veenhuis M, Konings WN, Driessen AJ. δ-(L-α- aminoadipyl)-L-cysteinyl-D-valine synthetase, that mediates the first committed step in penicillin biosynthesis, is a cytosolic enzyme. Fungal Genet Biol. 2002;37:49–55.
40. Ramos FR, López-Nieto MJ, Martín JF. Isopenicillin N synthetase of *Penicillium chrysogenum*, an enzyme that converts δ-(L-α-aminoadipyl)-L-cysteinyl-D-valine to isopenicillin N. Antimicrob Agents Chemother. 1985;27:380–7.
41. Perry D, Abraham EP, Baldwin JE. Factors affecting the isopenicillin N synthetase reaction. Biochem J. 1988;255:345–51.
42. Bainbridge ZA, Scott RI, Perry D. Oxygen utilisation by isopenicillin N synthase from *Penicillium chrysogenum*. J Chem Technol Biotechnol. 1992;55:233–8.
43. Cooper RD. The enzymes involved in biosynthesis of penicillin and cephalosporin; their structure and function. Bioorg Med Chem. 1993;1:1–17.
44. Abraham EP, Huddlestone JA, Jayatilake GS, O'Sullivan J, White R. Conversion of δ(L-α-aminoadipyl)-L-cysteinyl-D-valine to isopenicillin N in cell-free extracts of *Cephalosporium acremonium*. In: Gregory GI, editor. Recent advances in the chemistry of beta-lactam antibiotics. London: Royal Society of Chemistry; 1981. p. 125–34.
45. Sawada Y, Baldwin JE, Singh PD, Solomon NA, Demain AL. Cell-free cyclization of delta-(L-alpha-aminoadipyl)-L-cysteinyl-D-valine to isopenicillin N. Antimicrob Agents Chemother. 1980;18:465–70.

46. Carr LG, Skatrud PL, Scheetz ME, Queener SW, Ingolia TD. Cloning and expression of the isopenicillin N synthetase gene from *Penicillium chrysogenum*. Gene. 1986;48:257–66.
47. Barredo JL, Cantoral JM, Álvarez E, Díez B, Martín JF. Cloning, sequence analysis and transcriptional study of the isopenicillin N synthase of *Penicillium chrysogenum* AS-P-78. Mol Gen Genet. 1989;216:91–8.
48. Roach PL, Clifton IJ, Fülöp V, Harlos K, Barton GJ, Hajdu J, Andersson I, Schofield CJ, Baldwin JE. Crystal structure of isopenicillin N synthase is the first from a new structural family of enzymes. Nature. 1995;375:700–4.
49. Roach PL, Clifton IJ, Hensgens CMH, Shibata N, Schofield CJ, Hajdu J, Baldwin JE. Structure of isopenicillin N synthase complexed with substrate and the mechanism of penicillin formation. Nature. 1997;387:827–30.
50. Alvarez E, Meesschaert B, Montenegro E, Gutiérrez S, Díez B, Barredo JL, Martín JF. The isopenicillin-N acyltransferase of *Penicillium chrysogenum* has isopenicillin-N amidohydrolase, 6-aminopenicillanic acid acyltransferase and penicillin amidase activities, all of which are encoded by the single *penDE* gene. Eur J Biochem. 1993;215:323–32.
51. Queener SF, Neuss N. The biosynthesis of β-lactam antibiotics. In: Morin RB, Morgan N, editors. The chemistry and biology of β-lactam antibiotics. London: Academic; 1982. p. 1–81.
52. Barredo JL, van Solingen P, Díez B, Álvarez E, Cantoral JM, Kattevilder A, Smaal EB, Groenen MAM, Veenstra AE, Martín JF. Cloning and characterization of the acyl coenzyme A: 6 amino-penicilanic acid acyltransferase gene of *Penicillium chrysogenum*. Gene. 1989;83:291–300.
53. Whiteman PA, Abraham EP, Baldwin JE, Fleming MD, Schofield CJ, Sutherland JD, Willis AC. Acyl coenzyme A: 6-aminopenicillanic acid acyltransferase from *Penicillium chrysogenum* and *Aspergillus nidulans*. FEBS Lett. 1990;262:342–4.
54. Tobin MB, Fleming MD, Skatrud PL, Miller JR. Molecular characterization of the acylcoenzyme A:isopenicillin N acyltransferase gene (*penDE*) from *Penicillium chrysogenum* and *Aspergillus nidulans* and activity of recombinant enzyme in *Escherichia coli*. J Bacteriol. 1990;172:5908–14.
55. Tobin MB, Baldwin JE, Cole SCJ, Miller JR, Skatrud PL, Sutherland JD. The requirement for subunit interaction in the production of *Penicillium chrysogenum* acyl-coenzyme A: isopenicillin N acyltransferase in *Escherichia coli*. Gene. 1993;132:199–206.
56. Fernández FJ, Cardoza RE, Montenegro E, Velasco J, Gutiérrez S, Martín JF. The isopenicillin N acyltransferases of *Aspergillus nidulans* and *Penicillium chrysogenum* differ in their ability to maintain the 40-kDa alpha beta heterodimer in an undissociated form. Eur J Biochem. 2003;270:1958–68.
57. García-Estrada C, Vaca I, Fierro F, Sjollema K, Veenhuis M, Martín JF. The unprocessed preprotein form IAT^{C103S} of the isopenicillin N acyltransferase is transported inside peroxisomes and regulates its self-processing. Fungal Genet Biol. 2008;45:1043–52.
58. Müller WH, Bovenberg RA, Groothuis MH, Kattevilder F, Smaal EB, van der Voort LH, Verkleij AJ. Involvement of microbodies in penicillin biosynthesis. Biochim Biophys Acta. 1992;1116:210–13.
59. Müller WH, Essers J, Humbel BM, Verkleij AJ. Enrichment of *Penicillium chrysogenum* microbodies by isopycnic centrifugation in nycodenz as visualized with immunoelectron microscopy. Biochim Biophys Acta. 1995;1245:215–20.
60. Meijer WH, Gidijala L, Fekken S, Kiel JA, van den Berg MA, Lascaris R, Bovenberg RA, van der Klei IJ. Peroxisomes are required for efficient penicillin biosynthesis in *Penicillium chrysogenum*. Appl Environ Microbiol. 2010;76:5702–09.
61. Martín JF, García-Estrada C, Ullán RV. Transport of substrates into peroxisomes: the paradigm of β-lactam biosynthetic intermediates. Biomol Concepts. 2013;4:197–211.
62. Spröte P, Hynes MJ, Hortschansky P, Shelest E, Scharf DH, Wolke SM, Brakhage AA. Identification of the novel penicillin biosynthesis gene *aatB* of *Aspergillus nidulans* and its putative evolutionary relationship to this fungal secondary metabolism gene cluster. Mol Microbiol. 2008;70:445–61.
63. García-Estrada C, Vaca I, Ullán RV, van den Berg MA, Bovenberg RA, Martín JF. Molecular characterization of a fungal gene paralogue of the penicillin *penDE* gene of *Penicillium chrysogenum*. BMC Microbiol. 2009;9:104.

64. Lambalot RH, Gehring AM, Flugel RS, Zuber P, LaCelle M, Marahiel MA, Reid R, Khosla C, Walsh CT. A new enzyme superfamily - the phosphopantetheinyl transferases. Chem Biol. 1996;3:923–36.
65. Keszenman-Pereyra D, Lawrence S, Twfieg ME, Price J, Turner G. The *npgA/cfwA* gene encodes a putative 4′-phosphopantetheinyl transferase which is essential for penicillin biosynthesis in *Aspergillus nidulans*. Curr Genet. 2003;43:186–90.
66. Márquez-Fernández O, Trigos A, Ramos-Balderas JL, Viniegra-González G, Deising HB, Aguirre J. Phosphopantetheinyl transferase CfwA/NpgA is required for *Aspergillus nidulans* secondary metabolism and asexual development. Eukaryot Cell. 2007;6:710–20.
67. Cohen G, Argaman A, Schreiber R, Mislovati M, Aharonowitz Y. The thioredoxin system of *Penicillium chrysogenum* and its possible role in penicillin biosynthesis. J Bacteriol. 1994;176:973–84.
68. Gledhill L, Greaves PA, Griffin JP. Phenilacetyl-CoA ligase from *Penicillium chrysogenum*. Patent IPN WO97/02349, Smithkline Beecham, 1997.
69. Lamas-Maceiras M, Vaca I, Rodríguez E, Casqueiro J, Martín JF. Amplification and disruption of the phenylacetyl-CoA ligase gene of *Penicillium chrysogenum* encoding an arylcapping enzyme that supplies phenylacetic acid to the isopenicillin N acyltransferase. Biochem J. 2006;395:147–55.
70. Wang FQ, Liu J, Dai M, Ren ZH, Su CY, He JG. Molecular cloning and functional identification of a novel phenylacetyl-CoA ligase gene from *Penicillium chrysogenum*. Biochem Biophys Res Commun. 2007;360:453–8.
71. Koetsier MJ, Jekel PA, van den Berg MA, Bovenberg RA, Janssen DB. Characterization of a phenylacetate-CoA ligase from *Penicillium chrysogenum*. Biochem J. 2009;417:467–76.
72. Koetsier MJ, Gombert AK, Fekken S, Bovenberg RA, van den Berg MA, Kiel JA, Jekel PA, Janssen DB, Pronk JT, van der Klei IJ, Daran JM. The *Penicillium chrysogenum aclA* gene encodes a broad-substrate-specificity acyl-coenzyme A ligase involved in activation of adipic acid, a side-chain precursor for cephem antibiotics. Fungal Genet Biol. 2010;47:33–42.
73. Yu ZL, Liu J, Wang FQ, Dai M, Zhao BH, He JG, Zhang H. Cloning and characterization of a novel CoA-ligase gene from *Penicillium chrysogenum*. Folia Microbiol (Praha). 2011;56: 246–52.
74. Kubicek CP, Hönlinger CH, Jaklitsch WM, Affenzeller K, Mach R, Gerngross TU, Ying L. Regulation of lysine biosynthesis in the fungus *Penicillium chrysogenum*. In: Lubec G, Rozenthal GA, editors. Amino acids: chemistry, biology and medicine. Leiden: ESCOM; 1990.
75. Jaklitsch WM, Kubicek CP. Homocitrate synthase from *Penicillium chrysogenum*. Localization, purification of the cytosolic isoenzyme, and sensitivity to lysine. Biochem J. 1990;269:247–53.
76. Klionsky DJ, Herman PK, Emr SD. The fungal vacuole: composition, function, and biogenesis. Microbiol Rev. 1990;54:266–92.
77. Kubicek-Pranz EM, Kubicek CP. Production and biosynthesis of amino acids by fungi. In: Arora DK, Elander RP, Mukerji KG, editors. Handbook of applied mycology. vol. 4: fungal biotechnology. New York: Marcel Dekker; 1991. p. 313–56.
78. Horak J. Amino acid transport in eukaryotic microorganisms. Biochim Biophys Acta. 1986;864:223–56.
79. Bañuelos O, Casqueiro J, Gutiérrez S, Riaño J, Martín JF. The specific transport system for lysine is fully inhibited by ammonium in *Penicillium chrysogenum*: an ammonium-insensitive system allows uptake in carbon-starved cells. Antonie Van Leeuwenhoek. 2000;77:91–100.
80. Trip H, Evers ME, Konings WN, Driessen AJ. Cloning and characterization of an aromatic amino acid and leucine permease of *Penicillium chrysogenum*. Biochim Biophys Acta. 1992;1565:73–80.
81. Benko PV, Word TC, Segel IH. Specificity and regulation of methionine transport in filamentous fungi. Arch Biochem Biophys. 1967;122:783–804.
82. Skye GE, Segel IH. Independent regulation of cysteine and cystine transport in *Penicillium chrysogenum*. Arch Biochem Biophys. 1970;138:306–18.
83. Benko PV, Wood TC, Segel IH. Multiplicity and regulation of amino acid transport in *Penicillium chrysogenum*. Arch Biochem Biophys. 1969;129:498–508.

84. Hunter DR, Segel IH. Acidic and basic amino acid transport systems of *Penicillium chrysogenum*. Arch Biochem Biophys. 1971;144:168–83.
85. Evers ME, Trip H, van den Berg MA, Bovenberg RA, Driessen AJ. Compartmentalization and transport in β-lactam antibiotics biosynthesis. Adv Biochem Eng Biotechnol. 2004;88:111–35.
86. Fernández-Aguado M, Teijeira F, Martín JF, Ullán RV. A vacuolar membrane protein affects drastically the biosynthesis of the ACV tripeptide and the beta-lactam pathway of *Penicillium chrysogenum*. Appl Microbiol Biotechnol. 2013;97:795–808.
87. Eriksen SH, Jensen B, Schneider I, Kaasgaard S, Olsen J. Uptake of phenoxyacetic acid by *Penicillium chrysogenum*. Appl Microbiol Biotechnol. 1995;42:945–50.
88. Hillenga DJ, Versantvoort H, van der Molen S, Driessen A, Konings WN. *Penicillium chrysogenum* takes up the penicillin G precursor phenylacetic acid by passive diffusion. Appl Environ Microbiol. 1995;61:2589–95.
89. Fernández-Cañón JM, Reglero A, Martínez-Blanco H, Ferrero MA, Luengo JM. Phenylacetic acid transport system in *Penicillium chrysogenum* Wis 54-1255: molecular specificity of its induction. J Antibiot (Tokyo). 1989;42:1410–15.
90. Fernández-Cañón JM, Reglero A, Martínez-Blanco H, Luengo JM. Uptake of phenylacetic acid by *Penicillium chrysogenum* Wis 54-1255: a critical regulatory point in benzylpenicillin biosynthesis. J Antibiot (Tokyo). 1989;42:1398–409.
91. Martínez-Blanco H, Reglero A, Ferrero MA, Fernández-Cañón JM, Luengo JM. Repression of phenylacetic acid transport system in *Penicillium chrysogenum* Wis 54-1255 by free amino acids and ammonium salts. J Antibiot (Tokyo). 1989;42:1416–23.
92. Eriksen SH, Soderblom TB, Jensen B, Olsen J. Uptake of phenylacetic acid by two strains of *Penicillium chrysogenum*. Biotechnol Bioeng. 1998;60:310–6.
93. Fernández-Aguado M, Ullán RV, Teijeira F, Rodríguez-Castro R, Martín JF. The transport of phenylacetic acid across the peroxisomal membrane is mediated by the PaaT protein in *Penicillium chrysogenum*. Appl Microbiol Biotechnol. 2013;97:3073–84.
94. Yang J, Xu X, Liu G. Amplification of an MFS transporter encoding gene penT significantly stimulates penicillin production and enhances the sensitivity of *Penicillium chrysogenum* to phenylacetic acid. J Genet Genomics. 2012;39:593–602.
95. Demain AL. Biosynthesis of β-lactam antibiotics. In: Demain AL, Solomon NA, editors. Antibiotics containing the β-lactam structure. Berlin: Springer; 1983. p. 189–228.
96. López-Nieto MJ, Ramos FR, Luengo JM, Martín JF. Characterization of the biosynthesis in vivo of α-aminoadipyl-cysteinylvaline in *Penicillium chrysogenum*. Appl Microbiol Biotechnol. 1985;22:343–51.
97. Meister A, Anderson ME. Glutathione. Annu Rev Biochem. 1983;52:711–60.
98. del Carmen MR, Sánchez S. Transport of neutral amino acids and penicillin formation in *Penicillium chrysogenum*. J Gen Genet. 1990;254:417–26.
99. García-Estrada C, Vaca I, Lamas-Maceiras M, Martín JF. In vivo transport of the intermediates of the penicillin biosynthetic pathway in tailored strains of *Penicillium chrysogenum*. Appl Microbiol Biotechnol. 2007;76:169–82.
100. Ullán RV, Teijeira F, Guerra SM, Vaca I, Martín JF. Characterization of a novel peroxisome membrane protein essential for conversion of isopenicillin N into cephalosporin C. Biochem J. 2010;432:227–36.
101. Ullán RV, Teijeira F, Martín JF. Expression of the *Acremonium chrysogenum cefT* gene in *Penicillum chrysogenum* indicates that it encodes an hydrophilic beta-lactam transporter. Curr Genet. 2008;54:153–61.
102. Nijland JG, Kovalchuk A, van den Berg MA, Bovenberg RA, Driessen AJ. Expression of the transporter encoded by the *cefT* gene of *Acremonium chrysogenum* increases cephalosporin production in *Penicillium chrysogenum*. Fungal Genet Biol. 2008;45:1415–21.
103. Kirschbaum J. Penicillin G potassium. In: Florey K, editor. Analytical profiles of drugs substances, vol. 15. New York: Academic; 1986. p. 427–509.
104. Hillenga DJ, Versantvoor HJ, Driessen AJ, Konings WN. Structural and functional properties of plasma membranes from the filamentous fungus *Penicillium chrysogenum*. Eur J Biochem. 1994;224:581–7.

105. van de Kamp M, Driessen AJ, Konings WN. Compartmentalization and transport in β-lactam antibiotic biosynthesis by filamentous fungi. Antonie Van Leeuwenhoek. 1999;75:41–78.
106. Andrade AC, Van Nistelrooy JG, Peery RB, Skatrud PL, De Waard MA. The role of ABC transporters from *Aspergillus nidulans* in protection against cytotoxic agents and in antibiotic production. Mol Gen Genet. 2000;263:966–97.
107. Roze LV, Chanda A, Linz JE. Compartmentalization and molecular traffic in secondary metabolism: a new understanding of established cellular processes. Fungal Genet Biol. 2010;48:35–48.
108. Bartoszewska M, Kiel JA, Bovenberg RA, Veenhuis M, van der Klei IJ. Autophagy deficiency promotes beta-lactam production in *Penicillium chrysogenum*. Appl Environ Microbiol. 2011;77:1413–22.
109. Brakhage AA. Molecular regulation of β-lactam biosynthesis in filamentous fungi. Microbiol Mol Biol Rev. 1998;62:547–85.
110. Martín JF, Casqueiro J, Kosalková K, Marcos AT, Gutiérrez S. Penicillin and cephalosporin biosynthesis: mechanism of carbon catabolite regulation of penicillin production. Antonie Van Leeuwenhoek. 1999;75:21–31.
111. Martín JF. Molecular control of expression of penicillin biosynthesis genes in fungi: regulatory proteins interact with a bidirectional promoter region. J Bacteriol. 2000;182:2355–62.
112. Brakhage AA, Spröte P, Al-Abdallah Q, Gehrke A, Plattner H, Tüncher A. Regulation of penicillin biosynthesis in filamentous fungi. Adv Biochem Eng Biotechnol. 2004;88:45–90.
113. Fierro F, García-Estrada C, Castillo NI, Rodríguez R, Velasco-Conde T, Martín JF. Transcriptional and bioinformatic analysis of the 56.8 Kb DNA region amplified in tandem repeats containing the penicillin gene cluster in *Penicillium chrysogenum*. Fungal Genet Biol. 2006;43:618–29.
114. van den Berg MA, Westerlaken I, Leeflang C, Kerkman R, Bovenberg RA. Functional characterization of the penicillin biosynthetic gene cluster of *Penicillium chrysogenum* Wisconsin 54-1255. Fungal Genet Biol. 2007;44:830–44.
115. Revilla G, Ramos FR, López-Nieto MJ, Álvarez E, Martín JF. Glucose represses formation of delta-(L-alpha-aminoadipyl)-L-cysteinyl-D-valine and isopenicillin N synthase but not penicillin acyltransferase in *Penicillium chrysogenum*. J Bacteriol. 1986;168:947–52.
116. Brakhage AA, Browne P, Turner G. Regulation of *Aspergillus nidulans* penicillin biosynthesis and penicillin biosynthesis genes *acvA* and *ipnA* by glucose. J Bacteriol. 1992;174: 3789–99.
117. Espeso EA, Peñalva MA. Carbon catabolite repression can account for the temporal pattern of expression of a penicillin biosynthetic gene in *Aspergillus nidulans*. Mol Microbiol. 1992;6:1457–65.
118. Espeso EA, Tilburn J, Arst Jr HN, Peñalva MA. pH regulation is a major determinant in expression of a fungal penicillin biosynthetic gene. EMBO J. 1993;12:3947–56.
119. Tilburn J, Sarkar S, Widdick DA, Espeso EA, Orejas M, Mungroo J, Peñalva MA, Arst Jr HN. The *Aspergillus* PacC zinc finger transcription factor mediates regulation of both acid- and alkaline-expressed genes by ambient pH. EMBO J. 1995;14:779–90.
120. Suárez T, Peñalva MA. Characterization of a *Penicillium chrysogenum* gene encoding a PacC transcription factor and its binding sites in the divergent *pcbAB-pcbC* promoter of the penicillin biosynthetic cluster. Mol Microbiol. 1996;20:529–40.
121. Orejas M, Espeso EA, Tilburn J, Sarkar S, Arst Jr HN, Peñalva MA. Activation of the Aspergillus PacC transcription factor in response to alkaline ambient pH requires proteolysis of the carboxy-terminal moiety. Genes Dev. 1995;13:1622–32.
122. Marzluf GA. Genetic regulation of nitrogen metabolism in the fungi. Microbiol Mol Biol Rev. 1997;61:17–32.
123. Haas H, Marzluf GA. NRE, the major nitrogen regulatory protein of *Penicillium chrysogenum*, binds specifically to elements in the intergenic promoter regions of nitrate assimilation and penicillin biosynthetic gene clusters. Curr Genet. 1995;28:177–83.
124. Kolar M, Holzmann K, Weber G, Leitner E, Schwab H. Molecular characterization and functional analysis in *Aspergillus nidulans* of the 5′-region of the *Penicillium chrysogenum* isopenicillin N synthetase gene. J Biotechnol. 1991;17:67–80.

125. Demain AL. Inhibition of penicillin formation by lysine. Arch Biochem Biophys. 1957;67:244–6.
126. Swartz RW. Penicillins. In: Blanch HW, Drew S, Wang DIC, editors. Comprehensive biotechnology. The principles, applications and regulations of biotechnology in industry, agriculture and medicine. The practice of biotechnology: Current commodity products, vol. 3. Oxford: Pergamon; 1985. p. 7–47.
127. Hilgendorf P, Heiser V, Diekmann H, Thoma M. Constant dissolved oxygen concentrations in cephalosporin C fermentation: applicability of different controllers and effect on fermentation parameters. Appl Microbiol Biotechnol. 1987;27:247–51.
128. Renno DV, Saunders G, Bull AT, Holt G. Transcript analysis of penicillin genes from *Penicillium chrysogenum*. Curr Genet. 1992;21:49–54.
129. Martín J, García-Estrada C, Rumbero A, Recio E, Albillos SM, Ullán RV, Martín JF. Characterization of an auto inducer of penicillin biosynthesis in *Penicillium chrysogenum*. Appl Environ Microbiol. 2011;77:5688–96.
130. Martín J, García-Estrada C, Kosalková K, Ullán RV, Albillos SM, Martín JF. The inducers 1,3-diaminopropane and spermidine produce a drastic increase in the expression of the penicillin biosynthetic genes for prolonged time, mediated by the LaeA regulator. Fungal Genet Biol. 2012;49:1004–13.
131. García-Estrada C, Barreiro C, Jami MS, Martín-González J, Martín JF. The inducers 1,3-diaminopropane and spermidine cause the reprogramming of metabolism in *Penicillium chrysogenum*, leading to multiple vesicles and penicillin overproduction. J Proteomics. 2013;85C:129–59.
132. Liggett RW, Koffler H. Corn steep liquor in microbiology. Bacteriol Rev. 1948;12:297–311.
133. Cove DJ. The induction and repression of nitrate reductase in the fungus *Aspergillus nidulans*. Biochim Biophys Acta. 1966;113:51–6.
134. Peñalva MA, Espeso E, Orejas M, Gómez-Pardo E. Evolution and control of gene expression of the *Aspergillus nidulans* isopenicillin N synthetase gene. In: Kleinkauf H, von Döhren H, editors. Fifty years of penicillin application. Czech Republic: PUBLIC Ltd; 1991. p. 224–30.
135. Peñalva MA, Espeso E, Orejas M, Gómez-Pardo E. Penicillin production by *Aspergillus nidulans*: studies on the regulation of expression of the IPNS gene. In: Stahl U, Tudzinsky P, editors. Molecular biology of filamentous fungi. Weinheim: VCH Verlagsgesellschaft; 1991. p. 217–27.
136. Bok JW, Keller NP. LaeA, a regulator of secondary metabolism in *Aspergillus* spp. Eukaryot Cell. 2004;3:527–35.
137. Kosalková K, García-Estrada C, Ullán RV, Godio RP, Feltrer R, Teijeira F, Mauriz E, Martín JF. The global regulator LaeA controls penicillin biosynthesis, pigmentation and sporulation, but not roquefortine C synthesis in *Penicillium chrysogenum*. Biochimie. 2009;91:214–25.
138. Kato N, Brooks W, Calvo AM. The expression of sterigmatocystin and penicillin genes in *Aspergillus nidulans* is controlled by *veA*, a gene required for sexual development. Eukaryot Cell. 2003;2:1178–86.
139. Spröte P, Brakhage AA. The light-dependent regulator velvet A of Aspergillus nidulans acts as a repressor of the penicillin biosynthesis. Arch Microbiol. 2007;188:69–79.
140. Hoff B, Kamerewerd J, Sigl C, Mitterbauer R, Zadra I, Kürnsteiner H, Kück U. Two components of a velvet-like complex control hyphal morphogenesis, conidiophore development, and penicillin biosynthesis in *Penicillium chrysogenum*. Eukaryot Cell. 2010;9:1236–50.
141. Kopke K, Hoff B, Bloemendal S, Katschorowski A, Kamerewerd J, Kück U. Members of the *Penicillium chrysogenum* velvet complex play functionally opposing roles in the regulation of penicillin biosynthesis and conidiation. Eukaryot Cell. 2013;12:299–310.
142. Domínguez-Santos R, Martín JF, Kosalková K, Prieto C, Ullán RV, García-Estrada C. The regulatory factor PcRFX1 controls the expression of the three genes of β-lactam biosynthesis in Penicillium chrysogenum. Fungal Genet Biol. 2012;49:866–81.
143. Bergh KT, Litzka O, Brakhage AA. Identification of a major cis-acting DNA element controlling the bidirectionally transcribed penicillin biosynthesis genes *acvA* (*pcbAB*) and *ipnA* (*pcbC*) of *Aspergillus nidulans*. J Bacteriol. 1996;178:3908–16.

144. Litzka O, Then Bergh K, Brakhage AA. The *Aspergillus nidulans* penicillin-biosynthesis gene *aat* (*penDE*) is controlled by a CCAAT-containing DNA element. Eur J Biochem. 1996;238:675–82.
145. Caruso ML, Litzka O, Martic G, Lottspeich F, Brakhage AA. Novel basic-region helix-loop-helix transcription factor (AnBH1) of *Aspergillus nidulans* counteracts the CCAAT-binding complex AnCF in the promoter of a penicillin biosynthesis gene. J Mol Biol. 2002;323:425–39.
146. Kosalková K, Marcos AT, Fierro F, Hernando-Rico V, Gutiérrez S, Martín JF. A novel heptameric sequence (TTAGTAA) is the binding site for a protein required for high level expression of *pcbAB*, the first gene of the penicillin biosynthesis in *Penicillium chrysogenum*. J Biol Chem. 2000;275:2423–30.
147. García-Rico RO, Fierro F, Mauriz E, Gómez A, Fernández-Bodega MA, Martín JF. The heterotrimeric Galpha protein pga1 regulates biosynthesis of penicillin, chrysogenin and roquefortine in *Penicillium chrysogenum*. Microbiology. 2008;154:3567–78.
148. Fierro F, Barredo JL, Díez B, Gutiérrez S, Fernández FJ, Martín JF. The penicillin gene cluster is amplified in tandem repeats linked by conserved hexanucleotide sequences. Proc Natl Acad Sci U S A. 1995;92:6200–4.
149. Newbert RW, Barton B, Greaves P, Harper J, Turner G. Analysis of a commercially improved *Penicillium chrysogenum* strain series: involvement of recombinogenic regions in amplification and deletion of the penicillin biosynthesis gene cluster. J Ind Microbiol Biotechnol. 1997;19:18–27.
150. Smith DJ, Bull JH, Edwards J, Turner G. Amplification of the isopenicillin N synthetase gene in a strain of *Penicillium chrysogenum* producing high levels of penicillin. Mol Gen Genet. 1989;216:492–7.
151. Nijland JG, Ebbendorf B, Woszczynska M, Boer R, Bovenberg RA, Driessen AJ. Nonlinear biosynthetic gene cluster dose effect on penicillin production by *Penicillium chrysogenum*. Appl Environ Microbiol. 2010;76:7109–15.
152. Smidák R, Jopcík M, Kralovicová M, Gajdosíková J, Kormanec J, Timko J, Turna J. Core promoters of the penicillin biosynthesis genes and quantitative RT-PCR analysis of these genes in high and low production strain of *Penicillium chrysogenum*. Folia Microbiol (Praha). 2010;55:126–32.
153. Fernández-Cañón JM, Peñalva MA. Overexpression of two penicillin structural genes in *Aspergillus nidulans*. Mol Gen Genet. 1995;246:110–18.
154. Mingot JM, Peñalva MA, Fernández-Cañón JM. Disruption of *phacA*, an *Aspergillus nidulans* gene encoding a novel cytochrome P450 monooxygenase catalyzing phenylacetate 2-hydroxylation, results in penicillin overproduction. J Biol Chem. 1999;274:14545–50.
155. Arias-Barrau E, Olivera ER, Luengo JM, Fernández C, Galán B, García JL, Díaz E, Miñambres B. The homogentisate pathway: a central catabolic pathway involved in the degradation of L-phenylalanine, L-tyrosine, and 3-hydroxyphenylacetate in *Pseudomonas putida*. J Bacteriol. 2004;186:5062–77.
156. Ferrer-Sevillano F, Fernández-Cañón JM. Novel *phacB*-encoded cytochrome P450 monooxygenase from *Aspergillus nidulans* with 3-hydroxyphenylacetate 6-hydroxylase and 3,4-dihydroxyphenylacetate 6-hydroxylase activities. Eukaryot Cell. 2007;6:514–20.
157. Rodríguez-Sáiz M, Barredo JL, Moreno MA, Fernández-Cañón JM, Peñalva MA, Díez B. Reduced function of a phenylacetate-oxidizing cytochrome P450 caused strong genetic improvement in early phylogeny of penicillin-producing strains. J Bacteriol. 2001;183:5465–71.
158. Rodríguez-Sáiz M, Díez B, Barredo JL. Why did the Fleming strain fail in penicillin industry? Fungal Genet Biol. 2005;42:464–70.
159. Kiel JA, van der Klei IJ, van den Berg MA, Bovenberg RA, Veenhuis M. Overproduction of a single protein, Pc-Pex11p, results in 2-fold enhanced penicillin production by *Penicillium chrysogenum*. Fungal Genet Biol. 2005;42:154–64.

Chapter 3
Cephalosporins

Sandra Bloemendal and Ulrich Kück

Introduction

Alexander Fleming's initial observation of the antibacterial effect of fungal secondary metabolites in 1928 was an important milestone in medicine and health. Before this finding, bacterial infections were often lethal, but the discovery and subsequent purification of penicillin paved the way for controlling diseases caused by Gram-positive bacteria. The discovery and development of other antibiotics soon followed, broadening the spectrum of activity. In 1945, Guiseppe Brotzu isolated the ascomycete *Acremonium chrysogenum* from Sardinian coastal seawater, and described the antibiotic effect of extracts generated from this fungus. This active compound was called cephalosporin C, and its structure was determined some years later [1, 2]. In contrast to penicillin, which is mainly active against Gram-positive bacteria, cephalosporin C is a broadband antibiotic that affects both Gram-positive and Gram-negative bacteria. However, the original natural product exerts only weak antibiotic activity. This activity has been gradually increased through the generation of semisynthetic derivatives, representing a good example of the effective chemical modification of natural products [3].

In addition to penicillin and cephalosporin C, other β-lactam antibiotics have now also been discovered. β-lactam antibiotics can be divided into five subgroups: penicillins, cephalosporins, clavams, carbapenems, and monobactams (for an overview see [4]). They are produced by a wide variety of microorganisms, including filamentous fungi such as *Penicillium chrysogenum* and *A. chrysogenum*, Gram-positive streptomycetes, and a small number of Gram-negative bacteria [4]. All of these organisms produce β-lactams through essentially the same biosynthesis pathway, which has been well characterized chemically and kinetically owing to the

S. Bloemendal, Ph.D. • U. Kück, Ph.D. (✉)
Christian Doppler Laboratory for Fungal Biotechnology, Lehrstuhl für Allgemeine und Molekulare Botanik, Ruhr-Universität Bochum, Bochum, Germany
e-mail: ulrich.kueck@ruhr-uni-bochum.de

J.-F. Martín et al. (eds.), *Biosynthesis and Molecular Genetics of Fungal Secondary Metabolites*, Fungal Biology, DOI 10.1007/978-1-4939-1191-2_3,

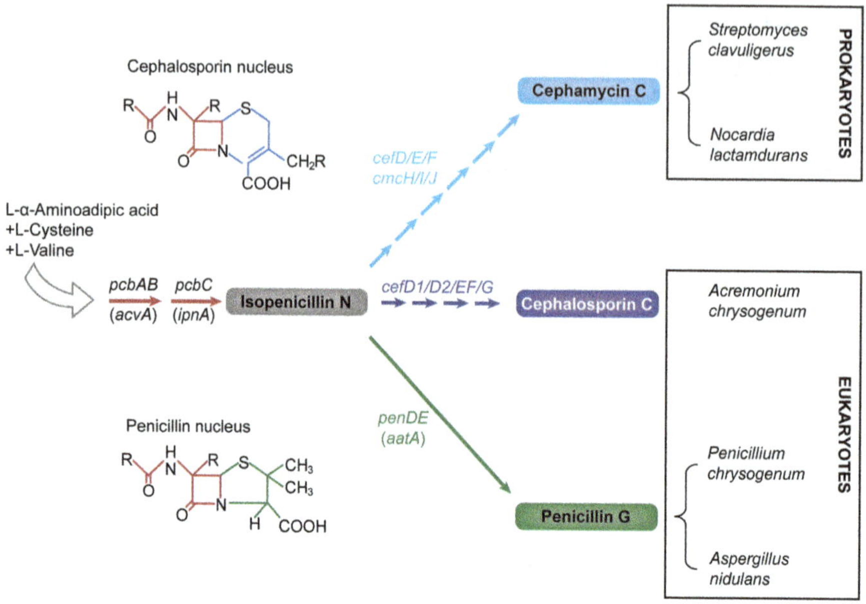

Fig. 3.1 β-lactam antibiotic biosynthesis in prokaryotes and eukaryotes. The first two steps of β-lactam biosynthesis are common among all producer organisms. The amino acid precursors L-α-aminoadipic acid, L-cysteine, and L-valine are combined to form the tripeptide δ-(L-α-aminoadipyl)-L-cysteinyl-D-valine (ACV). This step is catalyzed by the ACV synthase (ACVS), which is encoded by *pcbAB* (*acvA*). The second step is the cyclization of the linear ACV tripeptide, which is mediated by the IPNS, encoded by *pcbC* (*ipnA*). This step produces the first bioactive intermediate, isopenicillin N (IPN), which comprises a four-membered β-lactam ring fused to the five-membered thiazolidine ring. IPN formation is the branch point of β-lactam biosynthesis. In prokaryotes (e.g., *Streptomyces clavuligerus* and *Nocardia lactamdurans*), the subsequent step is the formation of cephamycin C. The eukaryote *Acremonium chrysogenum* produces the antibiotic compound cephalosporin C, which shows strong similarities to cephamycin C. Other eukaryotes, such as *Penicillium chrysogenum* and *Aspergillus nidulans*, form penicillin

considerable industrial potential of these antibiotics (Fig. 3.1) [5–7]. In particular, all β-lactam antibiotics share the same first two biosynthesis steps, in which a tripeptide is built from the amino acid precursors L-α-aminoadipic acid, L-cysteine, and L-valine, and then this tripeptide is converted into isopenicillin N (IPN), the first intermediate with an antibiotic effect and the characteristic β-lactam ring [4, 7, 8]. β-lactam antibiotics are the major anti-infective agents worldwide, having an estimated world market of about US $22 billion at the dosage form level [6], of which cephalosporin C and its semisynthetic derivatives hold a market share of 50 %.

In this chapter, we focus on the biosynthesis of cephalosporin C and its derivatives, and will further elucidate the molecular genetics of its sole industrial producer *A. chrysogenum*.

Biosynthesis Pathway

While the *P. chrysogenum* genes involved in penicillin biosynthesis are located on a single 17-kb cluster, cephalosporin C biosynthesis by *A. chrysogenum* requires two clusters of genes on two different chromosomes [9, 10]. Genes involved in the first four synthesis steps are contained in the so-called "early" cluster, which is located on either chromosome VI or VII [11–13]. Genes responsible for the last four steps of cephalosporin C production are found in the "late" cluster, which is located on either chromosome I or II (Fig. 3.2) [11, 12]. The chromosomal locations of the clusters differ depending on the industrial production strain being analyzed, indicating that chromosome rearrangements have occurred during strain development.

The first reaction of cephalosporin C biosynthesis involves the combination of three amino acid precursors—L-α-aminoadipic acid (an intermediate in the lysine biosynthetic pathway), L-cysteine, and L-valine—to form the tripeptide δ-(L-α-aminoadipyl)-L-cysteinyl-D-valine (ACV). This step is catalyzed by the *pcbAB*-encoded ACV synthetase (ACVS), a monomeric 420-kDa enzyme that binds all three amino acid precursors and activates them in an ATP-dependent step to generate the ACV peptide. The function of this enzyme is similar to that of other bacterial or fungal peptide synthetases in that it mediates the non-ribosomal synthesis of peptides via a multiple-carrier thiotemplate mechanism [14, 15]. The ACVS contains three repeated modules with conserved amino acid sequences, each including domains involved in activation, thiolation, and condensation of the amino acid precursors [16, 17]. The C-terminal region comprises the epimerase and the thioesterase domains that mediate the epimerization of LLL-ACV to LLD-ACV and the hydrolysis of the thioester bond [17].

The second step of cephalosporin C biosynthesis is the cyclization of the linear ACV tripeptide, which is mediated by the isopenicillin N synthase (IPNS), a small 38-kDa dioxygenase encoded by the *pcbC* gene. This step forms the first bioactive intermediate, IPN, comprising a four-membered β-lactam ring fused to the five-membered thiazolidine ring [18]. IPN formation is the branch point of penicillin

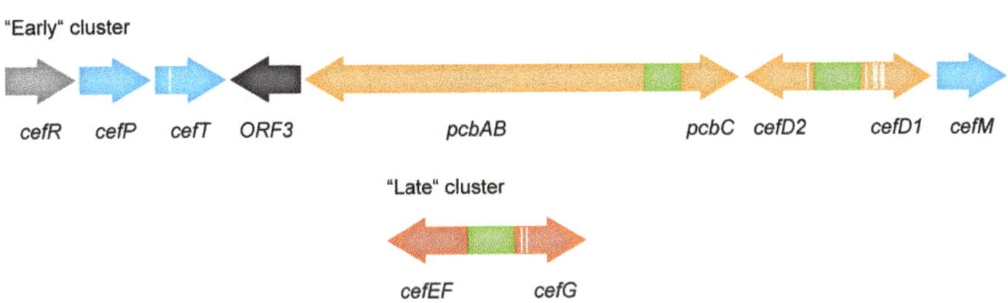

Fig. 3.2 Positions and orientations of genes in the "early" and "late" cephalosporin C clusters. Genes that encode enzymes of the biosynthesis pathway are shown in orange, whereas genes that encode transporters are depicted in blue. The regulatory gene *cefR* is presented in light gray. The gene pairs *pcbAB*/*pcbC*, *cefD1*/*cefD2*, and *cefEF*/*cefG* are controlled by bidirectional promoters, illustrated in green. Introns are shown in white

and cephalosporin C biosynthesis. Both *pcbAB* and *pcbC* are found in all β-lactam-producing microorganisms. Figure 3.2 shows the chromosomal organization of both genes, which share a bidirectional promoter and are intron-free, suggesting that they were passed from prokaryotes to eukaryotes by horizontal gene transfer [19]. The corresponding gene products have a cytoplasmic localization, indicating that the first two biosynthesis steps occur in the cytoplasm [20].

The third step of cephalosporin C biosynthesis involves an epimerization reaction that converts IPN to penicillin N, thereby initiating the pathway that is specific for cephalosporin synthesis. This is the last step that is catalyzed by products of genes in the "early" cluster, namely *cefD1* and *cefD2*. *cefD1* encodes an approximately 71-kDa protein with a high degree of similarity to other eukaryotic acyl-CoA synthetases, and *cefD2* encodes a 41-kDa protein with similarities to acyl-CoA racemases (Fig. 3.2) [21]. This two-component epimerization system found in *A. chrysogenum* differs from that in prokaryotes, in which this reaction only involves one enzyme encoded by the *cefD* gene [22]. The *cefD* gene is completely different from *cefD1* and *cefD2*, particularly in that the latter share a bidirectional promoter and have introns, supporting their eukaryotic origin. Putative peroxisomal targeting signals (PTSs) have been identified for both *A. chrysogenum* proteins, indicating that the epimerization of IPN to penicillin N takes place in peroxisomes [23]. This observation is supported by the identification of the *cefP* gene, located upstream of *pcbAB* in the "early" cephalosporin C cluster, which encodes a membrane protein with 11 transmembrane domains, and a Pex19-binding site that has been described in other proteins as mediating integration into the peroxisomal membrane [24]. A *cefP* disruption mutant is unable to convert IPN to penicillin N, and is therefore incapable of producing cephalosporin C [25].

The last steps of the cephalosporin C production pathway occur in the cytosol, and therefore penicillin N must be exported from the peroxisomes. The *cefM* gene was recently identified in the "early" cephalosporin C cluster downstream of *cefD1* (Fig. 3.2). The disrupted *cefM* mutant accumulates a significant amount of penicillin N, and is unable to synthesize cephalosporin C. Furthermore, the CefM protein localizes to peroxisomes, indicating its involvement in the export of penicillin N from peroxisomes to the cytoplasm [26].

Once in the cytoplasm, penicillin N is converted to deacetoxycephalosporin C (DAOC) through a two-step reaction in which the five-membered thiazolidine ring is expanded to the six-membered dihydrothiazine ring that is characteristic of cephalosporins, and then to deacetylcephalosporin C (DAC). In *A. chrysogenum*, this reaction is catalyzed by a single enzyme: an approximately 37-kDa DAOC/DAC synthetase that has both expandase and hydroxylase activities. This enzyme is encoded by the *cefEF* gene, which is one of the two genes present in the "late" cephalosporin C cluster and does not possess any intron sequences, suggesting a prokaryotic origin [27]. In contrast, antibiotic-producing streptomycetes carry two different genes, *cefE* and *cefF*, which are linked together in a single cluster by the *cefD* gene (for an overview see [19]).

The second gene in the "late" cluster in *A. chrysogenum* is *cefG*, which shares a bidirectional promoter with *cefEF* (Fig. 3.2). *cefG* encodes the last enzyme of the cephalosporin C production pathway, which is the approximately 50-kDa acetyl-CoA:DAC acetyltransferase [28, 29]. In the final reaction, an acetyl moiety is transferred from the acetyl coenzyme A to the hydroxyl group on the sulfur-containing ring of DAC, resulting in the final product cephalosporin C, which possesses high antibiotic activity (for an overview see [4, 7, 8]).

Cephalosporin Derivatives

Brotzu discovered *A. chrysogenum* and its secondary metabolite cephalosporin C in 1945, after which the compound was isolated and purified, leading to determination of the cephalosporin C structure in the late 1950s [30]. Cephalosporin C was active against both Gram-positive and Gram-negative bacteria; however, the antibiotic activity of the original compound was quite low. Therefore, right from the start, research was focused on the generation of semisynthetic derivatives with elevated antibiotic activity. There are two main groups of cephalosporin antibiotics: one derived from penicillin (G or V), the second from the chemical nucleus of cephalosporin C, 7-aminocephalosporanic acid (7-ACA). The 7-ACA compound was first isolated by Abraham and co-workers in 1959; however, sufficient quantities were not produced until some years later [31]. This advancement in production was shortly followed by the launch of the first semisynthetic cephalosporin C antibiotic, cefalotin, in 1964. To date, more than 50 derivatives exist, making cephalosporins by far the most successful group of antibiotics [32].

The first generation of cephalosporins display good activity against several Gram-positive bacteria and are relatively resistant to penicillinase; however, they have rather low activity against Gram-negative bacteria and enterococci. They are active against *E. coli* and *Klebsiella pneumoniae*, but they do not have activity against *Pseudomonas* [32, 33]. Three derivatives of the first generation are currently licensed in the European Union, namely cefazolin, cephalexin, and cefadroxil (Fig. 3.3). The limitations of the first generation of cephalosporins resulted in the development of the second generation, which includes cefaclor and cefuroxime (Fig. 3.3). This group is characterized by slightly reduced activity against Gram-positive species, but has the advantages of showing clearly increased effects against Gram-negative bacteria, including *Proteus* spp. and *Enterobacter* spp., as well as higher resistance towards β-lactamases [8].

The third generation of cephalosporins comprises cefotaxime, cefixime, and cefpodoxime (Fig. 3.3), and displays even further increases in activity against Gram-negative species, especially enterobacteria. Their reduced activity against staphylococci is compensated by strong activity against streptococci, and most members of the third generation also inhibit *Pseudomonas aeruginosa* [32, 33]. While the first generation shows good activity against Gram-positive bacteria, the third generation is characterized by increased effects against Gram-negative bacteria. Members of the fourth generation, such as cefepime (Fig. 3.3), exhibit the combination of both advantages [32]. Furthermore, pharmaceutical companies have recently announced

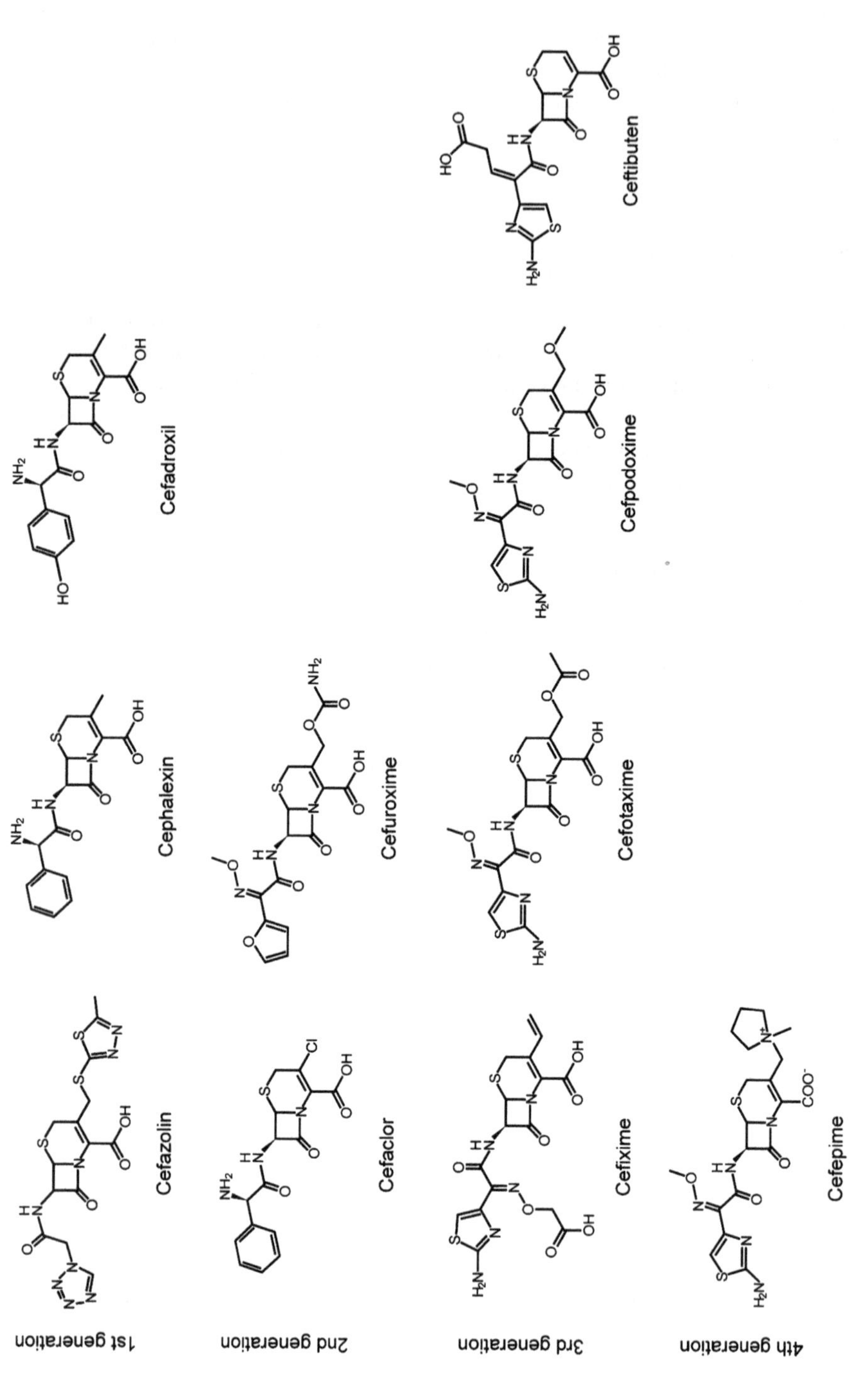

Fig. 3.3 Cephalosporin derivatives. Selection of derivatives that are currently licensed for treatment in human medicine within the EU [32, 102]

the development of fifth-generation cephalosporins that particularly focus on the increasing problem with methicillin-resistant *Staphylococcus aureus* (MRSA). Recently, ceftobiprole became the first broad-spectrum cephalosporin with activity against MRSA to be assessed in late-stage clinical trials [34, 35], thus indicating that cephalosporin derivatives remain extremely valuable antibiotics for current and future medical applications.

Acremonium chrysogenum: The Sole Producer of Cephalosporin C

Cephalosporin C is exclusively produced by the ascomycete *A. chrysogenum*. The term cephalosporin was derived from the original name of this fungus, *Cephalosporium acremonium*. The genus name *Cephalosporium* was formerly used for colorless molds with simple unbranched conidiophores and conidiogenous cells bearing a group or "head" of unicellular conidia at the tip [36]. However, more recent observations led to the integration of most *Cephalosporium* species into the *Acremonium* group, thus resulting in the renaming of the cephalosporin C producer as *Acremonium chrysogenum* [37].

Acremonium is a highly polyphyletic taxon with affiliations to at least three ascomycetous orders. Phylogenetic analyses have revealed that *A. chrysogenum* belongs to the order Hypocreales [38]. To date, no sexual life cycle has been identified for *A. chrysogenum* and it is only known to propagate asexually; therefore, it belongs to the so-called fungi imperfecti [39, 40], a group that comprises many industrially relevant fungi. However, a sexual cycle has eventually been discovered for some fungi of this group, such as *P. chrysogenum* and *Trichoderma reesei* [41, 42]. Despite the lack of a known sexual cycle, it was recently shown that *A. chrysogenum* carries a functional mating-type locus, indicating that a sexual cycle may exist [43]. Sexual recombination of *A. chrysogenum* strains under controlled laboratory conditions would be a valuable tool for strain improvement programs aiming to optimize cephalosporin C production for industrial applications [44].

On malt extract agar, *A. chrysogenum* exhibits a slow growth rate and forms yeast-like colonies of 8–15 mm in diameter. Vegetative sporulation is sparse and conidiophores are indistinguishable from the mycelium. The conidia are short or long ellipsoidal, with a flattened base and a size of about 5×2 μm (Fig. 3.4(a)) (for an overview see [36]). During cellular differentiation of *A. chrysogenum*, the direct fragmentation of swollen hyphae can be observed. Such fragmentation results in the formation of uni- or bicellular arthrospores, also called "yeast-like cells," which are clearly distinguishable from conidia based on their morphology and formation (Fig. 3.4(b)). While conidia arise from the ends of conidiophores that are formed laterally from vegetative hyphae, arthrospores are generated through hyphal fragmentation during prolonged cultivation under limited nutrient conditions [40, 45].

Arthrospores are metabolically active cells that are enriched with intracellular organelles and lipid-containing vacuoles [46]. During cultivation, arthrospore

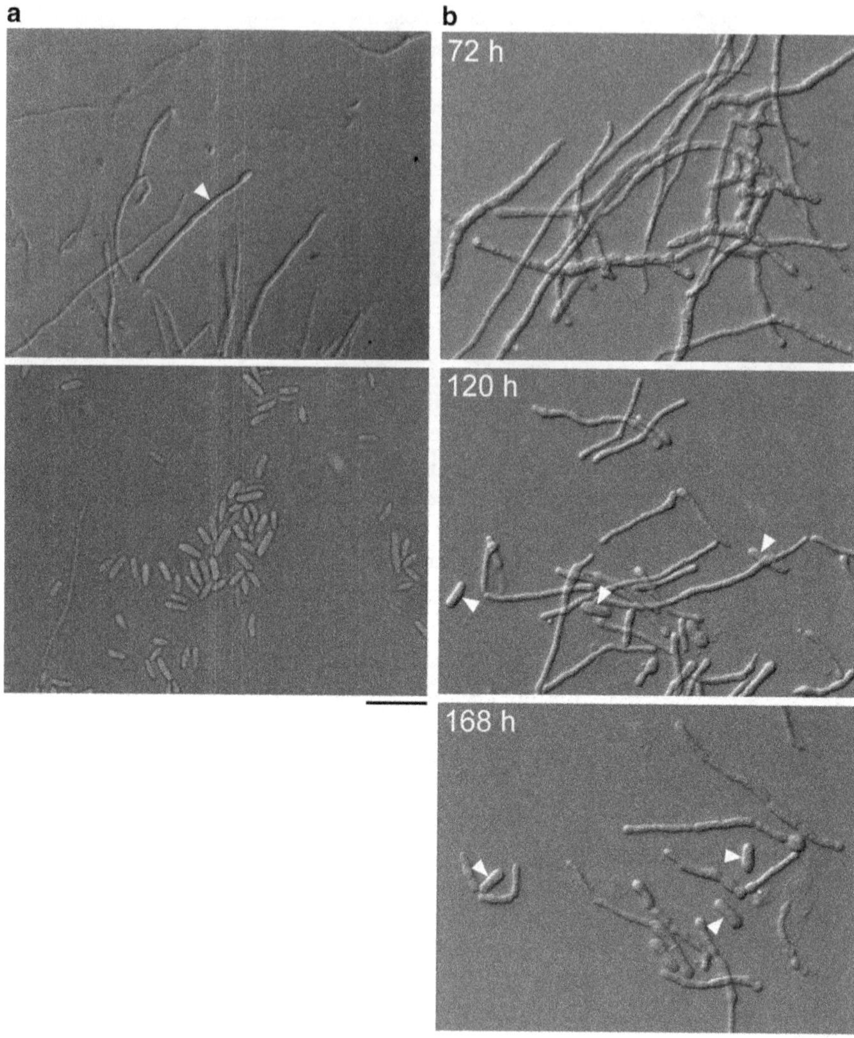

Fig. 3.4 Morphology of *A. chrysogenum*. (**a**) Conidiophore and conidia. *A. chrysogenum* was grown on agar plates for 168 h. The upper image shows a conidiophore (arrowhead) harboring a single conidium. The lower image shows several conidia. Calibration bar, 20 μm. (**b**) Arthrospore formation. *A. chrysogenum* was grown in liquid cultures and aliquots were taken at the indicated time-points. Fragmentation of hyphae and arthrospore formation starts at 72 h, and the mycelium is mainly fragmented after 168 h. Examples of arthrospores are indicated by arrowheads. Calibration bar, 20 μm

formation is stimulated by physiological changes, such as methionine addition or glucose depletion [47–51]. Furthermore, in *A. chrysogenum*, differentiation into arthrospores coincides with the maximum rate of cephalosporin C biosynthesis, with arthrospore formation seemingly correlated with high-yield cephalosporin C production [40, 46, 52]. Several proteins have been recently identified as being involved in *A. chrysogenum* arthrospore formation, including the winged helix transcription factor CPCR1, which controls hyphal fragmentation and arthrospore formation [39]. The putative septation protein AcSepH is also involved in this developmental process [53]. The section on *Regulation of Cephalosporin C Production* in this chapter discusses the role of these proteins and other regulators of *A. chrysogenum* in detail.

The Effect of Methionine on Cephalosporin C Production and Morphology

The morphological effect of methionine on arthrospore formation was first described by Caltrider and Niss [47]. Supplementation of culture medium with methionine resulted in swollen and fragmented hyphae, whereas growth of *A. chrysogenum* in medium lacking methionine was filamentous. Fifty years ago, it was demonstrated that methionine positively stimulates cephalosporin C biosynthesis [54]. Methionine-supplemented cultures of *A. chrysogenum* exhibited a two to threefold increase in cephalosporin C titers [55], thus suggesting that methionine is a promising target for research on strain improvement.

Several investigations found that methionine is directly involved in cephalosporin C biosynthesis through the reverse trans-sulfuration pathway (for an overview see [7]), in which the sulfur atom of methionine is transferred to L-cysteine, one of the three precursor amino acids in the biosynthesis pathway (Fig. 3.1) [47, 56]. However, further research revealed that the stimulating effect of methionine on the cephalosporin C yield is not predominantly based on its role as a sulfur donor. Other sulfur compounds—including other intermediates of the reverse trans-sulfuration pathway, such as homocysteine or cystathionine—were not able to stimulate cephalosporin C biosynthesis [57]. Additionally, it was found that the cephalosporin C titer is also enhanced by norleucine, a methionine analog that does not contain sulfur [58]. Norleucine also mimics the methionine effect on arthrospore formation, indicating that this role is also independent from the aspect of sulfur donation [48].

Sawada et al. [59] found that growth with either methionine or norleucine increases activity of both the IPNS and the DAOC/DAC synthetase, thus suggesting a possible regulatory role of these compounds. Similarly, Zhang et al. [60] investigated increased activity of the first enzyme in the biosynthesis pathway (the ACVS) in the presence of methionine, and found that it resulted in corresponding increases in β-lactam production in both a low-producing strain and a high-producing mutant. These results correlate with the observation that methionine increases the transcription levels of at least four cephalosporin C biosynthesis genes from both the "early" and "late" gene clusters, namely *pcbAB*, *pcbC*, *cefEF*, and *cefG* [55].

Other studies have revealed that the enzyme cystathionine-γ-lyase—which catalyzes the conversion of cystathionine to L-cysteine in the reverse trans-sulfuration—is crucial for the titer-enhancing effect of methionine [61]. The results of targeted inactivation of *mecB* (which encodes cystathionine-γ-lyase) indicate that the production of L-cysteine through the reverse trans-sulfuration pathway is required for high-level cephalosporin C production, but not for low-level biosynthesis [62]. This finding proves that the essential L-cysteine is obtained from both the autotrophic and the reverse trans-sulfuration pathways. However, *mecB* disruption did not affect the induction of cephalosporin C biosynthesis genes by methionine, thus indicating that their expression is not mediated by a putative regulatory role of cystathionine-γ-lyase. Instead, the induction of these genes may be triggered by methionine itself or by a methionine-derived catabolite.

Altogether, these data indicate that methionine has a dual effect on cephalosporin C biosynthesis in *A. chrysogenum*. On one hand, it seems to be the main supplier of L-cysteine via the reverse trans-sulfuration pathway and, on the other hand, it induces transcription of cephalosporin C biosynthesis genes (reviewed in [7, 63]).

Molecular Genetics of *A. chrysogenum*

Over the decades since the initial discovery of *A. chrysogenum* and its ability to produce the valuable broadband antibiotic cephalosporin C, strain improvement has been performed through random mutagenesis. Data from pulsed-field gel electrophoresis show that three related *A. chrysogenum* strains with differing rates of cephalosporin C biosynthesis also display different chromosome patterns, indicating that chromosome translocations in industrial strains may be responsible for increased β-lactam synthesis [64].

The ability to manipulate filamentous fungi using molecular genetic tools has advanced considerably, thus paving the way for targeted approaches to optimizing the production rate of industrial strains. In the 1980s, the development of a DNA-mediated transformation system allowed the ectopic integration of recombinant DNA into fungal genomic DNA, which was a prerequisite for targeted genetic approaches [65, 66]. The initial transformations were performed using a vector molecule carrying the bacterial phosphotransferase gene under control of the strong *pcbC* promoter [67].

The integration of DNA into a genome is generally performed via one of two different DNA repair mechanisms: ectopic integration by non-homologous end-joining (NHEJ), or locus-specific integration of the target DNA by homologous recombination (for the latter, see Fig. 3.5(a)). Unlike yeast, which has a highly efficient homologous recombination system [68, 69], DNA integration into the genome of filamentous fungi is predominantly performed through the NHEJ system, thus making these kinds of experiments very time consuming and laborious. This is also an issue in *A. chrysogenum*, with recent gene deletion experiments exhibiting a homologous recombination rate of between 0.1 and 1 % [70–72]. One very effective way to overcome this restriction is to delete a gene that encodes one of the essential

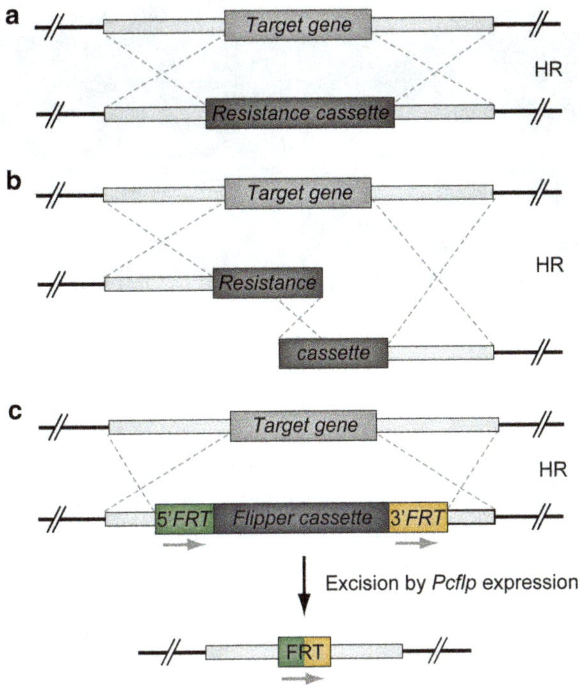

Fig. 3.5 Different gene replacement strategies established for *A. chrysogenum*. (**a**) Homologous recombination (HR). A target gene is replaced by a resistance cassette through a crossover of homologous flanking sites. (**b**) Split-marker system. The resistance cassette is split into two overlapping fragments, thus requiring three crossover events for replacement of the target gene with the resistance cassette. (**c**) FLP/*FRT* recombination system. The target gene is replaced by a flipper cassette flanked by *FRT* sites, which comprises the resistance cassette and, for a one-step approach, the gene encoding the Pcflp recombinase under control of an inducible promoter. After induction of the *Pcflp* gene, the whole cassette is excised, leaving only a single *FRT* site in the genome

NHEJ components—for example, Ku70 or Ku80 (reviewed in [73, 74]). The use of this approach recently resulted in the generation of an *A. chrysogenum* ΔAcku70 strain. Subsequent gene deletion experiments using this strain as a recipient have shown recombination rates of between 50 and 89 %,[1] in contrast to the 0.6 % rate observed for the deletion of the *Acku70* gene itself [75].

Another method for increasing the homologous recombination rate is the split-marker system. For transformation with this approach, two PCR fragments are transferred into the recipient strain: one comprising the 5′ flank of the target gene and part of a resistance cassette, and the other containing the 3′ flank and another part of the resistance cassette [76]. As shown in Fig. 3.5(b), three crossover events are necessary to generate a functional resistance cassette that substitutes the target

[1] D. Löper and U. Kück, unpublished.

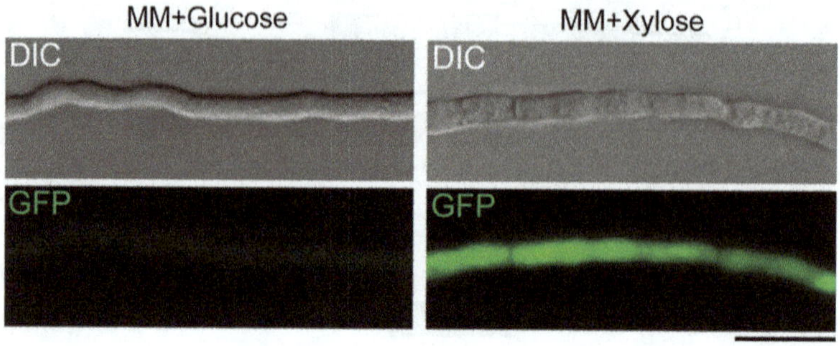

Fig. 3.6 Fluorescence microscopy to demonstrate *gfp* expression under control of the *Smxyl* promoter in *A. chrysogenum*. The strain was grown under inducing (MM + Xylose) and non-inducing (MM + Glucose) conditions. Calibration bar, 10 μm

gene by homologous recombination. This technique drastically increases the rate of homologous recombination; however, the transformation frequency is reduced [77].

Another limitation of gene deletion experiments is that both homologous recombination in a NHEJ-deficient strain and the split-marker system result in a residual resistance cassette in the genome of the deletion strain. For production strains, this has two disadvantages. First, the number of suitable resistance markers is limited for most filamentous fungi, thus limiting the ability to generate multiple gene deletions in a single strain. Second, the resistance cassette often includes foreign DNA, making it difficult to fulfill safety requirements needed for marketing approval. For *A. chrysogenum*, the latter concern has already been overcome by the use of a targeted approach to generate strains that lack foreign DNA. In this process, the fungus is subjected to DNA-mediated transformations using a vector without bacterial DNA sequences. Recombinant fungal strains have been generated that carry a mutated version of the homologous β-tubulin gene from *A. chrysogenum*, resulting in strains with a resistance against benomyl [78, 79].

Further advancing work with *A. chrysogenum*, the recently established FLP/*FRT* recombination system allows both the reuse of resistance marker and the generation of strains that are devoid of any foreign DNA. As shown in Fig. 3.5(c), in this process, the target gene is substituted by a so-called flipper cassette flanked by *FRT* sites. This cassette comprises the resistance gene and the codon-adapted gene for an Flp recombinase under an inducible promoter. After induction, the recombinase gene is expressed and the recombinase subsequently excises the complete flipper cassette, thus leaving only one 34-bp *FRT* site in the genome. This approach was recently used to generate a marker-free strain lacking the *Acku70* gene, representing a good recipient strain for further gene deletion experiments [75].

Establishment of this one-step FLP/*FRT* recombination system first required the identification of an inducible promoter for controlled expression of the recombinase gene in *A. chrysogenum*. For this, the heterologous xylose-inducible promoter *Smxyl* from *Sordaria macrospora* was used [75]. This promoter is a valuable tool for the recombination system as well as for other applications in which controlled gene expression is desirable. Figure 3.6 shows *A. chrysogenum* hyphae carrying the *gfp* gene under the control of the *Smxyl* promoter; strong fluorescence can be observed after growth under inducing conditions in medium with xylose. Additionally, the promoter of *mir1* (a putative siderophore transporter gene) has recently been developed as an endogenous inducible promoter for *A. chrysogenum*, called *mir1*^{P}. This promoter is inducible under iron starvation and thus allows the expression of genes encoding green fluorescence protein and phleomycin resistance [80].

For analyses of gene function, RNA silencing represents a good alternative to deletion experiments, especially when investigating essential genes that have a lethal deletion phenotype. The RNAi system used in *A. chrysogenum* was established using the DsRed protein as a reporter to monitor the silencing process in transformants. This system generates strains with high levels of the *DsRed* gene, resulting in red colonies. Subsequent transformation of a hairpin-expressing vector that carries fragments of the *DsRed* gene and the *pcbC* gene allows co-silencing of both genes. Transformant screening is facilitated by the colorless phenotype that is caused by downregulation of the *DsRed* gene. All colorless mutants additionally show downregulation of the *pcbC* gene, although at different levels [81].

Due to the stem-loop structure of the single-stranded DNA, the cloning of hairpin constructs is often rather time consuming and can be difficult to perform. However, these expression vectors have the advantages of stability and high efficiency compared to vectors containing dual-promoter constructs. Ullán et al. [82] used the *pcbC* and *cefEF* genes flanked by two divergently orientated strong promoters, and found that 15 % of the selected transformants were knockdown mutants with reduced cephalosporin C production, indicating that this process is effective. Other advantages of this expression system include the rather easy construction of the corresponding vector, and the high number of genes that can be tested. Figure 3.7 illustrates the similarities and differences between both RNA silencing approaches that have been established for *A. chrysogenum*.

Along with gene deletion or downregulation for functional characterization, the identification of interaction partners is another valuable approach to gain mechanistic insights into cephalosporin C regulation. Established methods for this include yeast one- and two-hybrid analyses for either protein–DNA interactions or protein–protein interactions. Both methods have been successfully applied in *A. chrysogenum*. Using an *A. chrysogenum* cDNA library as prey and the *pcbC* promoter as bait, the winged helix transcription factor CPCR1 was identified, and the binding of CPCR1 to the promoter region of *pcbC* was verified by electrophoretic mobility shift assay [83]. Furthermore, the same cDNA library was used in a two-hybrid screen with CPCR1 as bait, which identified the forkhead transcription factor AcFKH1 as an interaction partner of CPCR1. The interaction between CPCR1 and AcFKH1 was further verified through a GST pull-down assay, another in vitro approach to determine protein–protein interactions [84].

It must be noted that all of these techniques are performed in vitro, and thus harbor the possibility of false-positive or false-negative results. Bimolecular fluorescence complementation (BiFC) was established in *A. chrysogenum*, as an in vivo approach to determine protein–protein interactions [85]. This method is highly valuable, not only because the interaction takes place in vivo but also because localization of the

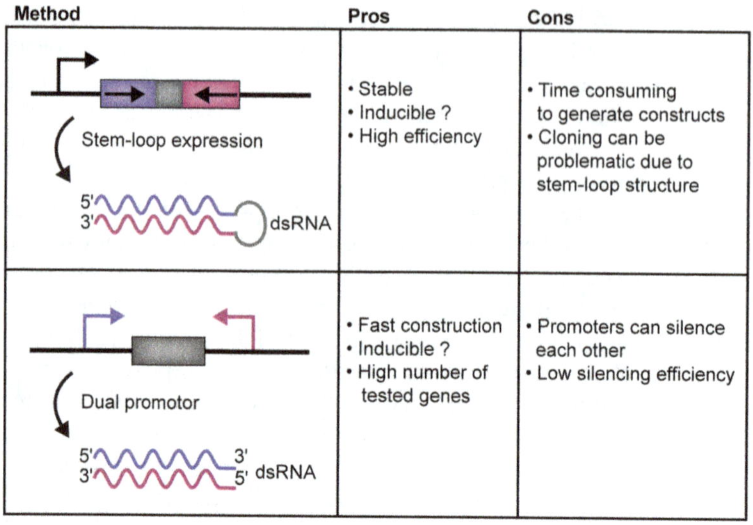

Fig. 3.7 Two RNAi vector systems: stem-loop expression versus dual promoter. The stem-loop expression RNAi system established for *A. chrysogenum* uses a vector that contains intron or spacer sequences between two inversely orientated target fragments to express double-stranded RNA (dsRNA) with a hairpin structure [81]. This system provides stable and high silencing frequencies; however, construction of stem-loop vectors containing two repeated sequences with an opposite orientation is very time consuming. An alternative is to use dual-promoter plasmids, such that a single target fragment is transcribed from both sites by divergently orientated fungal promoter sequences [82]

interaction can be visualized through live-cell imaging. For this approach, the gene for the enhanced yellow fluorescent protein (EYFP) is split into two parts, which are each fused then with one interaction partner of interest. Subsequently, fluorescence in the fungal cell is only observed when these two putative interaction partners come into close proximity.

Regulation of Cephalosporin C Production

The expressions of all cephalosporin C biosynthesis genes in *A. chrysogenum* are controlled by bidirectional promoters. In the "early" cluster, *pcbAB* and *pcbC* share a bidirectional promoter, as do *cefD1* and *cefD2*. In the "late" cluster, *cefEF* and *cefG* are equally controlled by a bidirectional promoter (Fig. 3.2). Interestingly, studies with the β-glucuronidase and β-galactosidase reporter genes fused to the bidirectional promoter in opposite orientation indicate that the *pcbC* promoter is at least five times stronger than the *pcbAB* promoter [86]. Furthermore, the *pcbC* gene from *A. chrysogenum* is differentially expressed in strains that show high or low cephalosporin C production. A synthetic gene encoding the thrombin inhibitor hirudin (from the leech *Hirudo medicinalis*) was fused with the 5′ and 3′ regions of the *pcbC* gene, and hirudin synthesis was determined with a thrombin-inhibition assay. Transformants from strains with high cephalosporin C production showed three to eightfold higher hirudin gene expression compared to low cephalosporin-C-producing strains, with the strongest signal after 3 days. The highest hirudin activity was detected after 4–5 days [87].

These bidirectional promoters of cephalosporin C biosynthesis genes represent a good starting point for identifying regulators that control cephalosporin C biosynthesis on the transcriptional level. For example, fusion of the sequential deletion derivatives of the *pcbC* promoter to the *lacZ* reporter gene has led to the identification of a fragment responsible for transcriptional activation of the *pcbC* gene. Interestingly, sequence analysis of this fragment revealed a consensus binding site for the fungal transcription factor PACC. Furthermore, an in vitro binding assay of the *A. chrysogenum* PACC protein revealed binding sites for the bidirectional *pcbAB-pcbC* promoter as well as for the *cefEF-cefG* promoter, indicating the involvement of PACC in regulating both "early" and "late" cephalosporin C biosynthesis genes [88]. PACC is a C_2H_2 zinc-finger transcription factor with three fingers, and it regulates pH-dependent gene expression in several filamentous fungi [89, 90]. Although filamentous fungi can grow in a wide pH range, from acidic to alkaline environments, the production of secondary metabolites varies under different pH conditions. Cephalosporin and penicillin production occur more strongly in alkaline conditions [7]. Interestingly, the expression of cephalosporin C biosynthesis genes was found to be optimum at a pH of 8 in a wild-type strain, but at a pH of 6 in a production strain [88]. This observation suggests that PACC-dependent regulation of cephalosporin C biosynthesis genes changed during production strain development.

It must be noted, however, that pH regulation of PACC is a general mechanism that affects a multitude of genes. Similarly, the glucose-dependent transcriptional regulator CRE1 is involved in carbon catabolite repression, and also affects expression of cephalosporin C biosynthesis genes. In *A. chrysogenum*, cephalosporin C biosynthesis is repressed by glucose, resulting in decreased expressions of both *pcbC* and *cefEF* in a wild-type strain grown in the presence of glucose. However, in a production strain with elevated cephalosporin C biosynthesis, the *pcbC* transcript level is not affected by the presence of glucose, indicating that strain improvement correlated with deregulation of glucose repression [91, 92]. Wild-type-like regulation of *pcbC* expression can be restored in the production strain through the introduction of additional copies of *cre1*, suggesting that *pcbC* expression is regulated by CRE1 [92].

The influence of global transcriptional regulators, such as PACC and CRE1, on the expression of cephalosporin C biosynthesis genes illustrates interplay between primary and secondary metabolism that enables sophisticated reactions to environmental conditions. However, more specialized regulators have also been identified. One example is velvet (VeA), which acts together with other velvet-like proteins and the methyltransferase LaeA to regulate secondary metabolism and morphology in several filamentous fungi (for a review see [93, 94]). In *A. chrysogenum*, AcVEA controls the transcriptional expression of all six cephalosporin C biosynthesis genes. Accordingly, the cephalosporin C titer was strongly reduced in *AcveA* disruption strains. Furthermore, compared to the wild-type strain, the *AcveA* disruption strains showed accelerated formation of arthrospores and hyperbranching of hyphal tips on osmotically nonstabilized media [71]. Data from *Aspergillus nidulans* suggest that velvet proteins act as transcriptional regulators by directly binding to DNA, which enables these proteins to regulate the transcription of different target genes [95]. Accordingly, velvet proteins have been shown to localize to the nucleus in several filamentous fungi [71, 96–98].

A concerted effect on both secondary metabolism and morphology is a ubiquitous phenomenon observed for several regulatory proteins. Another example from *A. chrysogenum* is the transcription factor CPCR1, a winged helix transcription factor that was initially identified in a yeast one-hybrid screen with the *pcbC* promoter as bait [83]. CPCR1 recognizes and binds at least two sequences in the bidirectional *pcbAB-pcbC* promoter. Although the overall cephalosporin C titers were not altered in deletion and overexpression mutants, the knock-out strains exhibited a >20 % reduction in biosynthesis of the intermediate penicillin N compared to in the wild-type strain [70]. Thus, CPCR1 seems to be involved in the regulation of early biosynthesis genes. Additionally, CPCR1 is required for hyphal fragmentation and thus arthrospore formation. Deletion of the *cpcR1* gene leads to a complete loss of arthrospore formation, whereas overexpression results in accelerated fragmentation into arthrospores [39]. Protein–protein interaction studies have revealed that CPCR1 forms a heterodimer with the forkhead transcription factor AcFKH1 [84]. AcFKH1 recognizes two forkhead consensus binding sites within the promoter region of the divergently orientated *pcbAB–pcbC* gene pair. However, in contrast to CPCR1, AcFKH1 is not directly involved in arthrospore formation. The *AcFKH1* deletion mutant still formed arthrospores, but these spores did not separate from each other, resulting in yeast-like pseudohyphal growth [39].

Another protein that was recently identified as a determinant of hyphal morphology is the septation protein AcSepH. In a mutant strain, development of vegetative hyphae into arthrospores was strongly disturbed, and only a few arthrospores were detectable. This phenotype occurred through obviously reduced septation of vegetative hyphae, which seems to be a prerequisite of arthrospore formation. Interestingly, AcSepH affects not only the morphology of *A. chrysogenum* but also the cephalosporin C biosynthesis. The disruption mutant exhibited delayed expression of *pcbC* and significantly decreased expressions of *cefEF*, *cefD1*, and *cefD2*, resulting in strong reduction of cephalosporin C production [53].

Regulatory proteins can also target genes other than those involved in cephalosporin C biosynthesis. The *cefR* gene was recently identified in the "early" cephalosporin C cluster, encoding the regulatory protein CefR. The predicted protein harbors a nuclear targeting signal and a "Fungal_trans" domain, indicating its action as a transcription factor. A *cefR* disruption mutant exhibited delayed expression of the *cefEF* gene, increased penicillin N secretion, and reduced cephalosporin C production. On the other hand, overexpression of *cefR* resulted in decreased penicillin N secretion and, consequently, increased cephalosporin C production. CefR was further shown to act as a repressor of the exporter CefT, which is responsible for the secretion of the intermediates IPN and penicillin N [99].

Conclusion

Cephalosporins are one of the most important classes of antibiotics, and further development of new cephalosporin derivatives harboring novel characteristic traits will help to ensure that future medical challenges are overcome. The enzymes of the cephalosporin C biosynthesis pathway have been known for quite a long time, and additional knowledge about the localization and transporters of these compounds has emerged more recently. Still, the exporter of cephalosporin C into the environment remains unknown [99]. Along with knowledge about the biosynthesis pathway itself, understanding its regulation is crucial to fully comprehending this process. In the last decade, several cephalosporin C biosynthesis regulatory proteins were identified, for example, the velvet protein AcVEA and the transcriptional regulator CefR [71, 99]. AcVEA also links secondary metabolism and morphology in *A. chrysogenum*, as was already shown in several other filamentous fungi [98, 100]. The relationship between these processes is even more obvious for *A. chrysogenum*, as the conversion into arthrospores coincides with the maximum rate of cephalosporin C biosynthesis [40, 46, 52].

The further development of molecular tools is a mandatory prerequisite to improving our understanding of the regulation of cephalosporin C biosynthesis and arthrospore formation in *A. chrysogenum*. The peculiar growth characteristics of this fungus—including its exclusively asexual reproduction, sporadic conidiospore production, and slow growth rate—makes this aim much more challenging than for other industrially relevant filamentous fungi [73, 101]. However, decent progress has been made in developing several molecular methods, including RNAi techniques and the FLP/*FRT* recombination system to facilitate knockdown or knockout experiments [75, 81, 82], and tools to analyze DNA–protein and protein–protein interactions, such as the one- or two-hybrid system and BiFC [83–85]. Together, the use of these existing molecular tools along with future innovations will continue to expand our understanding of how cephalosporin C biosynthesis is regulated and connected to arthrospore formation, thus allowing further strain optimization.

Acknowledgements The experimental work of the authors is funded by the Christian Doppler Society (Vienna) and Sandoz GmbH (Kundl). We are deeply thankful to Gabriele Frenßen-Schenkel for the artwork, and to MSc Janina Kluge and MSc Dominik Terfehr for their help with some figures. We further thank Isabel van der Kamp for her support in preparing the chapter about cephalosporin derivatives.

References

1. Abraham EP, Newton GGF, Olson BH, Schuurmans DM, Schenck JR, Hargie MP, Fisher MW, Fusari SA. Identity of cephalosporin N and synnematin B. Nature. 1955;176(4481):551.
2. Newton GGF, Abraham EP. Cephalosporin C, a new antibiotic containing sulphur and D-α-aminoadipic acid. Nature. 1955;175(4456):548.

3. Demain AL, Elander RP. The beta-lactam antibiotics: past, present, and future. Antonie Van Leeuwenhoek. 1999;75(1–2):5–19.
4. Brakhage AA, Spröte P, Al-Abdallah Q, Gehrke A, Plattner H, Tüncher A. Regulation of penicillin biosynthesis in filamentous fungi. Adv Biochem Eng Biotechnol. 2004;88:45–90.
5. Elander RP. Industrial production of β-lactam antibiotics. Appl Microbiol Biotechnol. 2003;61(5–6):385–92.
6. Demain AL. Antibiotics: natural products essential to human health. Med Res Rev. 2009;29(6):821–42.
7. Schmitt EK, Hoff B, Kück U. Regulation of cephalosporin biosynthesis. Adv Biochem Eng Biotechnol. 2004;88:1–43.
8. Ozcengiz G, Demain AL. Recent advances in the biosynthesis of penicillins, cephalosporins and clavams and its regulation. Biotechnol Adv. 2013;31(2):287–311.
9. Díez B, Gutiérrez S, Barredo JL, van Solingen P, van der Voort LH, Martín JF. The cluster of penicillin biosynthetic genes. Identification and characterization of the pcbAB gene encoding the alpha-aminoadipyl-cysteinyl-valine synthetase and linkage to the pcbC and penDE genes. J Biol Chem. 1990;265(27):16358–65.
10. Liras P, Martín JF. Gene clusters for beta-lactam antibiotics and control of their expression: why have clusters evolved, and from where did they originate? Int Microbiol. 2006;9(1):9–19.
11. Gutiérrez S, Fierro F, Casqueiro J, Martín JF. Gene organization and plasticity of the beta-lactam genes in different filamentous fungi. Antonie Van Leeuwenhoek. 1999;75(1–2):81–94.
12. Skatrud PL, Queener SW. An electrophoretic molecular karyotype for an industrial strain of *Cephalosporium acremonium*. Gene. 1989;78(2):331–8.
13. Smith AW, Collis K, Ramsden M, Fox HM, Peberdy JF. Chromosome rearrangements in improved cephalosporin C-producing strains of *Acremonium chrysogenum*. Curr Genet. 1991;19(3):235–7.
14. Byford MF, Baldwin JE, Shiau CY, Schofield CJ. The mechanism of ACV synthetase. Chem Rev. 1997;97(7):2631–50.
15. Aharonowitz Y, Bergmeyer J, Cantoral JM, Cohen G, Demain AL, Fink U, Kinghorn J, Kleinkauf H, MacCabe A, Palissa H, Pfeifer E, Schwecke T, van Liempt H, van Döhren H, Wolfe S, Zhang J. Delta-(L-alpha-aminoadipyl)-L-cysteinyl-D-valine synthetase, the multienzyme integrating the four primary reactions in beta-lactam biosynthesis, as a model peptide synthetase. Nat Biotechnol. 1993;11(7):807–10.
16. Gutiérrez S, Diéz B, Montenegro E, Martín JF. Characterization of the *Cephalosporium acremonium pcbAB* gene encoding alpha-aminoadipyl-cysteinyl-valine synthetase, a large multidomain peptide synthetase: linkage to the *pcbC* gene as a cluster of early cephalosporin biosynthetic genes and evidence of multiple functional domains. J Bacteriol. 1991;173(7):2354–65.
17. Wu X, García-Estrada C, Vaca I, Martín JF. Motifs in the C-terminal region of the *Penicillium chrysogenum* ACV synthetase are essential for valine epimerization and processivity of tripeptide formation. Biochimie. 2012;94(2):354–64.
18. Samson SM, Belagaje R, Blankenship DT, Chapman JL, Perry D, Skatrud PL, VanFrank RM, Abraham EP, Baldwin JE, Queener SW, Ingolia TD. Isolation, sequence determination and expression in *Escherichia coli* of the isopenicillin N synthetase gene from *Cephalosporium acremonium*. Nature. 1985;318(6042):191–4.
19. Aharonowitz Y, Cohen G, Martín JF. Penicillin and cephalosporin biosynthetic genes: structure, organization, regulation, and evolution. Annu Rev Microbiol. 1992;46:461–95.
20. van der Lende TR, van de Kamp M, van den Berg M, Sjollema K, Bovenberg RAL, Veenhuis M, Konings WN, Driessen AJM. δ-(L-α-aminoadipyl)-L-cysteinyl-D-valine synthetase, that mediates the first committed step in penicillin biosynthesis, is a cytosolic enzyme. Fungal Genet Biol. 2002;37(1):49–55.
21. Ullán RV, Casqueiro J, Bañuelos O, Fernandez FJ, Gutiérrez S, Martín JF. A novel epimerization system in fungal secondary metabolism involved in the conversion of isopenicillin N into penicillin N in *Acremonium chrysogenum*. J Biol Chem. 2002;277(48):46216–25.

22. Kovacevic S, Tobin MB, Miller JR. The beta-lactam biosynthesis genes for isopenicillin N epimerase and deacetoxycephalosporin C synthetase are expressed from a single transcript in *Streptomyces clavuligerus*. J Bacteriol. 1990;172(7):3952–8.
23. Martín JF, Ullán RV, García-Estrada C. Regulation and compartmentalization of beta-lactam biosynthesis. Microb Biotechnol. 2010;3(3):285–99.
24. Rottensteiner H, Kramer A, Lorenzen S, Stein K, Landgraf C, Volkmer-Engert R, Erdmann R. Peroxisomal membrane proteins contain common Pex19p-binding sites that are an integral part of their targeting signals. Mol Biol Cell. 2004;15(7):3406–17.
25. Ullán RV, Teijeira F, Guerra SM, Vaca I, Martín JF. Characterization of a novel peroxisome membrane protein essential for conversion of isopenicillin N into cephalosporin C. Biochem J. 2010;432(2):227–36.
26. Teijeira F, Ullán RV, Guerra SM, García-Estrada C, Vaca I, Martín JF. The transporter CefM involved in translocation of biosynthetic intermediates is essential for cephalosporin production. Biochem J. 2009;418(1):113–24.
27. Samson SM, Dotzlaf JE, Slisz ML, Becker GW, Van Frank RM, Veal LE, Yeh WK, Miller JR, Queener SW, Ingolia TD. Cloning and expression of the fungal expandase/hydroxylase gene involved in cephalosporin biosynthesis. Nat Biotechnol. 1987;5(11):1207–14.
28. Mathison L, Soliday C, Stepan T, Aldrich T, Rambosek J. Cloning, characterization, and use in strain improvement of the *Cephalosporium acremonium* gene *cefG* encoding acetyl transferase. Curr Genet. 1993;23(1):33–41.
29. Velasco J, Gutiérrez S, Campoy S, Martín JF. Molecular characterization of the *Acremonium chrysogenum cefG* gene product: the native deacetylcephalosporin C acetyltransferase is not processed into subunits. Biochem J. 1999;337:379–85.
30. Abraham EP, Newton GG. The structure of cephalosporin C. Biochem J. 1961;79:377–93.
31. Morin RB, Roeske RW, Flynn EH, Jackson BG. Chemistry of cephalosporin antibiotics I. 7-aminocephalosporanic acid from cephalosporin C. J Am Chem Soc. 1962;84(17):3400–1.
32. Barber MS, Giesecke U, Reichert A, Minas W. Industrial enzymatic production of cephalosporin-based β-lactams. Adv Biochem Eng Biotechnol. 2004;88:179–215.
33. Sonawane VC. Enzymatic modifications of cephalosporins by cephalosporin acylase and other enzymes. Crit Rev Biotechnol. 2006;26(2):95–120.
34. Noel GJ, Bush K, Bagchi P, Ianus J, Strauss RS. A randomized, double-blind trial comparing ceftobiprole medocaril with vancomycin plus ceftazidime for the treatment of patients with complicated skin and skin-structure infections. Clin Infect Dis. 2008;46(5):647–55.
35. Noel GJ, Strauss RS, Amsler K, Heep M, Pypstra R, Solomkin JS. Results of a double-blind, randomized trial of ceftobiprole treatment of complicated skin and skin structure infections caused by gram-positive bacteria. Antimicrob Agents Chemother. 2008;52(1):37–44.
36. Onions AHS, Brady BL. Taxonomy of *Penicillium* and *Acremonium*. In: Peberdy JF, editor. *Penicillium* and *Acremonium*. New York: Plenum; 1987. p. 1–36.
37. Gams W. *Cephalosporium*-artige Schimmelpilze (Hyphomycetes). Stuttgart: Gustav Fischer Verlag; 1971. German.
38. Glenn AE, Bacon CW, Price R, Hanlin RT. Molecular phylogeny of *Acremonium* and its taxonomic implications. Mycologia. 1996;88(3):369–83.
39. Hoff B, Schmitt EK, Kück U. CPCR1, but not its interacting transcription factor AcFKH1, controls fungal arthrospore formation in *Acremonium chrysogenum*. Mol Microbiol. 2005; 56(5):1220–33.
40. Nash CH, Huber FM. Antibiotic synthesis and morphological differentiation of *Cephalosporium acremonium*. Appl Microbiol. 1971;22(1):6–10.
41. Böhm J, Hoff B, O'Gorman CM, Wolfers S, Klix V, Binger D, Zadra I, Kürnsteiner H, Pöggeler S, Dyer PS, Kück U. Sexual reproduction and mating-type-mediated strain development in the penicillin-producing fungus *Penicillium chrysogenum*. Proc Natl Acad Sci U S A. 2013;110(4):1476–81.
42. Seidl V, Seibel C, Kubicek CP, Schmoll M. Sexual development in the industrial workhorse *Trichoderma reesei*. Proc Natl Acad Sci U S A. 2009;106(33):13909–14.

43. Pöggeler S, Hoff B, Kück U. Asexual cephalosporin C producer *Acremonium chrysogenum* carries a functional mating type locus. Appl Environ Microbiol. 2008;74(19):6006–16.
44. Pöggeler S. Mating-type genes for classical strain improvements of ascomycetes. Appl Microbiol Biotechnol. 2001;56(5–6):589–601.
45. Queener SW, Ellis LF. Differentiation of mutants of *Cephalosporium acremonium* in complex medium: the formation of unicellular arthrospores and their germination. Can J Microbiol. 1975;21(12):1981–96.
46. Bartoshevich YE, Zaslavskaya PL, Novak MJ, Yudina OD. *Acremonium chrysogenum* differentiation and biosynthesis of cephalosporin. J Basic Microb. 1990;30(5):313–20.
47. Caltrider PG, Niss HF. Role of methionine in cephalosporin synthesis. Appl Microbiol. 1966;14(5):746–53.
48. Drew SW, Winstanley DJ, Demain AL. Effect of norleucine on mycelial fragmentation in *Cephalosporium acremonium*. Appl Environ Microbiol. 1976;31(1):143–5.
49. Karaffa L, Sándor E, Kozma J, Szentirmai A. Methionine enhances sugar consumption, fragmentation, vacuolation and cephalosporin C production in *Acremonium chrysogenum*. Process Biochem. 1997;32(6):495–9.
50. Sándor E, Szentirmai A, Paul GC, Thomas CR, Pócsi I, Karaffa L. Analysis of the relationship between growth, cephalosporin C production, and fragmentation in *Acremonium chrysogenum*. Can J Microbiol. 2001;47(9):801–6.
51. Tollnick C, Seidel G, Beyer M, Schügerl K. Investigations of the production of cephalosporin C by *Acremonium chrysogenum*. Adv Biochem Eng Biotechnol. 2004;86:1–45.
52. Smith B, Warren SC, Newton GG, Abraham EP. Biosynthesis of penicillin N and cephalosporin C. Antibiotic production and other features of the metabolism of Cephalosporium sp. Biochem J. 1967;103(3):877–90.
53. Long LK, Wang Y, Yang J, Xu X, Liu G. A septation related gene *AcsepH* in *Acremonium chrysogenum* is involved in the cellular differentiation and cephalosporin production. Fungal Genet Biol. 2013;50:11–20.
54. Demain AL, Newkirk JF. Biosynthesis of cephalosporin C. Appl Microbiol. 1962;10(4): 321–5.
55. Velasco J, Gutiérrez S, Fernández FJ, Marcos AT, Arenós C, Martín JF. Exogenous methionine increases levels of mRNAs transcribed from *pcbAB*, *pcbC*, and *cefEF* genes, encoding enzymes of the cephalosporin biosynthetic pathway, in *Acremonium chrysogenum*. J Bacteriol. 1994;176(4):985–91.
56. Nüesch J, Treichler HJ, Liersch M. The biosynthesis of cephalosporin C. In: Vanek Z, Hostalek Z, Cudlin J, editors. Genetics of industrial microorganisms. Prague: Academia; 1973. p. 309–34.
57. Drew SW, Demain AL. Production of cephalosporin C by single and double sulfur auxotrophic mutants of *Cephalosporium acremonium*. Antimicrob Agents Chemother. 1975;8(1):5–10.
58. Drew SW, Demain AL. Methionine control of cephalosporin C formation. Biotechnol Bioeng. 1973;15(4):743–54.
59. Sawada Y, Konomi T, Solomon NA, Demain AL. Increase in activity of β-lactam synthetases after growth of Cephalosporium acremonium with methionine or norleucine. FEMS Microbiol Lett. 1980;9(4):281–4.
60. Zhang JY, Banko G, Wolfe S, Demain AL. Methionine induction of ACV synthetase in *Cephalosporium acremonium*. J Ind Microbiol. 1987;2(4):251–5.
61. Treichler HJ, Liersch M, Nüesch J, Döbeli H. Role of sulfur metabolism in cephalosporin C and penicillin biosynthesis. In: Sebek OK, Laskin AI, editors. Genetics of industrial microorganisms. Washington: American Society for Microbiology; 1979. p. 79–104.
62. Liu G, Casqueiro J, Bañuelos O, Cardoza RE, Gutiérrez S, Martín JF. Targeted inactivation of the *mecB* gene, encoding cystathionine-γ-lyase, shows that the reverse transsulfuration pathway is required for high-level cephalosporin biosynthesis in *Acremonium chrysogenum* C10 but not for methionine induction of the cephalosporin genes. J Bacteriol. 2001;183(5): 1765–72.

63. Martín JF, Demain AL. Unraveling the methionine-cephalosporin puzzle in *Acremonium chrysogenum*. Trends Biotechnol. 2002;20(12):502–7.
64. Walz M, Kück U. Polymorphic karyotypes in related *Acremonium* strains. Curr Genet. 1991;19(2):73–6.
65. Skatrud PL, Queener SW, Carr LG, Fisher DL. Efficient integrative transformation of *Cephalosporium acremonium*. Curr Genet. 1987;12(5):337–48.
66. Queener SW, Ingolia TD, Skatrud PL, Chapman JL, Kaster KR. A system for genetic transformation of *Cephalosporium acremonium*. In: Leive L, editor. Microbiology-1985. Washington: American Society for Microbiology; 1985. p. 468–72.
67. Kück U, Walz M, Mohr G, Mracek M. The 5′-sequence of the isopenicillin N-synthetase gene (*pcbC*) from *Cephalosporium acremonium* directs the expression of the prokaryotic hygromycin B phosphotransferase gene (*hph*) in *Aspergillus niger*. Appl Microbiol Biotechnol. 1989;31(4):358–65.
68. Takita Y, Takahara M, Nogami S, Anraku Y, Ohya Y. Applications of the long and accurate polymerase chain reaction method in yeast molecular biology: direct sequencing of the amplified DNA and its introduction into yeast. Yeast. 1997;13(8):763–8.
69. Wach A, Brachat A, Pohlmann R, Philippsen P. New heterologous modules for classical or PCR-based gene disruptions in *Saccharomyces cerevisiae*. Yeast. 1994;10(13):1793–808.
70. Schmitt EK, Bunse A, Janus D, Hoff B, Friedlin E, Kürnsteiner H, Kück U. Winged helix transcription factor CPCR1 is involved in regulation of β-lactam biosynthesis in the fungus *Acremonium chrysogenum*. Eukaryot Cell. 2004;3(1):121–34.
71. Dreyer J, Eichhorn H, Friedlin E, Kürnsteiner H, Kück U. A homologue of the *Aspergillus velvet* gene regulates both cephalosporin C biosynthesis and hyphal fragmentation in *Acremonium chrysogenum*. Appl Environ Microbiol. 2007;73(10):3412–22.
72. Walz M, Kück U. Targeted integration into the *Acremonium chrysogenum* genome: disruption of the *pcbC* gene. Curr Genet. 1993;24(5):421–7.
73. Kück U, Hoff B. New tools for the genetic manipulation of filamentous fungi. Appl Microbiol Biotechnol. 2010;86(1):51–62.
74. Krappmann S. Gene targeting in filamentous fungi: the benefits of impaired repair. Fungal Biol Rev. 2007;21(1):25–9.
75. Bloemendal S, Löper D, Terfehr D, Kopke K, Kluge J, Teichert I, Kück U. Tools for advanced and targeted genetic manipulation of the β-lactam antibiotic producer *Acremonium chrysogenum*. J Biotechnol. 2014;69:51–62.
76. Fairhead C, Llorente B, Denis F, Soler M, Dujon B. New vectors for combinatorial deletions in yeast chromosomes and for gap-repair cloning using 'split-marker' recombination. Yeast. 1996;12(14):1439–57.
77. Jeong JS, Mitchell TK, Dean RA. The *Magnaporthe grisea* snodprot1 homolog, MSP1, is required for virulence. FEMS Microbiol Lett. 2007;273(2):157–65.
78. Nowak C, Kück U. Development of an homologous transformation system for *Acremonium chrysogenum* based on the β-tubulin gene. Curr Genet. 1994;25(1):34–40.
79. Nowak C, Radzio R, Kück U. DNA-mediated transformation of a fungus employing a vector devoid of bacterial DNA sequences. Appl Microbiol Biotechnol. 1995;43(6):1077–81.
80. Gsaller F, Blatzer M, Abt B, Schrettl M, Lindner H, Haas H. The first promoter for conditional gene expression in *Acremonium chrysogenum*: iron starvation-inducible $mir1^P$. J Biotechnol. 2013;163(1):77–80.
81. Janus D, Hoff B, Hofmann E, Kück U. An efficient fungal RNA-silencing system using the *DsRed* reporter gene. Appl Environ Microbiol. 2007;73(3):962–70.
82. Ullán RV, Godio RP, Teijeira F, Vaca I, García-Estrada C, Feltrer R, Kosalkova K, Martín JF. RNA-silencing in *Penicillium chrysogenum* and *Acremonium chrysogenum*: validation studies using β-lactam genes expression. J Microbiol Methods. 2008;75(2):209–18.
83. Schmitt EK, Kück U. The fungal CPCR1 protein, which binds specifically to beta-lactam biosynthesis genes, is related to human regulatory factor X transcription factors. J Biol Chem. 2000;275(13):9348–57.

84. Schmitt EK, Hoff B, Kück U. AcFKH1, a novel member of the forkhead family, associates with the RFX transcription factor CPCR1 in the cephalosporin C-producing fungus *Acremonium chrysogenum*. Gene. 2004;342(2):269–81.
85. Hoff B, Kück U. Use of bimolecular fluorescence complementation to demonstrate transcription factor interaction in nuclei of living cells from the filamentous fungus *Acremonium chrysogenum*. Curr Genet. 2005;47(2):132–8.
86. Menne S, Walz M, Kück U. Expression studies with the bidirectional *pcbAB-pcbC* promoter region from *Acremonium chrysogenum* using reporter gene fusions. Appl Microbiol Biotechnol. 1994;42(1):57–66.
87. Radzio R, Kück U. Efficient synthesis of the blood-coagulation inhibitor hirudin in the filamentous fungus *Acremonium chrysogenum*. Appl Microbiol Biotechnol. 1997;48(1):58–65.
88. Schmitt EK, Kempken R, Kück U. Functional analysis of promoter sequences of cephalosporin C biosynthesis genes from *Acremonium chrysogenum*: specific DNA-protein interactions and characterization of the transcription factor PACC. Mol Genet Genomics. 2001;265(3): 508–18.
89. Shah AJ, Tilburn J, Adlard MW, Arst Jr HN. pH regulation of penicillin production in *Aspergillus nidulans*. FEMS Microbiol Lett. 1991;61(2–3):209–12.
90. Suárez T, Peñalva MA. Characterization of a *Penicillium chrysogenum* gene encoding a PacC transcription factor and its binding sites in the divergent *pcbAB-pcbC* promoter of the penicillin biosynthetic cluster. Mol Microbiol. 1996;20(3):529–40.
91. Jekosch K, Kück U. Glucose dependent transcriptional expression of the *cre1* gene in *Acremonium chrysogenum* strains showing different levels of cephalosporin C production. Curr Genet. 2000;37(6):388–95.
92. Jekosch K, Kück U. Loss of glucose repression in an *Acremonium chrysogenum* beta-lactam producer strain and its restoration by multiple copies of the *cre1* gene. Appl Microbiol Biotechnol. 2000;54(4):556–63.
93. Jain S, Keller N. Insights to fungal biology through LaeA sleuthing. Fungal Biol Rev. 2013;27(2):51–9.
94. Bayram Ö, Braus GH. Coordination of secondary metabolism and development in fungi: the velvet family of regulatory proteins. Fems Microbiol Rev. 2012;36(1):1–24.
95. Ni M, Yu JH. A novel regulator couples sporogenesis and trehalose biogenesis in *Aspergillus nidulans*. PLoS One. 2007;2(10):e970.
96. Bayram Ö, Krappmann S, Ni M, Bok JW, Helmstaedt K, Valerius O, Braus-Stromeyer S, Kwon NJ, Keller NP, Yu JH, Braus GH. VelB/VeA/LaeA complex coordinates light signal with fungal development and secondary metabolism. Science. 2008;320(5882):1504–6.
97. Stinnett SM, Espeso EA, Cobeno L, Araújo-Bazán L, Calvo AM. *Aspergillus nidulans* VeA subcellular localization is dependent on the importin α carrier and on light. Mol Microbiol. 2007;63(1):242–55.
98. Kopke K, Hoff B, Bloemendal S, Katschorowski A, Kamerewerd J, Kück U. Members of the *Penicillium chrysogenum* velvet complex play functionally opposing roles in the regulation of penicillin biosynthesis and conidiation. Eukaryot Cell. 2013;12(2):299–310.
99. Teijeira F, Ullán RV, Fernández-Aguado M, Martín JF. CefR modulates transporters of beta-lactam intermediates preventing the loss of penicillins to the broth and increases cephalosporin production in *Acremonium chrysogenum*. Metab Eng. 2011;13(5):532–43.
100. Sarikaya Bayram Ö, Bayram Ö, Valerius O, Park HS, Irniger S, Gerke J, Ni M, Han KH, Yu JH, Braus GH. LaeA control of velvet family regulatory proteins for light-dependent development and fungal cell-type specificity. PLoS Genet. 2010;6(12):e1001226.
101. Schuster A, Schmoll M. Biology and biotechnology of *Trichoderma*. Appl Microbiol Biotechnol. 2010;87(3):787–99.
102. Lüllmann H, Mohr K, Hein L. Pharmakologie und Toxikologie. 17th ed. Stuttgart: Thieme; 2010. German.

Chapter 4
Cyclosporines: Biosynthesis and Beyond

Tony Velkov and Alfons Lawen

Abbreviations

AdoMet	*S*-adenosyl-L-methionine
A-domain	Adenylation domain
Bmt	(4R)-4-[(E)-2-butyl]-4-methyl-L-threonine
CoA	Coenzyme A
C-domain	Condensation domain
CsA	Cyclosporine A
CySyn	Cyclosporine synthetase
D-Hiv	D-hydroxy isovaleric acid
E-domain	Epimerization domain
N-MTase	*N*-methyltransferase
NRPS	Non-ribosomal peptide synthetase
Ppant	4′-phosphopantetheine
PKS	Polyketide synthase
PCP	Peptidyl carrier protein
T-domain	Thiolation domain

T. Velkov, Ph.D.
Department of Pharmaceutics, Monash University, Parkville, VIC, Australia

A. Lawen, Dipl.-Chem., Dr. rer. nat. (✉)
Department of Biochemistry and Molecular Biology, School of Biomedical Sciences, Monash University, Melbourne, VIC, Australia
e-mail: Alfons.Lawen@monash.edu

J.-F. Martín et al. (eds.), *Biosynthesis and Molecular Genetics of Fungal Secondary Metabolites*, Fungal Biology, DOI 10.1007/978-1-4939-1191-2_4,

Cyclosporine A

Microbial peptide secondary metabolites often exhibit beneficial pharmacological properties. The penicillins, vancomycins, and the present topic, cyclosporine A (CsA; Sandimmune®, Neoral®), are notable examples. CsA was first isolated from *Tolypocladium inflatum* [1, 2]. The immunosuppressive properties of CsA have been widely exploited clinically, predominantly in bone marrow and organ transplantations and in the treatment of certain autoimmune diseases [3–5]. In addition to its predominant use in transplantation, CsA has also been indicated for a number of new clinical applications, some important examples include: the reversal of multidrug resistance, antimalarial, herpes virus infection, rheumatoid arthritis, type I diabetes, and also as a potent anti-human immunodeficiency virus 1 (HIV-1), anti-hepatitis C, and anticancer agent [6–15]. Unfortunately, dose-limiting nephrotoxicity and a number of other adverse side effects prevent the routine clinical implementation of CsA for the treatment of the aforementioned conditions. Thus, the search for novel cyclosporines that possess the pharmacological properties of CsA without the associated adverse side effects remain an important goal. CsA, is a cyclic undecapeptide that contains three non-proteinogenic amino acids (D-alanine, (4*R*)-4-[(*E*)-2-butyl]-4-methyl-L-threonine (Bmt), and L-2-aminobutyric acid) and seven *N*-methylated peptide bonds (Fig. 4.1) [16, 17]. The numbering of the amino acids in the CsA molecule corresponds to the order in which each residue was identified by sequential Edman degradation [2]. X-ray and NMR analyses showed that the backbone of the molecule between residues 11 and 7 forms a β(beta)-fragment consisting of an antiparallel β(beta)-sheet with a type II β-turn between residues 2 and 5. Residues 7–11 form an open loop structure [16]. The molecule exhibits a *cis*-amide bond between the *N*-methyl leucine residues at positions 9 and 10. Four intramolecular hydrogen bonds maintain the rigidity of the backbone structure (Fig. 4.1) [16]. Owing to the broad spectrum of bioactivity of CsA, considerable effort has been invested into identifying new and safer cyclosporines.

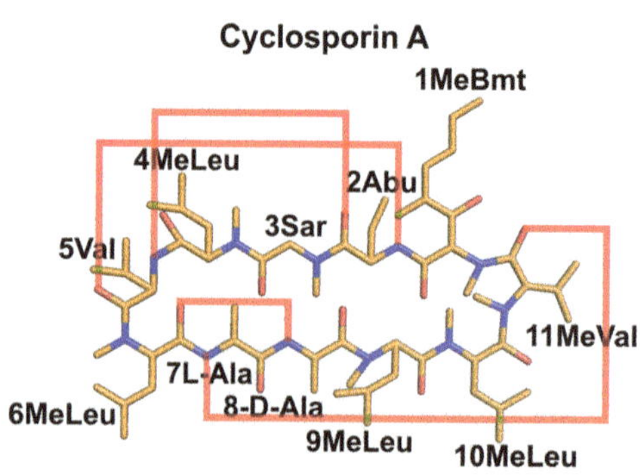

Fig. 4.1 The chemical structure of cyclosporine A. The three non-proteinogenic amino acids, are D-alanine, (4R)-4-[(E)-2-butyl]-4-methyl-L-threonine (Bmt), L-2-aminobutyric acid. The *red bars* indicate intramolecular hydrogen bonds

Cyclosporine Biosynthesis

Cyclosporine biosynthesis involves three different enzyme systems that cooperate in *trans* to generate and assemble the monomeric units of the cyclosporine molecule. The undecapeptide backbone is assembled on the multifunctional protein thiotemplate, cyclosporine synthetase (CySyn), a very complex high molecular weight non-ribosomal peptide synthetase (NRPS). *N*-methylation of specific amide positions in the cyclosporine backbone is critical for the complete assembly and cyclization of the cyclosporine peptide and is catalyzed by integral *N*-methyltransferase (*N*-MTase) domains during the assembly process prior to amide bond formation. The production and channeling of the non-proteinogenic precursor amino acids Bmt and D-alanine to CySyn is elaborated by specialized Bmt polyketide synthase (PKS) and alanine racemase enzymes. This section reviews the mechanistic functions and domain architecture of each of these key cyclosporine biosynthetic elements.

Cyclosporine Synthetase

The cyclosporines are products of CySyn, a very complex high molecular mass single polypeptide chain NRPS (Fig. 4.2) [18–20]. CySyn is one of the most complex NRPS systems known, consisting of a single polypeptide capable of catalyzing a total of 40 partial reaction steps in the synthesis of CsA via a protein template driven mechanism. The biosynthesis of cyclosporine A starts with the activation of D-alanine [21], thus the first module of CySyn is responsible for position 8 of the cyclosporine ring. A molecular mass of 1.7 MDa (15,281 amino acids), was delineated from the sequence of the CySyn gene, *simA*, which constitutes an intron-less

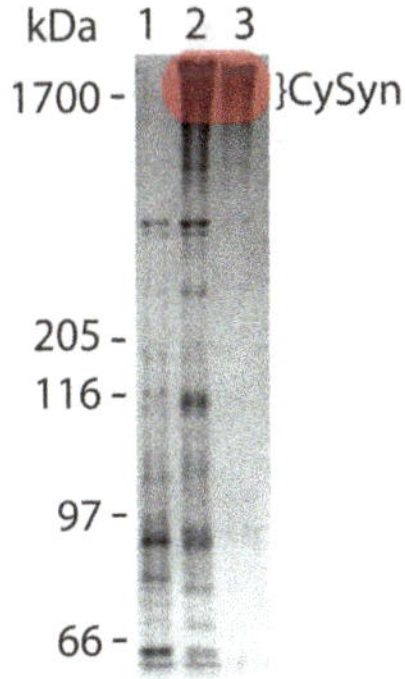

Fig. 4.2 SDS–PAGE analysis of CySyn purification in a 15-2 % gradient polyacrylamide gel. *Lane 1*, flow-through material from propyl-agarose chromatography. *Lane 2*, pooled active fractions from propyl-agarose chromatography. *Lane 3*, pooled active fractions from DEAE ion exchange chromatography. The migration of molecular mass marker protein standards is indicated on the ordinate. The 1.7 MDa CySyn polypeptide band is *shaded red*

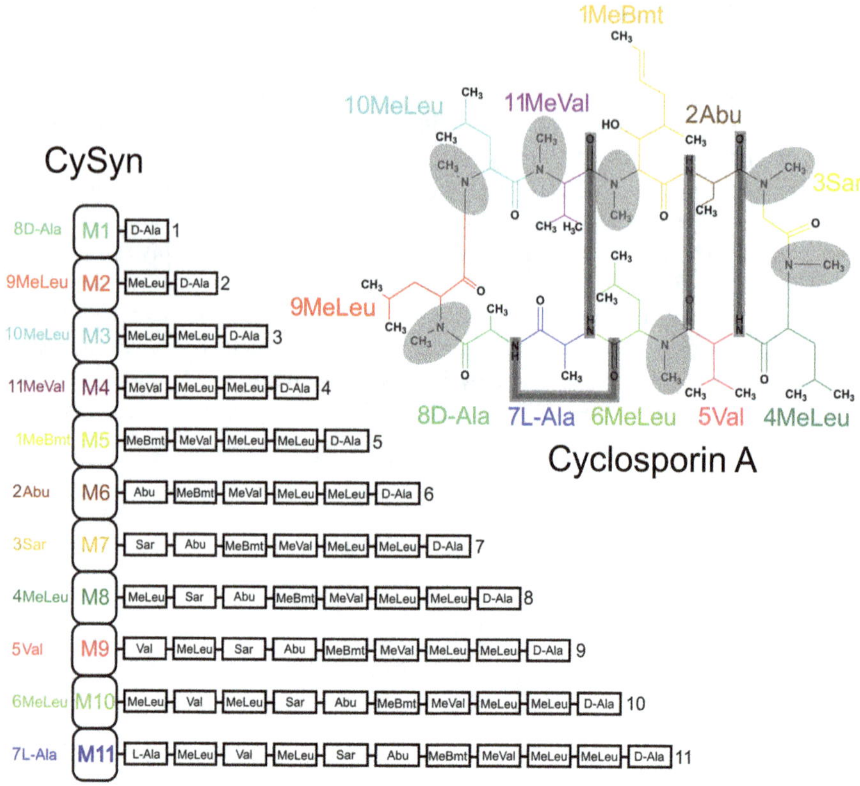

Fig. 4.3 Linear peptidyl-*S*-enzyme intermediate stages in CsA biosynthesis on the CySyn protein template. M1 to M11 represent the CySyn modular units responsible for binding and incorporation of individual amino acid components of CsA (the cognate amino acid substrates are indicated to the left of each module). The color-coding of each module corresponds to the particular amino acid it incorporates into the CsA structure (*top right panel*)

genomic open reading frame of 45.8 kb [22, 23]. The hydrodynamic shape of CySyn was first studied by sedimentation velocity ultracentrifugation, the findings suggested that CySyn is an oblate ellipsoid structure with a diameter of about 300 Å, with thickness of around 46 Å and a central opening of 50–60 Å, potentially to allow for peptide elongation [24, 25]. Transmission electron micrographs of negatively stained CySyn macromolecules revealed two distinct structures: one appeared to be a large globular structure of 25 nm whereas the second appeared to be a long chain of globular elements [26].

The CySyn polypeptide can be divided into eleven semiautonomous modular units, which are composed of homologous domains responsible for the activation, modification, and polymerization of the constituent amino acids of CsA [19, 20, 22]. The order of these modules is collinear with the sequence of the undecapeptide product, such that the repeating series of modules forms an ordered macromolecular assembly line (Fig. 4.3) [19–22]. The basic modular unit consists of catalytic domains responsible for substrate amino acid activation (A-domain), a peptidyl carrier protein

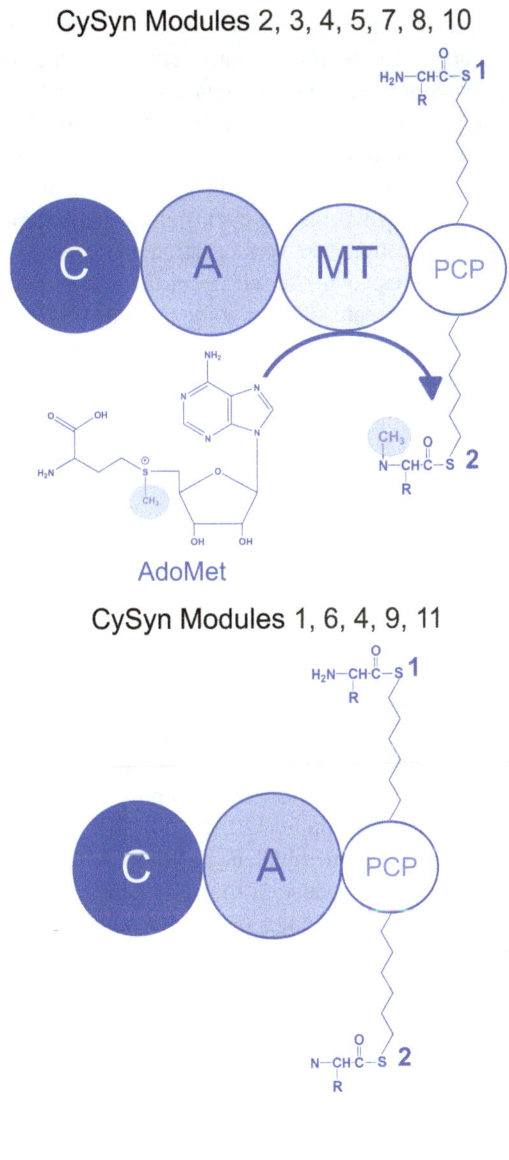

Fig. 4.4 Spatial organization of the functional domains in CySyn modules. A module (1,100–1,500 aa) consists of all of the functional domains required to incorporate a residue into the peptide chain. All CySyn modules contain domains responsible for the activation (A; ~550 aa), thiolation (PCP; ~100 aa), and condensation (C; ~450 aa) reactions of peptide assembly. *Top panel*, Domain organization of modules responsible for the incorporation of *N*-methylated amino acids. In addition to this essential catalytic domain triad, *N*-methylating modules exhibit an *N*-MTase (MT; ~430 aa) inserted between the A- and PCP-domains, that is responsible for the transfer of the methyl moiety of AdoMet to the cognate amino acid of the module, whereas it is covalently tethered to the respective modular 4′-Ppant prosthetic cofactor. *Bottom panel*, Domain organization of modules responsible for the incorporation of *des*-methyl amino acids

(PCP) domain (synonymous with the thiolation [T] domain) that acts to translocate the 4′-phosphopantetheine (4′-Ppant) covalently tethered peptidyl chain intermediate between modular active sites, and a condensation (C-domain) responsible for amide bond formation [27–30]. The activation of CySyn to the catalytically active form occurs posttranslationally by the covalent attachment of 11 4′-Ppant prosthetic cofactors and is catalyzed by a specific 4′-Ppant transferase. CySyn contains four (C)-(A)-(PCP) modules (modules 1, 6, 9, 11) and seven modules that display an additional *N*-MTase domain insert between the A- and PCP-domains (modules 2, 3, 4, 5, 7, 8, 10) of about 430 amino acids, presenting a modular domain order of (C)-(A)-(*N*-MTase)-(PCP) (Fig. 4.4). First, the adenylation domain of each module acts to activate its cognate amino acid substrate as an aminoadenylate. The activated amino acid is then covalently tethered to the free thiol of the 4′-Ppant prosthetic cofactor of the cognate modular PCP domain to yield aminoacyl-S-PCP [21, 31, 32]. The 4′-Ppant acts as a swinging arm that translocates the nascent peptidyl (or aminoacyl) intermediate to the following downstream aminoacyl-S-4′-PCP as peptide bond formation ensues by the action of the intervening C-domain [21, 28, 30, 31]. Thus, the growth of the cyclosporine peptide occurs by an ordered succession of transpeptidation and condensation reactions. The consummation of cyclosporine assembly results in the release of the mature peptide by ring closure catalyzed by a unique C-domain situated at the extreme C terminus of the last module [22].

Overall, it can be said that the functional characterization of the CySyn polypeptide and the encoding gene *simA*, represents a landmark achievement towards the understanding of the complex cyclosporine biosynthetic pathway.

N-*methylation*

Methyltransferases (MTase) are enzymes that catalyze the transfer of the methyl group from *S*-adenosyl-L-methionine (AdoMet) to nitrogen, oxygen, sulfur, or carbon atoms of a range of small molecules and macromolecular species. These enzymes can be divided into families according to four major substrate classes: protein, RNA, DNA, and small molecules [33–37]. *N*-methylation of amide bonds is a characteristic property of many therapeutic non-ribosomally synthesized peptides, such as the cyclosporines, enniatins, streptogramin Bs, pyochelins, and some cyanobacterial peptides [20, 38–43]. In CsA, seven of the eleven amino acid amide nitrogens are *N*-methylated (positions 1, 3, 4, 6, 9, 10, and 11). Sequence analysis of the cDNA sequence of the *simA* gene encoding the CySyn polypeptide revealed that 7 of the 11 modules display an additional *N*-MTase domain insert of approximately 430 amino acids between the A- and PCP-domains (Fig. 4.4) [22]. The *N*-MTase domains of CySyn catalyze the bimolecular transfer of the *S*-methyl group of AdoMet to the α-nitrogen of the thioesterified amino acid, releasing *S*-adenosyl-L-homocysteine as a reaction product [44]. Accordingly, the *N*-MTase domains of CySyn can be assigned as small molecule MTases. The presence of functional *N*-MTase activity in the CySyn polypeptide was first demonstrated by photoaffinity labeling with [*methyl*-^{14}C] AdoMet and by the ability of the purified enzyme to transfer the sulfonium methyl group from [*methyl*-^{14}C] AdoMet to CsA [18, 45]. Most AdoMet-dependent MTases have a bilobial structure and are organized into an AdoMet binding domain, which provides cofactor binding contacts and catalytic residues, and a second domain mainly responsible for conferring substrate specificity [35–37]. Despite the poor overall sequence identity across AdoMet-dependent MTases and varied substrate specificity, these enzymes display a structurally conserved cofactor binding domain termed the AdoMet binding fold [35–37]. A structure-guided sequence alignment of several structurally characterized AdoMet-dependent MTases revealed that the key residues governing the interactions with the AdoMet cofactor molecule are localized within four noncontiguous sequence motifs, I–IV [35–37]. Motifs I and II participate directly in cofactor binding, whereas motif III serves a structural role, forming the core of β-strand 5 of the AdoMet binding fold. Our group employed a highly specific photoaffinity labeling procedure, using ^{14}C-labeled AdoMet to define the chemical structure of the AdoMet binding centers of CySyn. Stoichiometric photoaffinity labeling demonstrated that CySyn is photolabeled with AdoMet in a molar ratio of ~7:1 (AdoMet:CySyn [45]), which is in good agreement with the seven *N*-MTase domains identified in the CySyn cDNA sequence [22]. The specificity of photolabeling was demonstrated by competitive displacement with nonradioactive AdoMet and its inhibitory analogs *S*-adenosyl-L-ethionine, *S*-adenosyl-L-homocysteine, and sinefungin [44, 46]. Photolabeling was only tenable with native CySyn and was not detected with denatured enzyme or with proteins, which do not bind AdoMet [46]. Notably, the photolabeling of the CySyn AdoMet binding sites displayed homotropic negative cooperativity, characterized by a curvilinear Scatchard plot with upward concavity [46]. Peptide mapping by tryptic digestion of purified CySyn photolabeled with either [*methyl*-^{14}C] AdoMet or [*carboxyl*-^{14}C] AdoMet yielded the sequence H_2N-Asn-Asp-Gly-Leu-Glu-Ser-Tyr-Val-Gly-Ile-Glu-Pro-Ser-Arg-COOH (residues 10,644–10,657), situated within the *N*-MTase domain of module 8 [45]. Radio-sequencing detected Glu10654 and Pro10655 as the major sites of derivatization. [*carboxyl*-^{14}C] AdoMet in addition labeled Tyr10650. Chymotryptic digestions generated the radiolabeled peptide H_2N-Ile-Gly-Leu-Glu-Pro-Ser-Gln-Ser-Ala-Val-Gln-Phe-COOH, corresponding to amino acids 2,125–2,136 of the *N*-MTase domain of module 2. The radiolabeled amino acids were identified as Glu2128 and Pro2129, which are equivalent in position and function to the modified residues identified with tryptic digestions in module 8. The modified sequence regions correspond to the motif II consensus sequence element, which is involved in directly complexing the adenine and ribose components of AdoMet. Homology modeling of the *N*-MTase domains of CySyn indicated that these regions conserve the consensus topology of the AdoMet binding fold and consensus cofactor interactions seen in structurally characterized AdoMet-dependent MTase [45]. Overall it appears that the AdoMet binding to the *N*-MTase domains of CySyn obey the consensus cofactor interactions observed among most structurally characterized AdoMet-dependent MTases.

N-methylation of specific amide positions in the cyclosporine backbone is critical for the complete assembly and cyclization of the cyclosporine peptide [44]. In in vitro biosynthetic reactions with purified CySyn, the main reaction product is CsA, as by-products, the desmethyl cyclosporines CsU ($MeLeu^6 \rightarrow Leu$) and CsQ ($MeLeu^4 \rightarrow Val$) are also observed, with CsA > CsU > CsQ in abundance [47, 48]. When the AdoMet inhibitors *S*-adenosyl-L-ethionine, *S*-adenosyl-L-homocysteine, and sinefungin were introduced at sub-inhibitory concentrations (10 μM), cyclosporine biosynthesis was preferentially shifted from CsA to cyclosporines that exhibit 5–6 *N*-methylated amides, with the following abundance: CsU ≥ CsL ($MeBmt^1 \rightarrow Bmt$) > CsE ($MeVal^{11} \rightarrow Val$) ≥ CsQ > CsA > CsR ($MeLeu^{6,10} \rightarrow Leu$) ≫ CsT ($MeLeu^{10} \rightarrow Leu$) [44]. The overall yield of cyclosporine products obtained in these reactions was low compared to the control CsA biosynthetic reaction mixture, suggesting biosynthesis of cyclosporines with <7 *N*-methyl amides is not a favorable process. This observation is in line with the frequency and abundance (compared to CsA) of desmethylated cyclosporines observed across the 32 cyclosporines isolated from *T. inflatum* nutrient broths [24]. Importantly, the desmethyl cyclosporine profile observed in the presence of sub-inhibitory concentrations of *N*-MTase inhibitors indicates that the enzyme has the ability to "skip" amide *N*-methylations at certain backbone positions. In order to determine the importance of amide *N*-methylation for the progression of peptide assembly and the mechanism whereby *N*-MTase inhibitors stall CsA biosynthesis, our group characterized the peptidyl-*S*-intermediates formed during the course of *N*-MTase inhibited reactions [44]. Omission of AdoMet or the inclusion of saturating levels of an inhibitory AdoMet analog from the otherwise complete reaction mixture stopped chain elongation at the D-Ala-L-Leu dipeptidyl stage, observed as the *cyclo*(D-alanyl-leucine diketopiperazine), a reaction product also observed with a mutant and with damaged enzyme [49]. This indicates that *N*-methylation of amide bonds appears to be critical for linear chain elongation beyond the two residue stages.

Molecular modeling and NMR analysis indicated that *N*-methylation of specific amide bond positions in the cyclosporine backbone is mandatory for the formation of a product-like conformation of the linear peptidyl-*S*-intermediate and for recognition by the acceptor site of the downstream peptide bond forming C-domain [44].

Condensation domains act as the gatekeepers of peptide bond formation and display rigid substrate selectivity for the incoming acceptor amino acid, thereby preserving the directionality of elongation and preventing mis-initiation at internal modules [50–57]. A widely established mechanism for why non-ribosomal peptide assembly stalls when a tailoring domain is deleted, mutated, or exchanged into another module is that the upstream C-domain does not recognize the nonnative amino acyl-*S*-PCP acceptor nucleophile [28, 58–66]. Crystallographic studies have revealed NRPS C-domains have a V-shaped architecture with a central canyon-like groove into which the upstream and downstream PCP-bound condensation substrates can be positioned from the opposing donor and acceptor sites, respectively, [55–57]. In the tyrocidine A NRPS, the PCP-C di-domain structure of TycC5-C6, the A/H conformation [32, 67] of the PCP domain interacts with the C-domain donor site in a catalytically unproductive orientation for peptide bond formation [56]. Whereas, in the structure of the termination module of surfactin A synthetase, SrfA [57], the A/H conformation of the PCP is stalled in the C-domain acceptor site with its 4′-Ppant binding site situated 16 Å away from the HHxxxDG core catalytic motif [28, 68], indicative of a catalytically productive orientation. We have modeled the C-domain of module 2 of CySyn (where CsA elongation stalls when *N*-methylation is inhibited or in the absence of AdoMet) with the upstream (donor) and downstream (acceptor) PCP domains (in the A/H state), and docked them into their respective C-domain donor and acceptor sites according to the productive PCP-C-domain contacts defined by mutagenesis studies [69–71], and observed in the SrfA termination module structure [57]. Our group's docking simulations with the 4′-Ppant donor and acceptor substrates suggested one of the mechanisms that stall CsA assembly at the dipeptidyl stage upon *N*-MTase inhibition is the inability of the acceptor site of the C-domain to recognize the unnatural desmethyl peptidyl-S-PCP acceptor [44].

CsA and the partially *des*-methylated analogs CsE ($MeVal^{11} \rightarrow Val$), CsQ ($MeLeu^{4} \rightarrow Val$), CsU ($MeLeu^{6} \rightarrow Leu$) and CsT ($MeLeu^{10} \rightarrow Leu$) were studied by 1D and 2D ^{1}H-NMR spectroscopy with the focus on the effect of *des*-methylation on backbone conformation and intra-chain hydrogen bonding [44]. The NMR data indicated that *des*-methylation at even a single amide position significantly affects the hydrogen bonding and backbone conformation of cyclosporines. In CsA, the backbone is constrained by *N*-methylation of specific amide bonds that limit the hydrogen bonding potential (Fig. 4.1). The removal of an *N*-methyl from position 11 as in CsE results in changes over the loop region formed by residues 6–11 with the formation of a hydrogen bond Ala^{7}-CO–HN-Val^{11}. Similarly, in CsU the loss of the position 6 amide *N*-methyl group ($MeLeu^{6} \rightarrow Leu^{6}$) leads to the formation of an additional hydrogen bond $MeBmt^{1}$-CO–HN-Leu^{6}.

Molecular models of the 3–11 amino acid stages of fully *N*-methylated and desmethyl CsA peptidyl-*S*-PCP intermediates suggested that the desmethyl backbone folds in on itself into a tight cyclic structure with the termini almost in contact [44]. In comparison, with the fully *N*-methylated form the simulation suggests that the amide *N*-methylation pattern of CsA only allows for specific intra-chain hydrogen bonds that may operate to stabilize the open undecapeptide chain into a product-like conformation, thereby bringing the amino and carboxyl termini of position 7 and 8 together to assist cyclization. Pre-organization of the growing peptidyl-*S*-intermediate into a product-like conformation is an emerging consensus mechanism across NRPS systems [72–76]. It appears that structural modifications to the peptide backbone introduced during assembly, such as amide *N*-methylation, help maintain the growing peptide in a product-like conformation that ensures that *trans*-peptidation and cyclization reactions proceed unhindered by futile intramolecular bonding events. In the assembly of the peptide tyrocidine A, the peptide substrate is pre-organized for cyclization via intramolecular backbone hydrogen bonds similar to those in the product to allow for the proper presentation of the termini to the thioesterase domain for ring closure [74, 75]. The overall rate of product cyclization/release was shown to be dependent on the rate of substrate pre-organization [74, 75]. Structural modifications such as *N*-methylation that inhibit backbone hydrogen bonding are therefore likely to influence the pre-organization of the cyclosporine peptide backbone for cyclization. Thus, the slower rate of biosynthesis of cyclosporines with five to six backbone *N*-methylations may be attributable to a less facile pre-organization rate. Coincidently, *N*-methylation of a specific amide bond position was shown to be mandatory for a high substrate turnover in the biosynthesis of streptogramin B antibiotics [41]. In the biosynthesis of tyrocidine A, key residues near the N- and C-termini are involved in the formation of intramolecular hydrogen bonds to allow for pre-organization of the linear peptide backbone, such that the N- and C-termini are presented in the correct orientation for macrocyclization [73]. In the CsA structure, the hydrogen bond between the amide nitrogen of D-Ala^{8} and the carbonyl oxygen of L-Ala^{7} helps to bring the ends of the molecule together for cyclization. The small side chains of the N- and C-terminal alanine residues of CsA also facilitate cyclization by allowing the ends to come close together into a suitable orientation for condensation. The D-configuration of the C-terminal alanine facilitates ring formation over side reactions as has been shown with other examples in the literature [77]. This is coincident with the strict selectivity of the A-domain of the last module incorporating for an L-amino acid and the strict D-epimer selectivity of the first A-domain module incorporating D-Ala^8 [48]. Similarly, N- and C-terminal amino acid side-chain stereochemistry and size constraints appear to be in place for other non-ribosomal peptide cyclization reactions [76]. Chemical cyclization of the synthetic CsA undecapeptide between positions 7 and 8, analogous to the enzymatic cyclization reaction, was the most efficient route of total chemical synthesis of cyclosporine A [78]. Moreover, L-Ala^7 and D-Ala^8 are the only consecutive pair of *N*-desmethyl amino acids in the CsA molecule. This further facilitates ring closure, as chemical amide bond formation between *N*-methylated amino acid residues is more difficult [79, 80]. Additional structural features of the CsA molecule that facilitate the pre-organizational folding process include the presence of the invariant *N*-methyl glycine (sarcosine) in the middle of the open peptide that serves as a β-turn forming element. A β-sheet appears to be a critical substructure requisite for proper pre-organization for cyclization and release of non-ribosomal peptides [72, 73]. Accordingly, backbone *N*-methylation is crucial for conservation of the proper hydrogen bonding pattern that preserves this β-sheet structure of CsA.

In summary, *N*-methylation of specific amide positions in the backbone of CsA is critical for the efficient progression of peptide assembly to the mature undecapeptidyl stage. It is evident that *N*-methylation of cyclosporines is most conserved at positions 3, 9, and 10, whereas loss of methylated peptide bonds is most frequently observed at positions 1, 4, 6, and 11. However, no more than two desmethyl positions are tolerated before peptide assembly stalls, and these never occur in direct succession. The *N*-methylation of enzyme-bound intermediates is of general importance in the synthesis of many peptide and depsipeptide antibiotics. In future, it is hoped that *N*-MTase tailoring functions can be routinely exploited in the genetic engineering of hybrid synthetases for the rational design of peptide antibiotics.

Alanine Racemase

Apart from D-alanine, the remaining amino acids in the CsA sequence are all of the L-form. Labeling experiments with ^{13}C-glucose revealed that the proteinogenic L-amino acids in the CsA sequence, namely L-valine, L-leucine, L-alanine, and glycine derived from classical biosynthetic pathways [81]. Whereas the non-proteinogenic L-2-aminobutyric acid is derived via the Krebs cycle from oxaloacetate via acetyl-coenzyme A [81]. The biosynthesis of Bmt is detailed in the following section. CySyn, which is unable to catalyze the isomerization of L-alanine, relies on an external fungal racemase enzyme for the provision of the D-epimer to initiate cyclosporine assembly. Hoffmann et al. [82] first reported the purification of a pyridoxyl phosphate cofactor-dependent oligomeric alanine racemase from *T. inflatum* strain 7939/45 that is responsible for the supply of D-alanine to CySyn. Interestingly, the alanine racemase was found to be capable of catalyzing the isomerization of several other amino acids (% relative to L-alanine conversion efficiency, [100 %]), including L-serine (23 %), 2-aminobutyric acid (15 %), and L-leucine (13 %) [82]. Coincidentally, precursor-directed in vitro synthesis with L-serine in vitro using enriched enzyme extracts or in vivo precursor feeding experiments of fungal cultures with L-serine, yielded D-Ser^8 [CsA] [83, 84]. Biochemical studies with isolated CySyn showed that in the absence of D-alanine the formation of peptidyl-*S*-intermediates could not be detected; suggesting D-alanine acts as the initiator amino acid for cyclosporine assembly [48]. This finding is coincident with the strict specificity of module 1 for the D-epimer of alanine and the absence of an integral epimerase domain within module 1 as inferred from sequence analysis of the *simA* gene [48].

Not surprisingly, when L-alanine and L-leucine was provided as the sole amino acid precursors, the co-incubation of the isolated alanine racemase with CySyn led to the formation of *cyclo*(D-alanyl-*N*-methylleucine), whereas, CySyn per se was incapable of diketopiperazine formation [82].

The 41 kDa *T. inflatum* alanine racemase is encoded by the *cssB* gene (1,149 bp). Notably, the *T. inflatum* alanine racemase shares high sequence identity with serine hydroxymethyltransferase and threonine aldolase, two other pyridoxyl phosphate dependent enzymes, with similar catalytic promiscuity and overlapping catalytic properties [23, 85, 86]. A recent report described a protocol for the recombinant expression and purification *T. inflatum* alanine racemase in *Escherichia coli* [86]. The authors noted some difficulty with obtaining a soluble enzyme, which is consistent with the localization of the enzyme on the outside of the vacuolar membrane [26].

Subcellular fractionation, together with immuno-electron microscopy, indicates CsA is localized within the fungal vacuole, with the CySyn enzyme and the cognate alanine racemase associated with the vacuolar membrane [26]. Due to the low concentration of D-alanine in the fungal cell milieu, it is possible that the alanine racemase channels D-alanine directly to the CySyn loading module via a direct protein–protein interaction. It is tenable to imagine that the cyclosporine biosynthetic machinery operates as a metabolon comprised of the CySyn, Bmt PKS, and the alanine racemase.

The production of cyclosporines appears to be regulated by the available flux of D-alanine from the racemase. The cyclosporine high-producing strain *T. inflatum* 7939/45 produces ~60 g/L of CsA, in comparison with medium producer strains, which produce ~1.5 g/L of culture medium [20]. This 40-fold difference in CsA production cannot be accounted for by the twofold difference in CySyn levels detected between the two strains, nor due to kinetic differences between the CySyn from each strain as the rate of CsA formation was about the same [20]. The cellular concentration of the alanine racemase in the fungal cell as estimated from the activity of crude extracts is ~0.01 % of the total cytosolic protein [82]. The high K_m of the alanine racemase for L-alanine, likely provides a sufficiently higher rate of substrate turnover compared to CySyn (which needs to perform 40 reaction steps in order to assemble one CsA molecule) to supply ample levels of D-alanine in the fungal cell to support CsA biosynthesis. Notably, feeding experiments with 3-fluoro-D-alanine, a known inhibitor of prokaryotic alanine racemases, produced a marked inhibition of cyclosporine production in *T. inflatum* cultures [87]. The transformation of a low CsA producer *T. inflatum* strain, which harbors a mutation in its *cssB* alanine racemase gene, with a plasmid vector containing the wild-type *cssB* gene restored CsA productivity to high levels [88].

Clearly the D-alanine racemase plays a pivotal role in CsA biosynthesis and may act as the dominant rate-limiting factor in the cyclosporine biosynthetic pathway in vivo.

Bmt Polyketide Synthase

Labeling experiments with ^{13}C-labeled acetate and glucose first pointed to a polyketide origin for the Bmt backbone [81, 89]. In line with our contemporary understanding of type II polyketide biosynthesis [90–95], Bmt biosynthesis is most likely elaborated over two stages: (1) the basic assembly of the polyketide backbone via the head-to-tail condensation of four acetate units, which involves reduction and dehydration and methylation reactions; (2) the transformation process into the final Bmt product, which involves incorporation of the amino group. Offenzeller et al. [96] performed detailed in vivo labeling experiments and in vitro polyketide synthesis assays with enriched enzyme fractions from *T. inflatum* strain NRRL 8044, which identified 3(*R*)-hydroxy-4(*R*)-*methyl*-6(*E*)-octenoyl-CoA as the end-product of the basic assembly process. The substrates for this basic assembly reaction were identified as acetyl-CoA, malonyl-CoA, NADPH, and AdoMet [96]. Notably, Offenzeller et al. [96] observed that the 3(*R*)-hydroxy-4(*R*)-*methyl*-6(*E*)-octenoyl-CoA Bmt backbone is released from the Bmt PKS as the CoA thioester, indicating that the ensuing transformation reactions most likely occur on this key intermediate [97]. The isolated Bmt PKS displays an apparent molecular mass of 600 kDa as estimated via SDS-PAGE [88]. Sequence analysis of the Bmt PKS encoding gene *cssC* indicated that it is a type II PKS [23, 88]. The Bmt PKS polypeptide appears to harbor all of the enzymatic activities required for the completion of the basic assembly stage of the Bmt biosynthetic route [97]. Once the activated building blocks are activated, they remain bound to the Bmt PKS during the reaction cycle [97]. The first elongation cycle together with the second condensation reaction and the methylation reaction appear to follow a processive mechanism as per macrolactone synthase [97]. Methylation takes place exclusively during the second backbone elongation cycle on the enzyme-bound 3-oxo-4-hexenoic acid intermediate [97].

T. inflatum strains with mutations in the Bmt PKS gene are defective in cyclosporine production [88]. So it follows that similar to the alanine racemase, the Bmt PKS plays a critical role in cyclosporine biosynthetic pathway by channeling the essential Bmt precursor to CySyn.

Directed Biosynthesis

The large spectrum of pharmacological actions of CsA has placed great importance on the search for derivatives that possess specific properties without the adverse effects. CySyn exhibits a low precursor amino acid specificity, which presents the opportunity for the directed biosynthesis of cyclosporine analogs via the utility of both in vitro and in vivo biosynthetic platforms.

In Vivo Directed Biosynthesis

Cyclosporines isolated from cultures of *T. inflatum* differ from the CsA sequence by one to two amino acids, displaying positional variations predominantly at position 2 (Abu) and by the presence or absence of 1–2 *N*-methyl groups at certain amino acid positions [84]. In view of the relaxed selectivity of some of the A-domains of CySyn, precursor feeding of fungal cultures has proved to be a most valuable method for directing the biosynthesis toward specific cyclosporine analogs [87, 98–100].

Cyclosporines are produced by fungi imperfecti. CsA and its naturally occurring analogs have been isolated from at least 17 fungal taxa [2, 88, 99, 101–104]. In addition to the main product CsA, more than 25 cyclosporines have been isolated from submerged cultures of *T. inflatum* [84, 87, 98–100, 105–108]. Many of these fungi exhibit a varied pattern of cyclosporine biosynthesis compared to the originally described cyclosporine producing fungus, *T. inflatum*. One notable example that we have investigated in our laboratory is the fungus *Cylindrotrichum oligospermum* (Corda) Bonorden that produces the peptolide SDZ 214-103 ([Thr^2, Leu^5, D-Hiv^8, Leu^{10}]-cyclosporine [109, 110]). Given the similarities between CySyn and peptolide SDZ 214-103 synthetase [109, 110], it should be possible to adapt the methods established for CySyn for the large-scale biosynthesis of peptolide SDZ 214-103 analogs. Moreover, two novel cyclosporines [Thr^2, Leu^5, Ala^{10}] CsA and [Thr^2, Ile^5] CsA have been isolated from strain F/88-3089/11 of *Acremonium luzulae* (Fuckel) WGams and strain F/93-4641/04 of the *Leptostroma anamorph* of *Hypoderma eucalypti* Cooke and Harkin [84]. The sequencing and annotation of the *T. inflatum* genome identified a total of 14 NRPSs, 20 PKSs, 4 Hybrid PKS-NRPSs, 11 putative NRPS-like enzymes, and 5 putative PKS-like enzymes [23]. Accordingly, this fungal species represents a rich source of NRPS genes with differing substrate specificities that can be employed for the construction recombinant synthases.

In Vitro Directed Biosynthesis

Although precursor-directed biosynthesis provides a useful and efficient means to produce certain cyclosporine analogs not occurring naturally, cell-free systems using isolated CySyn enzyme are superior as they avoid interference from fungal metabolism. The establishment of efficient industrial-scale purification procedures for the isolation of intact CySyn from cyclosporine producing fungi allows for the overproduction of CsA analogs with improved properties. To this end, we have explored enzyme purification strategies suited to large-scale processing and have presented a chromatographic sequence that serves as a pilot model for the industrial-scale preparation of CySyn from cyclosporine producing fungi [111]. A chromatographic sequence consisting of ammonium sulfate precipitation → gel filtration → hydrophobic interaction chromatography → anion exchange chromatography, yielded an electrophoretically almost homogenous CySyn preparation (Fig. 4.2). CySyn exhibited an optimal temperature range of 24–29 °C and a pH optimum of 7.6 [111]. The native enzyme displayed a pI of 5.7, as determined by isoelectric focusing. Presently, industrial processes for large-scale cyclosporine production involve a number of in vivo systems, predominantly submerged or solid state fermentation [88]. In such fermentations a complex mixture of cyclosporines is produced. Therefore, substantial chromatographic operations are required to achieve sufficient purity of the desired product; this in turn leads to decreased yields. The advantage of an in vitro system using isolated CySyn preparations is that a defined major product is obtainable, circumventing the need for the separation of complex peptide mixtures and providing higher yields. In vitro enzymatic biosynthesis avoids the inconsistent secondary metabolic physiology of the fungus and inconsistent cyclosporine production observed with some strains. Moreover, the industrial implementation of an in vitro biosynthetic approach could potentially prove useful for the production of important therapeutic cyclosporines, which occur as only minor fermentation by-products.

In addition to the naturally occurring CsA homologues, CySyn is capable of producing cyclosporines not obtainable in vivo. The in vitro enzymatic biosynthesis of cyclosporines allows for directed modifications of the peptide ring in positions 1, 2, 5, 7, 8, and 11 simply by varying the amino acid composition presented to the enzyme [47, 48, 109, 110, 112]. We have employed the isolated CySyn enzyme platform for the in vitro biosynthesis of more than 40 CsA analogs [19, 47, 48, 109, 110, 112]. A number of these in vitro synthesized cyclosporine analogs displayed clinically important bioactivities and therefore served as good candidates for pharmacological testing. One notable example that exhibited improved properties in comparison to the progenitor compound CsA was CsG ([norvaline^{2}] CsA), which exhibits potent immunosuppression with reduced nephrotoxicity. Two other important examples include the non-immunosuppressive analogs [MeIle^{4}] CsA *syn.* SDZ NIM 811, which displays a very potent anti-HIV-1 activity [113–117], and SDZ 214-103 which shows potent drug resistance reversion properties [9]. [MeIle^{4}] CsA (also designated SDZ NIM 811) was initially detected in cultures that had been supplemented with D-threonine. However, the large-scale production of this analog in vivo is impractical because of the requirement for feeding of D-threonine, which would be far too costly. With in vitro biosynthesis the substrate pool can be carefully controlled, providing an almost exclusive production of the desired analog. [MeIle^{4}] CsA was the first cyclosporine analog reported that lacked immunosuppressive activity while maintaining strong cyclophilin binding and peptidyl-prolyl-cis/*trans*-isomerase inhibiting activity [113]. It shows strong anti-immunodeficiency virus type 1 activity [118], antimalarial activity [119], and anti-hepatitis C activity [120, 121]. Furthermore, [MeIle[4]] CsA and its homologues proved valuable research tools for the discrimination between the effects of CsA on calcineurin and on cyclophilin (e.g., [122–124]).

The high biotechnological significance of an in vitro production system for cyclosporines accentuates the need for the development of stable in vitro biosynthetic platforms with isolated CySyn enzymes. With the use of purified CySyn preparations we have demonstrated the in vitro production of CsA can be sustained provided the continual removal of products and replenishment of substrates. It is tenable to imagine this process could be adapted to a bioreactor-type of system, which will allow for the maintenance of sustainable reaction conditions. An up-scaled version of this in vitro process will allow for the systematic synthesis of CsA analogs in sufficient quantities for preclinical evaluation.

Beyond Biosynthesis

Despite the versatility of the aforementioned biosynthetic strategies, we should not ignore the great potential of emerging in toto and recombinant synthetic platforms for the generation of novel and safer cyclosporines.

In Toto Synthetic Strategies

The routine total solid-phase synthesis of cyclosporines is complicated by the difficulty of coupling *N*-methylated amino acids and is also limited by the lack of commercial availability of Bmt due to its difficult synthesis [125–127]. Wenger [78, 125, 128] from Sandoz Ltd. described the first total synthesis of the CsA molecule using a fragment-condensation technique. Subsequently, the Sandoz team reported the solid-phase synthesis of CsA and several of its analogs [129]. More recently, the Danishefsky laboratory reported a versatile isonitrile fragment coupling route for the total synthesis of cyclosporines [130]. The ever-advancing peptide synthesis technologies, together with the commercial availability of protected forms of unusual non-proteinogenic amino acids such as Bmt, will undoubtedly result in the emergence of in toto synthetic platforms for the exploration of cyclosporine structure-activity-relationships as part of future drug discovery programs.

Recombinant Synthetic Strategies

One of the most exciting facets of NRPS research is the potential for the production of chimeras that are capable of synthesizing novel peptides. Over the years a variety of molecular biology strategies have been established for the cloning of peptide synthetase genes [53, 131–133]. Together with the recent advances in automated sequencing technology and bioinformatics, we are well equipped to implement combinatorial design projects to produce altered non-ribosomal products. Genetic level combinatorial approaches for the rational design of hybrid NRPSs include module/domain exchange, insertion or deletion, and the alteration of the substrate specificity of the A-domain via site-directed mutagenesis [27, 53, 132, 134]. Most of these combinatorial techniques have been successfully applied to PKS systems that synthesize its analogs of therapeutically important compounds, such as the erythromycin antibiotics [135–137]. The exchange of CySyn modules that display a stringent substrate selectivity (such as module 7, which displays a high specificity for Gly) with modules, which possess a broader substrate specificity (such as module 6 [Abu]), should allow for the production of a broader spectrum of cyclosporines via precursor-directed synthesis, than is accessible with the native enzyme. The work from Biochimie Kundl has demonstrated that the combinatorial biosynthesis of cyclosporines is tenable. A plasmid transformation system has been developed for *T. inflatum*, consisting of the promoter element derived from genomic cyclophilin fused to a bacterial hygromycin phosphotransferase gene [138]. Using this system, *T. inflatum* protoplasts were transformed at high frequency with plasmid constructs containing internal fragments of the CySyn gene. A successful homologous crossover event between the cloned fragment and the genomic CySyn DNA was evident by the detection of cyclosporine non-producing transformants. The insertions were also verified by Southern blot hybridization. The high frequency of CySyn knockout transformants obtained suggests that *T. inflatum* possesses a single genomic copy of the CySyn gene. These CySyn knockout mutants could also serve as hosts for CySyn genes mutated in vitro or for the fermentative production of D-alanine and Bmt precursors. Leitner et al. [139, 140] described the generation of a recombinant vector containing the entire CySyn DNA sequence and the transformation of *T. inflatum* cells with this construct. The implementation of this transformation system for *T. inflatum* will also allow for the selection of transformants with multiple inserts of the CySyn gene. In this manner, strains with enhanced cyclosporine production capabilities can be generated expediently compared to conventional mutagenesis and strain selection methodologies [139, 140]. The authors have also employed the CySyn DNA for screening of microbes for CySyn genes that may have been overlooked in product screening tests due to the inactive state of their CySyn gene. The CySyn DNA of *T. inflatum* can be recombined with the heterologous CySyn DNA isolated from a different cyclosporine producing fungi to construct a hybrid synthetase with the desired product profile.

Conclusion

Perspective

Many peptide secondary metabolites of microbial origin possess clinically useful pharmacological activities due to their expanded structural spectrum of non-proteinogenic amino- and hydroxy-acid monomeric units, together with unique structural modifications such as *N*-methylation of their peptide backbone. The enormous structural and functional diversity of these low-molecular weight peptides is attributable to their non-ribosomal mode of biosynthesis on NRPS protein templates. In addition to their ability to utilize a broad spectrum of proteinogenic and non-proteinogenic monomeric units, NRPS introduce further structural diversity into their products via auxiliary modifying functions that reside in either externally associated or integral enzyme activities. With the low hanging fruit such as the β(beta)-lactam antibiotics and CsA, which were harvested from the highly successful natural product discovery programs of the 1970s, progressing past their clinical usefulness, we must turn to recombinant NRPS systems to feed the insatiable drug discovery pipeline. The highly versatile cyclosporine biosynthetic system will unquestionably prove to be a valuable source of NRPS genes for the recombinant engineering of biosynthetic machinery for the production of the next generation of "natural product" pharmaceuticals.

Acknowledgements T. Velkov is a National Health and Medical Research Council of Australia CDA1 Industry Fellow (1003836).

References

1. Dreyfuss M, Härri E, Hofmann H, Kobel H, Pache W, Tcherter H. Cyclosporine-A and C new metabolites from *Trichoderma-polysporum* (Link Ex Pers) *rifai*. Eur J Appl Microbiol. 1976;3:125–33.
2. Rüegger A, Kuhn M, Lichti H, Loosli H-R, Huguenin R, Quiquerez C, Von Wartburg A. Cyclosporine A, ein immunsuppressiv wirksamer Peptidmetabolit aus *Trichoderma polysporum* (LINK ex PERS.) Rifai. Helv Chim Acta. 1976;59:1075–92.
3. Calne RY, White DJ, Thiru S, Evans DB, Mcmaster P, Dunn DC, Craddock GN, Pentlow BD, Rolles K. Cyclosporine A in patients receiving renal allografts from cadaver donors. Lancet. 1978;2:1323–7.
4. Morris PJ. The impact of cyclosporine A on transplantation. Adv Surg. 1984;17:99–127.
5. Borel JF, Baumann G, Chapman I, Donatsch P, Fahr A, Mueller EA, Vigouret JM. In vivo pharmacological effects of ciclosporin and some analogues. Adv Pharmacol. 1996;35:115–246.
6. Murphy JR, Baqar S, Baker RH, Roberts E, Nickell SP, Cole GA. Stage-selective inhibition of rodent malaria by cyclosporine. Antimicrob Agents Chemother. 1988;32:462–6.
7. Gaveriaux C, Boesch D, Boelsterli JJ, Bollinger P, Eberle MK, Hiestand P, Payne T, Traber R, Wenger R, Loor F. Overcoming multidrug resistance in Chinese hamster ovary cells in vitro by cyclosporine A (Sandimmune) and non-immunosuppressive derivatives. Br J Cancer. 1989;60:867–71.
8. Chase HP, Butler-Simon N, Garg SK, Hayward A, Klingensmith GJ, Hamman RF, O'brien D. Cyclosporine A for the treatment of new-onset insulin-dependent diabetes mellitus. Pediatrics. 1990;85:241–5.
9. Schwabe K, Steinheider G, Lawen A, Traber R, Hildebrandt A. Reversal of multidrug resistance by novel cyclosporine A analogues and the cyclopeptolide SDZ 214-103 biosynthesized in vitro. J Cancer Res Clin Oncol. 1995;121:407–12.
10. Stewart T, Tsai SC, Grayson H, Henderson R, Opelz G. Incidence of de-novo breast cancer in women chronically immunosuppressed after organ transplantation. Lancet. 1995;346:796–8.
11. Stein CM. Cyclosporine in the treatment of rheumatoid arthritis. Bull Rheum Dis. 1995;44:1–4.

12. Franke EK, Luban J. Inhibition of HIV-1 replication by cyclosporine A or related compounds correlates with the ability to disrupt the Gag-cyclophilin A interaction. Virology. 1996;222:279–82.
13. Loor F, Tiberghien F, Wenandy T, Didier A, Traber R. Cyclosporines: structure-activity relationships for the inhibition of the human MDR1 P-glycoprotein ABC transporter. J Med Chem. 2002;45:4598–612.
14. Rao SN. Treatment of herpes simplex virus stromal keratitis unresponsive to topical prednisolone 1 % with topical cyclosporine 0.05 %. Am J Ophthalmol. 2006;141:771–2.
15. Lee J. Cyclophilin A as a new therapeutic target for Hepatitis C virus induced hepatocellular carcinoma. Korean J Physiol Pharmacol. 2013;17:375–83.
16. Von Wartburg A, Traber R. Chemistry of the natural cyclosporine metabolites. Prog Allergy. 1986;38:28–45.
17. Sanglier JJ, Traber R, Buck RH, Hofmann H, Kobel H. Isolation of (4*R*)-4-[(*E*)-2-butenyl]-4-methyl-L-threonine, the characteristic structural element of cyclosporines, from a blocked mutant of *Tolypocladium inflatum*. J Antibiot (Tokyo). 1990;43:707–14.
18. Lawen A, Zocher R. Cyclosporine synthetase. The most complex peptide synthesizing multienzyme polypeptide so far described. J Biol Chem. 1990;265:11355–60.
19. Lawen A. Biosynthesis and mechanism of action of cyclosporines. Prog Med Chem. 1996;33:53–97.
20. Velkov T, Lawen A. Non-ribosomal peptide synthetases as technological platforms for the synthesis of highly modified peptide bioeffectors—cyclosporine synthetase as a complex example. Biotechnol Annu Rev. 2003;9:151–97.
21. Dittmann J, Wenger RM, Kleinkauf H, Lawen A. Mechanism of cyclosporine A biosynthesis. Evidence for synthesis via a single linear undecapeptide precursor. J Biol Chem. 1994;269:2841–6.
22. Weber G, Schorgendorfer K, Schneider-Scherzer E, Leitner E. The peptide synthetase catalyzing cyclosporine production in *Tolypocladium niveum* is encoded by a giant 45.8-kilobase open reading frame. Curr Genet. 1994;26:120–5.
23. Bushley KE, Raja R, Jaiswal P, Cumbie JS, Nonogaki M, Boyd AE, Owensby CA, Knaus BJ, Elser J, Miller D, Di Y, Mcphail KL, Spatafora JW. The genome of *Tolypocladium inflatum*: evolution, organization, and expression of the cyclosporine biosynthetic gene cluster. PLoS Genet. 2013;9:e1003496.
24. Lawen A, Dittmann J, Schmidt B, Riesner D, Kleinkauf H. Enzymatic biosynthesis of cyclosporine A and analogues. Biochimie. 1992;74:511–6.
25. Schmidt B, Riesner D, Lawen A, Kleinkauf H. Cyclosporine synthetase is a 1.4 MDa multienzyme polypeptide. Re-evaluation of the molecular mass of various peptide synthetases. FEBS Lett. 1992;307:355–60.
26. Hoppert M, Gentzsch C, Schörgendorfer K. Structure and localization of cyclosporine synthetase, the key enzyme of cyclosporine biosynthesis in *Tolypocladium inflatum*. Arch Microbiol. 2001;176:285–93.
27. Stachelhaus T, Schneider A, Marahiel MA. Rational design of peptide antibiotics by targeted replacement of bacterial and fungal domains. Science. 1995;269:69–72.
28. Stachelhaus T, Mootz HD, Bergendahl V, Marahiel MA. Peptide bond formation in nonribosomal peptide biosynthesis. Catalytic role of the condensation domain. J Biol Chem. 1998;273:22773–81.
29. Stachelhaus T, Marahiel MA. Modular structure of genes encoding multifunctional peptide synthetases required for non-ribosomal peptide synthesis. FEMS Microbiol Lett. 1995;125:3–14.
30. Stachelhaus T, Marahiel MA. Modular structure of peptide synthetases revealed by dissection of the multifunctional enzyme GrsA. J Biol Chem. 1995;270:6163–9.
31. Stachelhaus T, Huser A, Marahiel MA. Biochemical characterization of peptidyl carrier protein (PCP), the thiolation domain of multifunctional peptide synthetases. Chem Biol. 1996;3:913–21.

32. Weber T, Baumgartner R, Renner C, Marahiel MA, Holak TA. Solution structure of PCP, a prototype for the peptidyl carrier domains of modular peptide synthetases. Structure. 2000;8:407–18.
33. Borchardt RT. S-Adenosyl-L-methionine-dependent macromolecule methyltransferases: potential targets for the design of chemotherapeutic agents. J Med Chem. 1980;23:347–57.
34. Copeland RA, Solomon ME, Richon VM. Protein methyltransferases as a target class for drug discovery. Nat Rev Drug Discov. 2009;8:724–32.
35. Jurkowska RZ, Jurkowski TP, Jeltsch A. Structure and function of mammalian DNA methyltransferases. Chembiochem. 2011;12:206–22.
36. Struck AW, Thompson ML, Wong LS, Micklefield J. S-adenosyl-methionine-dependent methyltransferases: highly versatile enzymes in biocatalysis, biosynthesis and other biotechnological applications. Chembiochem. 2012;13:2642–55.
37. Liscombe DK, Louie GV, Noel JP. Architectures, mechanisms and molecular evolution of natural product methyltransferases. Nat Prod Rep. 2012;29:1238–50.
38. Burmester J, Haese A, Zocher R. Highly conserved *N*-methyltransferases as an integral part of peptide synthetases. Biochem Mol Biol Int. 1995;37:201–7.
39. Hacker C, Glinski M, Hornbogen T, Doller A, Zocher R. Mutational analysis of the *N*-methyltransferase domain of the multifunctional enzyme enniatin synthetase. J Biol Chem. 2000;275:30826–32.
40. Patel HM, Walsh CT. In vitro reconstitution of the Pseudomonas aeruginosa nonribosomal peptide synthesis of pyochelin: characterization of backbone tailoring thiazoline reductase and *N*-methyltransferase activities. Biochemistry. 2001;40:9023–31.
41. Mahlert C, Sieber SA, Grűnewald J, Marahiel MA. Chemoenzymatic approach to enantiopure streptogramin B variants: characterization of stereoselective pristinamycin I cyclase from Streptomyces pristinaespiralis. J Am Chem Soc. 2005;127:9571–80.
42. Welker M, Von Döhren H. Cyanobacterial peptides—nature's own combinatorial biosynthesis. FEMS Microbiol Rev. 2006;30:530–63.
43. Hornbogen T, Riechers SP, Prinz B, Schultchen J, Lang C, Schmidt S, Műgge C, Turkanovic S, Sűssmuth RD, Tauberger E, Zocher R. Functional characterization of the recombinant *N*-methyltransferase domain from the multienzyme enniatin synthetase. Chembiochem. 2007;8:1048–54.
44. Velkov T, Horne J, Scanlon MJ, Capuano B, Yuriev E, Lawen A. Characterization of the *N*-methyltransferase activities of the multifunctional polypeptide cyclosporine synthetase. Chem Biol. 2011;18:464–75.
45. Velkov T, Lawen A. Mapping and molecular modeling of S-adenosyl-L-methionine binding sites in *N*-methyltransferase domains of the multifunctional polypeptide cyclosporine synthetase. J Biol Chem. 2003;278:1137–48.
46. Velkov T, Lawen A. Photoaffinity labeling of the *N*-methyltransferase domains of cyclosporine synthetase. Photochem Photobiol. 2003;77:129–37.
47. Lawen A, Traber R, Geyl D, Zocher R, Kleinkauf H. Cell-free biosynthesis of new cyclosporines. J Antibiot (Tokyo). 1989;42:1283–9.
48. Lawen A, Traber R. Substrate specificities of cyclosporine synthetase and peptolide SDZ 214-103 synthetase. Comparison of the substrate specificities of the related multifunctional polypeptides. J Biol Chem. 1993;268:20452–65.
49. Dittmann J, Lawen A, Zocher R, Kleinkauf H. Isolation and partial characterization of cyclosporine synthetase from a cyclosporine non-producing mutant of *Beauveria nivea*. Biol Chem Hoppe Seyler. 1990;371:829–34.
50. Belshaw PJ, Walsh CT, Stachelhaus T. Aminoacyl-CoAs as probes of condensation domain selectivity in nonribosomal peptide synthesis. Science. 1999;284:486–9.
51. Linne U, Marahiel MA. Control of directionality in nonribosomal peptide synthesis: role of the condensation domain in preventing misinitiation and timing of epimerization. Biochemistry. 2000;39:10439–47.
52. Doekel S, Marahiel MA. Dipeptide formation on engineered hybrid peptide synthetases. Chem Biol. 2000;7:373–84.

53. Mootz HD, Schwarzer D, Marahiel MA. Construction of hybrid peptide synthetases by module and domain fusions. Proc Natl Acad Sci U S A. 2000;97:5848–53.
54. Ehmann DE, Trauger JW, Stachelhaus T, Walsh CT. Aminoacyl-SNACs as small-molecule substrates for the condensation domains of nonribosomal peptide synthetases. Chem Biol. 2000;7:765–72.
55. Keating TA, Marshall CG, Walsh CT, Keating AE. The structure of VibH represents nonribosomal peptide synthetase condensation, cyclization and epimerization domains. Nat Struct Biol. 2002;9:522–6.
56. Samel SA, Schoenafinger G, Knappe TA, Marahiel MA, Essen LO. Structural and functional insights into a peptide bond-forming bidomain from a nonribosomal peptide synthetase. Structure. 2007;15:781–92.
57. Tanovic A, Samel SA, Essen LO, Marahiel MA. Crystal structure of the termination module of a nonribosomal peptide synthetase. Science. 2008;321:659–63.
58. Stachelhaus T, Walsh CT. Mutational analysis of the epimerization domain in the initiation module PheATE of gramicidin S synthetase. Biochemistry. 2000;39:5775–87.
59. Schauwecker F, Pfennig F, Grammel N, Keller U. Construction and in vitro analysis of a new bi-modular polypeptide synthetase for synthesis of *N*-methylated acyl peptides. Chem Biol. 2000;7:287–97.
60. Linne U, Doekel S, Marahiel MA. Portability of epimerization domain and role of peptidyl carrier protein on epimerization activity in nonribosomal peptide synthetases. Biochemistry. 2001;40:15824–34.
61. Luo L, Burkart MD, Stachelhaus T, Walsh CT. Substrate recognition and selection by the initiation module PheATE of gramicidin S synthetase. J Am Chem Soc. 2001;123:11208–18.
62. Luo L, Walsh CT. Kinetic analysis of three activated phenylalanyl intermediates generated by the initiation module PheATE of gramicidin S synthetase. Biochemistry. 2001;40:5329–37.
63. Luo L, Kohli RM, Onishi M, Linne U, Marahiel MA, Walsh CT. Timing of epimerization and condensation reactions in nonribosomal peptide assembly lines: kinetic analysis of phenylalanine activating elongation modules of tyrocidine synthetase B. Biochemistry. 2002;41:9184–96.
64. Clugston SL, Sieber SA, Marahiel MA, Walsh CT. Chirality of peptide bond-forming condensation domains in nonribosomal peptide synthetases: the C5 domain of tyrocidine synthetase is a (D)C(L) catalyst. Biochemistry. 2003;42:12095–104.
65. Stein DB, Linne U, Marahiel MA. Utility of epimerization domains for the redesign of nonribosomal peptide synthetases. FEBS J. 2005;272:4506–20.
66. Stein DB, Linne U, Hahn M, Marahiel MA. Impact of epimerization domains on the intermodular transfer of enzyme-bound intermediates in nonribosomal peptide synthesis. Chembiochem. 2006;7:1807–14.
67. Koglin A, Mofid MR, Löhr F, Schäfer B, Rogov VV, Blum MM, Mittag T, Marahiel MA, Bernhard F, Dötsch V. Conformational switches modulate protein interactions in peptide antibiotic synthetases. Science. 2006;312:273–6.
68. Bergendahl V, Linne U, Marahiel MA. Mutational analysis of the C-domain in nonribosomal peptide synthesis. Eur J Biochem. 2002;269:620–9.
69. Finking R, Mofid MR, Marahiel MA. Mutational analysis of peptidyl carrier protein and acyl carrier protein synthase unveils residues involved in protein-protein recognition. Biochemistry. 2004;43:8946–56.
70. Lai JR, Fischbach MA, Liu DR, Walsh CT. A protein interaction surface in nonribosomal peptide synthesis mapped by combinatorial mutagenesis and selection. Proc Natl Acad Sci U S A. 2006;103:5314–9.
71. Lai JR, Koglin A, Walsh CT. Carrier protein structure and recognition in polyketide and nonribosomal peptide biosynthesis. Biochemistry. 2006;45:14869–79.
72. Trauger JW, Kohli RM, Mootz HD, Marahiel MA, Walsh CT. Peptide cyclization catalysed by the thioesterase domain of tyrocidine synthetase. Nature. 2000;407:215–8.
73. Trauger JW, Kohli RM, Walsh CT. Cyclization of backbone-substituted peptides catalyzed by the thioesterase domain from the tyrocidine nonribosomal peptide synthetase. Biochemistry. 2001;40:7092–8.

74. Kohli RM, Trauger JW, Schwarzer D, Marahiel MA, Walsh CT. Generality of peptide cyclization catalyzed by isolated thioesterase domains of nonribosomal peptide synthetases. Biochemistry. 2001;40:7099–108.
75. Kohli RM, Takagi J, Walsh CT. The thioesterase domain from a nonribosomal peptide synthetase as a cyclization catalyst for integrin binding peptides. Proc Natl Acad Sci U S A. 2002;99:1247–52.
76. Tseng CC, Bruner SD, Kohli RM, Marahiel MA, Walsh CT, Sieber SA. Characterization of the surfactin synthetase C-terminal thioesterase domain as a cyclic depsipeptide synthase. Biochemistry. 2002;41:13350–9.
77. Brady SF, Varga SL, Freidinger RM, Schwenk DA, Mendlowski M, Holly FW, Veber DF. Practical synthesis of cyclic peptides, with an example of dependence of cyclization yield upon linear sequence. J Org Chem. 1979;44:3101–5.
78. Wenger RM. Synthesis of cyclosporine: total syntheses of 'cyclosporine A' and 'cyclosporine H', two fungal metabolites isolated from species *Tolypocladium inflatum* Gams. Helv Chim Acta. 1984;67:502–25.
79. Nitecki DE, Halpern B, Westley JW. Simple route to sterically pure dioxopiperazines. J Org Chem. 1968;33:864–6.
80. Gund P, Veber DF. Ease of base-catalyzed epimerization of *N*-methylated peptides and diketopiperazines. J Am Chem Soc. 1979;101:1885–7.
81. Senn H, Weber C, Kobel H, Traber R. Selective 13C-labelling of cyclosporine A. Eur J Biochem. 1991;199:653–8.
82. Hoffmann K, Schneider-Scherzer E, Kleinkauf H, Zocher R. Purification and characterization of eucaryotic alanine racemase acting as key enzyme in cyclosporine biosynthesis. J Biol Chem. 1994;269:12710–4.
83. Billich A, Zocher R. Enzymatic synthesis of cyclosporine A. J Biol Chem. 1987;262: 17258–9.
84. Traber R. Biosynthesis of cyclosporines. In: Strohl WR, editor. Biotechnology of antibiotics. 2nd ed. New York: Marcel Dekker; 1997. p. 279–314. Chapter 9.
85. Kocher HP, Schneider-Scherzer E, Schörgendorfer K, Weber G. Recombinant Alanine Racemase and GAPDH from Tolypocladium. European Patent WO9425606; 1996.
86. Di Salvo ML, Florio R, Paiardini A, Vivoli M, D'aguanno S, Contestabile R. Alanine racemase from Tolypocladium inflatum: a key PLP-dependent enzyme in cyclosporine biosynthesis and a model of catalytic promiscuity. Arch Biochem Biophys. 2013;529:55–65.
87. Hensens OD, White RF, Goegelman RT, Inamine ES, Patchett AA. The preparation of [2-deutero-3-fluoro-D-Ala^8] cyclosporine A by directed biosynthesis. J Antibiot (Tokyo). 1992;45:133–5.
88. Kürnsteiner H, Zinner M, Kück U. Immunosuppressants. In: Osiewacz HD, editors. Esser K, Bennet JW, Series editors. The Mycota. A comprehensive treatise on fungi as experimental systems for basic and applied research. Volume X, Industrial Applications. Volume ed., Berlin: Springer; 2002. pp. 129–156.
89. Kobel H, Loosli HR, Voges R. Contribution to knowledge of the biosynthesis of cyclosporine A. Experientia. 1983;39:873–6.
90. Cane DE. Introduction: polyketide and nonribosomal polypeptide biosynthesis. From Collie to Coli. Chem Rev. 1997;97:2463–4.
91. Shen B, Kwon HJ. Macrotetrolide biosynthesis: a novel type II polyketide synthase. Chem Rec. 2002;2:389–96.
92. Shen B. Polyketide biosynthesis beyond the type I, II and III polyketide synthase paradigms. Curr Opin Chem Biol. 2003;7:285–95.
93. Walsh CT. Polyketide and nonribosomal peptide antibiotics: modularity and versatility. Science. 2004;303:1805–10.
94. Hertweck C, Luzhetskyy A, Rebets Y, Bechthold A. Type II polyketide synthases: gaining a deeper insight into enzymatic teamwork. Nat Prod Rep. 2007;24:162–90.
95. Olano C. Hutchinson's legacy: keeping on polyketide biosynthesis. J Antibiot (Tokyo). 2011;64:51–7.

96. Offenzeller M, Su Z, Santer G, Moser H, Traber R, Memmert K, Schneider-Scherzer E. Biosynthesis of the unusual amino acid (4R)-4-[(E)-2-butenyl]-4-methyl-L-threonine of cyclosporine A. Identification of 3(R)-hydroxy-4(R)-methyl-6(E)-octenoic acid as a key intermediate by enzymatic in vitro synthesis and by in vivo labeling techniques. J Biol Chem. 1993;268:26127–34.
97. Offenzeller M, Santer G, Totschnig K, Su Z, Moser H, Traber R, Schneider-Scherzer E. Biosynthesis of the unusual amino acid (4R)-4-[(E)-2-butenyl]-4-methyl-L-threonine of cyclosporine A: enzymatic analysis of the reaction sequence including identification of the methylation precursor in a polyketide pathway. Biochemistry. 1996;35:8401–12.
98. Kobel H, Traber R. Directed biosynthesis of Cyclosporines. Eur J Appl Microbiol Biotechnol. 1982;14:237–40.
99. Traber R, Hofmann H, Kobel H. Cyclosporines—new analogues by precursor directed biosynthesis. J Antibiot (Tokyo). 1989;42:591–7.
100. Von Wartburg A, Traber R. Cyclosporines, fungal metabolites with immunosuppressive activities. Prog Med Chem. 1988;25:1–33.
101. Nakajima H, Hamasaki T, Tanaka K, Kimura Y, Udagawa S. Production of cyclosporine by fungi belonging to the genus Neocosmospora. Agric Biol Chem. 1989;53:2291–2.
102. Moussaïf M, Jacques P, Schaarwächter P, Budzikiewicz H, Thonart P. Cyclosporine C is the main antifungal compound produced by *Acremonium luzulae*. Appl Environ Microbiol. 1990;63:1739–43.
103. Traber R, Dreyfuss MM. Occurrence of cyclosporines and cyclosporine-like peptolides in fungi. J Ind Microbiol Biotechnol. 1996;17:397–401.
104. Sallam LA, El-Refai AM, Hamdy AH, El-Minofi HA, Abdel-Salam IS. Role of some fermentation parameters on cyclosporine A production by a new isolate of Aspergillus terreus. J Gen Appl Microbiol. 2003;49:321–8.
105. Traber R, Kuhn M, Loosli H-R, Pache W, Von Wartburg A. Neue Cyclopeptide aus *Trichoderma polysporum* (LINK EX PERS.) RIFAI: Die Cyclosporine B, D und E. Helv Chim Acta. 1977;60:1568–78.
106. Traber R, Kuhn M, Rűegger A, Lichti H, Loosli R-H, Von Wartburg A. Die Structur von Cyclosporine C. Helv Chim Acta. 1977;60:1247–55.
107. Traber R, Hofmann H, Loosli H-R, Ponelle M, Von Wartburg A. Neue Cyclosporine aus *Tolypocladium inflatum*. Die Cyclosporine K–Z. Helv Chim Acta. 1987;70:13–36.
108. Traber R, Loosli H-R, Hofmann H, Kuhn M, Von Wartburg A. Isolierung und Strukturermittlung der neuen Cyclosporine E, F, G, H und I. Helv Chim Acta. 1982;65:1655–77.
109. Lawen A, Traber R, Geyl D. In vitro biosynthesis of [Thr^2, Leu^5, D-Hiv^8, Leu^{10}] cyclosporine, a cyclosporine-related immunosuppressive peptolide. Biomed Biochim Acta. 1991;50:S260–3.
110. Lawen A, Traber R, Geyl D. In vitro biosynthesis of [Thr^2, Leu^5, D-Hiv^8, Leu^{10}]-cyclosporine, a cyclosporine-related peptolide, with immunosuppressive activity by a multienzyme polypeptide. J Biol Chem. 1991;266:15567–70.
111. Velkov T, Singaretnam LG, Lawen A. An improved purification procedure for cyclosporine synthetase. Protein Expr Purif. 2006;45:275–87.
112. Lawen A, Traber R, Reuille R, Ponelle M. In vitro biosynthesis of ring-extended cyclosporines. Biochem J. 1994;300(Pt 2):395–9.
113. Traber R, Kobel H, Loosli H-R, Senn H, Rosenwirth B, Lawen A. [MeIle4] Cyclosporine, a novel natural cyclosporine with anti-HIV activity: structure elucidation and biosynthesis. Antivir Chem Chemother. 1994;5:331–9.
114. Billich A, Hammerschmid F, Peichl P, Wenger R, Zenke G, Quesniaux V, Rosenwirth B. Mode of action of SDZ NIM 811, a nonimmunosuppressive cyclosporine A analog with activity against human immunodeficiency virus (HIV) type 1: interference with HIV protein-cyclophilin A interactions. J Virol. 1995;69:2451–61.
115. Steinkasserer A, Harrison R, Billich A, Hammerschmid F, Werner G, Wolff B, Peichl P, Palfi G, Schnitzel W, Mlynar E, Rosenwirth B. Mode of action of SDZ NIM 811, a nonimmunosuppressive cyclosporine A analog with activity against human immunodeficiency virus type 1 (HIV-1): interference with early and late events in HIV-1 replication. J Virol. 1995;69:814–24.

116. Borvak J, Chou CS, Van Dyke G, Rosenwirth B, Vitetta ES, Ramilo O. The use of cyclosporine, FK506, and SDZ NIM811 to prevent CD25- quiescent peripheral blood mononuclear cells from producing human immunodeficiency virus. J Infect Dis. 1996;174:850–3.
117. Mlynar E, Bevec D, Billich A, Rosenwirth B, Steinkasserer A. The non-immunosuppressive cyclosporine A analogue SDZ NIM 811 inhibits cyclophilin A incorporation into virions and virus replication in human immunodeficiency virus type 1-infected primary and growth-arrested T cells. J Gen Virol. 1997;78(Pt 4):825–35.
118. Rosenwirth B, Billich A, Datema R, Donatsch P, Hammerschmid F, Harrison R, Hiestand P, Jaksche H, Mayer P, Peichl P, Quesniaux V, Schatz F, Schuurman H-J, Traber R, Wenger R, Wolff B, Zenke G, Zurini M. Inhibition of human immunodeficiency virus type 1 replication by SDZ NIM 811, a nonimmunosuppressive cyclosporine analog. Antimicrob Agents Chemother. 1994;38:1763–72.
119. Kocken CH, Van Der Wel A, Rosenwirth B, Thomas AW. *Plasmodium vivax*: in vitro anti-parasitic effect of cyclosporines. Exp Parasitol. 1996;84:439–43.
120. Goto K, Watashi K, Murata T, Hishiki T, Hijikata M, Shimotohno K. Evaluation of the anti-hepatitis C virus effects of cyclophilin inhibitors, cyclosporine A, and NIM811. Biochem Biophys Res Commun. 1996;343:879–84.
121. Gallay PA. Cyclophilin inhibitors. Clin Liver Dis. 2009;13:403–17.
122. Wolvetang EJ, Larm JA, Moutsoulas P, Lawen A. Apoptosis induced by inhibitors of the plasma membrane NADH-oxidase involves Bcl-2 and calcineurin. Cell Growth Differ. 1996;7:1315–25.
123. Bennett PC, Singaretnam LG, Zhao W-Q, Lawen A, Ng KT. Peptidyl-prolyl-*cis/trans-isomerase* activity may be necessary for memory formation. FEBS Lett. 1998;431:386–90.
124. Halestrap AP. A pore way to die: the role of mitochondria in reperfusion injury and cardio-protection. Biochem Soc Trans. 2010;38:841–60.
125. Wenger RM. Synthesis of Cyclosporine. I. Synthesis of enantiomerically pure (2*S*,3*R*,4*R*,6*E*)-3-hydroxy-4-methyl-2-methylamino-6-octenoic acid starting from tartaric acid. Helv Chim Acta. 1983;66:2308–21.
126. Togni A, Pastor SD, Rihs G. Application of the gold(I)-Catalyzed aldol reaction to a stereoselective synthesis of (2S, 3R, 4R, 6E)-3-Hydroxy-4-methyl-2-(methylamino)oct-6-enoic Acid (= MeBmt), Cyclosporine's unusual amino acid. Helv Chim Acta. 1989;72:1471–8.
127. Durand J-O, Genet J-P. Synthesis of MeBmt C9 amino acid present in cyclosporine, and analogues. Bull Chem Soc. 1994;131:612–9.
128. Wenger RM. Synthesis of Cyclosporine. Part II. Synthesis of Boc-D-Ala-MeLeu-MeLeu-MeVal-OH, a part of the peptide sequence of cyclosporine, by different strategic ways and synthesis of its isomers Boc-D-Ala-MeLeu-D-MeLeu-MeVal-OH, Boc-D-MeLeu-D-MeVal-OH, and Boc-D-Ala-MeLeu-MeLeu-D-MeVal-OH as reference compounds. Helv Chim Acta. 1983;66:2672–702.
129. Ko SY, Wenger RM. Solid-phase total synthesis of cyclosporine analogues. Helv Chim Acta. 1997;80:695–705.
130. Wu X, Stockdill JL, Wang P, Danishefsky SJ. Total synthesis of cyclosporine: access to *N*-methylated peptides via isonitrile coupling reactions. J Am Chem Soc. 2010;132: 4098–100.
131. Borchert S, Patil SS, Marahiel MA. Identification of putative multifunctional peptide synthetase genes using highly conserved oligonucleotide sequences derived from known synthetases. FEMS Microbiol Lett. 1992;71:175–80.
132. Mootz HD, Marahiel MA. Design and application of multimodular peptide synthetases. Curr Opin Biotechnol. 1999;10:341–8.
133. Turgay K, Marahiel MA. A general approach for identifying and cloning peptide synthetase genes. Pept Res. 1994;7:238–41.
134. Schneider A, Stachelhaus T, Marahiel MA. Targeted alteration of the substrate specificity of peptide synthetases by rational module swapping. Mol Gen Genet. 1998;257:308–18.
135. Jacobsen JR, Hutchinson CR, Cane DE, Khosla C. Precursor-directed biosynthesis of erythromycin analogs by an engineered polyketide synthase. Science. 1997;277:367–9.

136. Cane DE, Walsh CT, Khosla C. Harnessing the biosynthetic code: combinations, permutations, and mutations. Science. 1998;282:63–8.
137. Walsh CT. Combinatorial biosynthesis of antibiotics: challenges and opportunities. Chembiochem. 2002;3:125–34.
138. Weber G, Leitner E. Disruption of the cyclosporine synthetase gene of *Tolypocladium niveum*. Curr Genet. 1994;26:461–7.
139. Leitner E, Schneider-Scherzer E, Schörgendorfer K, Weber G. Cyclosporine synthetase. European patent application EP0578616 A2; 1994.
140. Leitner E, Schneider-Scherzer E, Schörgendorfer K, Weber G. Cyclosporine synthetase. U. S. patent 5,827,706; 1998.

Chapter 5
Aflatoxin Biosynthesis: Regulation and Subcellular Localization

John E. Linz, Josephine M. Wee, and Ludmila V. Roze

Aflatoxin Biosynthesis

Health and Economic Impacts of Aflatoxin Contamination of Crops

It is estimated that up to 4.5 billion people worldwide may be exposed to aflatoxin in their diet or as part of particles in the air they breathe [1]. The level of dietary aflatoxin exposure is strongly associated with the incidences of liver/lung cancer, HIV/AIDS, and stunted growth in Asia, Africa, and many areas of the world [1–3]. In the United States, human exposure to low levels of aflatoxin in food and feed crops is reported to impact susceptibility to a variety of infectious agents and to increase risk for liver and lung cancer [4]. Aflatoxin contamination of crops also results in a huge economic burden in the United States, especially on farmers [4]. Due to negative health impacts, the occurrence of aflatoxin in crops is tightly regulated in developed countries. Maximum tolerated levels (MTL) in human food are 20 ppb in the United States and 5 ppb in Europe [4].

J.E. Linz, M.S., Ph.D. (✉)
Department of Food Science and Human Nutrition, Microbiology and Molecular Genetics, Michigan State University, East Lansing, MI, USA
e-mail: jlinz@msu.edu

J.M. Wee, B.Sc.
Department of Food Science and Human Nutrition, Center for Integrative Toxicology, Michigan State University, East Lansing, MI, USA

L.V. Roze, Ph.D.
Department of Plant Biology, Michigan State University, East Lansing, MI, USA

J.-F. Martín et al. (eds.), *Biosynthesis and Molecular Genetics of Fungal Secondary Metabolites*, Fungal Biology, DOI 10.1007/978-1-4939-1191-2_5,

In the developing countries of Asia and Africa, the staple crops corn (maize) and peanuts (groundnuts) are often heavily contaminated with aflatoxins. Ingestion of contaminated crops, especially as a weaning food by children, is a frequent occurrence because food is scarce, and this situation is exacerbated during periods of drought. In these developing countries, aflatoxin exposure poses a severe health risk. Even in developed countries including the United States, aflatoxin contamination poses a significant health risk [2, 5]. Other affected crops include cotton seed, tree nuts, spices, as well as meat and dairy products derived from animals that consume contaminated feed [4].

Hepatocellular carcinoma (HCC) is the third leading cause of cancer deaths worldwide and its prevalence is up to 32 times higher in developing countries [2]. AFB_1, B_2, G_1, and G_2 are the most common aflatoxins synthesized in the laboratory and in the field; mixtures of aflatoxins in foods have been classified as a Group 1 human carcinogen by the International Agency for Research on Cancer [6]. AFB_1 is the most abundant of the aflatoxins and is the most potent naturally occurring carcinogen known [5, 7]. Experimental evidence demonstrates that aflatoxin is a potent carcinogen in a variety of animal species [4, 7]. There is also a strong epidemiological link between human aflatoxin exposure and HCC [2, 8]. Of the estimated 550,000–600,000 new HCC cases each year worldwide, up to 155,000 (28 %) may be attributable to aflatoxin exposure [2]. Aflatoxin exposure and hepatitis B and C virus infection are reported to be widespread and act synergistically to generate human HCC in large areas of Africa and Asia [8]. So, food shortage/drought, hepatitis B/C virus infection, and aflatoxin exposure generate a large risk for HCC in the developing world. Human aflatoxin exposure also has been linked to lung cancer [9].

A growing body of evidence suggests that aflatoxin exposure is strongly associated with impaired growth (stunting) in developing countries [10] and this may directly impact long-term health and mental development. Given that aflatoxin is a common contaminant of weaning foods in parts of the world where childhood stunting is prevalent, it is important to reduce aflatoxin exposure in foods consumed by children [10]. Experimental evidence also demonstrates that aflatoxins are immunosuppressive in animals and there is significant epidemiological evidence that human aflatoxin exposure is associated with increased incidences of HIV/AIDS and other infectious diseases in Africa [8].

Aflatoxin contamination in the United States results in up to $500,000,000 in losses each year [4] associated with screening, decontamination and/or destruction of contaminated crops. Farmers bear a disproportionate economic burden due to aflatoxin contamination [4]. Inability to export contaminated commodities to developed countries that enforce MTLs can result in export of contaminated commodities into markets in developing countries [11] where there are no import standards or where existing standards are not enforced.

Although many remediation strategies have been proposed to control human aflatoxin exposure, these have not been successfully applied on a large scale. Three promising remediation technologies under development include biocontrol, dietary adsorbents, and genetically engineered crops (Bt toxin) and these strategies have been reviewed related to their potential efficacy and cost [12–16].

The Proposed Role of Aflatoxin in Fungal Biology

Fungal secondary metabolites are most frequently synthesized after growth ceases or during a transition from active growth to stationary phase. Secondary metabolic pathways use intermediates/end products and energy supplied by primary metabolic pathways and their function does not appear to be essential for growth. For these reasons, secondary metabolism was proposed to provide an "evolutionary playing field" in which mutations in genes would provide ability to produce a wide variety of metabolites without a strong negative impact on growth [17]. Historically, other roles proposed for secondary metabolites include suppression of competitors in the natural environment, detoxification of toxic primary metabolites, and regulation of cellular differentiation [17].

Fungi represent an important food source for soil-dwelling organisms including insects and worms. Rohlfs et al. grew two strains of *Aspergillus nidulans* under standard laboratory conditions and then measured the feeding preference of the fungivorous springtail *Folsomia candida* (a soil arthropod that feeds on fungi) [18]. The two *A. nidulans* strains differed only in that one was a deletion mutant of the gene *laeA*, an important regulator of secondary metabolism. The Δ(Delta)*laeA* strain was unable to synthesize a variety of secondary metabolites including sterigmatocystin, a late intermediate in the aflatoxin biosynthetic pathway. The Δ(Delta)*laeA* strain was the preferred food source as compared to the wild-type strain that could synthesize all of the secondary metabolites. These data suggested that ability to synthesize secondary metabolites could influence food choices of soil predators of filamentous fungi and this process could provide selective pressure to maintain the ability to produce secondary metabolites in the soil environment.

Biochemistry of Aflatoxin Biosynthesis and Toxicity of Pathway Intermediates

Aflatoxin is a mycotoxin (toxin produced by a mold) that is synthesized as a family of related compounds predominantly by the filamentous fungi *Aspergillus parasiticus* and *A. flavus* as they grow on susceptible crops including peanuts, corn, tree nuts, and cottonseed [4, 19]. The molecular biology and biochemistry of aflatoxin biosynthesis have been reviewed recently [20–23]. Here we provide a brief overview of key enzymes and reactions involved in aflatoxin synthesis to provide a framework for the discussion presented later:

Aflatoxin B_1 (AFB_1), B_2 (AFB_2), G_1 (AFG_1), and G_2 (AFG_2) are the primary aflatoxins synthesized by *A. parasiticus* in culture and on susceptible plants. Of these, AFB_1 is the most abundant and most toxic. *A. flavus* synthesizes predominantly AFB_1 and B_2 while *A. nidulans* synthesizes sterigmatocystin, the next to last aflatoxin pathway intermediate. Aflatoxin biosynthesis is catalyzed by 25 or more enzymes encoded by up to 30 genes clustered on chromosome 3 in *A. flavus* and

A. parasiticus [24]. A similar gene cluster is also observed in *A. nidulans.* Primary substrates include acetyl CoA and methyl groups donated by *S*-adenosyl methionine. In the initial phase of aflatoxin synthesis, a specialized 2 subunit fatty acid synthetase catalyzes condensation of one molecule of acetyl CoA and two molecules of malonyl CoA to form hexanoyl-CoA [23]. A polyketide synthase (PksA) extends this 6-carbon molecule to form the 20-carbon decaketide, norosolorinic acid (NA), the first stable polyketide pathway intermediate. NA is converted via a series of enzymes to versicolorin A (VA), the first toxic pathway intermediate. VA is the first pathway intermediate to contain a bisfuran ring structure with a double bond. This double bond is oxidized to a short-lived epoxide primarily by specific cytochrome p450s in the liver of exposed animals and this "activation" is thought to be responsible for ability of aflatoxin to form adducts with cellular proteins and nucleic acids [7]. Adduct formation with cellular macromolecules drives much of the toxicity associated with aflatoxin exposure [7]. Aflatoxin B_1 may be the most toxic of the aflatoxins because its flat conformation enables close association with ring nitrogen 7 in guanine residues in the DNA double helix. This enables adduct formation. Repair of aflatoxin adducts in DNA can result in mutations and this event generates potentially large impacts on gene expression and gene function. In the final phase of aflatoxin synthesis, VA is chemically modified to generate ring structures characteristic of the aflatoxin B and G families [23].

Aflatoxin Biosynthesis and Export is Mediated by Vesicles and Endosomes

The Role of Endosomes in Cell Biology

Vacuoles serve many cellular functions including protein turnover, maintenance of intracellular pH, as well as storage of calcium, amino acids, cofactors, and toxic materials [25, 26]. Much is known about the structure, function, and biogenesis of fungal vacuoles [25–28]. The yeast vacuole is a single-membrane-bound organelle that arises by fusion of transport vesicles and endosomes [26, 29–31]. Transport vesicles carry a double-layer membrane and they transport proteins to vacuoles from a variety of intracellular sites. Transport from the Golgi occurs via the secretory pathway, transport from the cytoplasmic membrane occurs via the endocytosis pathway, and transport from the cytoplasm occurs via the autophagy and cytoplasm to vacuole transport (Cvt) pathways [25, 26, 29–32]. Vesicles can fuse directly with the growing vacuole, or they first can fuse with each other to generate late endosomes (multivesicular bodies), which fuse with the vacuole [26, 32]. Major steps in vesicle transport include *budding* from the source membrane, *transport* to the target site, *tethering* to the target membrane, and *fusion* [29].

Multi-subunit tethering complexes promote recognition between endosomes and target membranes prior to fusion. In yeast, the CORVET tethering complex facilitates protein transport from the Golgi to the endosome and from the endosome to the Golgi and the HOPS tethering complex facilitates protein transport from the endosome to the vacuole [29]. Rab7 (Ypt7, AvaA) is a small GTPase in the HOPS complex in yeast [29]. *Aspergillus nidulans avaA* encodes a protein with high identity to Ypt7; *avaA* disruption in *A. nidulans* resulted in a "fragmented vacuole morphology" and accumulation of a large number of small membrane-bound compartments [33]. HOPS tethering complex proteins Vps11, Vps16, Vps18, and Vps33 also contribute to CORVET function, suggesting that transport machinery is shared among transport pathways [29]. The close structural and functional relationship between the Cvt and autophagy pathways in yeast supports this idea [32].

The Role of Endosomes in Synthesis, Storage, and Export of Aflatoxin

In submerged culture in an aflatoxin inducing growth medium (yeast extract-sucrose [YES]), aflatoxin synthesis in *A. parasiticus* initiates at 24 h, reaches maximum rates between 30 and 48 h, and then declines [34]. The aflatoxin enzymes Nor-1 (early), Ver-1 (middle), and OmtA (late) are synthesized in the cytoplasm beginning at approximately 24 h [34], but later these enzymes localize to endosomes (30–48 h) and then to vacuoles (72 h) [34–37]. We observed that the number of endosomes, but not vacuoles, increased dramatically between 24 and 30 h, but this did not occur in a non-aflatoxin inducing growth medium (yeast extract-peptone [YEP]) [35]. Deletion of *veA* (Δ[Delta]*veA*), a global regulator of secondary metabolism and development, blocked the accumulation of endosomes, blocked the synthesis of aflatoxin and aflatoxin enzymes, and blocked the generation of asexual conidiospores in the dark. In the wild-type strain SU-1, we also demonstrated that VeA downregulates Vps16 and AvaA transcript accumulation as active growth ceases and this results in endosome accumulation and the accumulation, storage, and export of aflatoxin [35].

In contrast to *veA*, deletion of *avaA* (*rab7, ypt7*), a key component of the HOPS complex (mediates fusion of late endosomes, or multivesicular bodies to vacuoles), or inhibition of Vps16 with sortin3 [38] increased endosome accumulation, and increased aflatoxin synthesis, storage, and export [35]. Vps16 is a key component of both the CORVET and the HOPS tethering complexes [29], which mediate fusion of early and late endosomes, respectively—strongly suggesting that downregulation of tethering complex activity mediates the large increase in endosome accumulation observed as aflatoxin synthesis initiates.

Several aflatoxin enzymes carry an N terminal amphipathic helix but not glycosylation motifs. Nor-1, Ver-1, and OmtA, are cleaved at one or both ends as aflatoxin synthesis initiates and the timing and extent of cleavage correlated closely with endosome localization and enzyme activation [34, 39] strongly suggesting that most aflatoxin enzymes are transported to endosomes by Cvt vesicles [26, 30]. In contrast, Vbs is glycosylated [40] and localizes to the Golgi [39] suggesting that it alone is transported via secretory vesicles and the classical secretion pathway. We speculate that transport of Vbs by an alternative pathway assures that the timing and level of synthesis of VA, the first toxin pathway intermediate, is tightly regulated.

A Proposed Role for the Vps34 Homolog (Class III PI3 Kinase) in Aflatoxin Biosynthesis

We recently demonstrated that 3-methyladenine (3MA) treatment increases aflatoxin synthesis and export up to 15-fold. Others previously reported that 3MA is a specific inhibitor of Class III PI3-kinases [41, 42]. Vps34, a Class III PI3-kinase and key component of the yeast early endosome membrane, is known to activate the small GTPase protein Ypt7 (Rab7, AvaA) [43], a component of the HOPS tethering complex. Vps34 in yeast was also reported to recruit the CORVET tethering complex, which mediates fusion of early endosomes to late endosomes [29]. Knockdown of *vps34* (the gene that encodes Vps34) in *A. parasiticus* conducted in our laboratory up-regulated conidiospore number and pigment; the data also suggest that Vps34 knockdown increased the synthesis and export of aflatoxin. The phenotype of the Vps34 knockdown strains was nearly identical to the *avaA* knockout strains and to the phenotype of cells treated with sortin 3 as described previously. Together, these data argue that blocking HOPS and CORVET tethering complex activity in *A. parasiticus* downregulates endosome maturation and fusion. This results in an increase in endosome accumulation, aflatoxin synthesis, and aflatoxin export.

Subcellular Localization of Secondary Metabolism and Stress Response Enzymes to Endosomes

Because of their central role in aflatoxin synthesis, we developed a simple and effective procedure to purify transport vesicles, endosomes, and vacuoles (V fraction) from *A. parasiticus* and demonstrated that V fraction synthesizes aflatoxin in vitro [35, 44]. Feeding middle (vesicolorin A) and late (sterigmatocystin) pathway intermediates enhanced aflatoxin synthesis [35, 45] in V fraction up to tenfold. V fraction purity was >90 % based on the absence of marker enzymes for cytoplasm and mitochondria, enrichment for α(alpha)-mannosidase (a Cvt vesicle and vacuole marker) [26], and confocal laser scanning microscopy using stains for vesicle/vacuole membranes (MDY 64) and lumen (CMAC) [45]. We then purified V fraction under aflatoxin inducing (YES) and non-inducing (YEP) growth conditions and subjected these samples to gel electrophoresis, liquid chromatography, and mass spectrum (GE LC-MS/MS) analyses. More than 250 proteins were identified in V fraction [46] including: (1) sixteen aflatoxin enzymes; (2) enzymes associated with stress response including three SODs (Fe, Mn, Cu/Zn) and two catalase isozymes (encoded by *catA* and *cat1*); (3) proteins involved in heat stress and osmotic stress; and (4) protein markers associated with transport vesicles and endosomes (coatomer, clathrin, aminopeptidases, metallopeptidase, SNARE domain protein, and a Rab family GTPase). When we subjected the organelles in V fraction to centrifugation on a fresh sucrose gradient, all enzymes listed previously were enriched in the highly purified V fraction, strengthening their association with transport vesicles and endosomes. Chanda et al. reported that endosomes mediate aflatoxin export to the cell exterior likely via exocytosis [47]. In support of these data, Menke et al. demonstrated that at least part of the biosynthetic machinery for synthesis of the isoprenoid trichothecene mycotoxins is associated with "toxisomes" suggesting that vesicles may be a broadly utilized mechanism to synthesize and export mycotoxins [48].

Regulation of Secondary Metabolism and Aflatoxin Biosynthesis

Aflatoxin gene expression and enzyme activity are tightly regulated by growth conditions [19]. High levels of aflatoxin synthesis in *A. parasiticus* grown in culture require an easily metabolized sugar-like glucose or sucrose, temperatures between 25 and 30 °C, ammonia as a nitrogen source, and absence of light. We utilize defined (glucose mineral salts [GMS]) or rich (YES) growth media to induce aflatoxin synthesis in liquid submerged culture or on the surface of agar plates. These same media with peptone as carbon source are used as non-aflatoxin inducing media. It is clear that Aspergillus nuclei communicate with the growth environment by at least four key signal transduction pathways and this regulates expression and/or activity of several transcription factors [49]. Our recent data suggest that these signaling pathways and transcription factors regulate Branch 1 and 2 of aflatoxin synthesis (Fig. 5.1a–c [21, 29, 35, 45–47, 49–65]), expression of stress response genes, synthesis of other secondary metabolites, and conidiospore development [35, 66].

Signal Transduction Pathways

1. *cAMP/PKA signal transduction and nutrient availability*. In response to carbon source availability, protein kinase A (PKA) regulates AflR, one positive regulator of aflatoxin gene expression and aflatoxin synthesis [67]. During the transition from active growth to stationary phase, intracellular cAMP declines as carbon becomes limiting, PKA activity declines, and AflR transcription and activity increase [50, 67–69]. AflR activates aflatoxin gene expression and aflatoxin accumulation. The G protein FadA activates PKA signaling [50, 70]. A parallel pathway mediated by a Class I/II PI3-kinase (see later) also regulates intracellular cAMP/PKA activity, and aflatoxin gene expression [43, 71]. In *Saccharomyces cerevisiae*, PKA signaling regulates two pathways of relevance to this review: (1) it regulates the transition from Cvt transport to autophagy upon nutrient limitation, and (2) it regulates initiation of stress response [72]. In actively growing cells, PKA phosphorylates Atg1 and Atg13 [31]. Under nutrient limitation, PKA activity is downregulated, phosphorylation of Atg1 and Atg13 declines, and these proteins physically interact. The active Atg1::Atg13 complex downregulates Cvt vesicle biogenesis and activates autophagosome biogenesis [31].
2. *PI3 kinase*. Phosphatidylinositol 3-kinase signaling cascades play fundamental roles in eukaryotic cell growth, differentiation, survival, and motility [73, 74]. Abnormal PI3-kinase signaling results in diseases such as cancer and autoimmunity illustrating the central importance of these enzymes in cell biology.

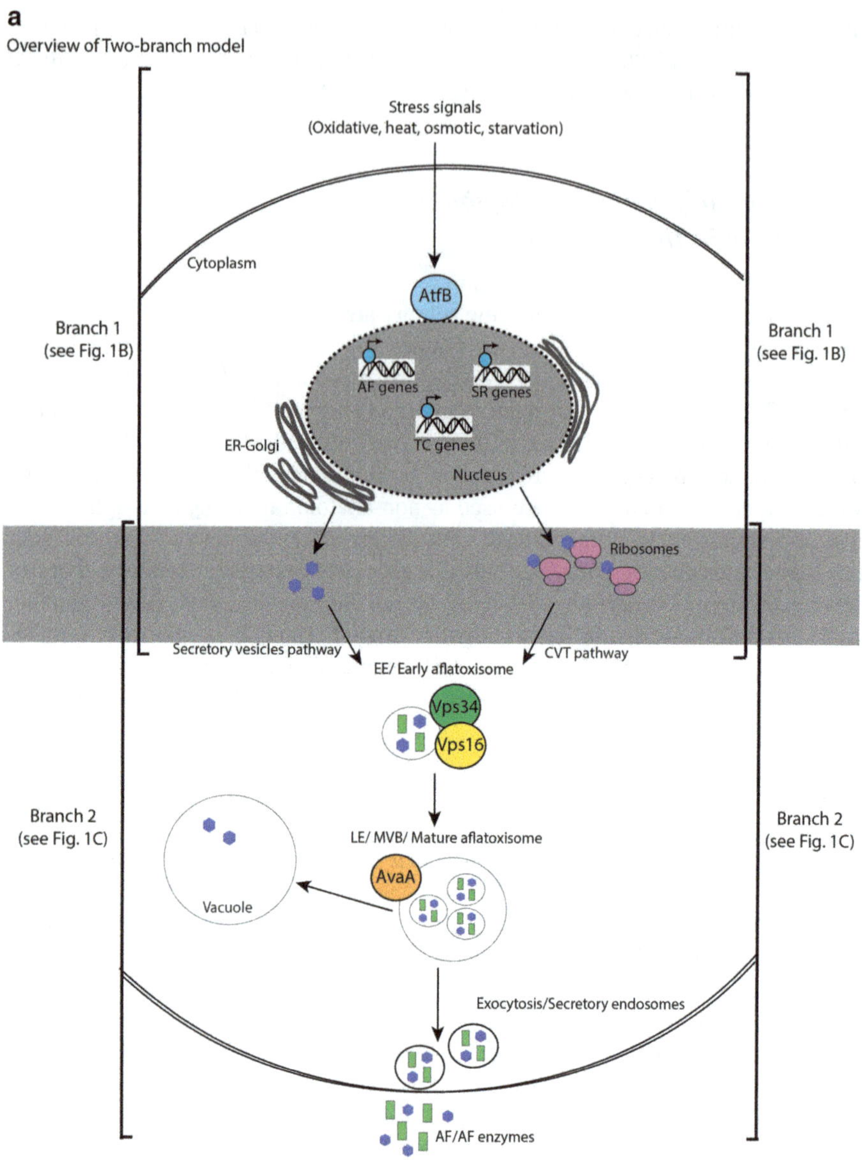

Fig. 5.1 The expanded "2 branch model": Regulation of Branch 1 and Branch 2 of aflatoxin biosynthesis. (**a**) *Overview of the two-branch aflatoxin biosynthesis model*. Branch 1 encompasses transcriptional regulation of coordinate expression of genes involved in aflatoxin biosynthesis (AF), stress response (SR), and tethering complex (TC) in response to stress/starvation as well as oxidative, heat, or osmotic stress. According to the model, a stress signal is sensed by a bZIP transcription factor AtfB and is transduced to the target genes defining the timing and levels of their expression—more detail can be observed in (**b**). Branch 2 encompasses protein traffic from the location of translation through CVT and secretory pathways to "aflatoxisomes" in which aflatoxin biosynthesis and exocytosis is proposed to occur. Early endosomes/aflatoxisomes containing aflatoxin biosynthetic enzymes are formed as a result of fusion of CVT vesicles and secretory vesicles bringing aflatoxin synthetic machinery together. Surface tethering complex proteins including Vps34, Vps16, and AvaA participate in biogenesis of endosomes/aflatoxisomes from early to mature states. The model proposes that

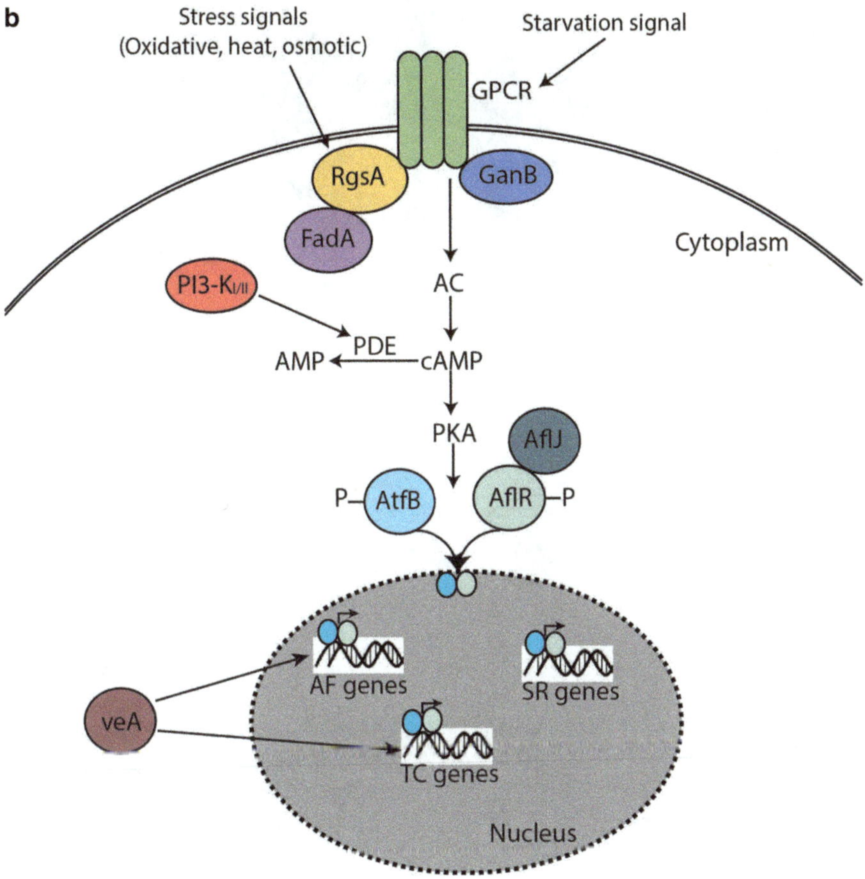

Fig. 5.1 (continued) aflatoxisomes have two intracellular destinations: they fuse to the vacuole for protein degradation and fuse to the cytoplasmic membrane for export of the toxin. During their function in the cell, Branch 1 and 2 are thought to exhibit concurrent, sequential and partially overlapping temporal patterns. (**b**, **c**) *Detailed schematic of Branch 1 and 2. Growth pathway*: When *Aspergillus parasiticus* is cultured under standard laboratory growth conditions, high glucose or sucrose is present, the G protein coupled receptor (GPCR) is occupied, and adenylate cyclase (AC) is activated by GanB and/or FadA (G proteins). This results in high internal cAMP levels and activation of protein kinase A (PKA) [35, 49, 50]. The transcription factors AflR and AtfB (if expressed) are phosphorylated and do not enter the nucleus or bind promoters of aflatoxin (AF), stress response (SR), or secondary metabolism (SM) genes; these genes are not expressed under rapid growth phase conditions [21, 49]. During active growth, vacuole biogenesis occurs via early endosome (EE) and late endosome (LE) maturation and fusion [29, 51] as follows. RabB (Rab5) in the EE membrane recruits and activates the Class III PI3-kinase Vps34 (CIII PI3K) [52], PIP increases, and Vps45, Vps19, and Vps27 are recruited to the EE. The CORVET tethering complex is recruited, which allows EE to fuse to generate LE. Then, the HOPS tethering complex replaces CORVET in the LE membrane and this stimulates LE fusion to generate vacuoles (V) [29]. Both HOPs and CORVETTE carry Vps16 while CORVETTE alone carries AvaA. *Glucose limitation* (*starvation*): As glucose declines during growth, the GPCR is occupied at lower frequency, and growth pathway activity is reduced. AflR and AtfB are not phosphorylated by PKA. They are transported to the nucleus and are active [53]. AtfB and AflR bind to AF, SR, and SM promoters and the genes are expressed (Branch 1). We recently generated additional evidence for a direct role for AtfB in activation of SM (see below). VeA is required for AtfB to bind to promoters (*green arrow*) [21]. Aflatoxin enzymes are synthesized on cytoplasmic

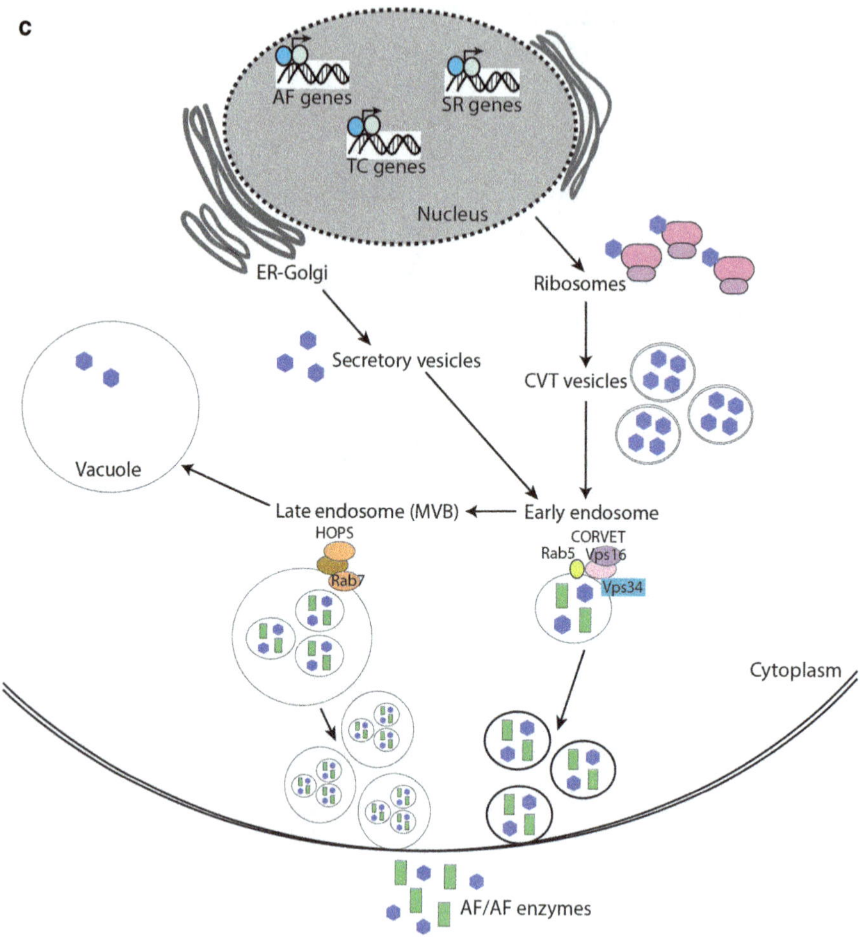

Fig. 5.1 (continued) ribosomes and packaged in double membrane-bound CVT vesicles, which fuse to form EE. Transport vesicles from the golgi carry the middle aflatoxin enzyme versicolorin B synthase (VBS) and transport vesicles from peroxisomes carry the early aflatoxin enzymes Nor-1, PksA, and Fas, as well as the important substrate acetyl CoA. These vesicles fuse with EE that carry other AF enzymes (Branch 2) [35, 45, 46, 54]. Meanwhile, VeA downregulates expression of *vps16* and *avaA* (*red arrow*), important components of HOPS and CORVET tethering complexes—so VeA directly regulates endosome accumulation. Vps16 and AvaA (Ypt7, Rab7) activities decline, EE and LE fail to fuse, and they accumulate in the cytoplasm [35, 45]. EE and/or LE migrate to and bind target receptors on the cytoplasmic membrane (CM). EE and LE fuse with the CM and empty their contents into the external environment most likely via exocytosis [47]. Our recent data suggest that expression of most SR and SM proteins is regulated similarly [55]. *Oxidative stress*: Oxidative stress from the external environment or due to cellular metabolism (respiration, reactive oxygen species [ROS] generated via pathway oxidases during aflatoxin synthesis) activates RgsA in the GPCR complex [49]. RgsA activates a wortmannin-sensitive Class I/II PI3-kinase (CI/II PI3K), which activates phosphodiesterase (PDE); this reduces cAMP and reinforces the starvation pathway. PDE decreases cAMP resulting in even higher levels of expression of AF, SR, and SM genes [21]. The remainder of the starvation pathway operates as described above. Oxidative stress in the environment can, at least in part, override the growth pathway by stimulating RgsA. This triggers precocious (early) expression of AF, SR, and SM genes. Redox signaling in endosomes has been studied in some detail in higher eukaryotes [56–65]. Despite this body of work, details of the mechanisms by which cells control the magnitude of the redox signal and trigger downstream targets are not clear

PI3-kinases are divided into Classes I, II, and IIII based on structural and functional differences. The Class I PI3-kinases are well-characterized while relatively little is known about the Class III kinases [74]. Our group added the PI3 kinase inhibitor wortmannin to *A. parasiticus* cultures and observed a 15-fold decrease in aflatoxin synthesis accompanied by a significant decrease in transcription of *nor-1* and *ver-1*, early and middle structural genes in the aflatoxin biosynthetic pathway [71]. Wortmannin predominantly inhibits Class I/II PI3-kinases but recently has been shown to influence activity of Class III PI3-kinases [75]. In contrast, we added 3 methyladenine (3MA) to *A. parasiticus* cultures and observed a 10- to 15-fold increase in aflatoxin synthesis. 3MA specifically inhibits Class III PI3 kinases [76].

To determine the specific role of Vps34, a Class III PI3 kinase, in aflatoxin biosynthesis, we conducted a knockdown experiment and transformants carrying the *vps34* gene knockdown construct were analyzed. Preliminary data suggest that Vps34 knockdown transformants accumulated higher levels of aflatoxin, conidiospores, and the conidiospore pigment than the control strain (SU-1), supporting the idea that Vps34 downregulates endosome maturation and fusion. In agreement with our model presented later (Fig. 5.1), as immature endosomes accumulate, aflatoxin synthesis and export also increase. Additional work is required to confirm these initial observations.

3. *SPK/MAPK and stress response.* A variety of stressors in the environment (osmotic, heat, UV, and oxidative stress) activate analogous protein kinase cascades in yeast and filamentous fungi [55] in order to trigger the cellular stress response. These kinase cascades activate transcription factors in the nucleus that target stress response genes and modulate their expression. Of these transcription factors, this review focuses predominantly on the bZIP factor AtfB, and this is described in more detail later.
4. *VeA and light.* VeA controls expression of a wide array of genes [77]. We demonstrated that deletion of *veA* in *A. parasiticus* blocks aflatoxin gene expression and AtfB binding to aflatoxin gene promoters in the dark [21]. Calvo reported that VeA is transported to the nucleus and this process is blocked by light [78]. Our recent data demonstrate that VeA likely modifies transcription factors or chromatin structure. Bayram et al. demonstrated that VeA interacts with several proteins in a complex including VelB, LaeA, and a red light sensor, FphA [79]. LaeA exhibits identity to protein methyltransferases, suggesting that VeA could influence gene expression by controlling methylation of proteins in the transcription apparatus or histones in chromatin [79]. Conidiospore development in Aspergillus is also controlled by VeA [79, 80], a pathway specific transcription factor BrlA [81, 82], and endogenous oleic- and linoleic-acid-derived oxylipins collectively called precocious sexual inducer or psi factor [83–86]. Absence of light is a key inducer of development and VeA up-regulates asexual conidiospore development in the dark [79, 85]. In the nucleus, VeA up-regulates expression and activity of the transcription factor BrlA, which then activates *wetA* and *abaA*. BrlA, WetA, and AbA control the timing of asexual spore development [81, 87, 88]. So VeA is a master regulator that controls secondary metabolism and conidiospore development in response to light.

Transcription Factors

1. *AflR, MsnA, and SrrA*. AflR is a binuclear zinc cluster transcription factor that binds specifically to promoters of many aflatoxin structural genes and helps to activate their expression [24]. AflR is regulated at the transcriptional and posttranslational level. AflR is one target of the PKA cascade and the protein is phosphorylated under rapid growth phase conditions and this prevents nuclear localization and promoter binding [67]. Under aflatoxin inducing conditions, AflR carries only low levels of phosphorylation and this activates nuclear localization and transcriptional activation of the aflatoxin genes. Co-immunoprecipitation using highly specific rabbit polyclonal antibodies to AflR suggested that AflR physically interacts with AtfB (see later), which may assist in recruiting AflR to aflatoxin gene promoters [89].

   Evidence for participation of MsnA, another binuclear zinc cluster transcription factor, in aflatoxin gene regulation was recently obtained by electrophoretic mobility shift analysis (EMSA) of specific aflatoxin and stress response promoters [90]. MsnA was previously demonstrated to bind to stress response elements (SRE; 5′ AGGGG 3″) in yeast promoters [91]. Use of a SRE competitor in competition EMSA greatly reduced DNA protein complex formation at aflatoxin and stress response promoters supporting a role for MsnA in their regulation [90]. Chang et al. recently conducted gene disruption of MsnA in *A. parasiticus* and *A. flavus* resulting in increased expression of *catA* and superoxide dismutase (SOD) as well as accumulation of aflatoxin, conidiospores, and kojic acid [92]. These data support our EMSA data and strongly suggest that MsnA is a negative regulator of secondary metabolism and asexual development.

   Evidence that SrrA may assist in gene regulation also was obtained in the study by Hong et al. [90]. A SrrA binding motif (5′ AAGCC 3′) was identified in aflatoxin and stress response gene promoters adjacent to cAMP response element (CRE) binding sites. Based on these data, the authors proposed a model in which AflR, AtfB, MsnA, and SrrA participate in a transcription factor regulatory network to regulate expression of genes involved in stress response, secondary metabolism, and asexual spore development [55]. SrrA has been demonstrated to be a two-component oxidative stress response regulator in bacteria and fungi [93] and SrrA binding to this motif and its proposed activity in secondary metabolism and stress response in *A. parasiticus* must be confirmed in future work.
2. *The bZIP transcription factors AtfA, AtfB, and AP-1 regulate fungal response to oxidative stress*. bZIP transcription factors regulate stress response in yeast [94, 95] and in the filamentous fungi Neurospora [96] and Aspergillus [68, 97–100]. AtfA, AtfB, and AP-1 in Aspergillus share several characteristics of bZIP family proteins: (1) they carry a leucine zipper that promotes homo- and heterodimer formation with bZIP factors; (2) they carry a "DNA basic domain" that mediates bZIP binding to cAMP-response elements (CRE) in DNA (promoters); (3) they mediate cellular response to cAMP signaling; and (4) they promote cellular response to oxidative stress. Confirming their role in stress response, disruption of AtfA or AtfB significantly reduced conidiospore tolerance to oxidative and heat stress

in *A. nidulans* [97] and *A. oryzae* [99, 100]. Disruption of *AP-1* (*APyapA*) in *A. parasiticus* [98, 101, 102] resulted in premature accumulation of aflatoxin, premature initiation of conidiospore development, and premature accumulation of intracellular reactive oxygen species (ROS), strongly suggesting a negative regulatory role for AP-1 in these cellular processes. In Neurospora, AP-1 nuclear localization and activation depend on the oxidation state of specific amino acid residues within the AP-1 protein [96]. These data demonstrate that AtfA and AtfB are positive regulators of stress response in Aspergilli and suggest that AP-1 is a negative regulator of oxidative stress, aflatoxin biosynthesis, and conidiospore development. Competition and shift inhibition EMSA analyses demonstrated that AtfB participates in formation of specific DNA/protein complexes in promoters of genes involved in aflatoxin biosynthesis and oxidative stress response in *A. parasiticus* and *A. flavus*. Although others demonstrated that AtfB helps regulate stress response [99, 100], little was known about the regulatory network by which AtfB co-regulates expression of genes involved in aflatoxin synthesis and stress response. The regulatory link between AtfB and aflatoxin biosynthesis is a novel discovery by our laboratory and our work is making important contributions to our understanding of the global AtfB regulatory network.

3. *Evidence that AtfB regulates secondary metabolism and stress response*. A strong association between oxidative stress and induction of aflatoxin biosynthesis has been previously reported [103–105], however the underlying mechanisms were not clear. Our data, presented as follows, provide a direct mechanistic link between the stress response regulator AtfB and up-regulation of aflatoxin biosynthetic genes in *A. parasiticus* and *A. flavus*.

   (a) *EMSA*. We utilized specific anti-AtfB rabbit polyclonal antibodies (anti-AtfB) to conduct shift inhibition, EMSA on enriched cellular extracts from *A. parasiticus SU-1* [90]. The data demonstrated that AtfB binds to CRE sites in promoters of *A. parasiticus* SU-1 genes involved in early, middle, and late steps in aflatoxin synthesis [66] as well as genes involved in oxidative stress response [55, 66], but not to genes that lack a CRE site (*vbs*, *laeA*). We recently confirmed formation of AtfB complexes in the same promoters in *A. flavus* 3357 strongly suggesting that the AtfB regulatory network is similar in these two key aflatoxin producers.

   (b) *RNA Seq*. To characterize the AtfB regulatory network in more detail, we conducted RNA sequence (RNA Seq) analysis of *A. parasiticus* and *A. flavus* under aflatoxin inducing (YES) and non-inducing conditions (YEP). A complete genome sequence for *A. flavus* 3357 was already available. However, we recently completed next-generation sequence (NGS) analysis of the *A. parasiticus* SU-1 genome to assist in RNA Seq data analysis. RNA Seq data demonstrated that the transcription factors AtfB and AflR are significantly up-regulated in YES (aflatoxin inducing growth conditions) as compared to YEP. In support of the shift inhibition EMSA data described previously, a large number of structural genes involved in aflatoxin synthesis as well as key genes involved in oxidative stress response are also up-regulated in YES. Of interest, several genes involved in other secondary metabolic pathways as well as a key regulator of conidiospore development are up-regulated in YES, suggesting that the AtfB regulatory network is more extensive than previously thought. Our goal now is to directly connect specific changes in gene expression to AtfB function and to map the extent of the AtfB regulatory network.

(c) *ChIP Seq. Analysis of the AtfB regulatory network in A. flavus and A. parasiticus.*[1] As a parallel approach to determine the extent of the AtfB regulatory network, we conducted chromatin immunoprecipitation/next-generation sequence (ChIP Seq) analysis of *A. flavus* and *A. parasiticus* grown in aflatoxin inducing and non-inducing conditions using highly specific, anti-AtfB rabbit polyclonal antibodies. Candidate target genes identified in the AtfB network in this preliminary screen included the aflatoxin genes *nor-1* and *fas1* (shown previously to bind AtfB in vitro using EMSA), the stress response genes *catA* and MnSOD (shown previously to bind AtfB in vitro using EMSA), and transcriptional regulators of conidispore development including Medusa and RosA. In addition, several polyketide synthases as well as a gene encoding a nonribosomal peptide synthase thought to be involved in gliotoxin biosynthesis were tentatively identified strongly suggesting that AtfB regulates many secondary metabolite gene clusters in *A. parasiticus*. These data must be confirmed by ChIP Seq analysis of the AtfB knockdown strains JW12 and JW13 (see later) and by shift inhibition and competition EMSA of each of these promoters. We will also delete the CRE sites in these promoters to help confirm a role for AtfB in their regulation. Together these data could provide additional support for a role for AtfB in regulation of these cellular processes.

(d) *Nucleotide sequence analysis; AtfB and Vps34.*[2] We recently completed next-generation sequence (NGS) analysis of the complete *A. parasiticus* genome as well as Southern blot analysis of *atfB* and *vps34*. These data confirmed high sequence identity between AtfB and Vps34 in *A. parasiticus* and homologous genes in *A. flavus*, *A. oryzae*, and *A. nidulans*. The data confirmed the presence of 2 *atfB* gene copies in *A. flavus* and *A. parasiticus* and a single gene copy of vps34 in both of these organisms.

(e) *Gene knockdown of atfB in A. parasiticus.*[3] We transformed *atfB* and *vps34* gene disruption constructs together with a selectable marker (*niaD*) into *A. parasiticus* NR1 (*niaD* –, aflatoxin +). PCR and Southern blot analysis of *niaD* + transformants identified several in which the disruption construct integrated into the *A. parasiticus* genome. The data suggested that multiple copies of the *atfB* construct integrated at ectopic sites in transformants. Preliminary semiquantitative RT PCR analysis determined that the *atfB* tran-

[1] Wee et al. (2014), in preparation.

[2] Linz JE, Wee J, Roze LV Aspergillus parasiticus SU-1 genome sequence, predicted chromosome structure, and comparative gene expression under aflatoxin inducing conditions: Evidence that differential expression contributes to species phenotype. Euk Cell 2014;13:000.

[3] Wee et al. (2014), in preparation.

scripts were downregulated in these transformants. Moreover, we were able to detect a relatively high level of transcripts derived from the gene disruption constructs using primer pairs specific for the gene disruption construct.

Two transformants carrying multiple copies of the *atfB* gene disruption construct integrated at ectopic sites were single spore isolated and analyzed in greater detail. We determined that even though the gene knockdown construct in JW12 and JW13 integrated at different ectopic locations, the phenotype of these transformants was nearly identical. They each accumulated several-fold lower levels of aflatoxin, asexual conidiospores, and the *atfB* transcript than the wild-type control strain (SU-1). RNA SEQ and semiquantitative RT PCR analyses of JW12 and JW13 suggested that the aflatoxin genes *aflR* and *nor-1* were also downregulated. Together the data suggest that we successfully knocked down *atfB* expression in JW12 and JW13 (probably by RNAi) and this resulted in the observed reductions in aflatoxin accumulation and conidiospore development.

(f) *CoIP. Possible physical interaction between AflR, AtfB, and AP-1.* Yeast 2 hybrid analysis performed by our collaborator Dr. Jeff Cary at USDA in New Orleans strongly suggests that AtfB and AP-1 physically interact. We previously used co-immunoprecipitation (CoIP) to demonstrate a physical interaction between AtfB and AflR [106]. Together the data suggest that AtfB and AflR interact to up-regulate aflatoxin biosynthesis but interaction of AtfB and AP-1 downregulates aflatoxin biosynthesis.

The Possible Role of Aflatoxin Biosynthesis in Stress Response

Recent data from several laboratories now add important perspective to the potential role of aflatoxin in fungal biology. These new data suggest that the ability to synthesize aflatoxin may contribute to ability of Aspergilli to respond to oxidative stress. For example, Narasaiah and colleagues observed a positive relationship between levels of aflatoxin synthesis and levels of ROS in several strains *A. parasiticus* blocked at different locations in the aflatoxin pathway [105]. Because the transcription factor, activating protein 1 (AP-1) has been demonstrated to play an important role in response to oxidative stress in other organisms, Reverberi conducted a gene disruption of the *A. parasiticus* AP-1 homolog, Ap*yapA* [98]. The disruption strain exhibited precocious production of aflatoxin, asexual conidiospores, and ROS as well increased susceptibility of conidiospores to extracellular oxidants. Complementation of the mutant restored normal levels and timing of ROS and aflatoxin synthesis. These observations suggest that AP-1 is a negative regulator of aflatoxin synthesis, conidiospore development, and response to oxidative stress. Sakamoto demonstrated that AtfB helps determine conidiospore resistance to oxidative stress response in Aspergillus [99, 100]. Disruption of AtfB in *A. oryzae* decreased ability of conidiospores to germinate and survive under hydrogen peroxide treatment. Our group demonstrated that mutants of *A. parasiticus* that are genetically blocked in the aflatoxin biosynthetic pathway are less resistant to hydrogen peroxide treatment than wild-type control strain (SU-1) and also accumulate less intracellular ROS.[4] In contrast, mutants that overproduce aflatoxin showed increased conidiospore resistance to oxidative stress and higher levels of intracellular ROS. Hong et al. demonstrated that AtfB binds to several aflatoxin gene promoters as well as promoters of genes involved in oxidative stress response [90]. Wee et al.[5] demonstrated that knockdown of AtfB also downregulates aflatoxin biosynthesis and conidiospore development. These observations demonstrate that AfB impacts response to oxidative stress, secondary metabolism, and fungal development.

Key Conclusions

Based on data from our laboratory we present the following conclusions regarding the role of AtfB and Vps34 in aflatoxin biosynthesis in *A. parasiticus*:

1. *AtfB is a key positive regulator of a network that controls secondary metabolism, stress response, and conidiospore development.*
2. *Vps34 is a negative regulator of conidiospore development.* Preliminary data suggest that Vps34 also plays a key role in regulating subcellular localization of the enzymes involved in aflatoxin biosynthesis.

An Expanded 2 Branch Model for Aflatoxin Biosynthesis and the Role of VPS34 and ATFB

Based on data previously described, we present a model for how *A. parasiticus* regulates secondary metabolism, stress response, and asexual conidiospore development (Fig. 5.1a–c). Based on this model we propose the following hypotheses that provide the framework for future work in this area:

1. *AtfB transcriptional activity is modulated by oxidative stress response.* Passage of AtfB through the endosome exposes it to ROS and this promotes transport to the nucleus. Subsequent reduction in the nucleus promotes promoter binding, which initiates gene expression. Oxidation and reduction of AP-1 have already been established to play an important role in transcriptional activation. This hypothesis is based on work presented in Linz et al. [46].

---

[4] Roze LV, Lavenieks M, Hong SY, Wee J, Wong SS, Vanos B, Awad D, Ehrlich K, Linz JE Aflatoxin biosynthesis is a novel source of reactive oxygen species – a potential redox mechanism to initiate resistance to oxidative stress.

[5] Wee et al. (2014), in preparation.

2. *AtfB interacts with AflR, SrrA, and MsnA to mediate activation of promoters involved in aflatoxin synthesis, stress response, and conidiospore development.* This hypothesis is based on work presented in Hong et al. [90].
3. *Vps34 functions predominantly as a regulator of Branch 2. Disruption of Vps34 will have no effect on aflatoxin gene expression or AtfB/AflR transcriptional activity*. This hypothesis is based on work presented in Wee et al.[6]

References

1. Liu Y, Chang CC, Marsh GM, Wu F. Population attributable risk of aflatoxin-related liver cancer: systematic review and meta-analysis. Eur J Cancer. 2012;48:2125–36.
2. Liu Y, Wu F. Global burden of aflatoxin–induced hepatocellular carcinoma: a risk assessment. Environ Health Perspect. 2010;118:818–24.
3. Wagacha JM, Muthomi JW. Mycotoxin problem in Africa: current status, implications to food safety, and health and possible management strategies. Int J Food Microbiol. 2008;124:1–12.
4. CAST (Council for Agricultural Science and Technology) Task Force Report #139. Mycotoxins: Risks in plant, animal, and human systems. Ames, Iowa. 2003.
5. Cullen JM, Newberne PM. Experimental toxicology of aflatoxins. In: Eaton DL, Groopman JD, editors. The toxicology of aflatoxins: human health, veterinary, and agricultural significance. San Diego: Academic; 1994.
6. IARC (International Agency for Research on Cancer). Some herbal medicines, some mycotoxins, naphthalene, and styrene. Monogr Eval Carcinog Risks Hum. 2002;82:171–300.
7. Roebuck BD, Maxuitenko YY. Biochemical mechanisms and biological implications of the toxicity of aflatoxins are related to aflatoxin carcinogenesis. In: Eaton DL, Groopman JD, editors. The toxicology of aflatoxins. San Diego: Academic; 1994.
8. Wild CP, Gong YY. Mycotoxins and human disease: a largely ignored global health issue. Carcinogenesis. 2010;32:71–82.
9. He XY, Tang L, Wang SL, Cai QS, Wang JS, Hong JY. Efficient activation of aflatoxin B1 by cytochrome P450 2A13, an enzyme predominantly expressed in human respiratory tract. Int J Cancer. 2006;118:2665–71.
10. Khlangwiset P, Shepard GS, Wu F. Aflatoxin and growth impairment: a review. Crit Rev Toxicol. 2011;41:740–55.
11. Wu F. Mycotoxin risk assessment for the purpose of setting international regulatory standards. Environ Sci Technol. 2004;38:4049–55.
12. Chulze SN. Strategies to reduce mycotoxin levels in maize during storage: a review. Food Addit Contam. 2010;27:651–7.
13. Khlangwiset P, Wu F. Costs and efficacy of public health interventions to reduce aflatoxin-induced human disease. Food Addit Contam. 2010;27:998–1014.
14. Wu F. Mycotoxin reduction in Bt corn: potential economic, health, and regulatory impacts. Transgenic Res. 2006;15:277–89.
15. Wu F, Khlangwiset P. Health economic impacts and cost effectiveness of aflatoxin—reduction in Africa: case studies in biocontrol and post-harvest interventions. Food Addit Contam. 2010;27:496–509.
16. Wu F, Khlangwiset P. Evaluating the technical feasibility of aflatoxin risk reduction strategies in Africa. Food Addit Contam. 2010;27:658–76.

[6] Linz JE, Wee J, Roze LV Aspergillus parasiticus SU-1 genome sequence, predicted chromosome structure, and comparative gene expression under aflatoxin inducing conditions: Evidence that differential expression contributes to species phenotype. Euk Cell 2014;13:000.

17. Zahner H, Anke H, Anke T. Evolution and secondary pathways. In: Bennett JW, Ciegler A, editors. Secondary metabolism and differentiation in fungi, Mycology series, volume 5. New York: Marcel Dekker; 1983. p. 153–71. Chapter 6.
18. Rolfs M, Albert M, Keller NP, Kempken F. Secondary chemicals protect mould from fungivory. Biol Lett. 2007;3:523–5.
19. Miller MJ, Linz JE. Genetic mechanisms involved in regulation of mycotoxin biosynthesis. In: Shetty K, editor. Food biotechnology. 2nd ed. Boca Raton: Taylor and Francis Group LLC; 2006.
20. Georgianna DR, Payne GA. Genetic regulation of aflatoxin biosynthesis: from gene to genome. Fungal Genet Biol. 2009;46:113–25.
21. Roze LV, Chanda A, Linz JE. Compartmentalization and molecular traffic in secondary metabolisms: a new understanding of established cellular processes. Fungal Genet Biol. 2011;48:35–48.
22. Roze LV, Hong S-Y, Linz JE. Aflatoxin biosynthesis: current Frontiers. Annu Rev Food Sci Technol. 2013;4:293–311.
23. Yabe K, Nakajima H. Aflatoxin biosynthesis. Shokuhin Eiseigaku Zasshi. 2011;52:135–47. Japanese.
24. Georgianna DR, Fedorova ND, Burroughs JL, Dolezal AL, Bok JW, Horowitz-Brown S, Woloshuk CP, Yu J, Keller NP, Payne G. Beyond aflatoxin; four distinct expression patterns and functional roles associated with *Aspergillus flavus* secondary metabolism gene clusters. Mol Plant Pathol. 2010;11:213–26.
25. Klionsky DJ, Herman PK, Emr SD. The fungal vacuole—composition, function, and biogenesis. Microbiol Rev. 1990;54:266–92.
26. Klionsky DJ. Nonclassical protein sorting to the yeast vacuole. J Biol Chem. 1998;273:10807–10.
27. Sarry JE, Chen S, Collum RP, Liang S, Peng M, Lang A, Naumann B, Dzierszinski F, Yuan C-X, Hippler M, Rea P. Analysis of the vacuolar luminal proteome of *Saccharomyces cerevisiae*. FEBS J. 2007;274:4287–305.
28. Wiederhold E, Gandhi T, Permentier HP, Breitling R, Poolman B, Slotboom DJ. The yeast vacuolar membrane proteome. Mol Cell Proteomics. 2009;8:380–92.
29. Cai H, Reinisch K, Ferro-Novik S. Coats, tethers, Rabs, and SNAREs work together to mediate the intracellular destination of a transport vesicle. Dev Cell. 2007;12:671–82.
30. Oda M, Scott S, Hefner-Gravnik A, Caffarelli A, Klionsky DJ. Identification of a cytoplasm to vacuole targeting determinant in aminopeptidase I. J Cell Biol. 1996;132:999–1010.
31. Yorimitsu T, Klionsky DJ. Autophagy: molecular machinery for self-eating. Cell Death Differ. 2005;12:1542–52.
32. Pollack JK, Harris SD, Marten MR. Autophagy in filamentous fungi. Fungal Genet Biol. 2009;46:1–8.
33. Ohsumi K, Arioka M, Nakajima H, Kitamoto K. Cloning and characterization of a gene (*avaA*) from *Aspergillus nidulans* encoding a small GTPase involved in vacuolar biogenesis. Gene. 2002;291:77–84.
34. Lee LW, Chiou CH, Klomparens K, Cary JW, Linz JE. Sub-cellular localization of aflatoxin biosynthetic enzymes Nor-1, Ver-1, and OmtA in time-dependent fractionated colonies of *Aspergillus parasiticus*. Arch Microbiol. 2004;181:204–14.
35. Chanda A, Roze LV, Kang S, Artymovich KA, Hicks GR, Raikel N. A key role for vesicles in fungal secondary metabolism. Proc Natl Acad Sci U S A. 2009;106:19533–8.
36. Hong SY, Linz JE. Functional expression and sub-cellular localization of the aflatoxin pathway enzyme Ver-1 fused to enhanced green fluorescent protein. Appl Environ Microbiol. 2008;74:6385–96.
37. Hong SY, Linz JE. Functional expression and sub-cellular localization of the early aflatoxin pathway enzyme Nor-1 in *Aspergillus parasiticus*. Mycol Res. 2009;113:591–601.
38. Zouhar J, Hicks GR, Raikhel N. Sorting inhibitors (sortins): chemical compounds to study vacuolar sorting in *Arabidopsis*. Proc Natl Acad Sci U S A. 2004;101:9497–501.

39. Chiou CH, Lee LW, Owens SA, Whallon JH, Klomparens KL, Townsend CA, Linz JE. Distribution and sub-cellular localization of the aflatoxin enzyme versicolorin B synthase in time-fractionated colonies of *Aspergillus parasiticus*. Arch Microbiol. 2004;182:67–79.
40. Silva JC, Townsend CA. Heterologous expression, isolation, and characterization of versicolorin B synthase from *Aspergillus parasiticus*. A key enzyme in the aflatoxin B_1 biosynthetic pathway. J Biol Chem. 1997;272:804–13.
41. Farkas T, Daugard M, Jaatela M. Identification of small molecule inhibitors of Phosphatidylinositol 3-kinase and autophagy. J Biol Chem. 2011;286:38904–12.
42. Park MR, Gupta MK, Lee HR, Das ZC, Uhm SJ, Lee HT. Possible involvement of Class III phosphatidylinositol-3-kinase in meiotic progression of porcine oocytes beyond germinal vesicle stage. Theriogenology. 2010;75:940–50.
43. Lindmo K, Stenmark H. Regulation of membrane traffic by phosphoinositide 3-kinases. J Cell Sci. 2005;119:605–14.
44. Chanda A, Roze LV, Linz JE. Purification of a vesicle-vacuole (V) fraction from Aspergillus. Methods Mol Biol. 2012;944:259–66.
45. Chanda A, Roze LV, Pastor A, Frame MK, Linz JE. Purification of a vesicle-vacuole fraction functionally linked to aflatoxin synthesis in *Aspergillus parasiticus*. J Microbiol Methods. 2009;78:28–33.
46. Linz JE, Chanda A, Whitten D, Roze LV. Proteomic and biochemical evidence support a role for membrane-bound, sub-cellular compartments in stress response and secondary metabolism in *Aspergillus parasiticus*. J Proteome Res. 2011;11:767–75.
47. Chanda A, Roze LV, Linz JE. A possible role for exocytosis in aflatoxin export in *Aspergillus parasiticus*. Eukaryot Cell. 2010;9:1724–7.
48. Menke J, Weber J, Broz K, Kistler HC. Cellular development associated with induced mycotoxin synthesis in the filamentous fungus fusarium graminearum. PLoS One. 2013;8(5):e63077.
49. Brodhagen M, Keller NP. Signaling pathways connecting mycotoxin production and sporulation. Mol Plant Pathol. 2006;7:285–301.
50. Roze LV, Beaudry RM, Keller NP, Linz JE. Regulation of aflatoxin synthesis by FadA/cAMP/protein kinase A signaling in *Aspergillus parasiticus*. Mycopathologia. 2004;158:219–32.
51. Abenza JF, Galindo A, Pantozopoulou A, Gil C, de los Rios V, Penalva MA. Aspergillus RabB (Rab5) integrates acquisition of degradative identity with the long distance movement of early endosomes. Mol Biol Cell. 2010;21:2756–69.
52. Slessareva JE, Rout SM, Temple B, Bankaitis VA, Dohlman HG. Activation of the phosphatidylinositol 3-kinase Vps34 by a G protein α subunit at the endosome. Cell. 2006;126:191–203.
53. Ehrlich KC, Mack BM, Wei Q, Li P, Roze LV, Dazzo F, Cary JW, Bhatnagar D, Wee J, Linz JE. Association with AflR in endosomes reveals new functions for AflJ in aflatoxin biosynthesis. Toxins. 2012;4:1582–600.
54. Maggio-Hall LA, Wilson RA, Keller NP. Fundamental contribution of beta-oxidation to polyketide mycotoxin production in planta. Mol Plant Microbe Interact. 2005;18:783–93.
55. Hong SY, Roze LV, Wee J, Linz JE. Evidence that a transcription factor regulatory network coordinates oxidative stress response and secondary metabolism. MicrobiologyOpen. 2013;2:144–60.
56. Elahi MM, Kong Y-X, Matata BM. Oxidative stress as a mediator of cardiovascular disease. Oxid Med Cell Longev. 2009;2:259–69.
57. Jackson MJ, McArdle A. Age-related changes in skeletal muscle reactive oxygen species generation and adaptive responses to reactive oxygen species. J Physiol. 2011;589:2139–45.
58. Janssen-Heininger YMW, Aesif SW, van der Velden J, Guala AS, Reiss JN, Roberson EC, Budd RC, Reynaert NL, Anathy V. Regulation of apoptosis through cysteine oxidation: implications for fibrotic lung disease. Ann N Y Acad Sci. 2010;1203:23–8.
59. Lee JL, Kim NY, Suh YA, Lee CH. Involvement of ROS in curcumin-induced autophagic cell death. Korean J Physiol Pharmacol. 2011;15:1–7.

60. Lewis DFV. Review. Oxidative stress and the role of cytochromes P450 in oxygen activation. J Chem Technol Biotechnol. 2002;77:1095–100.
61. Li X, Becker KA, Zhang Y. Ceramide in redox signalling and cardiovascular diseases. Cell Physiol Biochem. 2010;26:41–8.
62. Miaczymnski M, Bar-Sagi D. Signaling endosomes: seeing is believing. Curr Opin Cell Biol. 2010;22:535–40.
63. Muro S, Cui X, Gajewski C, Murciano J-C, Muzykantov VR, Koval M. Intracellular trafficking of catalase nanoparticle targeted to ICAM-1 protects endothelial cells from oxidative stress. Am J Physiol Cell Physiol. 2003;285:C1339–47.
64. Ushio-Fukai M. Localizing NADPH oxidase-derived ROS. Sci *Online.* 2006; *349/re8:* 1–8.
65. Ushio-Fukai M. Compartmentalization of redox signaling through NADPH oxidase-derived ROS. Antioxid Redox Signal. 2009;11:1289–99.
66. Roze LV, Chanda A, Laivenieks M, Wee J, Awad D, Linz JE. Stress-related transcription factor AtfB integrates secondary metabolism with oxidative stress response in Aspergilli. J Biol Chem. 2011;286:35137–48.
67. Shimizu K, Hicks JK, Huang T, Keller N. Pka, Ras and RGS protein interactions regulate activity of AflR, a Zn(II)Cys6 transcription factor in *Aspergillus nidulans*. Genetics. 2003;165:1095–104.
68. Bin W-B, Amaike S, Wohlback DJ, Gasch AP, Chiang YM, Wang CC, Bok J-W, Rohlfs M, Keller NP. An *Aspergillus nidulans* bZIP response pathway hardwired for defensive secondary metabolism operates through *aflR*. Mol Microbiol. 2012;83:1024–34.
69. Shimizu K, Keller NP. Genetic involvement of a cAMP-dependent protein kinase in a G protein signaling pathway regulating morphological and chemical transitions in *Aspergillus nidulans*. Genetics. 2001;157:591–600.
70. Tag A, Hicks J, Garifullina G, Ake CJ, Phillips TD, Beremand M, Keller N. G-protein signaling mediates differential production of toxic secondary metabolites. Mol Microbiol. 2000;38:658–65.
71. Lee JW, Roze LV, Linz JE. Evidence that a wortmannin-sensitive signal transduction pathway regulates aflatoxin biosynthesis. Mycologia. 2007;99:562–8.
72. Cebollero E, Reggiori F. Regulation of autophagy in yeast *Saccharomyces cerevisiae*. Biochim Biophys Acta. 2009;1793:1413–21.
73. McLean BA, Zhabyeyev P, Pituskin E, Paterson I, Haykowsky MJ, Oudit GY. PI3K inhibitors as novel cancer therapies. Implications for cardiovascular medicine. J Card Fail. 2013;19:268–82.
74. Kim EH, Suresh M. Role of PI3K/Akt signaling in memory CD8 T cell differentiation. Front Immunol. 2013;4:1–11.
75. Sprio DJ, Boll W, Kirchhausen T, Wessling-Resnick M. Wortmannin alters the transferrin receptor endocytic pathway in vivo and in vitro. Mol Biol Cell. 1996;7:355–67.
76. Wu YT, Tan HL, Shui G, Bauvy C, Huang Q, Wenk MR, Ong CN, Codongo P, Shen HM. Dual roles of 3-methyladenine in modulation of autophagy via different temporal patterns of inhibition of Class I and Class III phosphoinositide 3-kinase. J Biol Chem. 2010;285:10850–61.
77. Cary JW, Obrian GR, Nielsen DM, Nierman W, Harris-Coward P, Yu J, Bhatnagar D, Cleveland TE, Payne GA, Calvo AM. Elucidation of *veA*-dependent genes associated with aflatoxin and sclerotial production in *Aspergillus flavus* by functional genomics. Appl Microbiol Biotechnol. 2007;76:1107–18.
78. Calvo AM. The VeA regulatory system and its role in morphological and chemical development in fungi. Fungal Genet Biol. 2008;45:1053–61.
79. Bayram O, Krappman S, Ni M, Bok JW, Helmstaedt K, Valerius O, Braus-Stromeyer S, Kwon NJ, Keller NP, Yu JH, Braus G. VelB/VeA/LaeA complex coordinates light signal with fungal development and secondary metabolism. Science. 2008;320:1504–6.
80. Amaike S, Keller NP. Distinct roles for VeA and LaeA in development and pathogenesis of *Aspergillus flavus*. Eukaryot Cell. 2009;8:1051–60.
81. Adams TH, Deising H, Timberlake WE. *brlA* requires both zinc fingers to induce development. Mol Cell Biol. 1990;10:1815–7.

82. Timberlake WE. Molecular genetics of Aspergillus development. Annu Rev Genet. 1990;24:5–36.
83. Burow GB, Gardner HW, Keller NP. A peanut seed lipoxygenase responsive to Aspergillus colonization. Plant Mol Biol. 2000;42:689–701.
84. Burow GB, Nesbitt TC, Dunlap J, Keller NP. Seed lipoxygenase products modulate *Aspergillus* mycotoxin biosynthesis. Mol Plant Microbe Interact. 1997;10:380–7.
85. Calvo AM, Hinze LL, Gardner HW, Keller NP. Sporogenic effect of polyunsaturated fatty acids on development of *Aspergillus spp*. Appl Environ Microbiol. 1999;65:3668–73.
86. Tsitsiginis DI, Kowieski T, Zarnowski R, Keller NP. Three putative oxylipin biosynthetic genes integrate sexual and asexual development in *Aspergillus nidulans*. Microbiology. 2005;151:1809–21.
87. Andrianopoulos A, Timberlake WE. The *Aspergillus nidulans abaA* gene encodes a transcriptional activator that acts as a genetic switch to control development. Mol Cell Biol. 1994;14:2503–15.
88. Sewall TC, Mims CW, Timberlake WE. Conidium differentiation in *Aspergillus nidulans* wild-type and wet-white (*wetA*) mutant strains. Dev Biol. 1990;138:499–508.
89. Roze LV, Miller MJ, Rarick M, Mahanti N, Linz JE. A novel cAMP-response element, CRE1, modulates expression of *nor-1* in *Aspergillus parasiticus*. J Biol Chem. 2004;279:27428–39.
90. Hong SY, Roze L, Wee J, Linz JE. Evidence that a transcription factor regulatory network coordinates oxidative stress response and secondary metabolism in aspergilli. MicrobiologyOpen. 2012;2:144–60.
91. Martinez-Pastor MT, Marchler G, Schuller C, Marchler-Bauer A, Ruis H, Estruch F. The *Saccharomyces cerevisiae* zinc finger proteins Msn2p and Msn4p are required for transcriptional induction through the stress response element (STRE). EMBO J. 1996;15:2227–35.
92. Chang PK, Scharfenstein LL, Luo M, Mahoney N, Molyneaux RJ, Yu J, Brown R, Campbell BC. Loss of *msnA*, a putative stress regulatory gene, in *Aspergillus parasiticus* and *Aspergillus flavus* increased production of conidia, aflatoxins, and kojic acid. Toxins. 2010;10:82–104.
93. Vargas-Perez I, Sanchez O, Kawasaki L, Georgellis D, Aguirre J. Response regulators SrrA and SskA are central components of a phosphorelay system involved in stress signal transduction and asexual sporulation in Aspergillus nidulans. Eukaryot Cell. 2007;6:1570–83.
94. Eshaghi M, Lee JH, Lei Z, Poon SY, Li J, Cho KH, Chu Z, Karaturi KM, Liu J. Genomic binding profile of the fission yeast stress-activated MAPK Sty1 and the bZIP transcriptional activator Atf1 in response to H_2O_2. PLoS One. 2010;5:e11620.
95. Morgan BA, Banks GR, Toone WM, Raitt D, Kuge S, Johnston L. The Skn7 response regulator controls gene expression in the oxidative stress response of the budding yeast *Saccharomyces cerevisiae*. EMBO J. 1997;16:1035–44.
96. Tian C, Li J, Glass NL. Exploring the bZIP transcription factor regulatory network in *Neurospora crassa*. Microbiology. 2010;157:747–59.
97. Hagiwara D, Asano Y, Yamashino T, Mizuno T. Characterization of bZip-type transcription factor AtfA with reference to stress responses of conidia of *Aspergillus nidulans*. Biosci Biotechnol Biochem. 2008;72:2756–60.
98. Reverberi M, Zjalic S, Ricelli A, Punelli F, Camera E, Fabbri C, Picardo M, Fanelli C, Fabbri AA. Modulation of antioxidant defense in *Aspergillus parasiticus* is involved in aflatoxin biosynthesis: a role for the ApyapA gene. Eukaryot Cell. 2008;7:988–1000.
99. Sakamoto K, Arima T, Kazuhiro I, Yamada O, Gomi K. *Aspergillus oryzae atfB* encodes a transcription factor required for stress tolerance in conidia. Fungal Genet Biol. 2008;45:922–32.
100. Sakamoto K, Iwashita K, Yamada O, Kobayashi K, Mizuno A, Akita O, Mikami S, Shimoi H, Gomi K. *Aspergillus oryzae atfA* controls conidial germination and stress tolerance. Fungal Genet Biol. 2009;46:887–97.
101. Reverberi M, Zjalic S, Punelli F, Ricelli A, Fabbri AA, Fanelli C. Apyap1 affects aflatoxin biosynthesis during *Aspergillus parasiticus* growth in maize seeds. Food Addit Contam. 1994;24:1070–5.

102. Reverberi M, Ricelli F, Zjalic S, Fabbri A, Fanelli C. Natural functions of mycotoxins and control of their biosynthesis in fungi. Appl Microbiol Biotechnol. 2010;87:899–911.
103. Jayashree T, Subramanyam C. Oxidative stress as a prerequisite for aflatoxin production by *Aspergillus parasiticus*. Free Radic Biol Med. 2000;29:981–5.
104. Kim JH, Yu J, Mahoney N, Chan KL, Molyneux RJ, Varga J, Bhatnagar D, Cleveland TE, Nierman WC, Campbell BC. Elucidation of the functional genomics of antioxidant-based inhibition of aflatoxin biosynthesis. Int J Food Microbiol. 2008;122:49–60.
105. Narasaiah KV, Sashidar RB, Subramanyam C. Biochemical analysis of oxidative stress in the production of aflatoxin and its precursor intermediates. Mycopathologia. 2006;162:179–89.
106. Roze LV, Calvo AM, Gunterus A, Beaudry R, Kall M, Linz JE. Ethylene modulates development and toxin biosynthesis in aspergillus possibly via an ethylene sensor-mediated signaling pathway. J Food Prot. 2004;67(3):438–47.

Chapter 6
Roquefortine C and Related Prenylated Indole Alkaloids

Juan-Francisco Martín, Paloma Liras, and Carlos García-Estrada

Introduction

Many fungal metabolites, collectively designated as indole alkaloids contain in their structures a prenylated indole nucleus that derives from L-tryptophan and mevalonate. These metabolites include two large groups: (1) the ergot alkaloids produced by plant parasitic *Claviceps* species [1], and (2) the indole alkaloids produced by species of *Aspergillus*, *Penicillium*, and *Neosartorya* among others [2]. These alkaloids differ: (1) in the carbon atom of the indole molecule bearing the isopentenyl group, (2) in modifications of the diketopiperazine ring, and (3) in modifications of the N1 atom of indole, that are introduced by "late" modification enzymes encoded by additional genes in the clusters.

One of the best known indole alkaloid groups is that of the ergot alkaloids [1] that are reviewed in a separate chapter in this book. Another important group is that of roquefortine C and related indole alkaloids (glandicoline, meleagrin, neoxaline) [3, 4]. Several of these mycotoxins are produced by *Penicillium* species of the *Corymbifera* family, which infect onions, tulips, and other plant bulbs (Table 6.1) [5]. Others are produced by *Penicillium* species growing on cheese [6] or in contaminated food products and cereal grains. The acetylaszonalenin producer *N. fischeri* (formerly *Aspergillus fischeri*) is an opportunistic human pathogen closely related to the pathogenic *Aspergillus fumigatus* [7, 8].

J.-F. Martín, Ph.D. (✉) • P. Liras, Ph.D.
Department of Molecular Biology, Microbiology Section, University of León, León, Spain
e-mail: jf.martin@unileon.es

C. García-Estrada, D.V.M., Ph.D.
INBIOTEC (Institute of Biotechnology of León), Parque Científico de León, León, Spain

J.-F. Martín et al. (eds.), *Biosynthesis and Molecular Genetics of Fungal Secondary Metabolites*, Fungal Biology, DOI 10.1007/978-1-4939-1191-2_6,

Table 6.1 Roquefortine C, meleagrin, and neoxaline producer strains

Producer strains	Identified compounds
Penicillium roqueforti	Roquefortine C[a]
Penicillium crustosum	Roquefortine C[a]
Penicillium venetum	Roquefortine C, D[a]
Penicillium hordei	Roquefortine C, D
Penicillium hirsutum	Roquefortine C, D; glandicolines A, B
Penicillium albocoremium	Roquefortine C, glandicolines A, B; meleagrin
Penicillium alli	Roquefortine C, glandicolines A, B; meleagrin, neoxaline
Penicillium radiocolor	Roquefortine C, glandicolines A, B; meleagrin
Penicillium tulipae	Roquefortine C, D; glandicolines A, B; meleagrin and epineoxaline

[a]The lack of production of glandicolines A and B, and meleagrin may be due to lack of the enzymes converting roquefortine C to meleagrin (see text). The biosynthesis of neoxaline (epineoxaline) has only been reported in *P. tulipae*, but barely detectable amounts of neoxaline might be formed by other meleagrin producing strains

The roquefortine group contains compounds that are prenylated at carbon 3 (C-3) of the indole nucleus by "reverse prenylation"; i.e., by a bond of the isopentenyl group through its C-3′ carbon (see the prenylation reaction below). In this chapter, we limit our study to the compounds of this group.

Toxicity and Medical Interest

Indole alkaloids have very potent effects on humans and other animals. The medical effect of ergot alkaloids and the toxicity of many other indole alkaloids that are potent mycotoxins have been known for centuries. One of the indole alkaloids, fumitremorgin, is an inhibitor of the breast cancer resistance protein (BCRP) and is an important lead molecule for the development of advanced anticancer agents [9, 10].

The role in nature of these compounds is complex and still poorly known. Due to their toxicity some of those compounds are feeding deterrents for animals that discourage the consumption of feeds contaminated by the producer fungi [11]. Roquefortine C contamination of food and animal feedstuff is of great interest because of the well-known neurotoxicity of this mycotoxin [12, 13].

Early Studies on Roquefortine Production

Roquefortine C, a potent mycotoxin member of the DKP alkaloid family, was first isolated from cultures of *Penicillium roqueforti* [14] and later from cultures of other *Penicillium* species growing on contaminated feed grain and food wastes [15–17], onions [5], beer [18], and wine [19]. Roquefortine C is also synthesized by penicillin-producing strains of *Penicillium chrysogenum* derived from the Peoria strain *P. chrysogenum* NRRL1951 [4, 20].

The level of production and secretion of roquefortine C and related metabolites is highly dependent on the producer *Penicillium* species and the nutritional and environmental conditions [5, 21]. Interestingly, the production of roquefortine C is influenced by its extracellular concentration [22], which suggests a feedback control of the production, probably at the secretion level, as reported for some macrolide compounds [23].

Labeled ^{14}C-roquefortine C enters the cell and is incorporated into protein, suggesting that there is degradation and reutilization of the fragments as carbon and nitrogen sources [22, 24], but it is unclear whether roquefortine C enters the cell as an undegraded molecule. Re-entry into the cell of secreted secondary metabolites (e.g., benzylpenicillin in *P. chrysogenum*) [25] is very inefficient and most likely secondary metabolites are further oxidized and degraded by the producer strains before the component fragments of the molecules are taken up as nutrients.

Roquefortine C and related mycotoxins have been proposed to serve as nitrogen reserve nutrients [26], but taking into account the low production levels in nature (in the range of a few micrograms/mL) they will provide a limited nutrient amount. Therefore, it seems likely that roquefortine C in addition to be a possible reserve nutrient material, serves as a communication signal with other microorganisms living on the infected grains or bulbs' surface [21]. Indeed, roquefortine C has antimicrobial activity [27].

Precursors and Biosynthesis of the Prenylated Indole Alkaloids

Early studies on precursors incorporation led to the proposal that roquefortine C derives from tryptophan, histidine, and mevalonate [28, 29, 32]. Glandicolines A and B, meleagrin, and neoxaline were proposed to derive from roquefortine C [30–33] and this conversion was confirmed using ^{14}C-roquefortine as a precursor of those products [34]. However, the enzyme involved in the roquefortine C/meleagrin biosynthesis pathway and the genes that encode them have been only recently identified [4, 35].

It is accepted that the prenylated indole alkaloids derive from the component amino acids, which are easily deduced by analysis of their chemical structures (Fig. 6.1 and Table 6.2). However, there is very little experimental evidence confirming the nature of the precursors in the indole alkaloids of this group. Only in the cases of roquefortine C and acetylaszonalenin there are detailed studies on precursor incorporation. Early studies on roquefortine biosynthesis in cultures of *P. roqueforti* using labeled precursors showed that roquefortine C derives from tryptophan, histidine, and mevalonate [36, 37]. The addition of L-tryptophan stimulates the roquefortine C/meleagrin production in *P. chrysogenum* but, surprisingly, this stimulation is reversed by L-histidine when mixed with tryptophan [35].

However, the final structures of roquefortine C and meleagrin contain a molecule of dehydrohistidine, instead of histidine, and the exact origin of this dehydrohistidine has remained unclear for many years. The presence of dehydrohistidine (an antimetabolite of histidine) as a free compound in the cytosol is unreasonable due

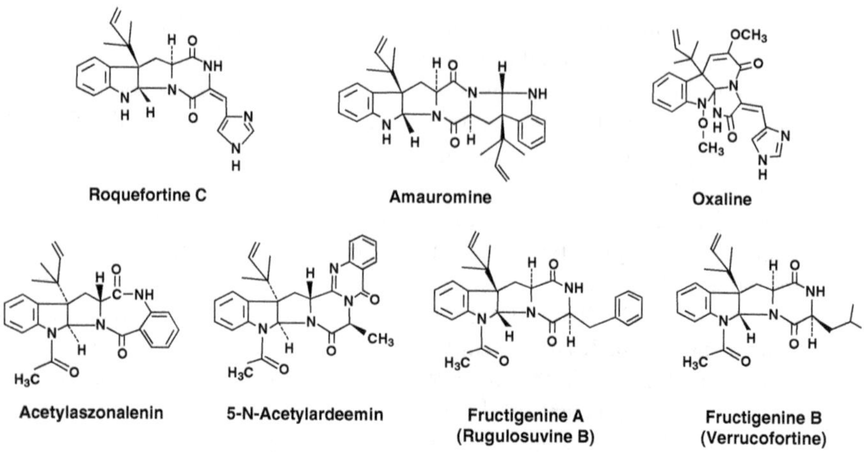

Fig. 6.1 Chemical structures of roquefortine C, acetylaszonalenin, and related members of the indole alkaloids prenylated at C-3 of the indole molecule

Table 6.2 Fungal diketopiperazines with a reverse prenylated indole at carbon C-3

Compound	Producer fungi	Amino acids forming the DKP nucleus	References
Roquefortine C	*P. roqueforti* *P. chrysogenum*	Tryptophan + histidine	[4–6, 17]
Glandicoline	*P. roqueforti, P. chrysogenum* *P. glandicola*	Tryptophan + histidine	[5]
Meleagrin	*P. roqueforti, P. chrysogenum*	Tryptophan + histidine	
Aszonalenin and acetylaszonalenin	*Aspergillus zonatus* *Aspergillus fumigatus* *Aspergillus carneus* *Aspergillus terreus* *Neosartorya fischeri*	Tryptophan + anthranilic acid	[57–60]
Epiaszonalenin	*Aspergillus novofumigatus*	Tryptophan + anthranilic acid	
Oxaline	*P. tulipae*	Tryptophan + histidine	[5]
Amauromine	*Amauroascus* sp.	Tryptophan + tryptophan[a]	[68]
Epi-amauromine	*Aspergillus ochraceus*	Tryptophan + tryptophan	
Fructigenine A	*Penicillium fructigenum*	Tryptophan + phenylalanine	[48]
= rugulosuvine	*Penicillium rugulosum*	Tryptophan + phenylalanine	
Verrucofortine	*Penicillium verrucosum*	Tryptophan + leucine	[69]
= fructigenine B	*Penicillium verrucosum* var. *cyclopium*	Tryptophan + alanine	[48]

[a]Each of the tryptophan moieties bears a 3′-isopentenyl group at C3 of the indole molecule

to its metabolic toxicity (see below enzymes putatively involved in the conversion of histidine to dehydrohistidine occurring in the roquefortine C cluster). Dehydrohistidine is also a precursor of phenylahistin produced by *Aspergillus ustus* [38].

Similarly, tryptophan was shown to be a precursor of acetylaszonalenin; the biosynthesis of this mycotoxin also involves the final incorporation of an acetyl group from acetyl-CoA [39–41].

The indole alkaloid cyclodipeptides are formed from unmodified L-tryptophan. Particularly relevant evidence confirming this fact is the incorporation of deuterium from L-(2,4,5,6,7)^{2}H-tryptophan in roquefortine C and acetylaszonalenin [40, 41]. Incorporation of this labeled tryptophan proceeds with retention of ^{2}H at all five positions of the tryptophan moiety thus excluding that prenylation (with removal of a ^{2}H atom) occurred transitorily at any of those positions. Indeed, it was confirmed that in the biosynthesis of aszonalenin the prenylation occurs directly at carbon 3 of the indole nucleus of tryptophan [42].

The anthranilic acid moiety of aszonalenin and acetylaszonalenin is likely to derive from the anthranilic acid pool, an intermediate in the tryptophan biosynthetic pathway, formed by the anthranilate synthase from the aromatic intermediate chorismic acid.

Formation of the Diketopiperazine Ring

A common structural feature of the indole alkaloids is the presence of a six-membered diketopiperazine ring that is formed by the head-to-head condensation of two amino acids (one of them tryptophan). The amino group of tryptophan is linked to the C-1 carboxyl group of the second amino acid and vice versa the amino group of the second amino acid is linked to the carboxyl group of tryptophan. This cyclization is catalyzed by a CDPS, a specific type of peptide synthetase that belongs to the large class of NRPSs [43]. The cyclization reaction appears to be catalyzed by the second condensation domain (C2) of the respective CDPS enzyme (see below).

Cyclodipeptide Synthetases

These enzymes consist of two modules, each containing an amino acid activation domain (A), an aminoacyl carrier (thiolation) domain (T) and a condensation domain (C) catalyzing peptide bond formation. The first condensation domain, C1, catalyzes dipeptide formation between two activated aminoacyl intermediates. A second condensation domain, C2, promotes cyclization to the diketopiperazine. The domain structure of these cyclodipeptide synthetases is generally ATCATC (see Table 6.3). These enzymes lack the integrated thioesterase domain that occurs at the C-terminus in the NRPSs of linear peptides such as the *P. chrysogenum* ACV synthetase [44, 45]. The roquefortine biosynthetic cluster (Fig. 6.2) includes a distinct dipeptide synthetase [4, 35]. A phylogenetic analysis of CDPSs of the type ATC_1ATC_2 shows an early divergence into: (1) indole alkaloid precursor-forming systems of the roquefortine/aszonalenin/fumitremorgin type, and (2) ETP (epi-polythiodioxopiperazine) precursor-forming systems (Fig. 6.3). Both types of CDPS genes are integrated in gene clusters with a variety of genes encoding cyclopeptide-modifying enzymes, particularly, a set of monooxygenases and

Table 6.3 Cyclopeptide synthetases in various ascomycetes

Organism	Gene (GI)	Size (aa)	Domain organization	Amino acid sequences of A1	Amino acid sequences of A2	Functions
Indole alkaloid group						
C. immitis	90306876	2454	ATCATC	DMLVCGFINK	DAQLICGICK	
T. atroviride	154060	2178	ATCATC	DTEDIGTIVK	DVSYVGSIWK	
G. zeae	42546236	2325	ATCATC	DVSNLSTFKK	DVLCVAWLSK	
A. fumigatus	224471203	2211	ATCATC	DVMFIGAVNK(Trp)	DVYFVGGICK(Pro)	cTrpPro, fumitremorgin
P. chrysogenum	Pc21g15480	2372	ATCATC	DSLELVAVVK(His)	DIAMIGSMYK(Trp)	cHisTrp, roquefortine
N. fischeri	119407561	2179	ATCATC	DSLELVAVVK(His)	DIAMVGSMYK(Trp)	cHisTrp, roquefortine?
P. chrysogenum	Pc21g12630	2382	ATCATC	DVRSVGAGIK	DIGLGAMVIK	
G. zeae	42555902	2381	ATCATC	DVRSVGAGIK	DIGLGCMVIK	
N. fischeri	119470744	2359	ATCATC	GALFFAAGVK(Ant)	DVMFVGEVAK(Trp)	cAntTrp, aszonalenin
A. terreus	114187602	2337	ATCATC	GALFFAAGVK(Ant)	DVMFVGEVAK(Trp)	cAntTrp, aszonalenin
A. fumigatus	Afu3g03350	2210	ATCATC	SARGTVSQLK	DVYFTGGVLK	
C. globosum	88184088	2499	ATCATC	SARETARQMK	DAQIVGVMVK	
A. fumigatus	70998751	2353	ATCATC	SARDVGSQLK	DGYNAGSICK	
C. globosum	88175466	2596	ATCATC	SARDTAAQVK	DAFMLCGILK	
P. anserina	171693316	2588	ATCATC	SARDTAAQVK	DAFMLCGILK	
ETP-group						
T. atroviride	77514	2149	ATCATCT	DAGAVGFIAK	DVVHGAFPIK	
T. reesei	41057	2146	ATCATC	DAAAMGVFAK	DVVYGAFPIK	
T. stipitatus	8102259	2151	ATCATC	DAGVVGFFAK	DVVYNAGPIK	
L. maculans	46403055	2176	ATCATC	DAGPFGCCTK	DCNMAAVISK	SirP, cProSer, sirodesmin
A. fumigatus	70999215	2306	ATCATC	DVGPFGACTK	DCKMTAGLIK	
A. fumigatus	159127346	2273	ATCATC	DVGPFGACTK	DCKMTAGLIK	
N. fischeri	119491024	2202	ATCATC	DVGPFGACTK	DCKMTAGLIK	
A. terreus	115433823	2152	ATCATC	DVGPFGSCTK	DCKMTAGLIK	
T. virens	70742	2038	ATCATC	DAGPFGCICK	DVAMLASIGK	
P. marneffei	7022587	2043	ATCATC	DAGPFGAISK	DVAMFAGIVK	
T. virens	78708	2032	ATCATC	DAGALGACAK(Phe)	DYNTYTAICK(Ser)	
A. fumigatus	GliP, 70992013	2135	ATCATCT	DAGALGACAK(Phe)	DYNTYTAICK(Ser)	GliP, cPheSer, gliotoxin
N. fischeri	119406235	2135	ATCATCT	DAIALGGCAK	DYNTYTAICK(Ser)	"GliP"
T. reesei	38441	2190	ATCATCT	DACAIGGSMK	DAETYTAISK	

Cyclodipeptide synthetases. Domain structure of the cyclodipeptide synthetases of indole alkaloids. Note that all contain the ATCATC domain structure. Amino acid code signature of the two A sites (indicated A1 and A2 sites) in the cyclodipeptide synthetases are shown. Note the total conservation of the amino acid code in the tryptophan and histidine activating domains of the roquefortine Rds and NFIA074300 (see text for details). The question marks indicate unconfirmed chemical compounds

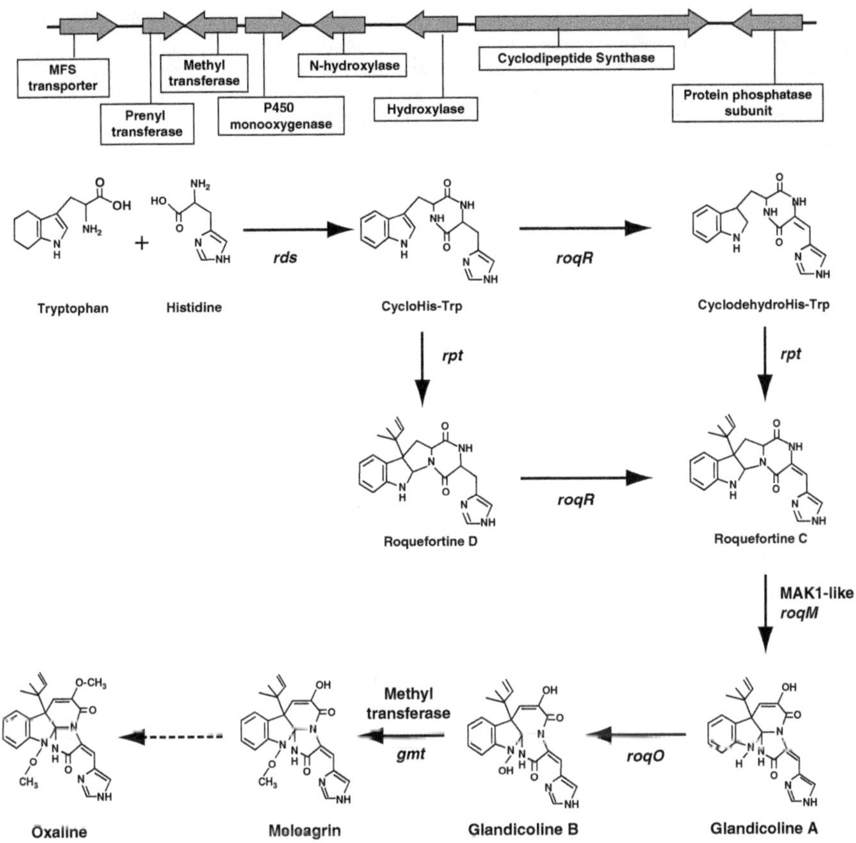

Fig. 6.2 Organization of the roquefortine C/meleagrin cluster in *P. chrysogenum* and biosynthetic pathway of these mycotoxins. The possible last step converting meleagrin to oxaline is indicated by a dotted arrow. The genes *rds*, *rpt*, and *gmt* encoding the roquefortine cyclodipeptide synthetase, roquefortine prenyltransferase, and glandicoline methyltransferase, respectively, are named as designated initially [4]. Other genes are indicated as designated by Ali et al. [35]. In P. *chrysogenum* glandicoline A is replaced by roquefortine L (see text)

oxydoreductases and different prenyltransferases (or dimethylallyltryptophan transferase, DMATs). ETP systems share a unique N-methyltransferase differing from integrated methyltransferases of other NRPSs.

The cyclodipeptide synthetase, Rds, of the roquefortine/meleagrin cluster (encoded by the *rds* gene) belongs phylogenetically to the indole alkaloid CDPS group, which diverged evolutively splitting into the fumitremorgin cluster and the roquefortine/aszonalenin cluster (Fig. 6.3).

Similarities of CDPS genes within the indole alkaloid clusters are fairly low, except for subclusters formed by orthologs (from different fungi). Thus, the closely related *N. fischeri* CDPS gene NFIA_074300, shows 59 % amino acid identity with the orthologous roquefortine Rds, while another *P. chrysogenum* CDPS (Pc21g12630) of unknown function, has only 31 % identity. As one of the two

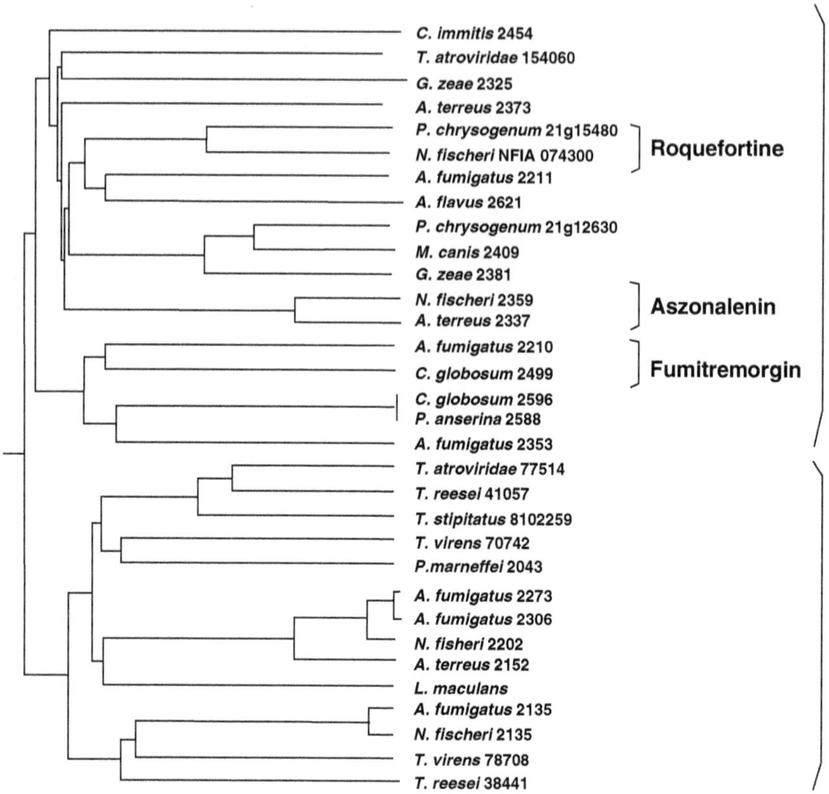

Fig. 6.3 Phylogenetic tree of the cyclodipeptide synthetases of indole alkaloids. Reproduced with permission of Cell Press from [4] García-Estrada C, Ullán RV, Albillos SM, Fernández-Bodega MÁ, Durek P, von Döhren H, Martín JF. A single cluster of coregulated genes encodes the biosynthesis of the mycotoxins roquefortine C and meleagrin in *Penicillium chrysogenum*. Chem Biol. 2011; 18:1499–1512

A-domains of this group of CDPS activates L-Trp, the study of substrate signature codes based on the Stachelhaus/Challis algorithm is useful for identifying Trp-activating domains (Table 6.3). This type of analysis confirmed that Rds and NFIA_074300 are orthologous genes. Indeed, the 34 amino acid "extended code" describing all side chains within 8 amstrongs of the binding pocket of *N. fischeri* NFIA_074300 modules 1 and 2, shows very high conservation to those of *P. chrysogenum* Rds (Pc21g15480) (only 2 and 4 residues, different respectively). Interestingly only low similarities have been found with other Trp-activating domains, indicating that the Trp-activation domain of Rds and NFIA_074300 belong to a phylogenetic subcluster with a unique Trp-binding site. BLAST searches comparing A-domains also failed to identify highly similar structured activation sites [4].

A comparison of the extended signature codes [46] divides the known Trp signatures into bacterial and fungal branches; the fungal branch is subdivided into the roquefortine and the aszonalenin/fumitremorgin type. The same holds true for the His activating domain of both Rds and NFIA_074300 CDPSs.

To elucidate the fitting of the substrates in the two adenylation domains, modeling was performed using structures based on the GrsA crystal structure [47]. The fitting of L-His and L-Trp side chain predicted that the charged His could only be accommodated in the first A site. As a result, the sequence of the initial biosynthetic reactions was established as formation of His and Trp-adenylates (A1 and A2) followed by thioester attachment (T1 and T2), formation of the His-Trp peptide bond (by C1), and cyclization to cycloHis-Trp (by C2) [4]. The involvement of the Rds enzyme in the formation of the cycloHis-Trp intermediate has been confirmed by genetic studies using RNAi silencing of *rds* and gene disruption [4, 35].

The Roquefortine Prenyltransferase Rpt

Fungal indole alkaloids are prenylated compounds in which an isopentenyl group is bound to one of the carbon atoms of the indole nucleus of tryptophan.

The prenylation of tryptophan-containing peptides is catalyzed by prenyltransferases (PT) that perform a Friedel-Crafts condensation using isopentenyl diphosphate (dimethylallyl-diphosphate, DMA-PP) as prenyl-donor [2]. In the normal type of condensation, the carbon-1 of dimethylallyl-diphosphate is linked to the tryptophan, whereas in the "reverse type" of condensation, the carbon-3′ of DMA-PP is bound to one of the carbon atoms of the indole nucleus. In plant parasitic fungi of the *Claviceps* genus producing ergot alkaloids, the prenylation is catalyzed by the so-called dimethylallyltryptophan synthase encoded by the *dmaW* gene [4] that uses L-tryptophan as substrate [48]. In other prenylated indole alkaloids, such as acetylaszonalenin, the substrates are cyclodipeptides containing tryptophan [2, 49]. Based on metabolic profiling Ali et al. [35] proposed that the Rpt enzyme prenylates either cyclohistidine-tryptophan (resulting in roquefortine D) or cyclo-dehydrohistidine-tryptophan resulting in roquefortine C. Further biochemical studies are needed to clarify the substrate specificity of Rpt. Several of the alkaloid prenyltransferases contain the signature sequence (N/D) DxxD. Lysine and other basic amino acid residues in the enzyme play an important role in substrate recognition [2, 50]. The reaction does not require magnesium. A few indole alkaloid prenyltransferases are now crystallized, notably those of *Claviceps* involved in ergotamine biosynthesis and FgaAT2 from *A. fumigatus* involved in the biosynthesis of fumigaclavine [49–51].

The great versatility of the alkaloid prenyltransferases is reflected by the presence in nature of cyclopeptides derived from prenylated tryptophan carrying the isopentenyl moiety at different carbon positions of the indole structure and even at the N1 atom of this molecule [2, 51]. Roquefortine C, acetylaszonalenin, and other related compounds reviewed in this article are 3-(3,3′) dimethylallyltryptophan derivatives (i.e., prenylated at C-3 of indole) (Fig. 6.1), but other indole alkaloids are prenylated at C-4, C-2, or N1 of the indole molecule.

A prenyltransferase gene (Pc21g15430), named *rpt*, containing an intron of 70 bp, was found in the roquefortine C gene cluster of *P. chrysogenum* (Fig. 6.2) [4]

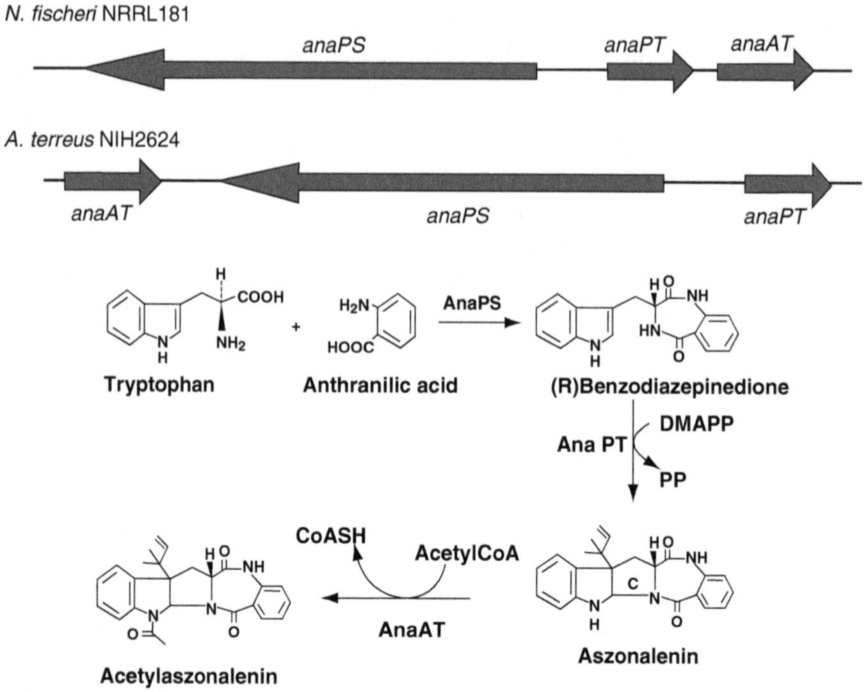

Fig. 6.4 Organization of the gene cluster for acetylaszonalenin in *N. fischeri* and *A. terreus* and the acetylaszonalenin biosynthetic pathway. Note the different locations of the *anaAT* gene in both clusters. The figure has been redrawn based on the results of Yin et al. [42]

and confirmed by Ali et al. [35]. The encoded protein shows high similarity (81 % homology, 67 % identity) to the putative cyclopeptide prenyltransferase protein from *N. fischeri* (NFIA_074280), and to a second DMAT synthase-like protein from *N. fischeri* (119483566, XP_001261686.1n, 70 % homology, 50 % identity). The roquefortine prenyltransferase, Rpt, is also similar to DMAT synthetases from *A. fumigatus* (146324647, XP_747143.2, and 159124027, EDP49146.1, 70 % homology, 50 % identity) and *Aspergillus clavatus* (121704507, XP_001270517.1, 57 % homology, 40 % identity). The involvement of the Rpt prenyltransferase of *P. chrysogenum* in roquefortine biosynthesis was established by RNAi silencing of the encoding gene [4] and confirmed by gene disruption [35]. Two mutants silenced in the Rpt gene (Pc21g15430) showed a drastic reduction in the production of roquefortine C and the deletion mutant lacked this mycotoxin. Thus, reverse prenylation of cycloHis-Trp in carbon 3 of indole is one of the early biosynthesis steps following the formation of the His-Trp or dehydroHis-Trp cyclodipeptide.

The prenylation is simultaneous with closure of the new ring C (Fig. 6.2) from position 2 of the indole ring to the N12-atom of the diketopiperazine ring, leading to the roquefortine D structure.

In the acetylaszonalenin system the formation of the fourth ring (Figs. 6.1 and 6.4) appears to be catalyzed by the prenyltransferase, since the in vitro product of the prenyltransferase reaction (aszonalenin) already contains this ring [42].

Conversion of Roquefortine D to Roquefortine C: A Grid in the Pathway

The presence of a dehydrohistidine molecule in roquefortine C is intriguing. Roquefortine D (3,12-dihydroroquefortine C) contains a histidine moiety in the cyclodipeptide instead of dehydrohistidine. Roquefortine D has been reported to be present in small amounts in cultures of *P. roqueforti* [28], *P. farinosum* [30] (Table 6.1) and other plant bulbs–infecting *Penicillium* species [5]. The conversion of roquefortine D to roquefortine C has been suggested [4, 52] and is likely to be carried out by a cytochrome P450 oxygenase (Pc21g15470) that catalyses a dehydrogenation or a hydroxylation/dehydration resulting in formation of a double bond in the histidine moiety (Fig. 6.2). Ali et al. [35] proposed that this enzyme also converts cycloHis-Trp to cyclodehydroHis-Trp, which results in a metabolic grid of the second and third steps of the pathway (Fig. 6.2).

Conversion of Roquefortine C to Meleagrin and Neoxaline

The related chemical structures of roquefortine C, meleagrin, and neoxaline suggest that these three metabolites derive from the same initial intermediates of the common pathway. Indeed, it was suggested that roquefortine C is converted to meleagrin in *Penicillium glandicola* [34]. This hypothesis was confirmed by García-Estrada et al. [4] using mutants silenced in different genes of the roquefortine/meleagrin pathway. Similarly, meleagrin may be converted to neoxaline (Fig. 6.2). These reactions are still poorly characterized.

The first step in the proposed roquefortine to meleagrin pathway (Fig. 6.2) involves a complex reorganization of the ring system in the roquefortine molecule. Roquefortine C is transformed to glandicoline A by a MAK1-related oxygenase and later to glandicoline B by oxygenation of the N1 atom of indole [4, 35]. Finally glandicoline B is converted to meleagrin by methylation of the N–OH group by the glandicoline B methyltransferase (Gmt) (Fig. 6.2). RNAi silencing of the roquefortine cluster genes showed that two oxidoreductase genes are involved in the biosynthesis of the roquefortine/meleagrin molecules, whereas the methyltransferase named Gmt, is specifically involved in the conversion of glandicoline B to meleagrin (Fig. 6.2) [4]. These results have been confirmed by metabolite profiling [35].

Recently Ries et al. [53] proposed a modification of the late steps of the pathway (conversion of roquefortine C to meleagrin and neoxaline). Based on chemical characterization of biosynthetic intermediates, these authors reported that glandicoline A does not exist in *P. chrysogenum* and is replaced by a new compound, roquefortine

L. Instead of the classical pathway proceeding from roquefortine C to glandicoline A, then to glandicoline B and finally to meleagrin, these authors proposed the pathway as follows: roquefortine C to roquefortine L, then to glandicoline B and to meleagrin. Other branches of those late steps lead to novel roquefortine products and to neoxaline.

Overy et al. [26] proposed that in *P. tulipae* the final product of the pathway is neoxaline, which contains two methylated hydroxyl groups instead of one (as in meleagrin). This raises the possibility that in this fungus there is a second methyltransferase, an O-methyltransferase, different from the Gmt, although it cannot be excluded that both methylations are catalyzed by the Gmt methyltransferase.

An MFS Transporter is Present in the Roquefortine/Meleagrin Gene Cluster

The roquefortine/meleagrin gene cluster of *P. chrysogenum* includes a gene (Pc21g15420) (Fig. 6.2) that encodes a metabolite transporter of the major facilitator superfamily (MFS) containing 12 transmembrane domains. This transporter is similar to the cercosporin facilitator protein (*cfp*) of *Cercosporium kikuchii*. The *cfp* gene encodes a 65-kDa MFS protein that is involved in cercosporin secretion [54]. The presence of a similar gene in the *P. chrysogenum* roquefortine/meleagrin cluster is consistent with the finding of MFS transporters in many secondary metabolite gene clusters [reviewed in [55, 56]]. MFS transporters show relatively wide substrate specificity. It is likely that the transporter encoded by Pc21g15420 is involved in the secretion of roquefortine C, meleagrin, and neoxaline. It has been reported that disruption of the MFS-encoding gene in *P. chrysogenum* only slightly reduces the secretion of meleagrin [35], suggesting that an alternative system exists for secretion of roquefortine C, meleagrin, and neoxaline.

A Roquefortine Orthologous Cluster in Neosartorya fischeri

The orthologous cluster of the roquefortine biosynthetic genes found in *N. fischeri*, consisting of NFIA_074280 (orthologous of Pc21g15430, 67 % identity to Rpt), NFIA_074290, (orthologous to Pc21g15470, 60 % identity to the cyclodipeptide dehydrogenase) and NFIA_074300 (orthologous to Pc21g15480, 59 % identity to Rds), is shorter than that of *P. chrysogenum*. This cluster lacks the three downstream genes involved in the conversion of roquefortine C to meleagrin. Due to the identical amino acid codes in the substrate binding site signatures of the cyclodipeptide synthetase NFIA_074300 and *Rds* (Table 6.3), it can be predicted that *N. fischeri* also produces roquefortine C or a closely related compound. In addition, this fungus produces acetylaszonalenin, another indole alkaloid of the same family. *N. fischeri* also contains clusters for verrucologen and fumitremorgin, compounds that are prenylated at the C-2 position of the indole ring. Indeed, *N. fischeri* contains several prenyltransferases-encoding genes that are associated to different secondary metabolite gene clusters [2].

The Biosynthesis of Acetylaszonalenin

Acetylaszonalenin is an indole alkaloid derived from L-tryptophan that is prenylated by reverse isopentenyldiphosphate at C-3 of the indole nucleus (Figs. 6.1 and 6.4). Aszonalenin and its acetylated derivative acetylaszonalenin are produced by *Aspergillus zonatus* [57], *A. fumigatus* [58], *A. carneus* [59], and *N. fischeri* NRRL 181 [60].

Using expression in *E. coli* of the putative aszonalenin biosynthesis genes from *N. fischeri* and in vitro transformation studies of the diketopiperazine intermediate, Yin et al. [42] identified a cluster of three genes that carry out the entire biosynthesis process. These genes are NFIA_055290 (*anaPS*), NFIA_055300 (*anaPT*), and NFIA_055310 (*anaAT*) (Fig. 6.4). The three genes are closely linked in a region of 13 kb in the *N. fischeri* genome.

The AnaPS enzyme is a dimodular CDPS with the domain structure A_1TCA_2TC that activates anthranilic acid and tryptophan. The high conservation of the amino acid code DVMFVGEVAK in the active center of the A_2 site of this CDPS with respect to the tryptophan-activating A_2 domain of the roquefortine synthetase (DIAMIGSMYK) (Table 6.3), indicates that this A_2 site activates tryptophan. Consistently, the A_1 site of the AnaPS (putatively for anthranilic acid activation) has a substrate pocket code clearly different from the A_1 site of the roquefortine Rds (which activates histidine). It seems that the first step of aszonalenin biosynthesis is the formation of a cycloanthranilyl-tryptophan dipeptide (Fig. 6.4).

The second gene of the aszonalenin cluster, *anaPT*, encodes a reverse prenyltransferase that introduces the isopentenyl group at the C-3 position of the indole nucleus by “reverse prenylation” [42, 50]. The protein encoded by this gene shares 30 % identity with the *P. chrysogenum* roquefortine prenyltransferase Rpt, which is encoded by the *rpt* gene (containing one intron of 70 bp). In a similar way, the *anaPT* gene contains one intron of 51 bp, unlike the cyclopeptide synthetase genes that typically lack introns presumably because the NRPS have a bacterial origin [61]. In vitro studies with the AnaPT enzyme confirmed that the reaction product is aszonalenin (the deacetylated intermediate of acetylaszonalenin). This product already contains the fourth ring of the aszonalenin structure indicating that the bond between the C-2 of the indole nucleus and the N12 atom of the diketopiperazine ring is formed by this enzyme.

The third gene in the aszonalenin cluster encodes an acetyltransferase (AnaAT) that has been shown in vitro to convert aszonalenin to acetylaszonalenin. The *anaAT* gene is absent from the roquefortine gene cluster; in roquefortine the N1 atom of the indole ring remains non-acetylated, although later it is hydroxylated during the reactions to form glandicoline B.

An acetylaszonalenin orthologous cluster occurs in the genome of *A. terreus* NIH2624n [42] (Fig. 6.3). The three genes of the *A. terreus* acetylaszonalenin cluster showed 76 %, 72 %, and 82 % identical amino acids to those of the AnaPS, AnaPT, and AnaAT of *N. fischeri* NRRL181, respectively.

The *anaPS* and *anaPT* genes in both *N. fischeri* and *A. terreus* are expressed from divergent promoters (in a bidirectional promoter region). Promoters with this arrangement are known to be subjected to common regulation by several transcriptional factors as occurs in a similar organization in the penicillin cluster (*pcbAB–pcbC* genes) [62–65]. On the other hand, the position of the third gene, *anaAT*, is different in the clusters of *N. fischeri* and *A. terreus* (Fig. 6.4).

Apart from the *ana*AT gene in the orthologous cluster of *A. terreus* there are no close relatives of the *anaAT* gene known in the sequenced genomes so far, suggesting that this indole N1-acetyltransferase is a rare acetylation system, specific for these indole alkaloids.

A Dimodular Cyclodipeptide Synthetase and a Prenyltransferase Involved in the Biosynthesis of Fumitremorgin

Two genes encoding a dimodular peptide synthetase with the ATCATC domain structure and a prenyltransferase are linked in the cluster of tryprostatin–fumitremorgin in *A. fumigatus*. Although fumitremorgin, a tremorgenic mycotoxin produced by *A. fumigatus* and related fungi does not belong strictly to the roquefortine C group, since it is not prenylated at C-3 of the indole, the initial steps of its biosynthesis closely resemble those involved in the biosynthesis of roquefortine C.

The dimodular CDPS encoded by the *ftmA* (*ftmPS*) gene is a cyclo-tryptophanyl-proline synthetase that forms a cyclodipeptide named breviamide F (also cyclo-L-Trp-L-Pro). Expression in *Aspergillus nidulans* and deletion of the gene of *A. fumigatus*, confirmed that this gene is involved in the biosynthesis of breviamide F that is later transformed into fumitremorgin [66].

The prenyltransferase FtmPT1 encoded by the *ftmPT1* gene, which is linked to the cyclodipeptide synthetase gene, was shown to prenylate breviamide F leading to the formation of tryprostatin [51]. Four other oxygenases and hydroxylases, one O-methyltransferase, and a second prenyltransferase (FtmPT2) are also included in the fumitremorgin cluster [10, 67] and perform a complex series of transformations converting tryprostatin to fumitremorgin.

Conclusion

The biosynthetic pathways of roquefortine C, meleagrin, and acetylaszonalenin have been largely elucidated. A good knowledge of the domains in the dimodular cyclodipeptide synthetases is available but the molecular mechanisms and sequence of formation of the two peptide bonds that give rise to the diketopiperazine ring are still obscure. Some of the indole prenyltransferases have been crystallized and this provides an insight into the mechanisms of the reaction of the "normal" and "reverse" prenyltransferases. Much less information is available on the oxidation reactions that catalyze the "late" transformation reactions that convert roquefortine C to meleagrin and neoxaline, which deserve further studies. Although one MFS transporter occurs in the roquefortine C/meleagrin gene cluster in *P. chrysogenum*, the functionality of this transporter remains obscure.

There is no information on the regulatory mechanisms and transcriptional factors controlling the expression of the roquefortine C/meleagrin biosynthetic genes. Furthermore, the subcellular localization of the biosynthetic enzymes and the possible involvement of vesicles or peroxisomes in the biosynthetic processes of these mycotoxins remain unexplored.

References

1. Tudzynsky P, Correia T, Keller U. Biotechnology and genetics of ergot alkaloids. Appl Microbiol Biotechnol. 2001;57:593–605.
2. Li SM. Evolution of aromatic prenyltransferases in the biosynthesis of indole derivatives. Phytochemistry. 2009;70:1746–57.
3. Sumarah MW, Miller JD, Blackwell BA. Isolation and metabolite production by *Penicillium roqueforti*, *P. paneum* and *P. crustosum* isolated in Canada. Mycopathologia. 2005;159: 571–7.
4. García-Estrada C, Ullán RV, Albillos SM, Fernández-Bodega MÁ, Durek P, von Döhren H, Martín JF. A single cluster of coregulated genes encodes the biosynthesis of the mycotoxins roquefortine C and meleagrin in *Penicillium chrysogenum*. Chem Biol. 2011;18:1499–512.
5. Overy DP, Frisvad JC, Steinmeier U, Thrane U. Clarification of the agents causing blue mold storage rot up on various flower and vegetable bulbs: implications for mycotoxin contamination. Postharvest Biol Technol. 2005;35:217–21.
6. Finoli C, Vecchio A, Galli A, Dragoni I. Roquefortine C occurrence in blue cheese. J Food Prot. 2001;64:246–51.
7. Fedorova ND, Khaldi N, Joardar VS, Maiti R, Amedeo P, Anderson MJ, Crabtree J, Silva JC, Badger JH, Albarraq A, Angiuoli S, Bussey H, Bowyer P, Cotty PJ, Dyer PS, Egan A, Galens K, Fraser-Liggett CM, Haas BJ, Inman JM, Kent R, Lemieux S, Malavazi I, Orvis J, Roemer T, Ronning CM, Sundaram JP, Sutton G, Turner G, Venter JC, White OR, Whitty BR, Youngman P, Wolfe KH, Goldman GH, Wortman JR, Jiang B, Denning DW, Nierman WC. Genomic islands in the pathogenic filamentous fungus *Aspergillus fumigatus*. PLoS Genet. 2008;4(4):e1000046.
8. Rydholm C, Szakacs G, Lutzoni F. Low genetic variation and no detectable population structure in *Aspergillus fumigatus* compared to closely related *Neosartorya* species. Eukaryot Cell. 2006;5:650–7.
9. Kato N, Suzuki H, Takagi H, Asami Y, Kakeya H, Uramoto M, Usui T, Takahashi S, Sugimoto Y, Osada H. Identification of cytochrome P450s required for fumitremorgin biosynthesis in *Aspergillus fumigatus*. Chembiochem. 2009;10:920–8.
10. Kato N, Suzuki H, Okumura H, Takahashi S, Osada H. A point mutation in *ftmD* blocks the fumitremorgin biosynthetic pathway in *Aspergillus fumigatus* strain Af293. Biosci Biotechnol Biochem. 2013;77:1061–7.
11. Rohlfs M, Albert M, Keller NP, Kempken F. Secondary chemicals protect mould from fungivory. Biol Lett. 2007;3:523–5.

12. Wagener RE, Davis ND, Diener UL. Penitrem A and roquefortine production by *Penicillium commune*. Appl Environ Microbiol. 1980;39:882–7.
13. Ohmomo S. Indole alkaloids produced by *Penicillium roqueforti*. J Antibact Antifung Agents. 1982;10:253–64.
14. Scott PM, Kennedy PC. Analysis of blue cheese for roquefortine and other alkaloids from *Penicillium roqueforti*. J Agric Food Chem. 1976;24:865–8.
15. Häggblom P. Isolation of roquefortine C from feed grain. Appl Environ Microbiol. 1990;56: 2924–6.
16. Ohmomo S, Kitamoto HK, Nakajima T. Detection of roquefortines in *Penicillium roqueforti* isolated from moulded maize silage. J Sci Food Agric. 1994;64:211–5.
17. Rundberget T, Skaar I, Flaoyen A. The presence of *Penicillium* and *Penicillium* mycotoxins in food wastes. Int J Food Microbiol. 2004;90:181–8.
18. Cole RJ, Dorner JW, Cox RH, Raymond LW. Two classes of alkaloid mycotoxins produced by *Penicillium crustosum* Thom isolated from contaminated beer. J Agric Food Chem. 1983;31: 655–7.
19. Möller T, Akerstrand K, Massoud T. Toxin-producing species of *Penicillium* and the development of mycotoxins in must and homemade wine. Nat Toxins. 1997;5:86–9.
20. García-Rico RO, Fierro F, Mauriz E, Gómez A, Fernández-Bodega MA, Martín JF. The heterotrimeric G alpha protein *pga1* regulates biosynthesis of penicillin, chrysogenin and roquefortine in *Penicillium chrysogenum*. Microbiology. 2008;154:3567–78.
21. Overy DP, Frisvad JC. New *Penicillium* species associated with bulbs and root vegetables. Syst Appl Microbiol. 2003;26:631–9.
22. Kulakovskaya TV, Reshetilova TA, Kuvichkina TN, Vinokurova NG. Roquefortine excretion and up take by *Penicillium crustosum* Thom VKM F-1746. Process Biochem. 1997;32:29–33.
23. Butler AR, Flint SA, Cundliffe E. Feedback control of polyketide metabolism during tylosin production. Microbiology. 2001;147:795–801.
24. Reshetilova TA, Kulakovskaya TV, Kuvichkina TN, Kozlovsky AG. Transport of alkaloid roquefortine into cells of the fungus *Penicillium farinosum*. Microbiology. 1994;63:230–2.
25. García-Estrada C, Vaca I, Lamas-Maceiras M, Martín JF. In vivo transport of the intermediates of the penicillin biosynthetic pathway in tailored strains of *Penicillium chrysogenum*. Appl Microbiol Biotechnol. 2007;76:169–82.
26. Overy DP, Nielsen KF, Smedsgaard J. Roquefortine/oxaline biosynthesis pathway metabolites in *Penicillium* ser. Corymbifera: in planta production and implications for competitive fitness. J Chem Ecol. 2005;31:2373–90.
27. Kopp-Holtwiesche B, Rehm HJ. Antimicrobial action of roquefortine. J Environ Pathol Toxicol Oncol. 1990;10:41–4.
28. Ohmomo S, Oguma K, Ohashi T, Abe M. Isolation of a new indole alkaloid, roquefortine D, from the cultures of *Penicillium roqueforti*. Agric Biol Chem. 1978;42:2387–9.
29. Kozlovsky AG, Vinokurova NG, Solovyeva TF, Buzilova IG. Microfungal nitrogen-containing secondary metabolites. Appl Biochem Microbiol. 1996;32:39Y48.
30. Kozlovsky AG, Solovieva TF, Reshetilova TA, Skryabin GK. Biosynthesis of roquefortine and 3,12-dihydroroquefortine by the culture *Penicillium farinosum*. Cell Mol Life Sci. 1981;37:472–3.
31. Kozlovsky AG, Vinokurova NG, Reshetilova TA, Sakharovsky VG, Baskunov BP, Seleznev SG. New metabolites of *Penicillium glandicola* var. glandicola: glandicoline A and glandicoline B. Appl Biochem Microbiol. 1994;30:334Y337.
32. Kawai K, Nozawa K, Nakajima S, Iitaka Y. Studies of fungal products. VII The structures of meleagrin and 9-O-p-bromobenzoylmeleagrin. Chem Pharm Bull. 1984;32:94Y98.
33. Nagel DW, Pachler KGR, Steyn PS. The chemistry and C NMR assignments of oxaline, a novel alkaloid from *Penicillium oxalicum*. Tetrahedron. 1976;32:2625Y2631.
34. Reshetilova TA, Vinokurova NG, Khmelenina VN, Kozlovsky AG. The role of roquefortine in the synthesis of alkaloids meleagrin, glandicolines A and B, and oxaline in fungi *Penicillium glandicola* and *P. atramentosum*. Microbiology. 1995;64:27–9.

35. Ali H, Ries MI, Nijland JG, Lankhorst PP, Hankemeier T, Bovenberg R, Vreeken RJ, Driesen AJM. A branched biosynthetic pathway is involved in production of roquefortine and related compounds in *Penicillium chrysogenum*. PloS One. 2013;8:e65328.
36. Barrow KD, Colley PW, Tribe DE. Biosynthesis of the neurotoxin alkaloid roquefortine. J Chem Soc Chem Commun. 1979;1979:225–6.
37. Gorst-Allman CP, Steyn PS, Vleggaar R. The biosynthesis of roquefortine: an investigation of acetate and mevalonate incorporation using field NMR spectroscopy. J Chem Soc Chem Commun. 1982;1982:652–3.
38. Kanoh K, Kohno S, Katada J, Takahashi J, Uno I. (-)-Phenylahistin arrests cells in mitosis by inhibiting tubulin polymerization. J Antibiot. 1999;52:134–41.
39. Williams RM, Stocking EM, Sanz-Cervera JF. Biosynthesis of prenylated alkaloids derived from tryptophan. Top Curr Chem. 2000;209:97–173. doi:10.1007/3-540-48146-X_3.
40. Bhat B, Harrison DM, Lamont HM. The biosynthesis of the tryptophan-derived mould metabolites roquefortine and aszonalenin. J Chem Soc Chem Commun. 1990;1990:1518–9.
41. Bhat B, Harrison DM, Lamont HM. The biosynthesis of the mold metabolites roquefortine and aszonalenin from L-[2,4,5,6,7-H-5(2)] tryptophan. Tetrahedron. 1993;49:10663–8.
42. Yin WB, Cheng J, Li SM. Stereospecific synthesis of aszonalenins by using two recombinant prenyltransferases. Org Biomol Chem. 2009;7:2202–7.
43. Von Döhren H. A survey of nonribosomal peptide synthetase (NRPS) genes in *Aspergillus nidulans*. Fungal Genet Biol. 2009;46:545–52.
44. Martín JF. α-Aminoadipyl-cysteinyl-valine synthetases in β-lactam producing organisms. From Abraham's discoveries to novel concepts of non-ribosomal peptide synthesis. J Antibiot. 2000;53:1008–21.
45. Wu X, García-Estrada C, Vaca I, Martín JF. Motifs in the C-terminal region of the *Penicillium chrysogenum* ACV synthetase are essential for valine epimerization and processivity of tripeptide formation. Biochimie. 2012;94:354–64.
46. Rausch C, Weber T, Kohlbacher O, Wohlleben W, Huson DH. Specificity prediction of adenylation domains in nonribosomal peptide synthetases (NRPS) using transductive support vector machines (TSVMs). Nucleic Acids Res. 2005;33:5799–808.
47. Schwecke T, Göttling K, Durek P, Dueñas I, Käufer NF, Zock-Emmenthal S, Staub E, Neuhof T, Dieckmann R, von Döhren H. Nonribosomal peptide synthesis in *Schizosaccharomyces pombe* and the architectures of ferrichrome-type siderophore synthetases in fungi. Chembiochem. 2006;7:612–22.
48. Arai K, Kimura K, Mushiroda T, Yamamoto Y. Structures of Fructigenines A and B, new alkaloids isolated from *Penicillium fructigenum* TAKEUCHI. Chem Pharm Bull. 1989;37:2937–9.
49. Metzger U, Schall C, Zocher G, Unsöld I, Stec E, Li SM, Heide L, Stehle T. The structure of dimethylallyl tryptophan synthase reveals a common architecture of aromatic prenyltransferases in fungi and bacteria. Proc Natl Acad Sci USA. 2009;106:14309–14.
50. Yin WB, Grundmann A, Cheng J, Li SM. Acetylaszonalenin biosynthesis in *Neosartorya fischeri*. Identification of the biosynthetic gene cluster by genomic mining and functional proof of the genes by biochemical investigation. J Biol Chem. 2009;284:100–9.
51. Grundmann A, Li SM. Overproduction, purification and characterization of FtmPT1, a brevianamide F prenyltransferase from *Aspergillus fumigatus*. Microbiology. 2005;151:2199–207.
52. Richard DJ, Schiavi B, Joullié MM. Synthetic studies of roquefortine C: synthesis of isoroquefortine C and a heterocycle. Proc Natl Acad Sci USA. 2004;101:11971–6.
53. Ries MI, Ali H, Lankhorst PP, Hankemeier T, Bovenberg RA, Driessen AJ, Vreeken RJ. Novel key metabolites reveal further branching of the roquefortine/meleagrin biosynthetic pathway. J Biol Chem. 2013;288:37289–95.
54. Callahan TM, Rose MS, Meade MJ, Ehrenshaft M, Upchurch RG. CFP, the putative cercosporin transporter of *Cercospora kikuchii*, is required for wild type cercosporin production, resistance, and virulence on soybean. Mol Plant Microbe Interact. 1999;12:901–10.
55. Martín JF, Casqueiro J, Liras P. Secretion systems for secondary metabolites: how producer cells send out messages of intercellular communication. Curr Opin Microbiol. 2005;8:282–93.

56. Martín JF, García-Estrada C, Ullan RV. Transport of substrates into peroxisomes. The paradigm of β-lactam biosynthetic intermediates. Biomol Concepts. 2013;4(2):197–203.
57. Kimura Y, Hamasaki T, Nakajima H, Isogai A. Structure of aszonalenin, a new metabolite of *Aspergillus zonatus*. Tetrahedron Lett. 1982;23:225–8.
58. Hayashi A, Fujioka S, Nukina M, Kawano T, Shimada A, Kimura Y. Fumiquinones A and B, nematicidal quinones produced by *Aspergillus fumigatus*. Biosci Biotechnol Biochem. 2007;71:1697–702.
59. Capon RJ, Skene C, Stewart M, Ford J, O_Hair RAJ, Williams L, Lacey E, Gill JH, Heiland K, Friedel T. Aspergillicins A-E: five novel depsipeptides from the marine-derived fungus *Aspergillus carneus*. Org Biomol Chem. 2003;1:1856–62.
60. Wakana D, Hosoe T, Itabashi T, Nozawa K, Okada K, de Campos Takaki GM, Yaguchi T, Fukushima K, Kawai KI. Isolation of isoterrein from *Neosartorya fischeri*. Mycotoxins. 2006;56:3–6.
61. Liras P, Martín JF. Gene clusters for β-lactam antibiotics and control of their expression: why have clusters been formed and where do they come from? Int Microbiol. 2006;9:9–19.
62. Díez B, Gutiérrez S, Barredo JL, van Solingen P, van der Voort LH, Martín JF. The cluster of penicillin biosynthetic genes. Identification and characterization of the pcbAB gene encoding the alpha-aminoadipyl-cysteinyl-valine synthetase and linkage to the pcbC and penDE genes. J Biol Chem. 1990;265:16358–65.
63. Kosalková K, Marcos AT, Fierro F, Hernando-Rico V, Gutiérrez S, Martín JF. A novel heptameric sequence (TTAGTAA) is the binding site for a protein required for high level expression of *pcb*AB, the first gene of the penicillin biosynthesis in *Penicillium chrysogenum*. J Biol Chem. 2000;275:2423–30.
64. Brakhage AA. Molecular regulation of beta-lactam biosynthesis in filamentous fungi. Microbiol Mol Biol Rev. 1998;62:547–85.
65. Martín JF. Molecular control of expression of penicillin biosynthesis genes in fungi: regulatory proteins interact with a bidirectional promoter region. J Bacteriol. 2004;182:2355–62.
66. Maiya S, Grundmann A, Li SM, Turner G. The fumitremorgin gene cluster of *Aspergillus fumigatus*: identification of a gene encoding brevianamide F synthetase. Chembiochem. 2006;7:1062–9.
67. Maiya S, Grundmann A, Li SM, Turner G. Improved tryprostatin B production by heterologous gene expression in *Aspergillus nidulans*. Fungal Genet Biol. 2009;46:436–40.
68. Takase S, Iwami M, Ando T, Okamoto M, Yoshida K, Horiai H, Kohsaka M, Aoki H, Imanaka H. Amauromine, a new vasodilator. Taxonomy, isolation and characterization. J Antibiot (Tokyo). 1984;37:1320–3.
69. Hodge RP, Harris CM, Harris TM. Verrucofortine, a major metabolite of *Penicillium verrucosum* var. *cyclopium*, the fungus that produces the mycotoxin verrucosidin. J Nat Prod. 1988;51: 66–73.

Chapter 7
Ochratoxin A and Related Mycotoxins

Massimo Reverberi, Anna Adele Fabbri, and Corrado Fanelli

Introduction

Some fungi of the genus *Aspergillus*, such as *A. niger*, *A. oryzae*, and *A. sojae*, are useful for industry by exploiting their metabolites or enzymes for food [1], while others synthesize hazardous secondary metabolites such as mycotoxins. Mycotoxins are products of fungal secondary metabolism, which now is considered an adaptation in response to different "environmental stressors" [2–5]. In fact, among the adaptation strategies of fungi, there is the capacity to modify not only their phenotypic characteristics (i.e., differentiating reproductive structures and spores from an undifferentiated vegetative mycelium) but also their "metabolic plasticity" (i.e., re-routing their enzymatic performance by activating alternative biochemical pathways—secondary metabolism) [6].

Ochratoxins are produced by some *Aspergillus* and *Penicillium* species found worldwide and known for their ability to adapt their metabolism to different environmental conditions [1]. Ochratoxin A (OTA) mainly is produced by *P. verrucosum* and *P. nordicum*, and by several *Aspergillus* species, notably *A. ochraceus*, *A. carbonarius*, and some isolates of *A. niger* [7]. Other *Aspergillus* species can also synthesize OTA: *A. westerdijkiae*, *A. wentii* [7], and *A. steynii* have recently been demonstrated to produce this toxin at considerable levels [8]. However, *A. ochraceus*, *P. verrucosum* and *Aspergillus* section Nigri (to which *A. carbonarius* belongs) can be considered the prevalent ochratoxigenic species. The mycotoxin synthesis and the production rate of the different ochratoxigenic species are influenced by the kind of substrate contaminated, by environmental conditions, and by the geographical regions considered [9]. The occurrence of OTA has been reported worldwide, from temperate to tropical climates. The first studies reported OTA produced by *P. verrucosum* and *A. ochraceus* mainly in

M. Reverberi, Ph.D. (✉) • A.A. Fabbri, Ph.D. • C. Fanelli, Ph.D.
Department of Environmental Biology, Università Sapienza, Rome, Italy
e-mail: massimo.reverberi@uniroma1.it

J.-F. Martín et al. (eds.), *Biosynthesis and Molecular Genetics of Fungal Secondary Metabolites*, Fungal Biology, DOI 10.1007/978-1-4939-1191-2_7,

cereals and derived products [10–12], but recently other species belonging to Black Aspergilli—i.e., *A. niger* and *A. carbonarius*—have been found to produce OTA in commodities such as wine, grapes, and raisins in different countries [9, 13].

Concerning the ochratoxin biosynthesis pathway, several steps have been clearly elucidated [14–17]. There is clear evidence of the involvement of a polyketide synthase in OTA production in *A. ochraceus* [18], *P. nordicum/verrucosum* [14], and *A. carbonarius* [19], while the involvement of a non-ribosomal peptide synthetase (*nrps*) has been unequivocally demonstrated in *A. carbonarius* only [20]. Several studies discuss the relation among the different forms of ochratoxins derived from the OTA pathway and OTA degradation [20–22]. However, the pathway scheme still is fragmentary and some genes are missing in the virtual overlap between the hypothesized biochemical pathway and the genes found in putative OTA clusters of both *Aspergillus* and *Penicillium* (Table 7.1) [21, 23–25]. A recent study of Gallo et al. [20] clarified the order of biosynthesis steps of the last part of the OTA pathway—i.e., the conversion of OTβ(beta) to OTA by passing through OTB—thereby "declassing" OTα(alpha) as degradation product, which thus now has to be considered out of the biosynthesis of OTA, at least in *A. carbonarius*.

Some genomic approaches such as cDNA-AFLP, SSH, and phage lambda genomic gene library have been performed and some models concerning the OTA pathway and its regulation have been proposed [14, 21, 23, 26]. However, the work performed hitherto did not result in an uncovering of all the genes necessary for the biosynthesis of this harmful toxin. Probably the recently published complete genome sequence of *A. carbonarius* in combination with RNA-seq or microarray driven experiments will soon provide more straight and clear indications on OTA biosynthesis in black Aspergilli.

To date, the huge work performed in previous studies allows us to suggest, at least in four fungi—namely *A. carbonarius*, *A. niger*, *A. ochraceus*, and *P. nordicum*—how OTA is synthesized (Table 7.1; Fig. 7.1) [20, 27] and which are the main factors driving its biosynthesis.

In this review, we report some recent peculiar aspects of OTA metabolism and some information about the ochratoxins of the group, notably OTB and OTC.

Environmental Factors Affecting OTA Production in *A. carbonarius*, *A. niger*, *A. ochraceus* and *Penicillium verrucosum*

Water Activity and Temperature Effect on Isolates Grown on Different Media

Some general considerations are necessary to introduce this multivariate subject. The knowledge on how environmental factors affect growth and mycotoxin biosynthesis of mycotoxigenic fungi is critical for the prevention of fungal contamination in the field and to hamper spoilage by fungi during feedstuff storage [28, 29].

Table 7.1 Putative genes for OTA biosynthesis in *A. carbonarius* (as proposed in Gallo et al. [20] and [27]) and its paralogs in *A. niger*, *A. ochraceus*, *P. nordicum*

Biosynthetic step	Biosynthetic step in Fig. 7.1	Enzymes	*A. carbonarius* (gene ID/Acc. No. NCBI/)	*A. niger* (gene ID/ Acc. No. NCBI/*A. niger* GBrowse)	*A. ochraceus* (gene ID/Acc. No. NCBI)	*P. nordicum* (gene ID/Acc. No. NCBI)
1 Acetyl CoA/malonyl CoA → mellein/7-methylmellein	1a	Polyketide synthase	*AcPKS*/AM944567	*PKS*/An15g07920	*PKS*/AY583208	*otapks*PN/AAP33839.2
7-methylmellein → 7-carboxymellein (OTβ[beta])	1b	Oxidase	ND/EST C099	ND/An15g07900	*p450-B03*/ DQ054596	ND
OTβ(beta) → OTB ethyl ester	2	Peptide synthase	*AcOTAnrps*/ND	ND/An15g07910	ND/AEN14499	*otanpsPN*/AAS98174
OTB ethyl ester → OTB	3	Esterase	ND	ND	ND	ND
OTB → OTA	4	Chloroperoxidase	ND	ND	ND	*otachl*PN/ND

ND not determined

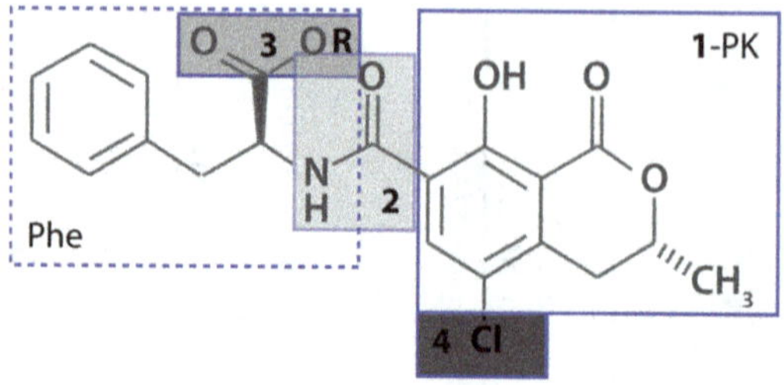

Fig. 7.1 Chemical structure of OTA (R=–OH) and OTC (R=ethyl ester). In OTB the chlorine present in OTA is substituted by a hydrogen. The isocoumarin nucleus of OTA is OTα(alpha) whereas its dechlorinated form is OTβ(beta) [20]. The *dotted box* represents the phenylalanine moiety deriving from the Shikimic acid pathway (Phe), the *white box* represents the isocoumarin nucleus (PK) which is synthesized through reactions 1a and 1b, the *light grey box* represents the peptide bond formed by non-ribosomal peptide synthases (2), the *middle grey box* highlights the site for de-esterification performed in reaction 3, whereas the *heavy grey box* highlights the chlorination step performed through reaction 4. The reaction numbering and the genes related to ochratoxin biosynthesis in *A. niger*, *A. carbonarius*, *A. ochraceus* and *P. nordicum* are indicated in Table 7.1

Temperature and water availability are key factors for the colonization of food commodities by fungi [30] and it is important to know the range in which water activity (a_w) and temperature affect the minimum and maximum level of fungal germination, growth, and toxin production. This is especially important for ochratoxigenic fungi; in fact, the ecological niches in which they develop are quite different (for example, between *A. carbonarius* and *P. verrucosum*, as detailed later). Further, it is worth mentioning that differences emerge about fungal growth and OTA biosynthesis in relation to the substrates from which the various strains were isolated, even if some species in some conditions do not show significant differences in growth and OTA profiles [31, 32]. Obviously genetic factors—i.e., natural genetic variation inside OTA-producing strains—drive the potential of OTA biosynthesis [1]. Moreover, environmental factors may drive secondary metabolism gene cluster expression in fungi at epigenetic level [33]. In addition, the studies on ochratoxigenic fungi are not easily comparable due to the different experimental conditions applied (i.e., synthetic media, in vivo on different substrates such as cereals, coffee, grapes, different storage conditions, preservatives, and so on). However, not always a direct correlation between fungal growth and OTA biosynthesis has been reported: Marin et al. showed that OTA was barely detected even when the growth of the four strains of *A. carbonarius* was maximum at 35 °C [32]. Aflatoxin production by *A. parasiticus* has also been reported to be inversely correlated to fungal growth [34]. In addition, the optimal conditions for fungal growth and mycotoxin production are not always the same [35]. Fungi may grow and proliferate on grains without producing mycotoxin and optimum OTA production can be observed when mycelial growth is not maximal. Water availability is probably the most important environmental factor affecting germination, growth and the fungal colonization of substrates [36].

In vitro studies reported an optimum growth temperature of 30–35 °C and a_w of 0.95–0.98 for *A. carbonarius* and an *A. niger* aggregate strain [1]. A study on *A. carbonarius* isolated from dried grapes with different a_w and temperatures showed that an inhibition of fungal growth occurred at 15 °C and a_w 0.82–0.90 [37]. *A. carbonarius* is considered the main species in grapes and wine responsible for OTA accumulation [12]. *A. niger*, as reported by Pitt and Hocking, grows at a minimum at 6–8 °C, at a maximum at 45–47 °C, and shows optimal growth at 35–37 °C [38]. Bellì et al. determined the temporal accumulation of OTA in vitro at different a_w (0.90, 0.93, 0.95, and 0.98) (simulating grape composition) by *A. carbonarius* and *A. niger* strains isolated from grapes [39]. They concluded that OTA production is favored at high a_w with a maximum reached after 5 days for *A. carbonarius* and from 7 to 13 days for *A. niger* when incubated at 20 °C. The authors suggested that these fungi may grow at a wide range of temperatures and water activities, with optimum conditions ranging from 25 to 35 °C and 0.95 to 0.99 a_w. Another study reported the effects of different fungal isolates (*Alternaria alternata*, *Cladosporium herbarum*, *Eurotium amstelodami*, *Trichoderma harzianum*, *Penicillium decumbens*, *P. janthinellum*, and *Candida* sp.) present on grapes on the capacity of *A. carbonarius* to synthesize OTA at different temperatures (20 and 30 °C) and a_w (0.92 and 0.97) in vitro when cultured in pairs [40]. At 0.97 a_w and 30 °C, OTA production was limited when *A. carbonarius* was grown in paired cultures. This result has been ascribed by the authors to (1) a limitation of *A. carbonarius* growth with a consequent decrease of OTA biosynthesis, (2) the utilization of nutrient sources by the different fungi being important for OTA biosynthesis, (3) OTA degradation by the other fungi as reported by Abrunhosa et al. [41]. At 0.97 a_w and 20 °C, the effect of the fungal interaction was not clear, whereas at 0.92 a_w OTA production by *A. carbonarius* was significantly inhibited. At 30 °C and 0.92 or 0.97 a_w reduced OTA production was found, which is also due to the interaction with the mycoflora present on dried vine fruit. Valero et al. suggested that conditions of 0.97 a_w and 20 °C, under which *A. carbonarius* shows good OTA production when not controlled by other fungi, have to be avoided to prevent OTA accumulation [40]. Other studies reported that conditions favorable for OTA production by different strains of *A. niger* aggregates were from 0.90 to 0.99 a_w depending on the strain and the culture medium assayed (Czapek yeast extract agar [CYA] and yeast extract sucrose [YES]). The a_w range for OTA production was narrower than that for growth [42]. Another study on *A. carbonarius* strains [43] reported the effect of a_w (0.78–0.99) and temperature (15 and 30 °C) on growth and production of ochratoxin A in synthetic media (CYA and YES) for 30 days. The results showed that *A. carbonarius* has the ability to grow and biosynthesize OTA in a wide range of water activities and also at high and low temperatures. Four isolates of *A. carbonarius* were studied as mixed inoculum on synthetic substrate (CYA) at 30 °C concerning growth rates and OTA biosynthesis [44]. No significant differences between growth rates of each single strain and the mixed inoculum were detected. The four strains differed in maximum OTA yield. When the inoculum was the mixture of the four *A. carbonarius* isolates, the fungi demonstrated a limiting water activity for OTA production at 0.87 and showed a xerotolerant behavior. In a study on the correlation between incubation time and different temperatures on the OTA-producing capacity and accumulation in four strains of *A. carbonarius* the following trend was reported: high temperature favors the earlier synthesis of OTA that then increases and in a few days becomes constant [32]. Garcia et al. assessed which environmental conditions (optimal: 0.98 a_w/25 °C; suboptimal: 0.90 a_w/25 °C and 0.98 a_w/37 °C) affected growth and OTA production of 30 *A. carbonarius* isolates [45]. By using ANOVA analysis they found that the coefficient variation (CV%) showed wider dispersion concerning growth and OTA synthesis at 0.98 a_w/37 °C, which can be considered "marginal" environmental conditions.

Equally interesting is the behavior of ochratoxigenic fungi in natural or semi-natural substrates. A lot of studies have been carried out by different authors to establish suitable environmental conditions for controlling ochratoxigenic fungal growth and OTA biosynthesis—some are reported in detail below. Pardo et al. reported *A. ochraceus*, *A. niger*, and *A. carbonarius* as being responsible of OTA accumulation in coffee with *A. ochraceus* presenting a high percentage (75–90 %) of isolates able to synthesize OTA [31]. These fungi have also been recognized as important contaminants and producers of OTA in grapes [39, 46, 47]. Different studies have been carried out on *A. niger* on corn grain, on *Aspergillus* section Nigri on peanuts, maize kernels, dried grapes, and coffee, and on *A. niger* aggregate on irradiated peanut seeds trying to dissect the suitable conditions for storing these foodstuffs [48–50]. These studies are interesting since those species are often recovered on substrates for foods and feeds in Argentina. Pardo et al. studied the effect on OTA synthesis in different strains of *A. ochraceus* on irradiated barley grains at 0.80–0.99 a_w and 10, 20, and 30 °C [51, 52]. The minimum a_w for OTA biosynthesis was 0.90 and the maximum level was 0.99. In addition, *A. ochraceus* produced lower levels of OTA on coffee-based media compared to wheat-based medium. Less OTA was produced by *P. verrucosum* and *P. nordicum* under the same culture conditions [53]. The authors further reported that growth was not indicative of OTA production, another case in which fungal growth and OTA synthesis are not correlated. When *A. carbonarius* and *A. niger* were grown on maize kernels at different a_w (0.92–0.98) and temperatures (5–45 °C) both fungi grew and produced OTA from 5 days of incubation. At 0.92 a_w colony diameter was significantly lower than at a_w 0.96 or 0.98. OTA was produced by *A. niger* in a temperature range of 15–45 °C and by *A. carbonarius* from 15 to 35 °C. The maximum OTA level was obtained from both fungi at 15 °C [54]. Medina et al. investigated the influence of carbon and nitrogen sources on OTA production by some ochratoxigenic strains from grapes (*A. niger*, *A. carbonarius*, *A. tubingensis*) [55]. Highest OTA levels were detected on arabinose (150 g/Lt) and phenylalanine (0.05 g/Lt) whereas there was no significant influence of the N source. The positive effect of phenylalanine on OTA biosynthesis was suggested to depend on a relation between the amino acid in the substrate and the phenylalanine in the structure of the OTA molecule.

P. verrucosum is the major producer of OTA in cereals (wheat and barley) in mainly temperate and cold climates [13], whereas *P. nordicum* was recovered from cheeses and products of dry-cured meat. Elmholt et al. reported the presence of *P. verrucosum* in more than 80 % of rye, barley, and wheat grains prior to storage and its presence can be indicative for OTA contamination [56]. *P. verrucosum* develops in cereals at temperatures below 30 °C and about 0.8 a_w, conditions typical of cool regions (Nord Europe). Schmidt-Heydt et al. in an in vivo study stored wheat remoistened and experimentally inoculated with *P. verrucosum* [57].

The grains were kept at different moisture content (24 %, 19 %, 14 %) and growth and OTA production were followed up to 6 months at ambient temperature. OTA was recovered, with a wave trend, only in the sample stored at 24 % moisture, whereas no fungal growth was detected at 14 and 19 %.

Different isolates of ochratoxigenic *P. verrucosum* were studied for growth at a_w 0.75–0.99 and 10–30 °C in seminatural medium (barley-meal extract agar). The authors reported 0.80–0.85 as the minimum a_w necessary for spore germination and fungal development and 0.90–0.99 a_w for maximum germination and mycelial growth [31]. Water activity and temperature significantly affected both mycelial growth and spore germination of OTA-producing *P. verrucosum* isolates. Moreover, different isolates may have slightly different responses to environmental factors. This information is important to assess predictive models toward OTA synthesis in barley that is correlated with fungal growth [31]. This study also revealed that a_w drives fungal behavior. In fact, conidial germination rates decreased when a_w was reduced, independently from temperature. These results are similar to those obtained with *Penicillium* spp. and *A. ochraceus* isolates on different cereals extract medium [52, 58]. In addition, an interesting correlation between the concentration of NaCl and OTA production has been shown in *P. nordicum* (isolated mainly from cheese and cured meat) and *P. verrucosum* (mainly isolated from cereals but also from brined olives) [59]. Both *P. nordicum* and *P. verrucosum* are able to produce OTA over a wide range of NaCl concentrations (5–100 g/L) and a mutant strain unable to synthesize OTA also showed a drastic reduction in growth when cultivated in the presence of a high concentration of NaCl. The authors proposed the biosynthesis of OTA as a metabolic mechanism of fungal adaptation to NaCl-rich foods, in fact OTA contains a chloride atom and the excretion of OTA by producer fungi can maintain a chloride homeostasis in the fungal cell. The same authors studied the physiological relationship among food preservatives (calcium propionate and potassium sorbate, 150, 300 ppm), environmental factors (temp. 25 °C; a_w 0.93, 0.95, 0.98), and *otapksPV* gene expression by *P. verrucosum* and demonstrated that OTA biosynthesis was activated mainly under optimal growth and weak stress conditions [57].

Finally, we want to report on "masked mycotoxins," which are currently studied in depth. "Masked mycotoxins" are derivatives of mycotoxins originating from plant metabolism that cannot be detected by conventional analytical instrumentations [60]. The first studies have been performed on cell cultures of maize and wheat using ^{14}C OTA, subsequently cell cultures of different plants were analyzed after the addition of OTA. The main metabolites recovered were ochratoxin α(alpha), which is considered nontoxic, 4-hydroxy-ochratoxin A, with immunosuppressant properties, and its β(beta)-glucosides. In general, the same derivatives have been recovered in other plant matrices [61]. Germinating cereals are also able to transform experimentally added OTA [62], although it is still not elucidated whether masked OTA is present also in naturally contaminated foods and whether its detection is related to the quantity of OTA produced by ochratoxigenic fungi, as seems to occur for some *Fusarium* toxins (DON) in wheat.[1]

[1] Reverberi et al., personal communication.

Light Effects on Growth and OTA Synthesis

Light is another environmental factor studied with the aim to propose some application to prevent or degrade OTA present in foodstuff. Recent studies by Schmidt-Heydt et al. provide hints on the role of light of different wavelengths on growth and OTA biosynthesis of different isolates (*A. carbonarius*, *A. niger*, *A. steynii*, *P. verrucosum*, and *P. nordicum*) [63–65]. It is reported that Penicillia are more affected by light than Aspergilli [63, 64]. OTA synthesis decreased by 20–30 % in Penicillia when incubated at 2800 lx, whereas at 1600 lx a lower decrease was observed and the fungi produced ochratoxin B in even higher amounts (five times) compared to growth in dark conditions (control). An unusual result was obtained when OTA had been added in the dark conditions to the substrate inoculated with Penicillia [63]. In this case, OTA was slightly toxic to the fungus and the toxicity was amplified by the exposure to light suggesting that *Penicillium* presented some adaptive strategy reducing OTA biosynthesis to control its toxicity [63]. In general, the extreme wavelengths of the spectrum, red (627 nm) and blue (470–455 nm), presented high inhibitory effects on growth and OTA biosynthesis. Nevertheless, *A. carbonarius* can produce high quantities of OTβ(beta) under blue light [64]. Not only the biosynthesis of OTA can be regulated by light but some wavelengths—notably white (470 nm) and blue (455 nm) light—are also able to contribute to OTA and OTB (and also citrinin) degradation in *Penicillium*. For application purposes it is interesting that OTA production was reduced by 50 % under blue light when wheat was experimentally contaminated with an ochratoxigenic strain of *P. verrucosum* [66]. The degradation of OTA and citrinin in the presence of light affects the physiology of the studied *Penicillium* strain. Under light conditions, the fungus shifts its metabolism towards other mycotoxin derivatives, probably to adapt its metabolism. The authors underline that these results have implications on the degradation of OTA in vivo.

pH Effect on OTA Biosynthesis

The pH of the substrates can affect OTA biosynthesis in different manners. O'Callaghan et al. showed that high quantities of OTA were produced by *A. ochraceus* in the lower pH range (pH 3.0–4.0) of the substrate, whereas OTA production was reduced at pH > 7.0 (up to 10.0) [23]. Transcript levels of *pks*, a gene related to OTA biosynthesis, paralleled the OTA production profile observed at the different pH values. However, the reason for the reduction in OTA biosynthesis at higher pH still is unclear. The authors suggested that the reduction of the transport of some nutrients important for OTA biosynthesis (i.e., glutamate) could affect the production of acetate molecules with consequent reduction in the synthesis of OTA. On the contrary, in *P. nordicum* OTA production and the expression of the *otapks*PN gene have been reported to be lower under acidic conditions (below pH 5.0) [67]. The highest expression of *otapksPN* gene was detected between pH 6–8. However, in general the expression of this gene followed the biosynthesis of OTA at different pH, that is, the genes related to OTA biosynthesis are up-regulated at an almost neutral pH and down-regulated at acidic pH at least in Penicillia. Different results have been obtained with six ochratoxigenic strains of *A. carbonarius* incubated at different temperatures (15 and 30 °C), in different culture media, at pH values ranging from 2.0 to 10.0. The results evidenced a high ability of this fungus to adapt its growth and OTA biosynthesis over a wide range of pH values at both the temperatures assayed [68].

Molecular Pathways Involved in Ochratoxin Synthesis

Ochratoxin Biosynthesis Pathway

The genome of *A. niger* CBS 513.88 has recently been sequenced and a putative ochratoxin cluster identified on the basis of a *pks* fragment found to be involved in ochratoxin biosynthesis in *A. ochraceus* [25]. Some *A. niger* strains have previously been reported to synthesize this toxin [69] but little is known about the biosynthetic pathway.

In general, OTA biosynthesis, even with interspecies/genus differences, is carried out through several steps by enzymes such as polyketide synthase, chloroperoxidase, reductase, esterase, dehydratase, and NRPS [16]. Of these, only a *pks* and an *nrps*-like gene were found in the putative OTA cluster of *A. niger* [25] (Table 7.2).

A suppression subtractive hybridization (SSH) approach has been used to isolate genes differentially expressed in a high and a low OTA-producing strain of *A. carbonarius*. Using the KEGG classification system [70] about 5 % of the differentially expressed genes could be identified as belonging to the secondary metabolism pathways [21]. In this gene set, two ESTs (C099 and C137) may encode genes related to

Table 7.2 Constituents of putative OTA cluster of *A. niger*

Gene ID	Protein function and similarity
An15g07860	Showing strong similarity to retinol dehydrogenase *Aspergillus kawachii* [IFO4308] (GAA83302) (e-value: $6e^{-141}$)
An15g07870	Representing isopropanol dehydrogenase XP_001397308.1
An15g07880	Representing radH flavin-dependent halogenase XP_001397309.2
An15g07890	Sharing similarity with the FBJ murine osteosarcoma viral oncogene homolog [*Xenopus (Silurana) tropicalis*] Sequence ID: ref|NP_001016200.1| (e-value: $3e^{-06}$)
An15g07900	Representing cytochrome P450 XP_001397311.1
An15g07910	Representing a NRPS with strong similarity to HC-toxin synthetase of *Metarhizium anisopliae* ARSEF 23 EFY97776.1 (e-value: 0.0)
An15g07920	Showing strong similarity to a PKS fragment of *A. ochraceus* involved in ochratoxin biosynthesis AAT92023.1 (e-value: 0.0)
An15g07930	Showing strong similarity to a nitric-oxide synthase of another strain of *A. niger* (ATCC 1015) EHA24963.1 (e-value: 0.0)
An16g00010	Showing similarity to a 6-phosphogluconolactonase of *Pedosphaera parvula* WP_007413006.1 (e-value: $7e^{-29}$)

the biosynthesis of mycotoxins. Notably, C099 showed homology to the gene *moxY* or *aflW*, which is a monooxygenase involved in the biosynthesis of aflatoxins [71], and C137 encodes a GAL4-like Zn_2Cys_6 binuclear cluster DNA-binding protein sharing similarity with AKtR-1, the main regulator of AK toxin biosynthesis in *Alternaria alternata* [72]. Other ESTs emerged by this study whose products are potentially involved in OTA synthesis are C086 and C058. The former is similar to a specific precursor of the acyl-CoA dehydrogenases (ACD), which are a family of mitochondrial enzymes oxidizing straight- or branched-chain acyl-CoAs in the metabolism of fatty acids or branched chain amino acids. Interestingly, part of the sequence of an acyl-CoA dehydrogenase-encoding gene, which resulted up-regulated in OTA biosynthesis stimulating conditions, has recently been used to develop a patented method for the detection of OTA-producing fungi [73]. C058 encodes a metal esteril oxidase involved in ergosterol biosynthesis, which is putatively correlated with OTA production [74]. Other molecular studies have focused on the identification of genes involved in OTA biosynthesis in *A. carbonarius*, with particular attention on the identification of the *pks* and *nrps* genes involved. In this respect, Atoui et al. [75] described the cloning of five different, highly diverse keto-synthase (KS) domain-encoding sequences of putative polyketide synthase genes in *A. carbonarius* and *A. ochraceus*. Recently, analysis of the *A. carbonarius* genome revealed about 13 *nrps* genes, one of which was identified as part of a putative OTA cluster based on homology to the predicted OTA biosynthetic cluster of *A. niger* [25]. In particular, *AcOTAnrps* encodes a protein characterized by one module with the typical three core domains and an additional adenylation domain similar to other NRPS involved in secondary metabolism. The product of this gene is needed for producing OTA in *A. carbonarius*: *nrps*-deleted mutant strains are unable to synthesize this toxin. Moreover, the absence of OTA, OTB, and OTα(alpha) and the concomitant increase of OTβ(beta) in the culture of Δ(Delta)Ac*OTAnrps* strains confirms that the bond between the phenylalanine and the polyketide dihydroisocoumarin, catalyzed by the synthetase, precedes the chlorination step, thereby clarifying the order of reactions in the ochratoxin biosynthetic pathway in *A. carbonarius* [20]. Similarly, it has been proposed that in *A. ochraceus* OTA is synthesized by a short gene cluster (<10-kb) composed of two putative P-450 monooxygenase-encoding genes together with the *pks* gene and a non-ribosomal peptide synthase-encoding gene [14, 23, 76]. *Pks* gene expression has been shown to correlate with OTA production in *A. ochraceus*, where in addition two putative P-450 monooxygenase-encoding genes being co-expressed with the *pks* gene are present. These also appear to be up-regulated during OTA production under different physiological conditions, indicating their possible role in its biosynthesis [23]. In *P. nordicum* and *P. verrucosum*, a putative OTA biosynthetic cluster has been identified containing the following biosynthetic genes:

- *otapks*PN encoding a PKS (AAP33839)
- *otanps*PN encoding a non-ribosomal peptide synthetase (NRPS) (AAS98174) putatively responsible for the formation of the peptide bond between the polyketide and the phenylalanine

- *otachl*PN encoding a protein with similarity to chlorinating enzymes, thought to be involved in the chlorination step
- *otatra*PN encoding a protein with high similarity to a predicted transporter protein (*Aspergillus clavatus* NRRL 1 putative MFS transporter XP_001273764.1; e-value: 1.4e^{-105}) hypothesized to be involved in OTA export
- *otaaspPN* (protein ID AAT65816.2), which codes for an alkaline serine proteinase [14, 24]

Expression of the genes contained in this cluster appears to be coordinated. Notably, peptide formation is performed by the product of the *otanps* gene, while the polyketide formation is catalyzed by the product of the *otapks* gene and the chlorination step by the activity of the chlorinating enzyme OTACHL. The *asp* and the *ntra* genes present at the boundaries of this short gene cluster are co-regulated with other genes related to OTA biosynthesis but their involvement in toxin synthesis is not yet known [65].

Molecular Factors Governing the Synthesis of Ochratoxins

A putative model for OTA biosynthesis regulation has been recently proposed by Botton et al. [26] exploiting cDNA-AFLP comparison in OTA high- and low-producing strains of *A. carbonarius*. Ochratoxin biosynthesis may be regulated by G protein signaling since many sequences found by cDNA-AFLP matched with elements involved in such signal transduction pathways, in the Ca^{2+}/calmodulin-dependent phosphorylation and in the dephoshorylation cascades.

In relation to this, some co-regulated genes encoding a calmodulin, a PI4 kinase, a protein involved in GTP metabolism, a PP2A phosphatase, a Gac1-like phosphatase, a Ser/Thr protein kinase, and a RgsA-like transcription factor have been identified by transcriptional studies carried out in *A. carbonarius* [21, 26]. The latter showed a high degree of similarity (70.2 %) with RgsA (regulator of G protein signaling A) of *A. nidulans* which is involved in the biosynthesis of sterigmatocystin and in asexual sporulation [77]. Differential expression of several MAPK-encoding genes was observed when comparing aflatoxin production at different temperatures [78]. The relevance of MAPK-mediated signal transduction during aflatoxin biosynthesis has previously been noted [79]. In addition, a cAMP-dependent protein is differentially regulated in high-OTA producing strains. cAMP is another important signaling molecule involved in aflatoxin production. At least ten genes examined so far from the aflatoxin cluster appear to have CRE1 (cAMP-response element) motifs in their promoters [80, 81]. From these studies a general scheme of the modulation of the OTA pathway similar to the one previously and clearly described in *Aspergillus* regarding sterigmatocystin (ST) and aflatoxin (AF) biosynthesis emerges [82, 83]. In *Aspergillus* species, the signal transduction proteins regulating ST/AF biosynthesis include: FlbA, an RGS (*R*egulator of *G* protein *S*ignaling) protein; FluG, an early-acting developmental regulator [84, 85]; FadA, the alpha subunit of a heterotrimeric G protein [86]; PkaA, the catalytic subunit of protein kinase A [82, 87].

Reverse-genetic approaches in *A. carbonarius* or other OTA-producing fungi shed some light on how OTA biosynthesis is actually driven by signal transduction pathways. Several of the EST found in the study of Botton et al. [26] and of Crespo-Sempere et al. [21] are related to the TOR complex. This complex works as a sensor for nutrients, energy, redox level, and controls protein synthesis in eukaryotic cells. The activity of this complex in mammals is stimulated by insulin, growth factors, serum, phosphatidic acid, amino acids (in particular leucine), and oxidative stress [88]. It has been recently proposed that inhibitors (resveratrol) of mammalian oxylipin biosynthesis, namely prostaglandins active in inflammatory responses, have Akt (mammalian target of rapamycin—mTOR) as cell target, leading to hypothesize a close link between these signals and the mTOR signaling pathway [89]. Fungal prostaglandin-like compounds, the Psi factors (i.e., fungal oxylipins), are active in filamentous fungi to drive development, virulence and secondary metabolism [90, 91]. In OTA-producing Aspergilli such as *A. ochraceus*, oxylipins are able to modulate these events too [92]. Notably, the disruption of an arachidonate 15-lipoxygenase-encoding gene (Ao*loxA*) in *A. ochraceus* results in a pleiotropic effect: the oxylipin-defective mutant displays altered development, does not trigger 9-oxylipins in contaminated wheat seeds and consequently the seeds do not support OTA biosynthesis [92]. In particular, Δ(Delta)Ao*loxA* strain delays conidia formation, shows copious sclerotia production, and hyphae distribution patterns that involve the whole seed's surface, i.e., the hyphal growth is not limited to the germ as it is in the wild-type strain. During host invasion, LOX activity of wheat seeds, a typical plant defense response, is stimulated after contact with *A. ochraceus* wild type and 9-HPODE formation is triggered. This reaction occurs in response to *A. ochraceus* contamination and is similar to that observed in the *A. nidulans*–maize interaction [93]. Thus, even in the OTA producer *A. ochraceus*, 9-HPODE was revealed as mycotoxin triggering factor suggesting a complex role for the oxylipin metabolism in the modulation of OTA biosynthesis in this fungus.

The interaction with the host and other environmental factors could thus generate signals that may modulate or have an influence on ochratoxin biosynthesis. CipC is a small protein with unknown function that was previously found to be up-regulated in an OTA high-producing strain of *A. carbonarius* [21]. CipC function is probably related to the perception of stress signals and modulation of cell responses such as antioxidant production. Nevertheless, CipC seems more related to a general stress response than displaying a specific role in OTA biosynthesis in *A. carbonarius* [94]. In relation to this, a close link between the onset of oxidative stress and OTA biosynthesis in *A. ochraceus* has been demonstrated [95] and it is likely that a similar stress perception mechanism and fate is ongoing in *A. carbonarius* too [96]. Over recent years, it has emerged that oxidative stress plays a pivotal role in controlling differentiation and secondary metabolism in fungi [95, 97–100]. The biosynthesis of different mycotoxins (such as aflatoxins, patulin, OTA, and some fusariotoxins) has a particular factor in common, namely, it can be affected by reactive oxygen species (ROS) [3, 95, 99, 101–103]. This straight link between the oxidative burst and OTA biosynthesis has been demonstrated by the inactivation of the paralog of

Saccharomyces cerevisiae Yap-1 in *A. ochraceus* (Ao*yap1*). Yap-1 is a cytosolic receptor/nuclear transcription factor able to promote gene expression targeted at restoring the oxidant/antioxidant balance in the cell [100, 104]. Accordingly, the Ao*yap1* gene contributes to controlling the redox balance in *A. ochraceus*. In fact, the Ao*yap1*-disrupted strain shows a higher quantity of unscavenged ROS, thus, the Δ(Delta)Ao*yap1* strain "experiences" a hyper-oxidant status over its entire lifecycle and produces a higher amount of OTA compared to the wild-type strain. The inability of the mutant to counteract oxidative processes is also demonstrated by the results obtained when treating the Δ(Delta)Ao*yap1* strain with the potent pro-oxidant CCl_4, which results in an inhibition of growth and OTA biosynthesis [105]. Thus, as emerged in other mycotoxigenic fungi [99, 102], ROS represent signals able to switch on/off secondary metabolism—namely OTA biosynthesis—in *A. ochraceus*, through a complex mechanism of cell signaling, which includes the oxidative stress-related transcription factor Ao*yap1* but probably also involves several additional regulatory factors. A recent study suggests that osmotic stress may modulate, through the HOG-like kinase cascade, the biosynthesis of OTA and that the chlorination step represents a way in which *P. nordicum* and *P. verrucosum* may adopt to ensure some kind of chloride homeostasis in the cell [65]. In fact, since chlorination and excretion seem to be at the very end of the ochratoxin A biosynthesis pathway, this step could represent a sort of self-protection mechanism. In relation to this, the synthesis of OTA on NaCl-rich substrates has the effect that chloride is permanently excreted out of the cell reducing its potential toxicity [65].

The expression of mycotoxin gene clusters is usually controlled by transcription factors such as the AflR and AflS proteins in the case of aflatoxin biosynthesis [83] or the genes *tri6* and *tri10* in the case of trichothecene biosynthesis [106], coding for zinc finger proteins. Several putative transcription factors were identified and their orthologs shown to be involved in the regulation of OTA biosynthesis and sexual/asexual sporulation. In *A. niger* also, a bZIP transcription factor (similar to the AK toxin regulator) has been found in the putative OTA cluster [25]. In *A. ochraceus* and in *Penicillium* these factors are still elusive. In *A. carbonarius*, Botton et al. [26] found that the expression of at least three different transcription factors is differentially regulated in an OTA producing strain compared to non-producing ones. Notably, a C_2H_2 zinc finger transcription factor, which is similar to transcription factors related to the regulation of sexual/asexual fungi development [107], was up-regulated in the OTA low-producing strain. Moreover, in parallel studies on OTA low- and high-producer strains of *A. carbonarius*, another two genes—namely, EST C073 and FD661669—encoding a C_2H_2 finger and a $Zn(II)_2Cys_6$ zinc finger domain transcription factor, respectively, were found to be differentially expressed.

Conclusion

It is more and more evident that some common factors driving the onset of mycotoxin synthesis are shared by different fungi. This is an important attainment in view of setting common preventive strategies to control different mycotoxins in foods and feeds at the same time. Among the various factors analyzed in this review and correlated with toxin production in fungi, oxidative stress is a crucial one. In fact, previous studies underline the use of external antioxidant to control the synthesis of different mycotoxins such as aflatoxins, OTA, *Fusarium* toxins and patulin [95, 102, 103, 108, 109]. It seems that the fungal adaptive behavior to extracellular stimuli and ROS could be considered an example of hormetic response. Hormesis is defined as an adaptive response of cells and organisms to slight and, in general, discontinuous stress. Hormesis is related to a biphasic dose response: stimulation of an adaptive response when a stimulus, present at a low dose, induces beneficial effects and toxic effects for the cell when present at high doses [110]. According to this metabolic hormetic response, a "waving" biphasic trend in adaptive responses to oxidative stress has been reported even in some toxigenic fungi [81, 99, 111].

The antioxidant response modulated at the molecular and physiological level following the perception of oxidative stress can be exploited not only to investigate the genetic potential of the different toxigenic fungi to respond with an hormetic strategy but also as a tool for controlling toxin biosynthesis by stimulating an intracellular reductive status.

References

1. Astoreca A, Magnoli CE, Dacero AM. Ecophysiology of *Aspergillus* section *Nigri* species potential ochratoxin A producers. Toxins. 2010;2:2593–605.
2. Calvo AM, Wilson RA, Bok JW, Keller NP. Relationship between secondary metabolism and fungal development. Microbiol Mol Biol Rev. 2002;66:447–59.
3. Reverberi M, Ricelli A, Zjalic S, Fabbri AA, Fanelli C. Natural functions of mycotoxins and control of their biosynthesis in fungi. Appl Microbiol Biotechnol. 2010;87:899–911.
4. Schmidt-Heydt M, Magan N, Geisen R. Stress induction of mycotoxin biosynthesis genes by abiotic factors. FEMS Microbiol Lett. 2008;284:142–9.
5. Hong SY, Roze LV, Wee J, Linz JE. Evidence that a transcription factor regulatory network coordinates oxidative stress response and secondary metabolism in aspergilli. Microbiologyopen. 2013;2(1):144–60.
6. Caddick MX. Perception and response: phenotypic plasticity in fungi. In: Jennings DH, editor. Stress tolerance of fungi. New York: Marcel Dekker; 1993. p. 13–44.
7. JEFCA. Ochratoxin A. Joint Food and Agriculture Organization/World Health Organization Expert Committee on Food Additives (JECFA 47, 2001). 2001. http://www.inchem.org/documents/jecfa/jecmono/v47je04.htm
8. Gil-Serna J, Vazquez C, Sardinas N, González-Jaén MT, Patiño B. Revision of ochratoxin a production capacity by the main species of *Aspergillus* section *Circumdati. Aspergillus steynii* revealed as the main risk of OTA contamination. Food Control. 2011;22:343–5.
9. Khalesi M, Khatib N. The effects of different ecophysiological factors on ochratoxin A production. Environ Toxicol Pharmacol. 2011;32:113–21.
10. Bau M, Bragulat MR, Abarca ML, Minguez S, Cabañes FJ. Ochratoxigenic species from Spanish wine grapes. Int J Food Microbiol. 2005;98:125–30.
11. Abarca ML, Accensi F, Bragulat MR, Bragi M, Castella G, Cabanes FJ. *Aspergillus carbonarius* as the main source of ochratoxin A contamination in dried vine fruits from the Spanish market. J Food Prot. 2003;66:504–6.
12. Cabañes FJ, Accensi F, Bragulat MR, Abarca ML, Castellá G, Minguez S. What is the source of ochratoxin A in wine? Int J Food Microbiol. 2002;79:213–5.

13. Cabañes FJ, Bragulat MR, Castellà G. Ochratoxin A producing species in the genus *Penicillium*. Toxins. 2010;2:1111–20.
14. Karolewiez A, Geisen R. Cloning a part of the ochratoxin A biosynthetic gene cluster of *Penicillium nordicum* and characterization of the ochratoxin polyketide synthase gene. Syst Appl Microbiol. 2005;28:588–95.
15. Harris JP, Mantle PG. Biosynthesis of ochratoxins by *Aspergillus ochraceus*. Phytochemistry. 2001;58:709–16.
16. Moss MO. Mode of formation of ochratoxin A. Food Addit Contam. 1996;13(Suppl):5–9.
17. Moss MO. Recent studies of mycotoxins. J Appl Microbiol. 1998;84(Suppl):62S–76.
18. O'Callaghan J, Caddick MX, Dobson ADW. A polyketide synthase gene required for ochratoxin A biosynthesis in *Aspergillus ochraceus*. Microbiology. 2003;149:3485–91.
19. Gallo A, Perrone G, Solfrizzo M, Epifani F, Abbas A, Dobson ADW. Characterisation of a *pks* gene which is expressed during ochratoxin A production by *Aspergillus carbonarius*. Int J Food Microbiol. 2009;129:8–15.
20. Gallo A, Bruno KS, Solfrizzo M, Perrone G, Mulè G, Visconti A, et al. New insight in the ochratoxin A biosynthetic pathway by deletion of an *nrps* 2 gene in *Aspergillus carbonarius*. Appl Environ Microbiol. 2012;78(23):8208–18.
21. Crespo-Sempere A, Gonzales-Candela L, Martinez-Culebras PV. Genes differentially expressed by *Aspergillus carbonarius* strains under ochratoxin A producing conditions. Int J Food Microbiol. 2010;142:170–9.
22. el Khoury A, Atoui A. Ochratoxin A: general overview and actual molecular status. Toxins. 2011;2:461–93.
23. O'Callaghan J, Stapleton PC, Dobson ADW. Ochratoxin A biosynthetic genes in *Aspergillus ochraceus* are differentially regulated by pH and nutritional stimuli. Fungal Genet Biol. 2006;43:213–21.
24. Geisen R, Schmidt-Heydt M, Karolewiez A. A gene cluster of the ochratoxin A biosynthetic genes in *Penicillium*. Mycotoxin Res. 2006;22:134–41.
25. Pel HJ, de Winde JH, Archer DB, Dyer PS, Hofmann G, Schaap PJ, et al. Genome sequencing and analysis of the versatile cell factory *Aspergillus niger* CBS 513.88. Nat Biotechnol. 2007;25(2):221–31.
26. Botton A, Ferrigo D, Scopel C, Causin R, Bonghi C, Ramina A. A cDNA-AFLP approach to study ochratoxin A production in *Aspergillus carbonarius*. Int J Food Microbiol. 2008;127:105–15.
27. Gallo A, Ferrara M, Perrone G. Phylogenetic study of polyketide synthases and nonribosomal peptide synthetases involved in the biosynthesis of mycotoxins. Toxins. 2013;5:717–42.
28. Garcia D, Ramos AJ, Sanchis V, Marin S. Predicting mycotoxins in foods: a review. Food Microbiol. 2009;26:757–69.
29. Garcia D, Ramos AJ, Sanchis V, Marin S. Modelling kinetics of aflatoxin production by *Aspergillus flavus* in maize-based medium and maize grain. Int J Food Microbiol. 2013;162:182–9.
30. Wilson DM, Abramson D. Mycotoxins. In: Sauer DB, editor. Storage of cereal grains and their products. Minnesota: Am Ass Cereal Chemist; 1992. p. 341–91.
31. Pardo E, Malet M, Marın S, Sanchis V, Ramos AJ. Effects of water activity and temperature on germination and growth profiles of ochratoxigenic *Penicillium verrucosum* isolates on barley meal extract agar. Int J Food Microbiol. 2006;106:25–31.
32. Marin S, Bellì N, Lasram S, Chebil S, Ramos AJ, Ghorbel A, et al. Kinetics of ochratoxin A production and accumulation by *Aspergillus carbonarius* on synthetic grape medium at different temperature levels. J Food Sci. 2006;71:196–9.
33. Strauss J, Reyes-Domiguez Y. Regulation of secondary metabolism by chromatin structure and epigenetic codes. Fungal Genet Biol. 2011;48:62–9.
34. Fabbri AA, Fanelli C, Panfili G, Passi S, Fasella P. Lipoperoxidation and aflatoxin biosynthesis by *Aspergillus parasiticus* and *Aspergillus flavus*. J Gen Microbiol. 1983;129:3447–52.
35. Sanchis V, Magan N. Environmental conditions affecting mycotoxins. In: Magan N, Olsen M, editors. Mycotoxins in food. Detection and control. Cambridge: Woodhead Publishing Limited; 2004. p. 72.

36. Magan N, Lacey J. Ecological determinants of mould growth in stored grain. Int J Food Microbiol. 1988;7:245–56.
37. Romero SM, Patriarca A, Fernandez Pinto V, Vaamonde G. Effect of water activity and temperature on growth of ochratoxigenic strains of *Aspergillus carbonarius* isolated from Argentinean dried vine fruits. Int J Food Microbiol. 2007;115:140–3.
38. Pitt JI, Hocking AD. Fungi and food spoilage. In: Sweigert BS, Stewart GF, editors. Food Science and Technology. Sydney: Academic. 1985; p. 413.
39. Bellì N, Marin S, Sanchis V, Ramos AJ. Influence of water activity and temperature on growth of isolates of *Aspergillus* section Nigri obtained from grapes. Int J Food Microbiol. 2004;96:19–27.
40. Valero A, Farrè JR, Sanchis V, Ramos AJ, Marín S. Effects of fungal interaction on ochratoxin A production by *Aspergillus carbonarius* at different temperature and water activity. Int J Food Microbiol. 2006;110:160–4.
41. Abrunhosa L, Serra R, Venâncio A. Biodegradation of ochratoxin by fungi isolated from grapes. J Agric Food Chem. 2002;50:7493–6.
42. Esteban A, Abarca ML, Bragulat MR, Cabañes FJ. Effect of water activity on ochratoxin A production by *Aspergillus niger* aggregate species. Int J Food Microbiol. 2006;108: 188–95.
43. Esteban A, Abarca ML, Bragulat MR, Cabañes FJ. Study on the effect of the effect water activity and temperature on ochratoxin A production by *Aspergillus carbonarius*. Food Microbiol. 2006;23:634–40.
44. Romero SM, Pinto VF, Patriarca A, Vaamonde G. Ochratoxin A production by a mixed inoculum of *Aspergillus carbonarius* at different conditions of water activity and temperature. Int J Food Microbiol. 2010;140:277–81.
45. Garcia D, Ramos AJ, Sanchis V, Marin S. Is intraspecific variability of growth and mycotoxin production dependent on environmental conditions? A study with *Aspergillus carbonarius* isolates. Int J Food Microbiol. 2011;144:432–9.
46. Battilani P, Logrieco A, Giorni P, Cozzi G, Bertuzzi T, Pietri A. Ochratoxin A production by *Aspergillus carbonarius* on some grape varieties grown in Italy. J Sci Food Agric. 2004;84:1736–40.
47. Da Rocha Rosa CA, Palacios V, Combinas M, Fraga ME, De Oliveira RA, Magnoli CE, et al. Potential ochratoxin A producers from wine grapes in Argentina and Brazil. Food Addit Contam. 2002;19:408–14.
48. Astoreca A, Magnoli C, Barberis C, Combina M, Dalcero A. Ochratoxin A production in relation to ecophysiological factors by *Aspergillus* section *Nigri* strains isolated from different substrates in Argentina. Sci Total Environ. 2007;388:16–23.
49. Astoreca A, Barberis C, Magnoli C, Combina M, Chiacchiera SM, Dalcero A. Ecophysiological factors effect on growth rate, lag phase and ochratoxin A production by *Aspergillus niger* aggregate strains on irradiated peanut seeds. Int J Food Microbiol. 2009;129:131–5.
50. Astoreca A, Barberis C, Magnoli C, Combina M, Dalcero A. Influence of ecophysiological factors effect on growth rate, lag phase and ochratoxin A production by *Aspergillus niger* aggregate strains in irradiated corn grains. Int J Food Microbiol. 2009;129:174–9.
51. Pardo E, Marın S, Solsona A, Sanchis V, Ramos AJ. Modelling of germination and growth of ochratoxigenic isolates of *Aspergillus ochraceus* as affected by water activity and temperature on a barley-based medium. Food Microbiol. 2004;21:267–74.
52. Pardo E, Marin S, Sanchis V, Ramos AJ. Prediction of fungal growth and ochratoxin A production by *Aspergillus ochraceus* on irradiated barley grains as influenced by temperature and water activity. Int J Food Microbiol. 2004;95:79–88.
53. Munoz K, Vega M, Rios G, Geisen R, Degen GH. Mycotoxin production by different ochratoxigenic *Aspergillus* and *Penicillium* species on coffee and wheat-based media. Mycotoxin Res. 2011;27:239–47.
54. Alborch L, Bragulat MR, Abarca ML, Cabañes FJ. Effects of water activity, temperature and incubation time on growth and ochratoxin production by *Aspergillus niger* and *Aspergillus carbonarius* on maize kernels. Int J Food Microbiol. 2011;147:53–7.

55. Medina A, Mateo EM, Valle-Algarra FM, Mateo F, Mateo R, Jiménez M. Influence of nitrogen and carbon sources on the production of ochratoxin A by ochratoxigenic strains of *Aspergillus* spp. isolated from grapes. Int J Food Microbiol. 2008;122:93–9.
56. Elmholt S, Rasmussen PH. *Penicillium verrucosum* occurrence and ochratoxin A contents in organically cultivated grain with special reference to ancient wheat types and drying practices. Mycopathology. 2005;159:421–32.
57. Schmidt-Heydt M, Baxter E, Geisen R, Geisen R, Magan N. Physiological relationship between food preservatives, environmental factors, ochratoxin and *otapks*PV gene expression by *Penicillium verrucosum*. Int Food Microbiol. 2007;119:277–83.
58. Marin S, Sanchis V, Saenz R, Ramos AJ, Vinas I, Magan N. Ecological determinants for germination and growth of some *Aspergillus* and *Penicillium* spp. from maize grain. J Appl Microbiol. 1998;84:25–36.
59. Schmidt-Heydt M, Graf E, Stoll D, Geisen R. The biosynthesis of ochratoxin A by *Penicillium* as one mechanism for adaptation to NaCl rich foods. Food Microbiol. 2012;29:233–41.
60. Tran ST, Smith TK. Determination of optimal conditions for hydrolysis of conjugated deoxynivalenol in corn and wheat with trifluoromethanesulfonic acid. Anim Feed Sci Technol. 2011;163:84–92.
61. Berthiller F, Crews C, Dall'Asta C, Saeger SD, Haesaert G, Karlovsky P, et al. Masked mycotoxins: a review. Mol Nutr Food Res. 2013;57(1):165–86.
62. Ruhland M, Engelhardt G, Wallnoefer PR. Transformation of the mycotoxin ochratoxin A in artificially contaminated vegetables and cereals. Mycotoxin Res. 1997;13:54–60.
63. Schmidt-Heydt M, Bode H, Raupp F, Raupp F, Geisen R. Influence of light on ochratoxin biosynthessis by *Penicillium*. Mycotoxin Res. 2010;26:1–8.
64. Schmidt-Heydt M, Rufer C, Raupp F, Bruchmann A, Perrone G, Geisen R. Influence of light on food relevant fungi with emphasis on ochratoxin producing species. Int J Food Microbiol. 2011;145:229–37.
65. Schmidt Heydt M, Graf E, Batzler J, Geisen R. The application of transcriptomics to understand the ecological reasons of ochratoxin a biosynthesis by *Penicillium nordicum* on sodium chloride rich dry cured foods. Trends Food Sci Technol. 2011;22:S39–48.
66. Schmidt-Heydt M, Cramer B, Graf I, Geisen R. Wavelength-dependent degradation of ochratoxin and citrinin by light in vitro and in vivo and its implications on *Penicillium*. Toxins. 2012;4:1535–51.
67. Geisen R. Molecular monitoring of environmental conditions influencing the induction of ochratoxin A biosynthesis genes in *Penicillium nordicum*. Mol Nutr Food Res. 2004;48:532–40.
68. Esteban A, Abarca ML, Bragulat MR, Cabanes FJ. Influence of pH and incubation time on ochratoxin A production by *Aspergillus carbonarius* in culture media. J Food Prot. 2005;68:1435–40.
69. Storari M, von Rohr R, Pertot I, Gessler C, Broggini GAL. Identification of ochratoxin A producing *Aspergillus carbonarius* and *A. niger* clade isolated from grapes using the loop-mediated isothermal amplification (LAMP) reaction. J Appl Microbiol. 2013;114:1193–200.
70. KEGG classification system. www.genome.ad.jp/kegg
71. Yu J, Chang PK, Ehrlich KC, Cary JW, Bhatnagar D, Cleveland TE, et al. Clustered pathway genes in aflatoxin biosynthesis. Appl Environ Microbiol. 2004;70:1253–62.
72. Tanaka A, Tsuge T. Structural and functional complexity of the genomic region controlling AK-toxin biosynthesis and pathogenicity in the Japanese pear pathotype of *Alternaria alternata*. Mol Plant Microbe Interact. 2000;13:975–86.
73. Dobson A, O'Callaghan J, inventors; University College Cork, National University of Ireland Cork, assignee. Detection of ochratoxin A producing fungi. US Patent US 7560234. 2009.
74. Saxena J, Munimbazi C, Bullerman LB. Relationship of mould count, ergosterol and ochratoxin A production. Int J Food Microbiol. 2001;71:29–34.
75. Atoui A, Dao HP, Mathieu F, Lebrihi A. Amplification and diversity analysis of ketosynthase domains of putative polyketide synthase genes in *Aspergillus ochraceus* and *Aspergillus carbonarius* producers of ochratoxin A. Mol Nutr Food Res. 2006;50:488–93.

76. Rodríguez A, Rodríguez M, Luque MI, Justesen AF, Córdoba JJ. Quantification of ochratoxin A-producing molds in food products by SYBR Green and TaqMan real-time PCR methods. Int J Food Microbiol. 2011;149:226–35.
77. Han KH, Seo JA, Yu JH. Regulators of G-protein signaling in *Aspergillus nidulans*: RgsA down-regulates stress response and stimulates asexual sporulation through attenuation of GanB (Gα) signaling. Mol Microbiol. 2004;53:529–40.
78. O'Brian GR, Georgianna DR, Wilkinson JR, Yu J, Abbas HK, Bhatnagar D, et al. The effect of elevated temperature on gene transcription and aflatoxin biosynthesis. Mycologia. 2007;99:232–9.
79. Juvvadi PR, Chivukula S. Putative calmodulin-binding domains in aflatoxin biosynthesis-regulatory proteins. Curr Microbiol. 2006;52:493–6.
80. Roze LV, Miller MJ, Rarick M, Mahanti N, Linz JE. A novel cAMP-response element, CRE1, modulates expression of nor-1 in *Aspergillus parasiticus*. J Biol Chem. 2004;26:27428–39.
81. Roze LV, Chanda A, Wee J, Awad D, Linz JE. Stress-related transcription factor Atfb integrates secondary metabolism with oxidative stress response in Aspergilli. J Biol Chem. 2011;286(40):35137–48.
82. Shimizu K, Keller NP. Genetic involvement of a cAMP-dependent protein kinase in a G protein signaling pathway regulating morphological and chemical transitions in *Aspergillus nidulans*. Genetics. 2001;157:591–600.
83. Georgianna DR, Payne GA. Genetic regulation of aflatoxin biosynthesis: from gene to genome. Fungal Genet Biol. 2009;46:113–25.
84. Dohlman HG, Thorner J. RGS proteins and signaling by heterotrimeric G proteins. J Biol Chem. 1997;272:3871–4.
85. Hicks JK, Yu JH, Keller NP, Adams TH. *Aspergillus* sporulation and mycotoxin production both require inactivation of the FadA G alpha protein dependent signaling pathway. EMBO J. 1997;16:4916–23.
86. Yu JH, Wieser J, Adams TH. The *Aspergillus* FlbA RGS domain protein antagonizes G protein signaling to block proliferation and allow development. EMBO J. 1996;15:5184–90.
87. Shimizu K, Hicks JK, Huang TP, Keller NP. Pka, Ras and RGS protein interactions regulate activity of AflR, a Zn(II)2Cys6 transcription factor in *Aspergillus nidulans*. Genetics. 2003;165:1095–104.
88. Foster KG, Fingar DC. Mammalian target of rapamycin (mTOR): conducting the cellular signaling symphony. J Biol Chem. 2010;285(19):14071–7.
89. Zhong LM, Zong Y, Sun L, Guo JZ, Zhang W, He Y, et al. Resveratrol inhibits inflammatory responses via the mammalian target of rapamycin signaling pathway in cultured LPS-stimulated microglial cells. PLoS One. 2012;7(2):e32195.
90. Tsitsigiannis DI, Keller NP. Oxylipins as developmental and host–fungal communication signals. Trends Microbiol. 2007;15:109–18.
91. Christensen SA, Kolomiets MV. The lipid language of plant-fungal interactions. Fungal Genet Biol. 2010;48(1):4–14.
92. Reverberi M, Punelli F, Scarpari M, Camera E, Zjalic S, Ricelli A, et al. Lipoperoxidation affects ochratoxin A biosynthesis in *Aspergillus ochraceus* and its interaction with wheat seeds. Appl Microbiol Biotechnol. 2010;85:1935–46.
93. Brodhagen M, Tsitsigiannis DI, Hornung E, Goebel C, Feussner I, Keller NP. Reciprocal oxylipin-mediated cross-talk in the *Aspergillus*-seed pathosystem. Mol Microbiol. 2008; 67(2):378–91.
94. Crespo-Sempere A, Selma-Lázaro C, Martínez-Culebras PV, González-Candelas L. Characterization and disruption of the *cipC* gene in the ochratoxigenic fungus *Aspergillus carbonarius*. Food Res Int. 2013;54(1):697–705.
95. Reverberi M, Gazzetti K, Punelli F, Scarpari M, Zjalic S, Ricelli A, et al. Ao*yap1* regulates OTA synthesis by controlling cell redox balance in *Aspergillus ochraceus*. Appl Microbiol Biotechnol. 2012;95(5):1293–304.
96. De Rossi P, Ricelli A, Reverberi M, Bello C, Fabbri AA, Fanelli C, et al. Grape variety related trans-resveratrol induction affects *Aspergillus carbonarius* growth and ochratoxin A biosynthesis. Int J Food Microbiol. 2012;156:127–32.

97. Aguirre J, Rios-Momberg M, Hewitt D, Hansberg W. Reactive oxygen species and development in microbial eukaryotes. Trends Microbiol. 2005;13:111–8.
98. Jayashree T, Subramanyam C. Oxidative stress as a prerequisite for aflatoxin production by *Aspergillus parasiticus*. Free Radic Biol Med. 2000;29:981–5.
99. Reverberi M, Zjalic S, Ricelli A, Punelli F, Camera E, Fabbri C, et al. Modulation of antioxidant defence in *Aspergillus parasiticus* is involved in aflatoxin biosynthesis: a role for the *ApyapA* gene. Eukariot Cell. 2008;7(6):988–1000.
100. Reverberi M, Fabbri AA, Fanelli C. Oxidative stress and oxylipins in plant-fungus interaction. In: Witzany G, editor. Biocommunication of Fungi. Dordrecht: Springer; 2012.
101. Narasaiah KV, Sashidar RB, Subramanyam C. Biochemical analysis of oxidative stress in the production of aflatoxin and its precursor intermediates. Mycopathology. 2006;162:179–89.
102. Ponts N, Pinson-Gadais L, Verdal-Bonnin MN, Barreau C, Richard-Forget F. Accumulation of deoxynivalenol and its 15-acetylated form is significantly modulated by oxidative stress in liquid cultures of *Fusarium graminearum*. FEMS Microbiol Lett. 2006;258:102–7.
103. Tolaini V, Zjalic S, Reverberi M, Fanelli C, Fabbri AA, Del Fiore A, et al. *Lentinula edodes* enhances the biocontrol activity of *Crypyptococcus laurentii* against *Penicillium expansum* contamination and patulin production in apple fruits. Int J Food Microbiol. 2010;138(3): 243–9.
104. Estruch F. Stress-controlled transcription factors, stress-induced genes and stress tolerance in budding yeast. FEMS Microbiol Rev. 2000;24:469–86.
105. Halliwell B, Gutteridge JMC, editors. Free radicals in biology and medicine. Oxford: University Press; 2007.
106. Peplow AW, Meek IB, Wiles MC, Phillips TD, Beremand MN. Tri16 is required for esterification of position C-8 during trichothecene mycotoxin production by *Fusarium sporotrichioides*. Appl Environ Microbiol. 2003;69:5935–40.
107. Coppin E. The FLE1 gene encoding a C2H2 zinc finger protein co-ordinates male and female sexual differentiation *in Podospora anserina*. Mol Microbiol. 2002;43:1255–68.
108. Fanelli C, Fabbri AA, Pieretti S, Finotti E, Passi S. Effect of different antioxidants and free radical scavengers on aflatoxin production. Mycol Res. 1985;1:65–9.
109. Reverberi M, Fabbri AA, Zjalic S, Ricelli A, Punelli F, Fanelli C. Antioxidant enzymes stimulation in *Aspergillus parasiticus* by *Lentinula edodes* inhibits aflatoxin production. Appl Microbiol Biotechnol. 2005;69:207–15.
110. Mattson MP. Hormesis defined. Ageing Res Rev. 2008;7:1–7.
111. Aguirre J, Hansberg W, Navarro R. Fungal responses to reactive oxygen species. Med Mycol. 2006;44(Suppl):101–7.

Chapter 8
Carotenoids

Javier Ávalos, Violeta Díaz-Sánchez, Jorge García-Martínez, Marta Castrillo, Macarena Ruger-Herreros, and M. Carmen Limón

Introduction

Carotenoids are a family of terpenoid pigments synthesized by all photosynthetic species, and by many fungi and non-photosynthetic bacteria [1–5]. In photosynthetic organisms the carotenoids provide bright yellow, orange, or reddish colors to many fruits and flowers, and fulfill essential roles in light harvesting and photoprotection of the photosynthetic machinery [6]. In addition, carotenoids are responsible for the bright colors of some animals, and play an essential role in vertebrates as a source of retinoids, such as the vision pigment retinal and the morphogen retinoic acid [7]. With some notable exception, explained by horizontal gene transfer from fungi [8], animals are unable to synthesize carotenoids and they obtain these pigments through dietary intake. Some carotenoids are also used in cosmetics and food industry, where fungi and algae are employed as biotechnological carotenoid producers [9, 10].

Biochemically, carotenoids are fat-soluble compounds consisting of an aliphatic polyene chain usually composed of eight isoprene units. Terpenoids are produced through two different biochemical pathways. In the first pathway, for many years the only one known by the biochemists, these compounds are produced from mevalonate, which derives from hydroxymethylglutaryl coenzyme A (HMG-CoA) (Fig. 8.1); later, an alternative pathway was discovered involving the condensation of hydroxyethyl-thiamin and glyceraldehyde 3-phosphate, via D-1-deoxyxylulose 1-phosphate [11]. In the cases investigated, plant and bacterial carotenoids derive from these compounds, while fungal carotenoids derive from HMG-CoA.

J. Ávalos, Ph.D. (✉) • V. Díaz-Sánchez, Ph.D. • J. García-Martínez • M. Castrillo
M. Ruger-Herreros, Pharm.D. • M.C. Limón, Ph.D.
Department of Genetics, Faculty of Biology, University of Seville, Sevilla, Spain
e-mail: avalos@us.es

J.-F. Martín et al. (eds.), *Biosynthesis and Molecular Genetics of Fungal Secondary Metabolites*, Fungal Biology, DOI 10.1007/978-1-4939-1191-2_8,

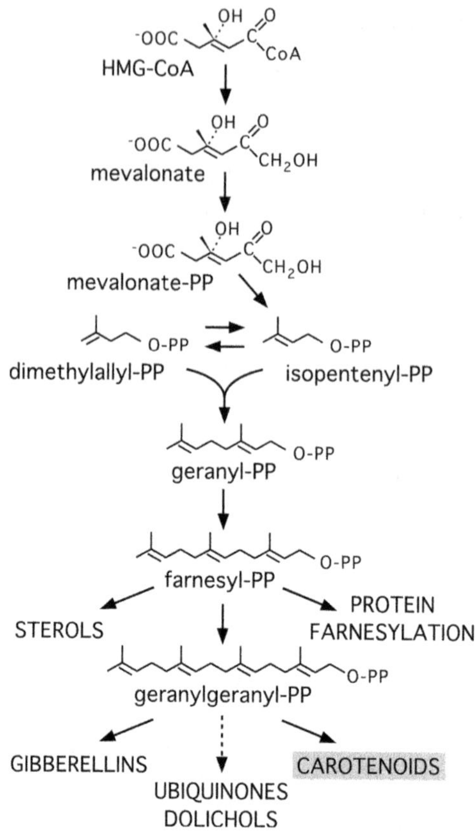

Fig. 8.1 First steps of the mevalonate terpenoid pathway. Some major terpenoid destinations are indicated in *capital letters*

Carotenoid biosynthesis proceeds through several condensation steps to yield geranyl pyrophosphate, farnesyl pyrophosphate (FPP, precursor of the sterols) and geranylgeranyl pyrophosphate (GGPP), precursor of carotenoids and other terpenoids (Fig. 8.1). The first molecule with a carotenoid structure is phytoene, produced by phytoene synthase through the condensation of two GGPP units (Fig. 8.2). Phytoene is not pigmented, but the subsequent introduction of conjugated double bonds by one or two dehydrogenases produces derivatives that absorb light of progressively longer wavelengths, providing the characteristic yellow, orange, or reddish colors of the carotenoids. The usual carotenoid structures are obtained by the introduction of α or β-ionone rings at either end of the molecules. Further chemical modifications, such as introduction of oxygenated radicals at the rings or oxidative cleavage reactions, may eventually be introduced and explain the large variety of the more than 750 carotenoids found in nature [2]. Oxygen-containing carotenoids are known as xanthophylls.

Carotenoid production is frequent in fungi, but it is not a universal trait. Some well-known fungal models, as *Aspergillus nidulans* and *Saccharomyces cerevisiae*, lack carotenoid pathways, but many other fungi produce carotenoids. In the following

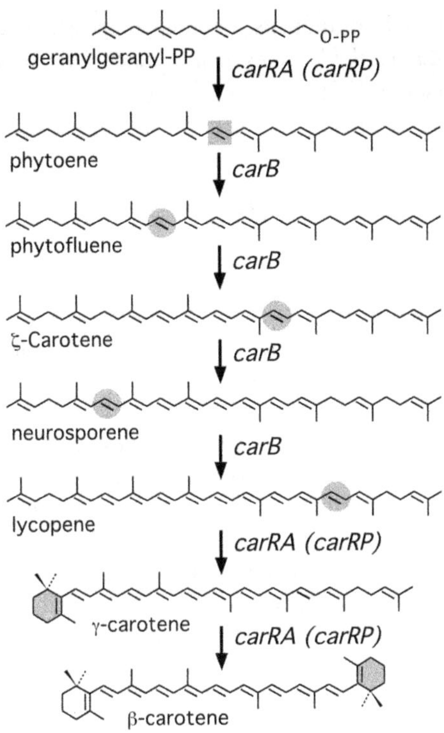

Fig. 8.2 β-carotene biosynthetic pathway. Genes responsible for each reaction in Mucorales are indicated close to the *arrows*. Reaction sites are shaded in the carotene product of each enzymatic step

sections we summarize the current knowledge on carotenoid production and its genetic basis in fungi, with special attention to those producing β-carotene and the xanthophyll neurosporaxanthin because of the considerable information available on their biochemistry and molecular biology.

Biosynthesis of β-carotene

β-carotene is one of the most widespread carotenes in nature. It is the major end product of the carotenoid pathway of different fungi, including several thoroughly investigated zygomycetes from the order Mucorales, such as *Phycomyces blakesleeanus* [12], *Blakeslea trispora* [9], and *Mucor circinelloides* [13, 14]. Moreover, β-carotene is an intermediary molecule in the synthesis of other carotenoids in many fungal species. In addition to these Mucorales, β-carotene production has also been described in fungi from other taxonomical groups, including basidiomycetes as *Rhodosporidium* sp. [15], *Sclerotium rolfsii* [16], *Sclerotinia sclerotiorum* [17], *Sporidiobolus pararoseus* [18], *Ustilago maydis* [19], and other *Ustilago* species [20, 21], ascomycetes as *Aspergillus giganteus* [22], *Cercospora nicotianae* [23], and *Penicillium* sp. [24], or the "imperfect fungus" *Aschersonia aleyroides* [25]. β-carotene production is also found in mutants of fungal species usually producing xanthophylls, such as *Xanthophyllomyces dendrorhous* [26], formerly *Phaffia rhodozyma*, or *Rhodotorula glutinis* [27].

Enzymatic Steps

The synthesis of β-carotene from phytoene involves four consecutive dehydrogenations to give the reddish lycopene, and two β-cyclizations (Fig. 8.2). The four desaturations are achieved by a single dehydrogenase, as indicated by the absence of mutants blocked in intermediate desaturation steps, and the finding of a leaky mutation in the phytoene dehydrogenase gene of *P. blakesleeanus* resulting in different catalytic efficiencies for each desaturation [28].

The genes for phytoene dehydrogenase in these fungi were identified because of its sequence similarity to *al-1* from *Neurospora crassa*, described in a later section. It was cloned and characterized in the Mucorales *P. blakesleeanus* [29], *M. circinelloides* [30], and *B. trispora* [31], where it was called *carB*, and in *C. nicotianae* [32]. In contrast to plants and bacteria, fungal phytoene synthase and cyclase enzymatic activities reside in two protein domains of a bifunctional enzyme encoded by a single gene. This dual activity was firstly discovered in the basidiomycete *X. dendrorhous* [33] and confirmed in *P. blakesleeanus* by mutation analysis [34] and in *M. circinelloides* by partial deletion analysis [35], by complementation of different *carRP* mutants and by heterologous expression in engineered *E. coli* strains [36]. Therefore, only two genes encoding a phytoene synthase/cyclase and a phytoene dehydrogenase are needed to produce β-carotene from GGPP. The phytoene synthase/cyclase gene is linked with *carB* and divergently transcribed from a common regulatory region in the three Mucorales investigated (*carRA* in *P. blakesleeanus* and *B. trispora* [31, 34], *carRP* in *M. circinelloides* [35]). Phytoene synthase (CarA) and lycopene cyclase (CarR) activities of the CarRA bifunctional polypeptide are separated later as independent proteins [37], probably through proteolytic cleavage in a conserved site in the boundary between both protein domains [34]. The physical separation of both domains was recently confirmed by western blot analyses with specific antisera against phytoene synthase and lycopene cyclase in *B. trispora* [38].

The molecular analyses of the genes for β-carotene biosynthesis were preceded in *P. blakesleeanus* by extensive genetic and biochemical research. Albino mutants either with no carotenoids (null *carRA* mutants) or accumulating phtyoene (*carB* mutants), and reddish mutants accumulating lycopene (affected in the cyclase segment of *carRA* or *carRP*) were first characterized in this fungus [12, 39] and later in *Mucor* [13, 40] and *Blakeslea* [41, 42]. In *P. blakesleeanus*, the *carB* mutants accumulate 15-*cis* phytoene [43], which is readily desaturated in vitro by a purified phytoene dehydrogenase [44]. Later intermediates up to β-carotene have the all-*trans* conformation, indicating that the isomerization step is achieved on phytoene. This was formerly suggested by genetic analyses that assigned the phytoene accumulating mutants to a single complementation group [45].

Some chemicals are specific inhibitors of either the dehydrogenase or the cyclase activities in Mucorales and mimic both *carB* and *carRA* mutant phenotypes. Diphenylamine efficiently blocks phytoene dehydrogenation in different species [46, 47], while nicotine, CFTA (2-[4-chlorophenylthio]-triethylamine), and other compounds inhibit the cyclization [43]. Quantitative studies of chemical inhibition in *P. blakesleeanus* showed the organization of the enzymes as an aggregate, formed by four dehydrogenases and two cyclases [48–50]. The multimeric organization of the dehydrogenases was further confirmed by interallelic complementation experiments with *carB* mutants [51]. The two identical cyclase units of the complex are responsible for the two sequential cyclizations converting lycopene to β-carotene, and are equally sensitive to chemical inhibitors. However, similar experiments in *B. trispora* indicated the occurrence of two different lycopene cyclases, one sensitive to inhibitors with a similar enzymatic role as that of *P. blakesleeanus*, and another one insensitive to inhibitors and responsible only for cyclization of lycopene to γ-carotene [52].

Regulation

A huge amount of information has been accumulated on the regulation of β-carotene biosynthesis in Mucorales [12, 53], more recently extended to mRNA levels of the structural genes. The three species more thoroughly investigated, *P. blakesleeanus*, *M. circinelloides*, and *B. trispora*, differ in their capacities to produce β-carotene under different culture conditions, indicating differences in the underlying regulatory mechanisms. Three major regulatory effects have been described in these fungi: induction by light, sexual stimulation, and a presumed feedback regulation, each one with different relevance in each species. Additionally, some chemicals stimulate differentially the pathway in different species, in some cases interacting with the mentioned regulatory mechanisms.

Activation by Light

Light induction of carotenoid biosynthesis is a frequent trait in carotenogenic fungi [54]. However, this is not a general rule: light stimulates β-carotene production in cultures of *P. blakesleeanus* [55, 56] and *M. circinelloides* [13, 40], but not in those of *B. trispora* [57] or *U. maydis* [19]. The photoinduction features have been investigated in special detail in *P. blakesleeanus*, where it coexists with photoregulated developmental stages. In this fungus, carotenogenesis responds to light in a specific period of its life cycle, preceding the formation of their vegetative reproductive structures, the sporangiophores [56]. Fluence response experiments revealed two levels of light sensitivity, with different thresholds and amplitudes. A minor but detectable increase of β-carotene content was produced after a weak illumination and a more robust response was obtained after strong illumination. The action spectra of the two components are not totally coincident, suggesting differences in their respective photoreception systems [56].

As found for other fungi (see next sections on regulation of neurosporaxanthin biosynthesis), regulation by light is achieved at transcriptional level. In *M. circinelloides*, fluence response curves for *carB* and *carRP* mRNA accumulation showed inductions up to 300-fold after 10 min illumination with blue light pulses compared to darkness levels, while no relevant induction was obtained for *carB* mRNA with red light [30, 35]. Slot blot analyses revealed a more modest photoinduction of *carB* mRNA in *P. blakesleeanus* [29, 58], later confirmed by northern blot experiments and extended also to *carRA* [59]. More detailed expression analyses found a biphasic response for *carRA* and *carB* transcripts following a brief light pulse, and the formation of a promoter binding complex, putatively involved in transcriptional *carRA/carB* repression [60]. A significant photoinduction of *carRA* and *carB* mRNA levels was also found in *B. trispora* despite its lower amount of β-carotene under continuous illumination [61], the later explained by the efficient photoadaptation of *carRA* and *carB* expression. However, a significant increase in its β-carotene content was found after illumination of a dark-grown culture.

Light stimulates carotene biosynthesis in the *carB* mutants of *P. blakesleeanus* [56], which accumulate the colorless precursor phytoene, indicating that β-carotene or other colored carotenoids are not components of the photoreceptor machinery. Such induction is supported by an increase of *carRA* and *carB* mRNA levels [59], also manifest in *carRA* mutants. However, the mutants of the *carRA* gene, either in its CarR or CarA domains, lack photoinduction in vivo [56], suggesting a regulatory role for the CarRA protein before its excision into CarR and CarA polypeptides.

Some regulatory mutants, as *madA* and *madB*, exhibit reduced photocarotenogenesis and are also affected in phototropism and photomorphogenesis, while the double *madA madB* mutant is completely blind [62, 63]. Carotenogenesis is also less sensitive to light in other mutants, such as *picA* and *picB* [62], specifically affected in this response, or *pimA*, altered as well in photomorphogenesis [64]. These mutants might be affected in a signal transduction cascade involving phosphorylation events [65], connecting light detection and biochemical or developmental responses.

The analysis of the genome sequence of *P. blakesleeanus*, together with former genetic data, revealed that *madA* en codes a White Collar 1 (WC-1)-like protein [66]. As found for WC complex in *N. crassa* [67], MadA interacts with a WC-2 partner, which turned out to be MADB [68]. The genome of *P. blakesleeanus* contains two and three additional WC-1 and WC-2 orthologs, respectively, called WcoA and WcoB, and WctB, WctC, and WctD [69]. The occurrence of these genes suggests the formation of different WC complexes; however, with the exception of MadA and MadB, no interaction could be detected between these proteins [68]. The genes *madA* and *madB* are not induced by light, but the MadA/B complex mediates the photoinduction of *wcoA*, *wcoB*, *wctB*, and *wctD*. Lack of stable transformation protocols for *P. blakesleeanus* has hindered the assignation of functions to the remaining WC genes, but the availability of efficient gene targeting procedures in *M. circinelloides* [70] allowed the functional analysis of its WC counterparts. Only the knockout mutants of one of them, Mcwc-1c (WcoA ortholog [69]), were defective in light induction of β-carotene biosynthesis, indicating its participation in the light transduction pathway that controls carotenogenesis, while those of Mcwc-1a (MadA ortholog) are affected in the phototropic responses of sporangiophores [71]. However, Mcwc-1a is involved in the regulation of Mcwc-1c by light, and the photoinduction of *carB* and *carRP* mRNA levels is drastically reduced in either the Mcwc-1a or the Mcwc-1c mutants, suggesting the participation of both proteins in the same regulatory network. On the other hand, no visible phenotype was found for knockout mutants of the WcoB ortholog, Mcwc-1b.

Mutational Deregulation

Mutations resulting in changes in the wild type pattern of β-carotene production are a powerful tool to identify regulatory genes. Deep-pigmented mutants, containing large amounts of β-carotene, have been described in different Mucorales. In *P. blakesleeanus*, the genetic analysis of β-carotene overproducing mutants assigned the mutations to three genes: *carS* [72], *carD* [73], and *carF* [74]. Deep-pigmented mutants have been also described in *B. trispora* [41, 42] and *M. circinelloides* [13, 14], but their three- to fourfold increases in carotene content were modest compared to the hundredfold increases in some *carS* and *carF* mutants of *P. blakesleeanus*. The molecular basis of the *carS* mutation was recently elucidated (see section "Downstream Metabolism of β-carotene") but the *carD* and *carF* genes remain to be identified. The large increases in β-carotene content in these mutants are not explained by the minor changes found in *carB* and *carRA* transcripts, suggesting the activation at posttranscriptional levels [59]. At least in the case of the *carS* and *carD* mutants, the higher carotenoid content correlates with enhanced enzymatic activities in vitro [75].

In *P. blakesleeanus*, genetic or chemical block of phytoene dehydrogenase or cyclase activities lead to the respective accumulation of phytoene or lycopene in much larger amounts than β-carotene in control strains, indicating the occurrence of a feedback regulation. The increase is less pronounced in mutants of the dehydrogenase of *M. circinelloides* [13, 40] and not detectable in those of *B. trispora* [41]. Some mutants of the *carA* segment of the bifunctional gene *carRA* contain low amounts of β-carotene that can be explained by leaky activity of the phytoene synthase domain [34]. However, the CarA mutation has pleiotropic effects that suggest additional regulatory roles for this protein: (1) they are affected in substrate transfer in the cyclization reactions [76], (2) they exhibit a photomorphogenetic defect [77], and (3) they are completely devoid of carotene photoinduction, the later also missing in the mutants of the *carR* domain [56]. Similarly, mutants of the *carC* gene in the same fungus contain less β-carotene than the wild type and are light yellow [78], but the molecular basis of this effect has not been elucidated. Carotenes are absent or present at reduced levels in other *B. trispora* [41] and *M. circinelloides* [13, 14, 40] mutants.

Carotene biosynthesis shares the first biosynthetic steps with other terpenoids (Fig. 8.1), such as the sterols. The different specific radioactivities found in ergosterol and β-carotene in the wild type and in a β-carotene superproducing mutant of *P. blakesleeanus*, grown in the presence of [^{14}C]mevalonate, indicated that both biosynthetic pathways occur in different cellular compartments and are independently regulated [79]. Accordingly, the variations in the β-carotene content of *P. blakesleeanus* mutants are not accompanied by changes in their sterol content. A similar approach led to a similar conclusion for the synthesis of ubiquinone and triacylglycerols, also made in different compartments from that of β-carotene [80]. These observations suggest the existence of regulatory mechanisms to control the targeting of the whole set of carotenogenic enzymes, from mevalonate to β-carotene, to the appropriate cell compartments.

The first gene involved in the regulation of β-carotene biosynthesis identified at molecular level in Mucorales was *crgA*, encoding a RING finger protein in *M. circinelloides* [81]. The mutation of *crgA* produces β-carotene overproduction as a result of a strong increase in the expression of *carRA* and *carB*, indicating that CrgA acts as a negative regulator [82, 83]. Independently of the enhanced β-carotene content, the *crgA* mutants conserve the photoinduction of the pathway, indicating that CrgA does not participate in light regulation. The CrgA protein contains several structural domains, including two RING-finger (RF) domains, two glutamine-rich regions, a LON domain, and a carboxy-terminal isoprenylation domain. At least two of these domains, RF-2 and a polyglutamine region, are essential for its regulatory function in carotenogenesis [83]. The RF domains have been associated to E3 ligase-type enzymes that mediate ubiquitylation of target proteins leading to their degradation by the proteasome. The essential role of RF-2 suggests that CrgA acts as an E3 ubiquitin ligase [83]; however, it has been shown that CrgA inactivates the WC protein Mcwc-1b by ubiquitylation-independent proteolysis [84]. In addition to suppressing the synthesis of β-carotene, CrgA controls other cellular processes, such as vegetative growth and sporulation, suggesting a role as a general regulator of the physiology of this fungus [85, 86]. This regulation is presumably conserved in other Mucorales, as indicates the ability of the CrgA ortholog from *B. trispora* to restore the wild type phenotype when introduced in a null *crgA* mutant of *M. circinelloides* [61].

Chemical Activation

Addition of some chemicals to several Mucorales results in notable increases in β-carotene content in a species-specific manner. A variety of examples are well known in *P. blakesleeanus* and *B. trispora*. Different compounds activate β-carotene production in *P. blakesleeanus* by at least two independent mechanisms [87]. A first set of chemicals containing β-ring in the molecule, with retinol and β-ionone as the most representative examples [88], presumably compete with β-carotene for binding to an enzyme or regulatory protein. It was proposed that β-carotene forms a complex with the *carS* and *carA* gene products to shut off the pathway [87], but this hypothesis remains to be demonstrated. As stated previously, the *carA* mutants are insensitive to light but are stimulated by retinol, indicating that both stimulatory mechanisms are independent. A second set of chemicals consists of different phenolic compounds, with no recognized common features and with variable activating efficacies [89]. Veratrol and dimethyl phthalate are the most potent activators, and their effects are additive to that of retinol when present simultaneously in the medium [87]. One of the most active phenolic compounds, cinnamic alcohol, also inhibits phytoene dehydrogenation by an independent mechanism [47].

Retinol addition results in a two- to threefold induction of β-carotene production in *B. trispora* [90], a minor activation compared to the 20-fold increase produced in *P. blakesleeanus* [87, 88], and no stimulatory effect has been described for phenols in this fungus. Because of the use of *B. trispora* for industrial carotene production, described in the next section, searches for stimulatory chemicals in this fungus are abundant in the literature. An early research found inducing effects by β-ionone and other related compounds as well as by a diversity of nitrogen-containing chemicals, including amides, imides, lactams, hydrazides, or substituted pyridines [91]. Recent reports have found other inducing chemicals: the surfactant sorbitan monolaurate duplicates the β-carotene content, possibly because it allows a more disperse mycelial morphology [90], and arachidonic acid results in a significant increase presumably through induction of *carRA* and *carB* transcription [92]. Finally, ketoconazole, an inhibitor of an enzymatic activity needed for ergosterol biosynthesis, produced a threefold stimulation of β-carotene biosynthesis that could be related with a compensatory response of the genes for early steps of the terpenoid pathway [93].

Sexual Activation and Industrial Production

A special group of chemical activators for the Mucorales are the trisporic acids, pheromones associated to the sexual cycle. Interaction of mycelia of opposite sex results in the induction of β-carotene production in different species [94]. This stimulation, which does not require a physical contact [95], is evidenced by the formation of a deep-pigmented band in the region where mycelia of opposite sex meet on solid medium (Fig. 8.3). In *P. blakesleeanus*, the mixed cultures of both mating types (mated cultures) contain about fivefold more β-carotene than the single cultures [96], exhibit higher in vitro carotenogenic activities [75], and contain higher *carRA* and *carB* transcript levels [59]. The response is triggered by the sexually induced production of trisporic acids, and it is usually accompanied by a sexual differentiation process [97, 98]. Variable degrees of induction may also be obtained by addition of different trisporoids to *P. blakesleeanus* [96] and other Mucorales [99].

The presence of the trisporic acids in the medium is sufficient to induce the response in single cultures, and in *P. blakesleeanus* this induction is additive with those produced by dimethyl phthalate, light or *carS* mutations [96]. In this fungus, the sexual induction is particularly efficient in intersexual heterokaryons, i.e., strains with nuclei of both mating types in the same cytoplasm [100]. Such heterokaryons may be stably maintained with appropriate selective markers, and their β-carotene

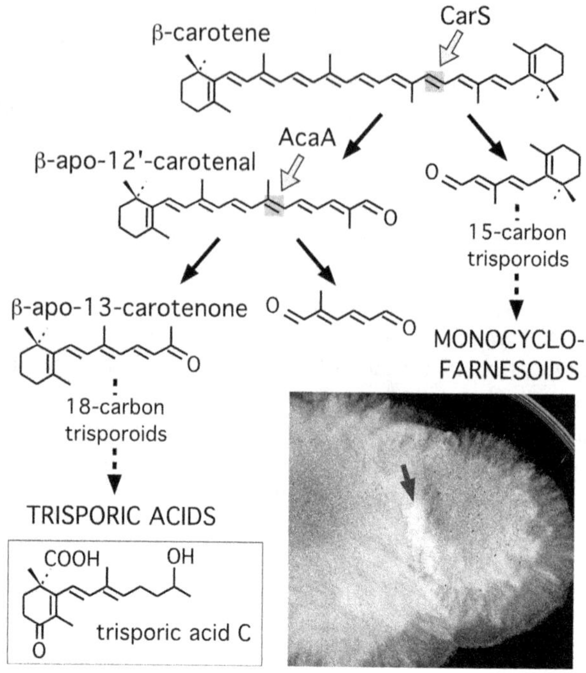

Fig. 8.3 Downstream metabolism of β-carotene. The *gray arrows* indicate cleavage sites for the oxygenases CarS and AcaA in *P. blakesleeanus*. The figure includes trisporic acid C as representative of the trisporic acids family. The picture (*lower right corner*) shows the meeting region of two colonies of opposite sex of the same fungus grown for 4 days at 16 °C on potato dextrose agar. The clearer band indicated by the *arrow* has a high β-carotene content (picture kindly provided by Lola Pérez de Camino)

content reach very high levels (2.5 % of dry weight) if the nuclei of opposite sex carry *carS* mutations. As described in the next section, trisporic acids derive from β-carotene [101], and strains unable to produce β-carotene do not stimulate β-carotene production in their sexual partners [102]. The sexual stimulation of carotenogenesis is drastically reduced in the presence of acetate in the medium; however, the formation of the sexual structures, the zygospores, is enhanced [103], indicating independent mechanisms for sexual activation of both processes.

The stimulation of β-carotene production in mated cultures is remarkably pronounced in *B. trispora* [9], and has been exploited for industrial production. As observed in *P. blakesleeanus*, the induction is accompanied by higher *carRA* and *carB* mRNA [104] and enzyme levels [38], and affects the expression of many other genes involved in different physiological processes [105]. Under laboratory conditions, β-carotene production yields are improved using appropriate mutant strains, and similar results are obtained with either mated cultures or sexual heterokaryons [42]. The efforts dedicated to set up the use of *B. trispora* for industrial β-carotene production are easily extrapolated for lycopene production by the use of appropriate mutants or cyclase inhibitors [42, 90].

The extensive knowledge on β-carotene biosynthesis in *P. blakesleeanus* and *M. circinelloides* makes these fungi promising alternatives for industrial purposes. Surface cultures of *P. blakesleeanus* intersexual heterokaryons in media containing subproducts from biological industries reached 2.5 % β-carotene in its dry mass [106]. Similar β-carotene levels were found in partial sexual diploids with a *carF* mutation, sufficiently stable for industrial purposes [107]. However, submerged or agitated liquid cultures were less productive than surface cultures [108], explaining the preference for *B. trispora* as a β-carotene biotechnological source. In the case of *M. circinelloides*, the availability of efficient gene manipulation protocols open new possibilities [70]. Transformants of this species with increased copy numbers of early genes of the terpenoid pathway led to increase up to fivefold the β-carotene content, indicating that some of the early enzymatic steps are bottlenecks in β-carotene production [109]. Furthermore, transformation tools allow the introduction of β-carotene modifying enzymes to produce xanthophylls of industrial interest, such as canthaxanthin or astaxanthin [109–111]. As in *B. trispora*, the genetic or chemical block of cyclase activity under high β-carotene production makes these species suitable sources for industrial lycopene production [42, 90, 112].

Downstream Metabolism of β-carotene

In Mucorales, β-carotene is the source for the synthesis of different chemical derivatives. As already stated, among them stand out the trisporic acids [101, 113], apocarotenoid-derived compounds that belong to a wider family, the trisporoids. The synthesis of these compounds has been thoroughly investigated in *B. trispora* and other Mucorales. The trisporic acids are classified in series A, B, C, D, and E according to the oxidation of the side chain and the position of substituents on the ring [101, 114, 115]. The refinement of the chemical analyses revealed an outstanding apocarotenoid complexity: culture media of *B. trispora* contain compounds from at least three groups: 18-carbon trisporoids, 15-carbon cyclofarnesoids, and 7-carbon methylhexanoids [116]. Similar compounds were found in the filtrates of *P. blakesleeanus* cultures, using as a negative control a *carB* mutant to elucidate the side-products of β-carotene metabolism [117]. Some apocarotenoids were not produced in the presence of acetate, providing an explanation for the inhibitory effect of this organic acid on sexual activation of carotenogenesis [103].

The analysis of the genome of *Rhizopus oryzae*, the first one available in Mucorales, led to the identification of two genes for carotenoid oxygenases, called *tsp3* and *tsp4* because of their presumed involvement in trisporic acid biosynthesis [118]. The same authors identified the *tsp3* ortholog in *B. trispora*, showed that *tsp3* transcript levels are activated in mated cultures in both species, and found that the *B. trispora* Tsp3 enzyme cleaves β-carotene in engineered *E. coli* cells. Additionally, the chemical analysis of the many apocarotenoids found in *P. blakesleeanus*, most particularly the finding of the 7-carbon compounds [119], suggested their origin from the sequential cleavage of β-carotene at its 11′,12′ and 12,13 double bonds (Fig. 8.3). Unexpectedly, the search for the gene *carS*, based on its genetic linkage to the genes *carRA* and *carB* in the *P. blakesleeanus* genome, revealed that it encodes a carotenoid oxygenase [120], which turned out to be the Tsp3 ortholog. The identity of the gene was confirmed by the finding of relevant mutations in six independent *carS* mutants. Expression of *carS* in β-carotene-producing *E. coli* cells showed its capacity to cleave this molecule to produce β-apo-12′-carotenal, the first intermediate in the apocarotenoid pathway in this fungus [121], and the same enzymatic reaction was later confirmed for Tsp3 [99]. Thus, *carS* does not encode a regulatory protein, as anticipated the β-carotene over-accumulation of its mutants, but a carotenoid cleavage oxygenase. This finding led to reinterpret the regulatory effect of the *carS* mutation as a result from the block of the pathway or the lack of a negative-acting apocarotenoid signal, involved in the feedback regulation. However, a direct regulatory role of the CarS enzyme cannot be discarded, as suggested by the albino phenotype exhibited by some double *carS* mutants [122].

The genome of *P. blakesleeanus* contains four additional genes for presumptive carotenoid cleavage oxygenases, two of them with the expected key residues required for the cleaving activity. The enzyme encoded by one of them, called AcaA, cleaves the CarS product β-apo-12′-carotenal to produce β-apo-13-carotenone [121]. Taken together, the available data indicate that first reactions in the apocarotenoid pathway of *P. blakesleeanus* are the cleavage of β-carotene at its C11′–C12′ double bond by CarS and the cleavage of the resulting C25-fragment at its C13–14 double bond by AcaA (Fig. 8.3). The 18-carbon product is then used for the synthesis of the trisporic acids and other trisporoids, while the 15-carbon product is the origin of the cyclofarnesoids. These early enzymatic steps are probably conserved in other Mucorales.

The genes and enzymes for trisporoid metabolism are currently under investigation. The first enzyme identified from this pathway was 4-dihydromethyltrisporate dehydrogenase, purified from *Mucor mucedo* cultures induced by trisporic acids [123]. The sequencing of protease-cleaved peptides from this enzyme allowed the identification of the encoding gene, called *tsp1*, which is expressed constitutively at transcriptional level and it is induced by trisporic acids at enzyme activity level [124]. A similar approach led to identify a second gene, *tsp2*, coding 4-dihydrotrisporin-dehydrogenase, in this case induced both at transcription and activity levels [125]. Tsp1 and Tsp2 belong to different enzymatic families, aldo/keto reductases and short chain dehydrogenases, despite they carry out the same enzymatic reaction on similar substrates. The availability of genomes sequences from other Mucorales in gene databases showed the occurrence of orthologs for these genes in different species [126], including *P. blakesleeanus* [127].

Biosynthesis of Neurosporaxanthin

Neurosporaxanthin is a carboxylic apocarotenoid discovered more than 50 years ago in *N. crassa*. The first biochemical analyses of *N. crassa* wild type cultures indicated the presence of mixture of carotenes and carotenoids that included acidic and neutral xanthophylls [128], tentatively identified as spirilloxanthin, and an acidic carotenoid [129]. These compounds were found with phytoene and γ-carotene, and minor amounts of other neutral carotenes. Later, the occurrence of spirilloxanthin was not confirmed and the acidic carotenoid fraction was found to contain a single component that was called neurosporaxanthin [130], chemically identified as the carboxylic carotenoid β-apo-4′-carotenoic acid [131]. Neurosporaxanthin is also produced by other fungal species, as those of the genus *Fusarium* [46, 132], *Verticillium* [133, 134], and *Podospora* [135]. More refined chemical studies identified other carotenes found in *N. crassa* in addition to neurosporaxanthin [136, 137], leading to postulate the biosynthetic pathway displayed in Fig. 8.4. Similar carotene to were also found in *F. fujikuroi* [138] and *F. aquaeductuum* [132], suggesting a similar biosynthetic pathway. Neurosporaxanthin may be subject to further chemical modifications, e.g., methyl ester and glycosyl ester derivatives were reported in *Verticillium agaricinum* [139] and in a marine *Fusarium* species [140], respectively.

Enzymatic Steps for Torulene Production

Neurosporaxanthin biosynthesis coincides with that of β-carotene in the initial steps, starting with the formation of phytoene from GGPP [141]. In *N. crassa*, carotenoid mixtures include 3,4-didehydrolycopene, a carotene resulting from the introduction of five sequential desaturations on the phytoene backbone. The cyclization of one end of the 3,4-didehydrolycopene molecule produces torulene, but the cyclization may be introduced in less desaturated precursors, as indicates the finding in *N. crassa* of γ-carotene or β-zeacarotene, resulting from the cyclization of neurosporene or lycopene, respectively (Fig. 8.4). In *Fusarium*, however, no 3,4-didehydrolycopene has been found [132, 138], suggesting a preference in this fungus to introduce the cyclization in earlier substrates.

The genetic analysis of albino mutants of *N. crassa* led to identify the first fungal genes involved in carotenoid biosynthesis. Recombination and complementation studies defined the genes *al-1* and *al-2* [142], to which a third gene, *al-3*, was subsequently added [143, 144]. Characterization of *al-3* showed that it encodes the prenyl transferase responsible for the synthesis of the phytoene precursor GGPP from FPP [145–147]. Only partially defective mutants have been isolated for this gene [148], as expected from the need of a minimal amount of GGPP to produce other essential terpenoids (Fig. 8.1).

The product of gene *al-1* was identified as phytoene dehydrogenase based on its similarity to formerly described phytoene dehydrogenases from bacteria [149], and the ability of *al-1* to complement a defective phytoene dehydrogenase mutant of *Rhodobacter capsulatus* [150]. The carotenoids produced in this bacterium indicated that AL-1 is able to catalyze at least four desaturations, unlike bacterial dehydrogenases that can only perform some of these reactions [151]. Expression of *al-1* in *E. coli* and in vitro studies with the purified enzyme showed that AL-1 is able to carry out the five desaturation steps of the neurosporaxanthin biosynthetic pathway as an assembly of identical AL-1 units, with the capacity to accept or exchange individual intermediates in and out of the complex [152]. The availability of the *al-1* sequence facilitated the cloning of the *F. fujikuroi* ortholog, *carB* [153, 154], coding the dehydrogenase responsible also for all desaturation steps in this fungus [155].

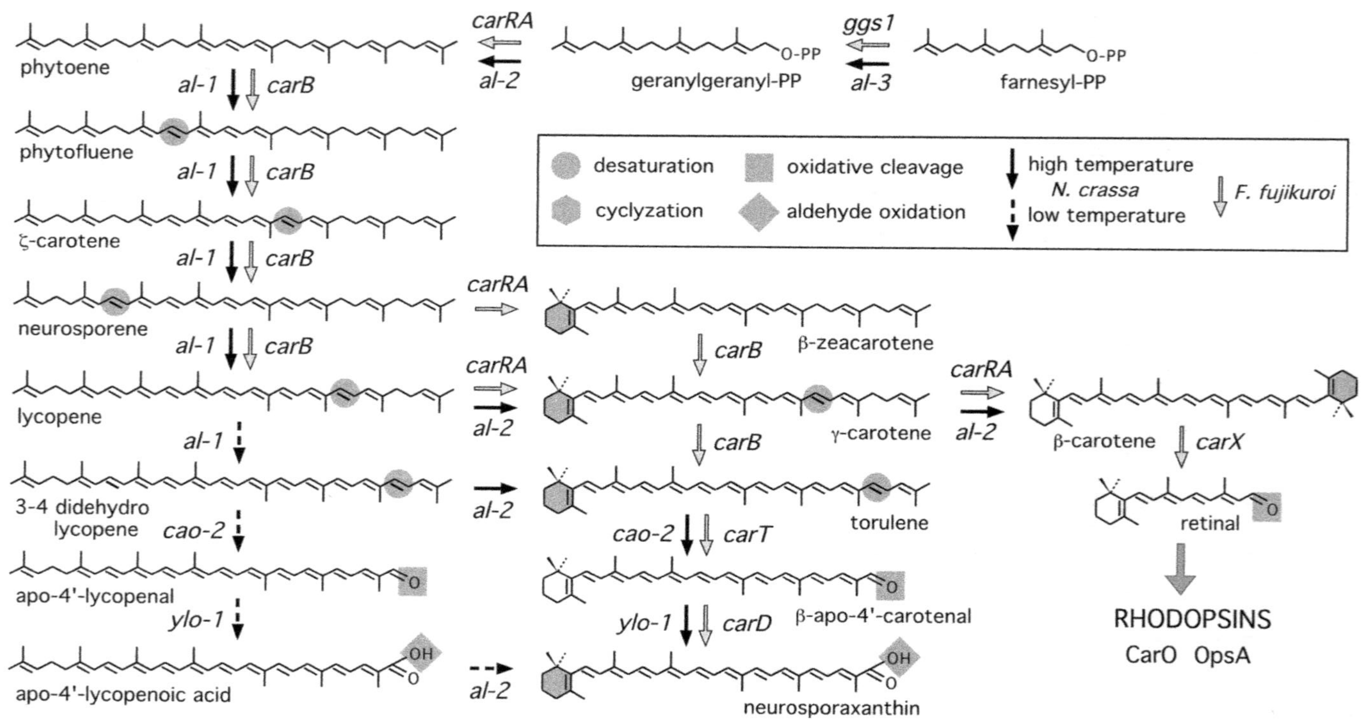

Fig. 8.4 Neurosporaxanthin biosynthetic pathway in *N. crassa and Fusarium* sp. Gene products responsible for each reaction are indicated close to the arrows (*al-1*, *al-2*, *cao-1* and *ylo-1* for *N. crassa*, and *carRA*, *carB*, *carT*, *carD* and *carX* for *Fusarium* sp.). Reaction sites are shaded in the carotenoid product of each enzymatic step. *Black arrows* indicate reaction steps in *N. crasssa* (*dotted arrows* for reactions that occur at low temperature) and *white arrows* indicate reactions in *Fusarium* sp.

The product of gene *al-2* was found to exhibit high sequence similarity to formerly investigated phytoene synthases, but it contained a long hydrophobic amino segment of unknown function, suggested to play a role in membrane binding [156]. As already stated, the finding of cyclase activity in the homologous segment of the AL-2 ortholog CrtYB of *X. dendrorhous* [33] led to assign also this function to the amino AL-2 segment, as confirmed the identification of mutations specifically affected in this enzymatic activity [157]. The dual activity of AL-2 was later reaffirmed through the analysis of further *al-2* alleles [158] and enzymatic assays of *al-2* cDNA expressed in appropriately engineered *E. coli* strains [159]. The latter approach showed that AL-2 preferentially carries out a single cyclization, fitting its participation in the synthesis of the monocyclic neurosporaxanthin. Its ortholog in *F. fujikuroi*, *carRA*, presumably encodes an equivalent bifunctional enzyme [154], but no mutants specifically affected in the cyclase domain of this protein have been described. The physical separation of the phytoene synthase and cyclase domains has not been demonstrated in *N. crassa* or in *F. fujikuroi*, but the scarcity of mutants specifically affected in the cyclase activity, and the finding in *N. crassa* of a mutant of the phytoene synthase domain lacking cyclase activity [158] suggest that both enzymatic functions are not split in separate polypeptides.

Biosynthetic Steps from Torulene to Neurosporaxanthin

The analysis of *Neurospora* and *Fusarium* genomes and the availability of appropriate mutant phenotypes promoted the identification of the enzymes involved in the late steps of neurosporaxanthin biosynthesis (Fig. 8.4). These genomes contain two genes encoding presumptive carotenoid oxygenases, a group of enzymes mediating oxidative cleavage reactions to produce apocarotenals [160]. One of them, called *carT*, was able to complement a torulene-accumulating mutant of *F. fujikuroi* [161], and the targeted mutation of its ortholog in *N. crassa*, *cao-2*, resulted in the production of torulene instead of neurosporaxanthin [162]. In vivo and in vitro analyses of *carT* and *cao-2* enzymatic activities and the finding of mutations in the *carT* or *cao-2* alleles of torulene-accumulating mutants demonstrated the role of this enzyme on torulene cleavage. CAO-2 cleaves torulene but not γ-carotene, indicating the need of the former five desaturations. The orthologous *carT* gene, together with the *carRA* and *carB* counterparts, have been also identified and confirmed by targeted gene disruption in the teleomorph of *Fusarium graminearum*, *Gibberella zeae* [163].

CarT and CAO-2 cleave the uncycled end of torulene (40 carbons) to yield β-apo-4′-carotenal (35 carbons). This molecule is converted to neurosporaxanthin by an aldehyde dehydrogenase, called YLO-1 in *N. crassa* [164] and CarD in *F. fujikuroi* [165]. The slant cultures of the *ylo-1* mutants exhibit a bright yellow pigmentation, strikingly different from the orange color of those of the wild type, indicating the accumulation of a different carotenoid mixture. The presence of ζ(zeta)-carotene, γ-carotene, and torulene and the absence of neurosporaxanthin in this mutant suggested a block at a late step of the pathway [144, 166]. Depending on the culture conditions, different carotenoids have been found in the *ylo-1* mutant, including apo-4′-lycopenal, apo-4′-lycopenol [164], apo-4′-γ-carotenal, and apo-4′-γ-carotenol [167], indicating a high reactivity of the aldehyde group. This was also shown by the identification of myristic acid esters of some of the carotenoids in the *ylo-1* mutant, with 4′-apolycopene-4′-myristate and 4′-apo-γ-carotene-4′-myristate as the predominant ones [167]. The *carD* mutant of *F. fujikuroi* also exhibits a complex biochemical phenotype: its young colonies are orange-yellowish because of the accumulation of β-apo-4′-carotenal, but their color turned gradually to yellow due to its conversion to β-apo-4′-carotenol and, possibly, to its fatty-acid esters [165].

Because of their lipophilic nature, the synthesis of carotenoids is expected to be associated to membranes. Analysis of subcellular fractions of *N. crassa* identified carotenoids in lipid globules and in fractions that were rich in membranes of the endoplasmic reticulum [168]. In support, the carotenogenic enzyme system is bound to membranes, the enzymes are partially solubilized by detergent treatments, and their activities are restored by lipid addition [169]. The sequence features of the proteins fit these predictions: the cyclase domain of AL-2 contains predicted transmembrane domains [157], as found for other fungal orthologs [34]. One of such domains is also found in the CarD/YLO-1 terminal region [164, 165], a rare feature in the usually soluble ALDH enzymes. The regulatory mechanisms used by fungi to control the cell compartment where carotenoids would be synthesized or stored are unknown. As formerly found in *P. blakesleeanus* [79], the syntheses of different terpenoids in *F. fujikuroi*, in this case sterols, gibberellins, and carotenoids, are located in different subcellular compartments and are provided by independent substrate pools [170].

A Side Branch in Neurosporaxanthin Biosynthesis

The carotenoid mixtures found in *N. crassa* and *F. fujikuroi* contain minor amounts of β-carotene [137, 138], indicating the occurrence of a side branch from the predominant neurosporaxanthin pathway involving a second cyclization step. The intermediate carotenes found in *F. fujikuroi* suggest that the branch diverts at the level of γ-carotene, which may be used as a substrate either by the cyclase or the dehydrogenase enzymes (Fig. 8.4). This also seems to be the case of *N. crassa*, where in vitro analyses showed that γ-carotene was not accepted as a substrate by the *Neurospora* dehydrogenase AL-1 [152], indicating that the formation of torulene must proceed exclusively through 3,4-didehydrolycopene and that γ-carotene can be only used for the synthesis of β-carotene (Fig. 8.4).

The occurrence of rhodopsin-like proteins in these fungi, NOP-1 in *N. crassa* [171] and CarO [172] and OpsA [173] in *F. fujikuroi*, indicates the need for the rhodopsin chromophore retinal. In support of retinal availability, retinal binding and photoactivity have been demonstrated for NOP-1 [174] and CarO (J. García-Martínez, unpublished). In animals, retinal is produced by the oxidative cleavage of β-carotene by a group of carotenoid oxygenase enzymes [175]. In *F. fujikuroi*, a carotenoid oxygenase encoded by the second gene from this family, *carX* [176], cleaves β-carotene to produce retinal [177]. This gene is located in a co-regulated gene cluster that includes the genes needed to produce β-carotene, *carRA* and *carB*, and the rhodopsin gene *carO*. In contrast, no retinal-forming enzymatic activity has been identified in *N. crassa*. The second predicted carotenoid oxygenase gene in the genome of this fungus, CAO-1, has no detectable activity on carotenoids, but cleaves the phytoalexin resveratrol [178].

Regulation

Activation by Light

In *N. crassa* and *Fusarium* sp., neurosporaxanthin biosynthesis is stimulated by light [179, 180]. Submerged cultures of *N. crassa* are not pigmented in the dark but accumulate neurosporaxanthin after exposure to light [129]. The response is detectable 1 h after light onset and reaches maximum levels after 6–12 h of illumination [129, 181]. The photoresponse is very sensitive, 1 min of light or less is enough to produce a detectable induction [182, 183], and it requires aerobic conditions [129], presumably to keep the photoreceptor in a proper oxidation state [184]. A similar response was found in *F. aquaeductuum* [132] and *F. fujikuroi* [185], but in these fungi 1 h of light exposure was needed for a significant response. Fluence response experiments showed a biphasic response in *N. crassa* photocarotenogenesis [183], reminding that observed in *P. blakesleeanus* [56], but the two steps were not observed in *Fusarium* sp. In *N. crassa*, after the photoresponse has been triggered by a light pulse, the mycelia become temporarily insensitive to a second exposure to light [186].

In *F. aquaeductuum*, the sulfhydryl reagents p-chloro- and p-hydroxymercuribenzoate [187], or the oxidative reagent hydrogen peroxide [188] result in sustained activation of neurosporaxanthin biosynthesis in the dark, suggesting that oxidation of –SH groups replaces the effect of light in this fungus. Moreover, strong reducing agents, such as dithionite and hydroxylamine [188], abolish the light response. The effect of oxidizing agents is additive with that of light, indicating independent mechanisms of action [189].

Action spectrum for neurosporaxanthin photoinduction in *N. crassa* expands from 400 to 500 nm [182], with peaks at ca. 450 and 480 nm [190], suggesting the participation of a flavin photoreceptor. This was further supported by the lower photoinduction exhibited by mutants with reduced flavin content [191], and by lack of restoration of normal photoinduction by addition of riboflavin or its analogs [192]. A similar action spectrum was also found in *F. aquaeductuum* [193] and red light proved to be ineffective in both species [194]. However, incubation of this fungus with some redox dyes, as methylene blue or toluidine blue, allows it to respond to red light [195], suggesting that they might act as artificial photoreceptors.

Neurosporaxanthin photoinduction is mediated in *N. crassa* by the White Collar complex, consisting of the flavin photoreceptor WC-1 and its partner WC-2 [67]. This was shown by the loss of light-induced phytoene synthetase activity in vitro [196], and carotenoid accumulation in vivo [197] in *wc-1* mutants compared to the wild type, and by the loss of transcriptional induction of the structural genes in *wc-1* and *wc-2* mutants [162]. The WC complex binds directly to the promoters of their target genes [198] and putative regulatory elements have been identified in the *al-3* promoter [199]. In contrast, mutants of the WC-1 gene of *F. fujikuroi wcoA* and *F. oxysporum wc1* maintained a significant carotenoid photoinduction [200, 201], indicating the participation of other photoreceptors for carotenogenesis in both species. Mutants of putative *F. fujikuroi* photoreceptors, such as the DASH cryptochrome CryD [202] or the rhodopsins CarO [172] and OpsA [173] also exhibited light-induced carotenogenesis. As formerly proposed for *V. agaricinum* photocarotenogenesis [134], the participation of more than one photoreceptor is not discarded. However, photoinduction was reduced in mutants of the adenylate cyclase gene *acyA* of *F. fujikuroi* [203] or the *MAT1-2-1* mating-type gene of *F. verticillioides* [204], pointing to direct or indirect involvement of different regulatory networks in the photoregulation of the carotenoid biosynthesis in these fungi.

In both *N. crassa* or *Fusarium* sp., the photoresponse is achieved through a remarkable increase in mRNA levels of most of the structural genes. This has been shown for *al-1*[149], *al-2* [156] and *cao-2* [162] in *N. crassa*, and for *carRA*, *carB* [154, 172], and *carT* [177] in *F. fujikuroi*. Similar results have been obtained in *F. oxysporum* [205] and *F. verticillioides* [204]. The photoinduction of transcripts of the *al* genes is very fast in *N. crassa*, reaching a maximum between 15 and 30 min after the illumination onset, and decreasing afterwards (photoadaptation), and requires fully active WC-1 and WC-2 proteins, while in *Fusarium* sp. the maximal levels are reached approximately after 1 h. Expression of *ylo-1*, responsible for the last reaction of the pathway, is not stimulated by light in *N. crassa* [164], but the prenyl transferase responsible for GGPP synthesis, *al-3*, is strongly photoinduced [145, 206], a regulation also exhibited by the *M. circinelloides* ortholog *carG* [207]. In *F. fujikuroi*, neither the *ylo-1* and *al-3* orthologs, *carD* and *ggs1*, nor the gene coding farnesyl pyrophosphate synthase are photoregulated [165, 208, 209]. This fungus contains a second GGPP synthase gene, *ggs2*, located in the gibberellin gene cluster [210], but the effect of light on its expression has not been investigated.

The photoadaptation mechanism allows adjusting the expression of the carotenoid genes to prolonged light exposures. In *N. crassa*, the photoadaptation involves changes in the activity of the WC complex, with the participation of other regulatory proteins [198]. One such protein is VIVID (VVD), a small photoreceptor with a flavin-binding domain required for efficient photoadaptation of light-regulated genes [211, 212]. The *vvd* mutants exhibit a deep pigmentation and accumulate about fivefold more carotenoids under light than the wild type [213, 214], a phenotypic trait explained by the sustained expression of the structural genes for carotenoid biosynthesis. Presumably, *vvd* is not the only gene involved in photoadaptation. New strains deficient in photoadaptation were identified through a search for antibiotic-resistant mutants under continuous illumination, using a strain carrying an antibiotic resistance gene under the control of a light-regulated promoter [214]. Some of them exhibited a *vvd*-like phenotype, with higher carotenoid content and sustained light induction of the carotenoid genes, but they were not allelic to *vvd*. Their molecular characterization suggested the occurrence of at least two further genes involved in photoadaptation, which remain to be investigated.

Developmental Regulation

N. crassa grown on solid media in the dark presents mycelia without colored carotenoids and pigmented neurosporaxanthin-containing conidia. The pigmented conidia are also produced by the *wc* mutants grown in the light, confirming that the accumulation of carotenoids in the conidia is independent of the light-inducing mechanism. Stimulation of conidiation by exposure of mycelia to air revealed that the genes *al-1* and *al-2* are induced after ca. 16 h under such conditions in the dark [215]. This WC-independent induction coincides with the formation of major constrictions during the conidiation process, and is not observed in non-conidiating regulatory mutants, such as *fluffy* or *fluffyoid*. The developmental induction was also observed under continuous illumination, when *al-1* and *al-2* mRNA levels were low because of photoadaptation. However, an earlier WC-dependent induction by light was also observed under these conditions at the start of the activation of conidiation [215], indicating a regulatory connection of the WC system with conidia formation.

The independent regulation of carotenogenesis by light and development is evidenced by the synthesis of specific *al-3* transcripts for each inducing condition: a 1.6 kb mRNA, *al-3*(*m*), in response to light, and a 2.2 kb mRNA, *al-3*(*c*), under conidiation [216, 217]. The different transcription start sites, confirmed by primer extension analyses, imply specific regulatory elements for each regulatory condition. The effects of different regulatory mutations draw attention to the complexity of *al-3* expression: *al-3*(*c*) formation is prevented in conidiation mutants, such as *fluffy* or *acon-2* [216], but not in *fluffyoid* [217], while it is still affected by light [216]. Additionally, alternative translational start sites were found in this gene, as demonstrated the residual activity found in a deletion mutant [218]. Gene expression analyses failed to detect *al-3* mRNA in mature conidia, indicating that the accumulation of carotenoids is achieved in the earlier stages of conidia formation. However, the *al-3* gene maintains the ability to respond to light in the conidia through the WC complex [219].

Neurosporaxanthin biosynthesis is also connected to development through other regulatory mechanisms, such as the cAMP-signaling pathway. Changes in the expression of gene *gna-1*, coding a Gα component of a heterotrimeric G complex, produced morphological and developmental alterations and exhibited a negative correlation with carotenoid levels [220]. Thus, strains with mutationally activated *gna-1* alleles developed longer aerial hyphae and less conidia, and contained less carotenoids in comparison to the wild type. Actually, former observations showed an inverse correlation between cAMP levels and carotenoid content in the dark in this fungus: (1) a transient induction of the cAMP content was observed during the lag phase that precedes carotenoid photoinduction [221], (2) mutants of the adenylyl-cyclase gene *cr-1* (*crisp-1*) contained more carotenoids in the dark [221], and (3) carotenoid photoinduction was reduced by addition of exogenous cAMP [222]. Interestingly, *al-3* mutants, devoid of carotenoids, contained more cAMP than the wild type [221]. More recently, it was shown that the mutation of the histidine kinase gene *dcc-1* results in enhanced conidia formation and carotenoid production [223], and that this effect is reversed by exogenous cAMP addition.

Dependence on Temperature and Nitrogen

Temperature has an unexpected influence in *Neurospora* carotenogenesis. Neurosporaxanthin accumulation in the mycelia is particularly efficient if dark-grown mycelia are illuminated at low temperature [224]. The response is very weak at 37 °C and the efficiency increases at lower temperatures, reaching the optimal response between 6 and 12 °C. This unusual effect has not been described in other fungi, and it is not a specific property of neurosporaxanthin production, e.g., in *F. aquaeductuum* the synthesis is less efficient as the temperature decreases [225]. The stimulatory effect of the lower temperature in *N. crassa* requires protein synthesis, as indicated by the loss of the response in the presence of cycloheximide [224].

The illumination of *N. crassa* at low temperature not only results in a higher accumulation of carotenoids, but also in a higher proportion of neurosporaxanthin among the total carotenoids. This circumstance facilitated the phenotypic characterization of mutants of late steps of the pathway, such as *cao-2* or *ylo-1* [162, 164]. Moreover, the analysis of the carotenoids accumulated by these mutants showed that the order of the reactions depends on the temperature of illumination [226]: at low temperature the CAO-2 oxidative cleavage step precedes the AL-2 cyclization, and apo-4′-lycopenal and apo-4′-lycopenoic acid are the neurosporaxanthin immediate precursors. Therefore, both the oxygenase CAO-2 and the cyclase activity of AL-2 compete for 3,4-didehydrolycopene as a substrate, prevailing one or the other as a function of the temperature (Fig. 8.4).

Neurosporaxanthin biosynthesis in *F. fujikuroi* is stimulated by nitrogen starvation: immobilized mycelia incubated under low nitrogen conditions produce carotenoids, but addition of nitrogen slows down the synthesis [227]. Moreover, the exchange of immobilized mycelia between media with different nitrogen contents results in induction or cessation of carotene synthesis, depending on the absence or presence of nitrogen, respectively. In the wild type, the induction by nitrogen is independent of the regulation by light, and it is also manifest in the *carS* overproducing strains [228], described in the next section. Transfer experiments from high to low nitrogen conditions showed that nitrogen starvation results in a transient increase of transcript levels of the structural genes of the pathway. Neurosporaxanthin biosynthesis is also regulated by nitrogen in *N. crassa*: incubation under nitrogen starvation increased the levels of *al-1* and *al-2* transcripts [229]. As in *F. fujikuroi*, this activation was independent of the regulation by light, as indicated the similar result obtained with *wc* mutants.

Neurosporaxanthin Overproduction

Deep-pigmented mutants, exhibiting a derepressed neurosporaxanthin production, have been described in *F. fujikuroi* and *F. oxysporum* [138, 205]. Such mutants, generically called *carS*, have been investigated in special detail in *F. fujikuroi*, where they accumulate large amounts of carotenoid irrespective of illumination, contain very high mRNA levels of *car* genes [161, 172, 176], and exhibit high carotenogenic activity in cell-free systems [230]. Despite the large mRNA amounts of the structural *car* genes in the dark, the *carS* mutants still exhibit a detectable mRNA photoinduction [161, 205]. The *F. fujikuroi carS* mutants are also affected in the production of other secondary metabolites like gibberellins or bikaverin [228, 231], suggesting regulatory connections between their respective biosynthetic pathways.

The gene affected in the *Fusarium carS* strains was identified through the characterization of T-DNA insertional mutants of *F. oxysporum* and was confirmed in both species by complementation and by identification of mutations in different *carS* alleles [205, 232]. The predicted CarS polypeptide has sequence similarity with the RF protein CrgA, whose mutation produces a similar carotenoid overproducing phenotype in *M. circinelloides* (*see* section "Mutational deregulation"). The sequence conservation between CarS and CrgA is rather low, but covers the whole length of the respective proteins and the most relevant CrgA proteins domains, suggesting a common but distant origin. The function of the CarS protein appears to be species-specific in the genus *Fusarium*: *carS*-like carotenoid overproducing mutants do not appear in mutagenesis screenings in *F. verticillioides* (J. García-Martínez and V. Díaz-Sánchez, unpublished observations), and have not been described in *F. aquaeductuum* or other *Fusarium species*. On the other hand, the *F. fujikuroi carS* gene restored partially the control of carotenogenesis in a *carS* mutant of *F. oxysporum*, but the restoration was less efficient at transcriptional level [232], indicating an unexpected specificity in the regulation of the pathway in close taxonomical relatives.

Mutants with a high carotenoid content in the dark are unknown in *N. crassa*, but two mutants producing significant amounts of carotenoids compared to the wild type, *ccb-1* and *ccb-2*, were identified through an ingenious selection system [233]. The *al-1* and *al-3* mRNA levels in these mutants in the dark were similar to those of the wild type, indicating a transcriptional independent regulatory mechanism to explain their increased carotenoid content. However, the *ccb-1* mutant contained more carotenoids also in the light, and exhibited a higher photoinduction of *al-1* and *al-3* transcripts. Other mutants, as the already mentioned *vvd*, were identified because of their higher carotenoid content in the light. This was also the case of the mutant called *ovc*, from "overaccumulator of carotenoids" [234]. The *ovc* strain accumulates about twofold more carotenoids than the wild type and exhibits osmotic sensitivity [213]. This pleiotropic phenotype is due to a 77-kb deletion that includes gene *cut-1*, which is involved in osmotic stress [235]. The deletion covers 20 additional presumed ORFs with no apparent connection with the regulation of carotenogenesis.

Neurosporaxanthin production in *N. crassa* is limited by substrate availability. The introduction of the catalytic domain of *S. cerevisiae* 3-hydroxy-3-methylglutaryl coenzyme A reductase, a key enzymatic step in the synthesis of isoprenoids [236], under the control of two different inducible promoters led to increases of up to threefold in the carotenoid content [237].

Biosynthesis of Other Xanthophylls

An economically important xanthophyll is astaxanthin—the pigment responsible for the pink color of salmonids, boiled crustaceans, and flamingoes—and is used as an additive in aquaculture feeds to provide the required pigmentation in the bred animals. Astaxanthin is synthesized by different microorganisms [238] through the introduction of keto and hydroxy groups in the β-rings of the β-carotene molecule. This book contains a specific chapter dedicated to the fungal production of this xanthophyll (see Chap. 9).

The production of other fungal xanthophylls has received attention from the researchers because of their potential biotechnological applications. Many searches have been focused on basidiomycetes species, especially yeasts. The best known example is torularhodin, a torulene derivative with a carboxylic group, actually a 40-carbon counterpart of neurosporaxanthin (Fig. 8.5). Torularhodin has been found in different species, and has been investigated in special detail in the genus *Rhodotorula* [239], where the factors affecting its production have been recently reviewed [240]. In addition to *Rhodotorula*, torularhodin has been also found mixed with other carotenoids in several yeasts, such as *Peniophora* [241], *Cystofilobasidium*, *Rhodosporidium*, *Sporobolomyces*, and *Sporidiobolus* [242–244]. A different torulene derivative, 16′-hydroxytorulene (Fig. 8.5), is produced by some

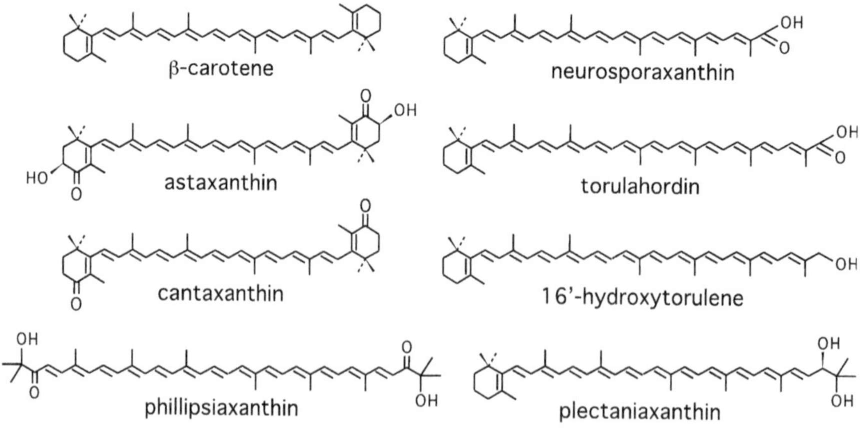

Fig. 8.5 Chemical structures of β-carotene, neurosporaxanthin, and fungal xanthophylls mentioned in the text

Cystofilobasidium species under certain culture conditions [245]. Other yeasts accumulate different xanthophylls (Fig. 8.5): the strong reddish color of *Cantharellus cinnabarinus* is due to the accumulation of canthaxanthin [246], an intermediary compound in astaxanthin biosynthesis, while that of the basidiomycete red yeast *Dioszegia* sp. is due to the production of plectaniaxanthin [247]. As a final example, phillipsiaxanthin, a lycopene derivative with hydroxy and keto groups, is found in the ascomycete *Phillipsia carminea* [248].

Biological Roles of Fungal Carotenoids

Carotenoids are not essential for fungi, as indicates the viability of albino mutants blocked in early steps of their carotenoid pathway in different species. Examples of such mutants are found in the three major taxonomic groups, e.g., the ascomycete *F. fujikuroi* [138], the zygomycete *P. blakesleeanus* [12], or the basidiomycete *X. dendrorhous* [26]. However, the prevalence of carotenoid pathways in a large variety of fungal species strongly point to relevant biological roles, not essential under laboratory conditions. This is clearly the case of the use of carotenoids as a source for apocarotenoid compounds, such as the trisporic acids or the rhodopsin chromophore, retinal (discussed in earlier sections). Putative rhodopsin-encoding genes are found in the genomes of many carotenoid-producing fungi [249], although they are absent in the available genomes from Mucorales. Many of the predicted rhodopsins are potentially photoactive, as indicates the presence of a highly conserved lysine residue required for covalent binding of the retinal chromophore. This is the case for NOP-1 from *N. crassa* [171], which binds retinal and undergoes a photochemical reaction cycle [174, 250]. Possible ion-pumping activities for the *F. fujikuroi* rhodopsins CarO and OpsA are currently under investigation (J. García-Martínez, personal communication).

The carotenoids are also known for their protective properties against oxidative stress [251]. Their protective roles in fungi are supported by many experimental observations: exposure to hydrogen peroxide results in enhanced production of β-carotene in *B. trispora* [252], astaxanthin in *X. dendrorhous* [253] and neurosporaxanthin in *N. crassa* [254] and *F. aquaeductuum* [132]. In *N. crassa*, light-dependent carotenogenesis was also enhanced in mutants of the superoxide dismutase gene *sod-1* [255], and mutants lacking a functional CAT-3 catalase accumulated more carotenoids than the wild type either in the light or in the dark [256]. A most revealing demonstration was provided by the higher resistance to hydrogen peroxide exhibited by *S. cerevisiae* cells, usually devoid of carotenoids, engineered to produce astaxanthin [257].

Further reports providing valuable information are available in the literature. Addition of $FeCl_3$, to *B. trispora* cultures augment the superoxide dismutase (SOD) and catalase levels without changing the β-carotene content [258]. However, elevation of dissolved oxygen concentrations [259] or addition of butylated hydroxytoluene [260] in shake cultures increase SOD and catalase activities as well as β-carotene levels, suggesting a protective role of β-carotene against oxidative stress. Similarly, mated cultures of *B. trispora* contain lower superoxide dismutase (SOD) and catalase activities than a single culture, and addition of SOD inhibitors result in increased β-carotene content [261]. A similar result was obtained inducing the oxidative stress by paraffin addition [262]. Despite these results, different data point to a lower efficiency of β-carotene as a protective agent compared to xanthophylls. The enhanced accumulation of astaxanthin produced by hydrogen peroxide in *X. dendrorhous* is accompanied by a β-carotene reduction [253], and in *Sporobolomyces roseus* an enhanced aeration results in a shift from the predominant β-carotene to torulene and torularhodin [242]. In the basidiomycete *Rhodotorula glutinis*, the chemical generation of singlet oxygen, superoxide anion radicals or peroxy radicals results in a marked increase in torulene and torularhodin but not in that of β-carotene [263]; and a mutant producing higher amounts of torularhodin exhibits a lower susceptibility to injury induced by active oxygen species than a β-carotene overproducing strain [264]. Finally, in this yeast torularhodin inhibits 2,5-diphenyl-3,4-benzofran decomposition by singlet oxygen quenching more efficiently than β-carotene [265].

The protective roles of carotenoids in *N. crassa* are presumably associated with their induction by light [254, 255]. In this fungus, illumination results in enhanced SOD and catalase activities [261]. Albino conidia were more sensitive to visible light than carotenoid-containing conidia in the presence of the photosensitizing agent methylene blue, or to long-wave UVA (380–400 nm) radiation in its absence. Similar results were obtained with blue light and toluidine blue [266], suggesting that carotenoids exert a protecting role by quenching damaging singlet molecular oxygen. Visible light negatively affects respiration of *N. crassa* hyphae, with no alteration in growth, and this effect is enhanced in an albino mutant, indicating a protective effect of carotenoids [267]. Additional evidence was provided by the correlation between carotenoid content in *Neurospora* wild type isolates and the latitude where they were isolated [268]. The strains found at higher latitudes contain less carotenoid, and their conidia are more sensitive to UV exposure. Accordingly, lack of carotenoids results in enhanced sensitivity to UV rays, but not to X-rays [269], although other authors found no difference in UV sensitivity between wild type and albino conidia [270]. An association between light, oxidative stress, and carotene accumulation also has been found in sclerotia-forming fungi. In the ascomycete *Penicillium* sp. [24, 271] and the basidiomycetes *S. rolfsii* [16] and *S. sclerotiorum* [17], sclerotia and β-carotene productions are enhanced by light and oxidative stress, and addition of external β-carotene counteracts the stress-induced sclerotia formation.

Conclusion

Since the identification of the first fungal gene for a specific enzyme of carotenogenesis, *al-1* of *Neurospora crassa*, enormous progress has been reached on the genetics and biochemistry of the synthesis of these pigments in model fungi from representative taxonomic groups. Currently all the genes involved in the synthesis of β-carotene, neurosporaxanthin, and astaxanthin (see Chap. 9), have been identified. The first steps are highly conserved in the investigated fungi and, as anticipated the chemical diversity, more variability has been found for the last enzymatic steps in xanthophyll-producing species. However, less information is available on the regulatory mechanisms. Although most carotenogenic fungi coincide in the induction of the pathway by light and by oxidative stress, the variability in regulatory mutations or the specific effects of certain chemical activators suggest substantial differences between the investigated species. Currently, considerable effort is dedicated to the identification of regulatory networks and genes controlling these pathways. Notably, very little information is available on the mechanisms that control physical location of synthesis and accumulation of carotenoids in the cell, an aspect of particular interest to understand their biological functions.

The carotenoid biosynthetic pathways are not essential in fungi (as shown by its absence in different species), but certainly they should provide adaptive advantages in their natural habitats. It is noteworthy the use of carotenoids as precursors of apocarotenoids with different biochemical or physiological roles, such as retinal and trisporic acids, and new biological functions in the synthesis of other compounds may be discovered in the future. On the other hand, the high carotenoid biosynthetic capacity of some species has made them ideal organisms for industrial production, able to compete with chemical synthesis methods or production by other organisms, such as algae. The vast amount of information accumulated on the genetics and biochemistry of these pathways provide the basis for future biotechnological improvements, and its application to fungi producing new xanthophylls opens promising perspectives.

Acknowledgments We thank Spanish Government (Ministerio de Ciencia y Tecnología, projects BIO2006-01323 and BIO2009-11131) and Andalusian Government (project P07-CVI-02813 and CTS-6638) for funding support.

References

1. Britton G, Liaaen-Jensen S, Pfander H. Carotenoids. Basel: Birkhäuser; 1998.
2. Britton G, Liaaen-Jensen S, Pfander H. Carotenoids: handbook. Boston: Birkhauser; 2004.
3. Sandmann G, Misawa N. Fungal carotenoids. In: Osiewacz HD, editor. The Mycota X industrial applications. Berlin: Springer; 2002. p. 247–62.
4. Sieiro C, Poza M, de Miguel T, Villa TG. Genetic basis of microbial carotenogenesis. Int Microbiol. 2003;6:11–6.
5. Fraser PD, Bramley PM. The biosynthesis and nutritional uses of carotenoids. Prog Lipid Res. 2004;43:228–65.
6. Domonkos I, Kis M, Gombos Z, Ughy B. Carotenoids, versatile components of oxygenic photosynthesis. Prog Lipid Res. 2013;52:539–61.
7. Blomhoff R, Blomhoff HK. Overview of retinoid metabolism and function. J Neurobiol. 2006;66:606–30.
8. Moran NA, Jarvik T. Lateral transfer of genes from fungi underlies carotenoid production in aphids. Science. 2010;328:624–7.
9. Avalos J, Cerdá-Olmedo E. Fungal carotenoid production. In: Arora DK, editor. Handbook of fungal biotechnology. 2nd ed. New York: Marcel Dekker, Inc.; 2004. p. 367–78.

10. Del Campo JA, García-González M, Guerrero MG. Outdoor cultivation of microalgae for carotenoid production: current state and perspectives. Appl Microbiol Biotechnol. 2007;74:1163–74.
11. Rohmer M, Knani M, Simonin P, Sutter B, Sahm H. Isoprenoid biosynthesis in bacteria: a novel pathway for the early steps leading to isopentenyl diphosphate. Biochem J. 1993;15: 517–24.
12. Cerdá-Olmedo E. Carotene. In: Cerdá-Olmedo E, Lipson ED, editors. Phycomyces. Cold Spring Harbor: Cold Spring Harbor Laboratory Press; 1987. p. 199–222.
13. Navarro E, Sandmann G, Torres-Martínez S. Mutants of the carotenoid biosynthetic pathway of *Mucor circinelloides*. Exp Mycol. 1995;19:186–90.
14. Fraser PD, Ruiz-Hidalgo MJ, López-Matas MA, Álvarez MI, Eslava AP, Bramley PM. Carotenoid biosynthesis in wild type and mutant strains of *Mucor circinelloides*. Biochim Biophys Acta. 1996;1289:203–8.
15. de Miguel T, Calo P, Díaz A, Villa TG. The genus *Rhodosporidium*: a potential source of β-carotene. Microbiologia. 1997;13:67–70.
16. Georgiou CD, Zervoudakis G, Tairis N, Kornaros M. β-Carotene production and its role in sclerotial differentiation of *Sclerotium rolfsii*. Fungal Genet Biol. 2001;34:11–20.
17. Georgiou CD, Tairis N, Polycratis A. Production of β-carotene by *Sclerotinia sclerotiorum* and its role in sclerotium differentiation. Mycol Res. 2001;105:1110–5.
18. Han M, He Q, Zhang WG. Carotenoids production in different culture conditions by *Sporidiobolus pararoseus*. Prep Biochem Biotechnol. 2012;42:293–303.
19. Estrada AF, Brefort T, Mengel C, Díaz-Sánchez V, Alder A, Al-Babili S, et al. *Ustilago maydis* accumulates β-carotene at levels determined by a retinal-forming carotenoid oxygenase. Fungal Genet Biol. 2010;46:803–13.
20. Will OH, Ruddat M, Garber ED, Kezdy FJ. Characterization of carotene accumulation in *Ustilago violacea* using high-performance liquid chromatography. Curr Microbiol. 1984;10:57–63.
21. Will OH, Ruddat M, Newland NA. Characterization of carotene accumulation in species of the fungal genus *Ustilago* using high-performance liquid chromatography. Bot Gaz. 1985;146:204–7.
22. El-Jack M, Mackenzie A, Bramley PM. The photoregulation of carotenoid biosynthesis in *Aspergillus giganteus* mut. *alba*. Planta. 1988;174:59–66.
23. Daub ME, Payne GA. The role of carotenoids in resistance of fungi to cercosporin. Phytopathology. 1989;79:180–5.
24. Han JR, Zhao WJ, Gao YY, Yuan JM. Effect of oxidative stress and exogenous β-carotene on sclerotial differentiation and carotenoid yield of *Penicillium* sp. PT95. Lett Appl Microbiol. 2005;40:412–7.
25. van Eijk GW, Mummery RS, Roeymans HJ, Valadon LR. A comparative study of carotenoids of *Aschersonia aleyroides* and *Aspergillus giganteus*. Antonie Van Leeuwenhoek. 1979;45:417–22.
26. Johnson EA. *Phaffia rhodozyma*: colorful odyssey. Int Microbiol. 2003;6:169–74.
27. Bhosale PB, Gadre RV. Production of β-carotene by a mutant of *Rhodotorula glutinis*. Appl Microbiol Biotechnol. 2001;55:423–7.
28. Bejarano ER, Govind NS, Cerdá-Olmedo E. ξ-Carotene and other carotenes in a *Phycomyces* mutant. Phytochemistry. 1987;26:2251–4.
29. Ruiz-Hidalgo MJ, Benito EP, Sandmann G, Eslava AP. The phytoene dehydrogenase gene of *Phycomyces*: regulation of its expression by blue light and vitamin A. Mol Gen Genet. 1997;253:734–44.
30. Velayos A, Blasco JL, Alvarez MI, Iturriaga EA, Eslava AP. Blue-light regulation of phytoene dehydrogenase (*carB*) gene expression in *Mucor circinelloides*. Planta. 2000;210:938–46.
31. Rodríguez-Saiz M, Paz B, De la Fuente JL, López-Nieto MJ, Cabri W, Barredo JL. *Blakeslea trispora* genes for carotene biosynthesis. Appl Environ Microbiol. 2004;70:5589–94.
32. Ehrenshaft M, Daub ME. Isolation, sequence, and characterization of the *Cercospora nicotianae* phytoene dehydrogenase gene. Appl Environ Microbiol. 1994;60:2766–71.

33. Verdoes JC, Krubasik KP, Sandmann G, van Ooyen AJ. Isolation and functional characterisation of a novel type of carotenoid biosynthetic gene from *Xanthophyllomyces dendrorhous*. Mol Gen Genet. 1999;262:453–61.
34. Arrach N, Fernández-Martín R, Cerdá-Olmedo E, Avalos J. A single gene for lycopene cyclase, phytoene synthase, and regulation of carotene biosynthesis in *Phycomyces*. Proc Natl Acad Sci USA. 2001;98:1687–92.
35. Velayos A, Eslava AP, Iturriaga EA. A bifunctional enzyme with lycopene cyclase and phytoene synthase activities is encoded by the *carRP* gene of *Mucor circinelloides*. Eur J Biochem. 2000;267:5509–19.
36. Sanz C, Velayos A, Alvarez MI, Benito EP, Eslava AP. Functional analysis of the *Phycomyces carRA* gene encoding the enzymes phytoene synthase and lycopene cyclase. PLoS One. 2011;6:e23102.
37. Torres-Martínez S, Murillo FJ, Cerdá-Olmedo E. Genetics of lycopene cyclization and substrate transfer in β-carotene biosynthesis in *Phycomyces*. Genet Res. 1980;36:299–309.
38. Breitenbach J, Fraser PD, Sandmann G. Carotenoid synthesis and phytoene synthase activity during mating of *Blakeslea trispora*. Phytochemistry. 2012;76:40–5.
39. Cerdá-Olmedo E. Carotene mutants of *Phycomyces*. Methods Enzymol. 1985;110:220–43.
40. Velayos A, López-Matas MA, Ruiz-Hidalgo MJ, Eslava AP. Complementation analysis of carotenogenic mutants of *Mucor circinelloides*. Fungal Genet Biol. 1997;22:19–27.
41. Mehta BJ, Cerdá-Olmedo E. Mutants of carotene production in *Blakeslea trispora*. Appl Microbiol Biotechnol. 1995;42:836–8.
42. Mehta BJ, Obraztsova IN, Cerdá-Olmedo E. Mutants and intersexual heterokaryons of *Blakeslea trispora* for production of β-carotene and lycopene. Appl Environ Microbiol. 2003;69:4043–8.
43. Goodwin TW. The biochemistry of the carotenoids. 2nd ed. London: Chapman & Hall; 1980.
44. Fraser PD, Bramley PM. The purification of phytoene dehydrogenase from *Phycomyces blakesleeanus*. Biochim Biophys Acta. 1994;1212:59–66.
45. Ootaki T, Crafts-Lighty A, Delbrück M, Hsu WJ. Complementation between mutants of *Phycomyces* deficient with respect to carotenogenesis. Mol Gen Genet. 1973;121: 57–70.
46. Avalos J, Cerdá-Olmedo E. Chemical modification of carotenogenesis in *Gibberella fujikuroi*. Phytochemistry. 1986;25:1837–41.
47. Bejarano ER, Cerdá-Olmedo E. Inhibition of phytoene dehydrogenation and activation of carotenogenesis in *Phycomyces*. Phytochemistry. 1989;28:1623–6.
48. De la Guardia MD, Aragón CM, Murillo FJ, Cerdá-Olmedo E. A carotenogenic enzyme aggregate in *Phycomyces*: evidence from quantitive complementation. Proc Natl Acad Sci USA. 1971;68:2012–5.
49. Aragón CM, Murillo FJ, de la Guardia MD, Cerdá-Olmedo E. An enzyme complex for the dehydrogenation of phytoene in *Phycomyces*. Eur J Biochem. 1976;63:71–5.
50. Candau R, Bejarano ER, Cerdá-Olmedo E. *In vivo* channeling of substrates in an enzyme aggregate for β-carotene biosynthesis. Proc Natl Acad Sci USA. 1991;88:4936–40.
51. Sanz C, Alvarez MI, Orejas M, Velayos A, Eslava AP, Benito EP. Interallelic complementation provides genetic evidence for the multimeric organization of the *Phycomyces blakesleeanus* phytoene dehydrogenase. Eur J Biochem. 2002;269:902–8.
52. Mehta BJ, Cerdá-Olmedo E. Lycopene cyclization in *Blakeslea trispora*. Mycoscience. 1999;40:307–10.
53. Lampila LE, Wallen SE, Bullerman LB. A review of factors affecting biosynthesis of carotenoids by the order Mucorales. Mycopathologia. 1985;90:65–80.
54. Avalos J, Bejarano ER, Cerdá-Olmedo E. Photoinduction of carotenoid biosynthesis. Methods Enzymol. 1993;214:283–94.
55. Bergman K, Eslava AP, Cerdá-Olmedo E. Mutants of *Phycomyces* with abnormal phototropism. Mol Gen Genet. 1973;123:1–16.
56. Bejarano ER, Avalos J, Lipson ED, Cerdá-Olmedo E. Photoinduced accumulation of carotene in *Phycomyces*. Planta. 1991;183:1–9.

57. Sutter RP. Effect of light on β-carotene accumulation in *Blakeslea trispora*. J Gen Microbiol. 1970;64:215–21.
58. Blasco JL, Roessink D, Iturriaga EA, Eslava AP, Galland P. Photocarotenogenesis in *Phycomyces*: expression of the *carB* gene encoding phytoene dehydrogenase. J Plant Res. 2001;114:25–31.
59. Almeida ER, Cerdá-Olmedo E. Gene expression in the regulation of carotene biosynthesis in *Phycomyces*. Curr Genet. 2008;53:129–37.
60. Sanz C, Benito EP, Orejas M, Alvarez MI, Eslava AP. Protein-DNA interactions in the promoter region of the *Phycomyces carB* and *carRA* genes correlate with the kinetics of their mRNA accumulation in response to light. Fungal Genet Biol. 2010;47:773–81.
61. Quiles-Rosillo MD, Ruiz-Vázquez RM, Torres-Martínez S, Garre V. Light induction of the carotenoid biosynthesis pathway in *Blakeslea trispora*. Fungal Genet Biol. 2005;42:141–53.
62. López-Díaz I, Cerdá-Olmedo E. Relationship of photocarotenogenesis to other behavioural and regulatory responses in *Phycomyces*. Planta. 1980;150:134–9.
63. Jayaram M, Leutwiler L, Delbrück M. Light-induced carotene synthesis in mutants of *Phycomyces* with abnormal phototropism. Photochem Photobiol. 1980;32:241–5.
64. Flores R, Cerdá-Olmedo E, Corrochano LM. Separate sensory pathways for photomorphogenesis in *Phycomyces*. Photochem Photobiol. 1998;67:467–72.
65. Tsolakis G, Parashi E, Galland P, Kotzabasis K. Blue light signaling chains in *Phycomyces*: phototransduction of carotenogenesis and morphogenesis involves distinct protein kinase/phosphatase elements. Fungal Genet Biol. 1999;28:201–13.
66. Idnurm A, Rodríguez-Romero J, Corrochano LM, Sanz C, Iturriaga EA, Eslava AP, et al. The *Phycomyces madA* gene encodes a blue-light photoreceptor for phototropism and other light responses. Proc Natl Acad Sci USA. 2006;103:4546–51.
67. Liu Y, He Q, Cheng P. Photoreception in *Neurospora*: a tale of two White Collar proteins. Cell Mol Life Sci. 2003;60:2131–8.
68. Sanz C, Rodríguez-Romero J, Idnurm A, Christie JM, Heitman J, Corrochano LM, et al. *Phycomyces* MADB interacts with MADA to form the primary photoreceptor complex for fungal phototropism. Proc Natl Acad Sci USA. 2009;106:7095–100.
69. Corrochano LM, Garre V. Photobiology in the Zygomycota: multiple photoreceptor genes for complex responses to light. Fungal Genet Biol. 2010;47:893–9.
70. Torres-Martínez S, Ruiz-Vázquez RM, Garre V, López-García S, Navarro E, Vila A. Molecular tools for carotenogenesis analysis in the zygomycete *Mucor circinelloides*. Methods Mol Biol. 2012;898:85–107.
71. Silva F, Torres-Martínez S, Garre V. Distinct *white collar-1* genes control specific light responses in *Mucor circinelloides*. Mol Microbiol. 2006;61:1023–37.
72. Murillo FJ, Cerdá-Olmedo E. Regulation of carotene synthesis in *Phycomyces*. Mol Gen Genet. 1976;148:19–24.
73. Salgado LM, Bejarano ER, Cerdá-Olmedo E. Carotene-superproducing mutants of *Phycomyces blakesleeanus*. Exp Mycol. 1989;13:332–6.
74. Mehta BJ, Salgado LM, Bejarano ER, Cerdá-Olmedo E. New mutants of *Phycomyces blakesleeanus* for β-carotene production. Appl Environ Microbiol. 1997;63:3657–61.
75. Salgado LM, Avalos J, Bejarano ER, Cerdá-Olmedo E. Correlation between *in vivo* and *in vitro* carotenogenesis in *Phycomyces*. Phytochemistry. 1991;30:2587–91.
76. Murillo FJ, Torres-Martínez S, Aragón CM, Cerdá-Olmedo E. Substrate transfer in carotene biosynthesis in *Phycomyces*. Eur J Biochem. 1981;119:511–6.
77. Corrochano LM, Cerdá-Olmedo E. Photomorphogenesis in behavioural and colour mutants of *Phycomyces*. J Photochem Photobiol B. 1990;6:325–35.
78. Revuelta JL, Eslava AP. A new gene (*carC*) involved in the regulation of carotenogenesis in *Phycomyces*. Mol Gen Genet. 1983;192:225–9.
79. Bejarano ER, Cerdá-Olmedo E. Independence of the carotene and sterol pathways of *Phycomyces*. FEBS Lett. 1992;306:209–12.
80. Kuzina V, Domenech C, Cerdá-Olmedo E. Relationships among the biosyntheses of ubiquinone, carotene, sterols, and triacylglycerols in zygomycetes. Arch Microbiol. 2006;186:485–93.

81. Navarro E, Ruiz-Pérez VL, Torres-Martínez S. Overexpression of the *crgA* gene abolishes light requirement for carotenoid biosynthesis in *Mucor circinelloides*. Eur J Biochem. 2000;267:800–7.
82. Navarro E, Lorca-Pascual JM, Quiles-Rosillo MD, Nicolás FE, Garre V, Torres-Martínez S, et al. A negative regulator of light-inducible carotenogenesis in *Mucor circinelloides*. Mol Genet Genomics. 2001;266:463–70.
83. Lorca-Pascual JM, Murcia-Flores L, Garre V, Torres-Martínez S, Ruiz-Vázquez RM. The RING-finger domain of the fungal repressor *crgA* is essential for accurate light regulation of carotenogenesis. Mol Microbiol. 2004;52:1463–74.
84. Silva F, Navarro E, Peñaranda A, Murcia-Flores L, Torres-Martínez S, Garre V. A RING-finger protein regulates carotenogenesis via proteolysis-independent ubiquitylation of a White Collar-1-like activator. Mol Microbiol. 2008;70:1026–36.
85. Quiles-Rosillo MD, Torres-Martínez S, Garre V. *cigA*, a light-inducible gene involved in vegetative growth in *Mucor circinelloides* is regulated by the carotenogenic repressor *crgA*. Fungal Genet Biol. 2003;38:122–32.
86. Murcia-Flores L, Lorca-Pascual JM, Garre V, Torres-Martínez S, Ruiz-Vázquez RM. Non-AUG translation initiation of a fungal RING finger repressor involved in photocarotenogenesis. J Biol Chem. 2007;282:15394–403.
87. Bejarano ER, Parra F, Murillo FJ, Cerdá-Olmedo E. End-product regulation of carotenogenesis in *Phycomyces*. Arch Microbiol. 1988;150:209–14.
88. Eslava AP, Alvarez MI, Cerdá-Olmedo E. Regulation of carotene biosynthesis in *Phycomyces* by vitamin A and β-ionone. Eur J Biochem. 1974;48:617–23.
89. Cerdá-Olmedo E, Hüttermann A. Förderung und Hemmung der Carotinsynthese bei *Phycomyces* durch Aromaten. Angew Botanik. 1986;60:59–70.
90. Choudhari SM, Ananthanarayan L, Singhal RS. Use of metabolic stimulators and inhibitors for enhanced production of β-carotene and lycopene by *Blakeslea trispora* NRRL 2895 and 2896. Bioresour Technol. 2008;99:3166–73.
91. Ninet L, Renaut J, Tissier R. Activation of the biosynthesis of carotenoids by *Blakeslea trispora*. Biotechnol Bioeng. 1969;11:1195–210.
92. Hu X, Sun J, Yuan Q. Improved β-carotene biosynthesis and gene transcription in *Blakeslea trispora* with arachidonic acid. Biotechnol Lett. 2012;34:2107–11.
93. Tang Q, Li Y, Yuan QP. Effects of an ergosterol synthesis inhibitor on gene transcription of terpenoid biosynthesis in *Blakeslea trispora*. Curr Microbiol. 2008;57:527–31.
94. Blakeslee AF. Sexual reproduction in the Mucorineae. Proc Am Acad Arts Sci. 1904;40:205–319.
95. Burgeff H. Untersuchungen über Sexualität und Parasitismus bei Mucorineen. Bot Abh. 1924;4:1–135.
96. Govind NS, Cerdá-Olmedo E. Sexual activation of carotenogenesis in *Phycomyces blakesleeanus*. J Gen Microbiol. 1986;132:2775–80.
97. Sutter RP, Harrison TL, Galasko G. Trisporic acid biosynthesis in *Blakeslea trispora* via mating type-specific precursors. J Biol Chem. 1974;249:2282–4.
98. Sutter RP. Sexual development. In: Cerdá-Olmedo E, Lipson ED, editors. Phycomyces. Cold Spring Harbor: Cold Spring Harbor Laboratory Press; 1987. p. 317–36.
99. Sahadevan Y, Richter M, Hoffmann K, Voigt K, Boland W. Early and late trisporoids differentially regulate β-carotene production and gene transcript levels in the mucoralean fungi *Blakeslea trispora* and *Mucor mucedo*. Appl Environ Microbiol. 2013;79(23):7466–75.
100. Murillo FJ, Calderón IL, López-Díaz I, Cerdá-Olmedo E. Carotene-superproducing strains of *Phycomyces*. Appl Environ Microbiol. 1978;36:639–42.
101. Austin DG, Bu'Lock JD, Winstanley DJ. Trisporic acid biosynthesis and carotenogenesis in *Blakesleea trispora*. Biochem J. 1969;113:34P.
102. Sutter RP. Mutations affecting sexual development in *Phycomyces blakesleeanus*. Proc Natl Acad Sci USA. 1975;72:127–30.
103. Kuzina V, Cerdá-Olmedo E. Modification of sexual development and carotene production by acetate and other small carboxylic acids in *Blakeslea trispora* and *Phycomyces blakesleeanus*. Appl Environ Microbiol. 2006;72:4917–22.

104. Schmidt AD, Heinekamp T, Matuschek M, Liebmann B, Bollschweiler C, Brakhage AA. Analysis of mating-dependent transcription of *Blakeslea trispora* carotenoid biosynthesis genes *carB* and *carRA* by quantitative real-time PCR. Appl Microbiol Biotechnol. 2005;67:549–55.
105. Kuzina V, Ramírez-Medina H, Visser H, van Ooyen AJ, Cerdá-Olmedo E, van den Berg JA. Genes involved in carotene synthesis and mating in *Blakeslea trispora*. Curr Genet. 2008;54:143–52.
106. Cerdá-Olmedo E. Production of carotenoids with fungi. In: Vandamme E, editor. Biotechnology of vitamin, growth factor and pigment production. London: Elsevier Applied Science; 1989. p. 27–42.
107. Mehta BJ, Cerdá-Olmedo E. Intersexual partial diploids of *Phycomyces*. Genetics. 2001;158:635–41.
108. Lilly VG, Barnett HL, Krause RF. The production of carotene by *Phycomyces blakesleeanus*. W Va Agric Exp Stn Bull. 1960;441T:1–80.
109. Csernetics A, Nagy G, Iturriaga EA, Szekeres A, Eslava AP, Vágvölgyi C, et al. Expression of three isoprenoid biosynthesis genes and their effects on the carotenoid production of the zygomycete *Mucor circinelloides*. Fungal Genet Biol. 2011;48:696–703.
110. Papp T, Velayos A, Bartok T, Eslava AP, Vagvolgyi C, Iturriaga EA. Heterologous expression of astaxanthin biosynthesis genes in *Mucor circinelloides*. Appl Microbiol Biotechnol. 2006;69:526–31.
111. Papp T, Csernetics A, Nagy G, Bencsik O, Iturriaga EA, Eslava AP, et al. Canthaxanthin production with modified *Mucor circinelloides* strains. Appl Microbiol Biotechnol. 2013;97:4937–50.
112. Nicolás-Molina FE, Navarro E, Ruiz-Vázquez RM. Lycopene over-accumulation by disruption of the negative regulator gene *crgA* in *Mucor circinelloides*. Appl Microbiol Biotechnol. 2008;78:131–7.
113. Austin DJ, Bu'Lock JD, Drake D. The biosynthesis of trisporic acids from β-carotene via retinal and trisporol. Experientia. 1970;26:348–9.
114. Caglioti L, Cainelli G, Camerino B, Mondelli R, Prieto A, Quilico A, et al. The structure of trisporic-C acid. Tetrahedron. 1966;22:175–87.
115. Sutter RP, Dadok J, Bothner-By AA, Smith RR, Mishra PK. Cultures of separated mating types of *Blakeslea trispora* make D and E forms of trisporic acids. Biochemistry. 1989;28:4060–6.
116. Barrero AF, Herrador MM, Arteaga P, Gil J, González JA, Alcalde E, et al. New apocarotenoids and β-carotene cleavage in *Blakeslea trispora*. Org Biomol Chem. 2011;9:7190–5.
117. Polaino S, González-Delgado JA, Arteaga P, Herrador MM, Barrero AF, Cerdá-Olmedo E. Apocarotenoids in the sexual interaction of *Phycomyces blakesleeanus*. Org Biomol Chem. 2012;10:3002–9.
118. Burmester A, Richter M, Schultze K, Voelz K, Schachtschabel D, Boland W, et al. Cleavage of β-carotene as the first step in sexual hormone synthesis in zygomycetes is mediated by a trisporic acid regulated β-carotene oxygenase. Fungal Genet Biol. 2007;44:1096–108.
119. Polaino S, Herrador MM, Cerdá-Olmedo E, Barrero AF. Splitting of β-carotene in the sexual interaction of *Phycomyces*. Org Biomol Chem. 2010;8:4229–31.
120. Tagua VG, Medina HR, Martín-Domínguez R, Eslava AP, Corrochano LM, Cerdá-Olmedo E, et al. A gene for carotene cleavage required for pheromone biosynthesis and carotene regulation in the fungus *Phycomyces blakesleeanus*. Fungal Genet Biol. 2012;49:398–404.
121. Medina HR, Cerdá-Olmedo E, Al-Babili S. Cleavage oxygenases for the biosynthesis of trisporoids and other apocarotenoids in *Phycomyces*. Mol Microbiol. 2011;82:199–208.
122. Salgado LM, Cerdá-Olmedo E. Genetic interactions in the regulation of carotenogenesis in *Phycomyces*. Curr Genet. 1992;21:67–71.
123. Czempinski K, Kruft V, Wöstemeyer J, Burmester A. 4-Dihydromethyltrisporate dehydrogenase from *Mucor mucedo*, an enzyme of the sexual hormone pathway: purification, and cloning of the corresponding gene. Microbiology. 1996;142:2647–54.

124. Schimek C, Petzold A, Schultze K, Wetzel J, Wolschendorf F, Burmester A, et al. 4-Dihydromethyltrisporate dehydrogenase, an enzyme of the sex hormone pathway in *Mucor mucedo*, is constitutively transcribed but its activity is differently regulated in (+) and (-) mating types. Fungal Genet Biol. 2005;42:804–12.
125. Wetzel J, Scheibner O, Burmester A, Schimek C, Wöstemeyer J. 4-dihydrotrisporin-dehydrogenase, an enzyme of the sex hormone pathway of *Mucor mucedo*: purification, cloning of the corresponding gene, and developmental expression. Eukaryot Cell. 2009;8:88–95.
126. Schimek C, Wöstemeyer J. Carotene derivatives in sexual communication of zygomycete fungi. Phytochemistry. 2009;70:1867–75.
127. Polaino S. Apocarotenoides en la interacción sexual de *Phycomyces*. PhD thesis, University of Seville; 2009.
128. Haxo F. Studies on the carotenoid pigments of *Neurospora*; composition of the pigment. Arch Biochem. 1949;20:400–21.
129. Zalokar M. Studies on biosynthesis of carotenoids in *Neurospora crassa*. Arch Biochem Biophys. 1954;50:71–80.
130. Zalokar M. Isolation of an acidic pigment in *Neurospora*. Arch Biochem Biophys. 1957;70:568–71.
131. Aasen AJ, Jensen SL. Fungal carotenoids II. The structure of the carotenoid acid neurosporaxanthin. Acta Chem Scand. 1965;19:1843–53.
132. Bindl E, Lang W, Rau W. Untersuchungen über die lichtabhängige Carotinoidsynthese. VI. Zeitlicher Verlauf der Synthese der einzelnen Carotinoide bei Fusarium aquaeductuum unter verschiedenen Induktionsbedingungen. Planta. 1970;94:156–74.
133. Valadon LRG, Mummery RS. Biosynthesis of neurosporaxanthin. Microbios. 1969;1A:3–8.
134. Valadon LRG, Osman M, Mummery RS, Jerebzoff-Quintin S, Jerebzoff S. The effect of monochromatic radiation in the range 350 to 750 nm on the carotenogenesis in *Verticillium agaricinum*. Physiol Plant. 1982;56:199–203.
135. Strobel I, Breitenbach J, Scheckhuber CQ, Osiewacz HD, Sandmann G. Carotenoids and carotenogenic genes in *Podospora anserina*: engineering of the carotenoid composition extends the life span of the mycelium. Curr Genet. 2009;55:175–84.
136. Harding RW, Huang PC, Mitchell HK. Photochemical studies of the carotenoid biosynthetic pathway in *Neurospora crassa*. Arch Biochem Biophys. 1969;129:696–707.
137. Mitzka U, Rau W. Composition and photoinduced biosynthesis of the carotenoids of a protoplast-like *Neurospora crassa* "slime" mutant. Arch Microbiol. 1977;111:261–3.
138. Avalos J, Cerdá-Olmedo E. Carotenoid mutants of *Gibberella fujikuroi*. Curr Genet. 1987;25:1837–41.
139. Valadon LRG, Mummery RS. Natural β-apo-4′-carotenoic acid methyl ester in the fungus *Verticillium agaricinum*. Phytochemistry. 1977;16:613–4.
140. Sakaki H, Kaneno H, Sumiya Y, Tsushima M, Miki W, Kishimoto N, et al. A new carotenoid glycosyl ester isolated from a marine microorganism, *Fusarium* strain T-1. J Nat Prod. 2002;65:1683–4.
141. Spurgeon SL, Turner RV, Harding RW. Biosynthesis of phytoene from isopentenyl pyrophosphate by a *Neurospora* enzyme system. Arch Biochem Biophys. 1979;195:23–9.
142. Huang PC. Recombination and complementation of albino mutants in *Neurospora*. Genetics. 1964;49:453–69.
143. Wang SS, Magill JM, Phillips RW. Auxotrophic and visible mutations in white spore (*ws-1*). Neurospora Newsl. 1971;18:16–7.
144. Goldie AH, Subden RE. The neutral carotenoids of wild-type and mutant strains of *Neurospora crassa*. Biochem Genet. 1973;10:275–84.
145. Nelson MA, Morelli G, Carattoli A, Romano N, Macino G. Molecular cloning of a *Neurospora crassa* carotenoid biosynthetic gene (*albino-3*) regulated by blue light and the products of the *white collar* genes. Mol Cell Biol. 1989;9:1271–6.
146. Carattoli A, Romano N, Ballario P, Morelli G, Macino G. The *Neurospora crassa* carotenoid biosynthetic gene (*albino 3*) reveals highly conserved regions among prenyltransferases. J Biol Chem. 1991;266:5854–9.

147. Sandmann G, Misawa N, Wiedemann M, Vittorioso P, Carattoli A, Morelli G, et al. Functional identification of *al-3* from *Neurospora crassa* as the gene for geranylgeranyl pyrophosphate synthase by complementation with *crt* genes, *in vitro* characterization of the gene product and mutant analysis. J Photochem Photobiol B. 1993;18:245–51.
148. Barbato C, Calissano M, Pickford A, Romano N, Sandmann G, Macino G. Mild RIP-an alternative method for in vivo mutagenesis of the *albino-3* gene in *Neurospora crassa*. Mol Gen Genet. 1996;252:353–61.
149. Schmidhauser TJ, Lauter FR, Russo VE, Yanofsky C. Cloning, sequence, and photoregulation of *al-1*, a carotenoid biosynthetic gene of *Neurospora crassa*. Mol Cell Biol. 1990;10:5064–70.
150. Bartley GE, Schmidhauser TJ, Yanofsky C, Scolnik PA. Carotenoid desaturases from *Rhodobacter capsulatus* and *Neurospora crassa* are structurally and functionally conserved and contain domains homologous to flavoprotein disulfide oxidoreductases. J Biol Chem. 1990;265:16020–4.
151. Sandmann G. Evolution of carotene desaturation: the complication of a simple pathway. Arch Biochem Biophys. 2009;483:169–74.
152. Hausmann A, Sandmann G. A single five-step desaturase is involved in the carotenoid biosynthesis pathway to β-carotene and torulene in *Neurospora crassa*. Fungal Genet Biol. 2000;30:147–53.
153. Fernández-Martín R, Cerdá-Olmedo E, Avalos J. Homologous recombination and allele replacement in transformants of *Fusarium fujikuroi*. Mol Gen Genet. 2000;263: 838–45.
154. Linnemannstöns P, Prado MM, Fernández-Martín R, Tudzynski B, Avalos J. A carotenoid biosynthesis gene cluster in *Fusarium fujikuroi*: the genes *carB* and *carRA*. Mol Genet Genomics. 2002;267:593–602.
155. Prado-Cabrero A, Schaub P, Díaz-Sánchez V, Estrada AF, Al-Babili S, Avalos J. Deviation of the neurosporaxanthin pathway towards β-carotene biosynthesis in *Fusarium fujikuroi* by a point mutation in the phytoene desaturase gene. FEBS J. 2009;276:4582–97.
156. Schmidhauser TJ, Lauter FR, Schumacher M, Zhou W, Russo VE, Yanofsky C. Characterization of *al-2*, the phytoene synthase gene of *Neurospora crassa*. Cloning, sequence analysis, and photoregulation. J Biol Chem. 1994;269:12060–6.
157. Arrach N, Schmidhauser TJ, Avalos J. Mutants of the carotene cyclase domain of *al-2* from *Neurospora crassa*. Mol Genet Genomics. 2002;266:914–21.
158. Díaz-Sánchez V, Estrada AF, Trautmann D, Limón MC, Al-Babili S, Avalos J. Analysis of *al-2* mutations in *Neurospora*. PLoS One. 2011;6:e21948.
159. Sandmann G, Zhu C, Krubasik P, Fraser PD. The biotechnological potential of the *al-2* gene from *Neurospora crassa* for the production of monocyclic keto hydroxy carotenoids. Biochim Biophys Acta. 2006;1761:1085–92.
160. Sui X, Kiser PD, Lintig JV, Palczewski K. Structural basis of carotenoid cleavage: from bacteria to mammals. Arch Biochem Biophys. 2013;539(2):203–13.
161. Prado-Cabrero A, Estrada AF, Al-Babili S, Avalos J. Identification and biochemical characterization of a novel carotenoid oxygenase: elucidation of the cleavage step in the *Fusarium* carotenoid pathway. Mol Microbiol. 2007;64:448–60.
162. Saelices L, Youssar L, Holdermann I, Al-Babili S, Avalos J. Identification of the gene responsible for torulene cleavage in the *Neurospora* carotenoid pathway. Mol Genet Genomics. 2007;278:527–37.
163. Jin JM, Lee J, Lee YW. Characterization of carotenoid biosynthetic genes in the ascomycete *Gibberella zeae*. FEMS Microbiol Lett. 2010;302:197–202.
164. Estrada AF, Youssar L, Scherzinger D, Al-Babili S, Avalos J. The *ylo-1* gene encodes an aldehyde dehydrogenase responsible for the last reaction in the *Neurospora* carotenoid pathway. Mol Microbiol. 2008;69:1207–20.
165. Díaz-Sánchez V, Estrada AF, Trautmann D, Al-Babili S, Avalos J. The gene *carD* encodes the aldehyde dehydrogenase responsible for neurosporaxanthin biosynthesis in *Fusarium fujikuroi*. FEBS J. 2011;278:3164–76.

166. Sandmann G. Photoregulation of carotenoid biosynthesis in mutants of *Neurospora crassa*: activities of enzymes involved in the synthesis and conversion of phytoene. Z Naturforsch. 1993;48c:570–4.
167. Sandmann G, Takaichi S, Fraser PD. C(35)-apocarotenoids in the yellow mutant *Neurospora crassa* YLO. Phytochemistry. 2008;69:2886–90.
168. Mitzka-Schnabel U, Rau W. The subcellular distribution of carotenoids in *Neurospora crassa*. Phytochemistry. 1980;19:1409–13.
169. Mitzka-Schnabel U. Carotenogenic enzymes from *Neurospora*. Pure Appl Chem. 1985;57: 667–9.
170. Domenech CE, Giordano W, Avalos J, Cerdá-Olmedo E. Separate compartments for the production of sterols, carotenoids and gibberellins in *Gibberella fujikuroi*. Eur J Biochem. 1996;239:720–5.
171. Bieszke JA, Braun EL, Bean LE, Kang S, Natvig DO, Borkovich KA. The *nop-1* gene of *Neurospora crassa* encodes a seven transmembrane helix retinal-binding protein homologous to archaeal rhodopsins. Proc Natl Acad Sci USA. 1999;96:8034–9.
172. Prado MM, Prado-Cabrero A, Fernández-Martín R, Avalos J. A gene of the opsin family in the carotenoid gene cluster of *Fusarium fujikuroi*. Curr Genet. 2004;46:47–58.
173. Estrada AF, Avalos J. Regulation and targeted mutation of *opsA*, coding for the NOP-1 opsin orthologue in *Fusarium fujikuroi*. J Mol Biol. 2009;387:59–73.
174. Bieszke JA, Spudich EN, Scott KL, Borkovich KA, Spudich JL. A eukaryotic protein, NOP-1, binds retinal to form an archaeal rhodopsin-like photochemically reactive pigment. Biochemistry. 1999;38:14138–45.
175. Wyss A. Carotene oxygenases: a new family of double bond cleavage enzymes. J Nutr. 2004;134:246S–50.
176. Thewes S, Prado-Cabrero A, Prado MM, Tudzynski B, Avalos J. Characterization of a gene in the *car* cluster of *Fusarium fujikuroi* that codes for a protein of the carotenoid oxygenase family. Mol Genet Genomics. 2005;274:217–28.
177. Prado-Cabrero A, Scherzinger D, Avalos J, Al-Babili S. Retinal biosynthesis in fungi: characterization of the carotenoid oxygenase CarX from *Fusarium fujikuroi*. Eukaryot Cell. 2007;6:650–7.
178. Díaz-Sánchez V, Estrada AF, Limón MC, Al-Babili S, Avalos J. The oxygenase CAO-1 of *Neurospora crassa* is a resveratrol cleavage enzyme. Eukaryot Cell. 2013;12:1305–14.
179. Avalos J, Estrada AF. Regulation by light in *Fusarium*. Fungal Genet Biol. 2010;47:930–8.
180. Avalos J, Corrochano LM. Carotenoid biosynthesis in *Neurospora*. In: Kasbekar DP, McCluskey K, editors. *Neurospora*: genomics and molecular biology. Norfolk: Caister Academic Press; 2013.
181. Rau W, Lindemann I, Rau-Hund A. Untersuchungen über die lichtabhängige Carotinoidsynthese. III Die Farbstoffbildung von Neurospora crassa in Submerskultur. Planta. 1968;80:309–16.
182. Zalokar M. Biosynthesis of carotenoids in *Neurospora*. Action spectrum of photoactivation. Arch Biochem Biophys. 1955;56:318–25.
183. Schrott EL. Fluence response relationship of carotenogenesis in *Neurospora crassa*. Planta. 1980;150:174–9.
184. Rau W. Untersuchungen über die lichtabhängige Carotinoidsynthese. IV. Die Rolle des Sauerstoffs bei der Lichtinduktion. Planta. 1969;84:30–42.
185. Avalos J, Schrott EL. Photoinduction of carotenoid biosynthesis in *Gibberella fujikuroi*. FEMS Microbiol Lett. 1990;66:295–8.
186. Schrott EL. The biphasic fluence response of carotenogenesis in *Neurospora crassa*: temporary insensitivity of the photoreceptor system. Planta. 1981;151:371–4.
187. Rau W, Feuser B, Rau-Hund A. Substitution of p-chloro- or p-hydroxymercuribenzoate for light in carotenoid synthesis by *Fusarium aquaeductuum*. Biochim Biophys Acta. 1967;136:589–90.
188. Theimer RR, Rau W. Untersuchungen über die lichtabhängige Carotinoidsynthese V. Aufhebung der Lichtinduktion dutch Reduktionsmittel und Ersatz des Lichts durch Wasserstoffperoxid. Planta. 1970;92:129–37.

189. Theimer RR, Rau W. Untersuchungen über die lichtabhängige Carotinoidsynthese VIII. Die unterschiedlichen Wirkungsmechanismen von Licht und Mercuribenzoat. Planta. 1972;106:331–43.
190. de Fabo EC, Harding RW, Shropshire Jr W. Action spectrum between 260 and 800 nanometers for the photoinduction of carotenoid biosynthesis in *Neurospora crassa*. Plant Physiol. 1976;57:440–5.
191. Paietta J, Sargent ML. Photoreception in *Neurospora crassa*: correlation of reduced light sensitivity with flavin deficiency. Proc Natl Acad Sci USA. 1981;78:5573–7.
192. Paietta J, Sargent ML. Modification of blue light photoresponses by riboflavin analogs in *Neurospora crassa*. Plant Physiol. 1983;72:764–6.
193. Rau W. Untersuchungen über die lichtabhängige Carotinoidsynthese. I. Das Wirkungsspektrum von Fusarium aquaeductuum. Planta. 1967;72:14–28.
194. Schrott EL, Huber-Willer A, Rau W. Is phytochrome involved in the light-mediated carotenogenesis in *Fusarium aquaeductuum* and *Neurospora crassa*? Photochem Photobiol. 1982;35:213–6.
195. Lang-Feulner J, Rau W. Redox dyes as artificial photoreceptors in light-dependent carotenoid synthesis. Photochem Photobiol. 1975;21:179–83.
196. Harding RW, Turner RV. Photoregulation of the carotenoid biosynthetic pathway in albino and white collar mutants of *Neurospora crassa*. Plant Physiol. 1981;68:745–9.
197. Degli-Innocenti F, Russo VE. Isolation of new white collar mutants of *Neurospora crassa* and studies on their behavior in the blue light-induced formation of protoperithecia. J Bacteriol. 1984;159:757–61.
198. He Q, Liu Y. Molecular mechanism of light responses in *Neurospora*: from light-induced transcription to photoadaptation. Genes Dev. 2005;19:2888–99.
199. Carattoli A, Cogoni C, Morelli G, Macino G. Molecular characterization of upstream regulatory sequences controlling the photoinduced expression of the *albino-3* gene of *Neurospora crassa*. Mol Microbiol. 1994;13:787–95.
200. Estrada AF, Avalos J. The White Collar protein WcoA of *Fusarium fujikuroi* is not essential for photocarotenogenesis, but is involved in the regulation of secondary metabolism and conidiation. Fungal Genet Biol. 2008;45:705–18.
201. Ruiz-Roldán MC, Garre V, Guarro J, Mariné M, Roncero MI. Role of the white collar 1 photoreceptor in carotenogenesis, UV resistance, hydrophobicity, and virulence of *Fusarium oxysporum*. Eukaryot Cell. 2008;7:1227–30.
202. Castrillo M, García-Martínez J, Avalos J. Light-dependent functions of the *Fusarium fujikuroi* CryD DASH cryptochrome in development and secondary metabolism. Appl Environ Microbiol. 2013;79:2777–88.
203. García-Martínez J, Ádám AL, Avalos J. Adenylyl cyclase plays a regulatory role in development, stress resistance and secondary metabolism in *Fusarium fujikuroi*. PLoS One. 2012;7:e28849.
204. Ádám AL, García-Martínez J, Szücs EP, Avalos J, Hornok L. The MAT1-2-1 mating-type gene upregulates photo-inducible carotenoid biosynthesis in *Fusarium verticillioides*. FEMS Microbiol Lett. 2011;318:76–83.
205. Rodríguez-Ortiz R, Michielse C, Rep M, Limón MC, Avalos J. Genetic basis of carotenoid overproduction in *Fusarium oxysporum*. Fungal Genet Biol. 2012;49:684–96.
206. Baima S, Macino G, Morelli G. Photoregulation of the *albino-3* gene in *Neurospora crassa*. J Photochem Photobiol B. 1991;11:107–15.
207. Velayos A, Papp T, Aguilar-Elena R, Fuentes-Vicente M, Eslava AP, Iturriaga EA, et al. Expression of the *carG* gene, encoding geranylgeranyl pyrophosphate synthase, is up-regulated by blue light in *Mucor circinelloides*. Curr Genet. 2003;43:112–20.
208. Homann V, Mende K, Arntz C, Ilardi V, Macino G, Morelli G, et al. The isoprenoid pathway: cloning and characterization of fungal FPPS genes. Curr Genet. 1996;30:232–9.
209. Mende K, Homann V, Tudzynski B. The geranylgeranyl diphosphate synthase gene of *Gibberella fujikuroi*: isolation and expression. Mol Gen Genet. 1997;255:96–105.
210. Tudzynski B, Holter K. Gibberellin biosynthetic pathway in *Gibberella fujikuroi*: evidence for a gene cluster. Fungal Genet Biol. 1998;25:157–70.

211. Shrode LB, Lewis ZA, White LD, Bell-Pedersen D, Ebbole DJ. *vvd* is required for light adaptation of conidiation-specific genes of *Neurospora crassa*, but not circadian conidiation. Fungal Genet Biol. 2001;32:169–81.
212. Schwerdtfeger C, Linden H. VIVID is a flavoprotein and serves as a fungal blue light photoreceptor for photoadaptation. EMBO J. 2003;22:4846–55.
213. Youssar L, Schmidhauser TJ, Avalos J. The *Neurospora crassa* gene responsible for the *cut* and *ovc* phenotypes encodes a protein of the haloacid dehalogenase family. Mol Microbiol. 2005;55:828–38.
214. Navarro-Sampedro L, Yanofsky C, Corrochano LM. A genetic selection for *Neurospora crassa* mutants altered in their light regulation of transcription. Genetics. 2008;178:171–83.
215. Li C, Schmidhauser TJ. Developmental and photoregulation of *al-1* and *al-2*, structural genes for two enzymes essential for carotenoid biosynthesis in *Neurospora*. Dev Biol. 1995;169: 90–5.
216. Arpaia G, Carattoli A, Macino G. Light and development regulate the expression of the *albino-3* gene in *Neurospora crassa*. Dev Biol. 1995;170:626–35.
217. Li C, Sachs MS, Schmidhauser TJ. Developmental and photoregulation of three *Neurospora crassa* carotenogenic genes during conidiation induced by desiccation. Fungal Genet Biol. 1997;21:101–8.
218. Vittorioso P, Carattoli A, Londei P, Macino G. Internal translational initiation in the mRNA from the *Neurospora crassa albino-3* gene. J Biol Chem. 1994;269:26650–4.
219. Baima S, Carattoli A, Macino G, Morelli G. Photoinduction of *albino-3* gene expression in *Neurospora crassa* conidia. J Photochem Photobiol B. 1992;15:233–8.
220. Yang Q, Borkovich KA. Mutational activation of a Gα_i causes uncontrolled proliferation of aerial hyphae and increased sensitivity to heat and oxidative stress in *Neurospora crassa*. Genetics. 1999;151:107–17.
221. Kritsky MS, Sokolovsky VY, Belozerskaya TA, Chernysheva EK. Relationship between cyclic AMP level and accumulation of carotenoid pigments in *Neurospora crassa*. Arch Microbiol. 1982;133:206–8.
222. Harding RW. Inhibition of conidiation and photoinduced carotenoid biosynthesis by cyclic AMP. Neurospora Newsl. 1973;20:20–1.
223. Barba-Ostria C, Lledias F, Georgellis D. The *Neurospora crassa* DCC-1 protein, a putative histidine kinase, is required for normal sexual and asexual development and carotenogenesis. Eukaryot Cell. 2011;10:1733–9.
224. Harding RW. The effect of temperature on photo-induced carotenoid biosynthesis in *Neurospora crassa*. Plant Physiol. 1974;54:142–7.
225. Rau W. Über den Einfluss der Temperatur auf die lichtabhängige Carotinoidbildung von *Fusarium aquaeductuum*. Planta. 1962;59:123–37.
226. Estrada AF, Maier D, Scherzinger D, Avalos J, Al-Babili S. Novel apocarotenoid intermediates in *Neurospora crassa* mutants imply a new biosynthetic reaction sequence leading to neurosporaxanthin formation. Fungal Genet Biol. 2008;45:1497–505.
227. Garbayo I, Vílchez C, Nava-Saucedo JE, Barbotin JN. Nitrogen, carbon and light-mediated regulation studies of carotenoid biosynthesis in immobilized mycelia of *Gibberella fujikuroi*. Enzyme Microb Technol. 2003;33:629–34.
228. Rodríguez-Ortiz R, Limón MC, Avalos J. Regulation of carotenogenesis and secondary metabolism by nitrogen in wild-type *Fusarium fujikuroi* and carotenoid-overproducing mutants. Appl Environ Microbiol. 2009;75:405–13.
229. Sokolovsky VY, Lauter FR, Müller-Röber B, Ricci M, Schmidhauser TJ, Russo VEA. Nitrogen regulation of blue light-inducible genes in *Neurospora crassa*. J Gen Microbiol. 1992;138:2045–9.
230. Avalos J, Mackenzie A, Nelki DS, Bramley PM. Terpenoid biosynthesis in cell-extracts of wild type and mutant strains of *Gibberella fujikuroi*. Biochim Biophys Acta. 1988;966: 257–65.
231. Candau R, Avalos J, Cerdá-Olmedo E. Gibberellins and carotenoids in the wild type and mutants of *Gibberella fujikuroi*. Appl Environ Microbiol. 1991;57:3378–82.

232. Rodríguez-Ortiz R, Limón MC, Avalos J. Functional analysis of the *carS* gene of *Fusarium fujikuroi*. Mol Genet Genomics. 2013;288:157–73.
233. Linden H, Rodríguez-Franco M, Macino G. Mutants of *Neurospora crassa* defective in regulation of blue light perception. Mol Gen Genet. 1997;254:111–8.
234. Harding RW, Philip DQ, Drozdowicz BZ, Williams NP. A *Neurospora crassa* mutant which overaccumulates carotenoid pigments. Neurospora newsl. 1984;31:23–5.
235. Youssar L, Avalos J. Genetic basis of the *ovc* phenotype of *Neurospora*: identification and analysis of a 77 kb deletion. Curr Genet. 2007;51:19–30.
236. Imblum RL, Rodwell VW. 3-Hydroxy-3-methylglutaryl CoA reductase and mevalonate kinase of *Neurospora crassa*. J Lipid Res. 1974;15:211–22.
237. Wang GY, Keasling JD. Amplification of HMG-CoA reductase production enhances carotenoid accumulation in *Neurospora crassa*. Metab Eng. 2002;4:193–201.
238. Bhosale P, Bernstein PS. Microbial xanthophylls. Appl Microbiol Biotechnol. 2005;68: 445–55.
239. Moliné M, Libkind D, van Broock M. Production of torularhodin, torulene, and β-carotene by *Rhodotorula* yeasts. Methods Mol Biol. 2012;898:275–83.
240. Frengova GI, Beshkova DM. Carotenoids from *Rhodotorula* and *Phaffia*: yeasts of biotechnological importance. J Ind Microbiol Biotechnol. 2009;36:163–80.
241. Arpin N, Lebreton P, Fiasson JL. Chemotaxonomic research on fungi. II. Carotenoids of *Peniophora aurantiaca* (Bres.) (Basidiomycete). Bull Soc Mycol Fr. 1966;82:450–9.
242. Davoli P, Mierau V, Weber RWS. Carotenoids and fatty acids in red yeasts *Sporobolomyces roseus* and *Rhodotorula glutinis*. Appl Biochem Microbiol. 2004;40:392–7.
243. Iurkov AM, Vustin MM, Tiaglov BV, Maksimova IA, Sineokii SP. Pigmented basidiomycete yeasts are a promising source of carotenoids and ubiquinone Q10. Mikrobiologiya. 2008;77:5–10.
244. Buzzini P, Innocenti M, Turchetti B, Libkind D, van Broock M, Mulinacci N. Carotenoid profiles of yeasts belonging to the genera *Rhodotorula*, *Rhodosporidium*, *Sporobolomyces*, and *Sporidiobolus*. Can J Microbiol. 2007;53:1024–31.
245. Herz S, Weber RW, Anke H, Mucci A, Davoli P. Intermediates in the oxidative pathway from torulene to torularhodin in the red yeasts *Cystofilobasidium infirmominiatum* and *C. capitatum* (Heterobasidiomycetes, Fungi). Phytochemistry. 2007;68:2503–11.
246. Haxo F. Carotenoids of the mushroom *Cantharellus cinnabarinus*. Bot Gaz. 1950;112: 228–32.
247. Madhour A, Anke H, Mucci A, Davoli P, Weber RWS. Biosynthesis of the xanthophyll plectaniaxanthin as a stress response in the red yeast *Dioszegia* (Tremellales, Heterobasidiomycetes, Fungi). Phytochemistry. 2005;66:2617–26.
248. Arpin N, Liaaen-Jensen S. Chemotaxonomic research on fungi. Fungal carotenoids. IV. Carotenoids of *Phillipsia carminea* (Pat.) Le Gal; isolation and identification of a new natural xanthophyll. Bull Soc Chim Biol. 1967;49:527–36.
249. Brown LS. Fungal rhodopsins and opsin-related proteins: eukaryotic homologues of bacteriorhodopsin with unknown functions. Photochem Photobiol Sci. 2004;3:555–65.
250. Brown LS, Dioumaev AK, Lanyi JK, Spudich EN, Spudich JL. Photochemical reaction cycle and proton transfers in *Neurospora* rhodopsin. J Biol Chem. 2001;276:32495–505.
251. Edge R, McGarvey DJ, Truscott TG. The carotenoids as anti-oxidants—a review. J Photochem Photobiol B. 1997;41:189–200.
252. Jeong JC, Lee IY, Kim SW, Park YH. Stimulation of β-carotene synthesis by hydrogen peroxide in *Blakeslea trispora*. Biotech Lett. 1999;21:683–6.
253. Liu YS, Wu JY. Hydrogen peroxide-induced astaxanthin biosynthesis and catalase activity in *Xanthophyllomyces dendrorhous*. Appl Microbiol Biotechnol. 2006;73:663–8.
254. Iigusa H, Yoshida Y, Hasunuma K. Oxygen and hydrogen peroxide enhance light-induced carotenoid synthesis in *Neurospora crassa*. FEBS Lett. 2005;579:4012–6.
255. Yoshida Y, Hasunuma K. Reactive oxygen species affect photomorphogenesis in *Neurospora crassa*. J Biol Chem. 2004;279:6986–93.

256. Michán S, Lledías F, Hansberg W. Asexual development is increased in *Neurospora crassa* *cat-3*-null mutant strains. Eukaryot Cell. 2003;2:798–808.
257. Ukibe K, Hashida K, Yoshida N, Takagi H. Metabolic engineering of *Saccharomyces cerevisiae* for astaxanthin production and oxidative stress tolerance. Appl Environ Microbiol. 2009;75:7205–11.
258. Nanou K, Roukas T. Oxidative stress response of *Blakeslea trispora* induced by iron ions during carotene production in shake flask culture. Appl Biochem Biotechnol. 2013;169: 2281–9.
259. Nanou K, Roukas T. Stimulation of the biosynthesis of carotenes by oxidative stress in *Blakeslea trispora* induced by elevated dissolved oxygen levels in the culture medium. Bioresour Technol. 2011;102:8159–64.
260. Nanou K, Roukas T. Oxidative stress response and morphological changes of *Blakeslea trispora* induced by butylated hydroxytoluene during carotene production. Appl Biochem Biotechnol. 2010;160:2415–23.
261. Gessler NN, Sokolov AV, Bykhovsky VY, Belozerskaya TA. Superoxide dismutase and catalase activities in carotenoid-synthesizing fungi *Blakeslea trispora* and *Neurospora crassa* in oxidative stress. Appl Biochem Microbiol. 2002;38:205–9.
262. Hu X, Ma X, Tang P, Yuan Q. Improved β-carotene production by oxidative stress in *Blakeslea trispora* induced by liquid paraffin. Biotechnol Lett. 2013;35:559–63.
263. Sakaki H, Nochide H, Komemushi S, Miki W. Effect of active oxygen species on the productivity of torularhodin by *Rhodotorula glutinis* No. 21. J Biosci Bioeng. 2002;93:338–40.
264. Sakaki H, Nakanishi T, Satonaka K, Miki W, Fujita T, Komemushi S. Properties of a high-torularhodin-producing mutant of *Rhodotorula glutinis* cultivated under oxidative stress. J Biosci Bioeng. 2000;89:203–5.
265. Sakaki H, Nakanishi T, Tada A, Miki W, Komemushi S. Activation of torularhodin production by *Rhodotorula glutinis* using weak white light irradiation. J Biosci Bioeng. 2001;92: 294–7.
266. Shimizu M, Egashira T, Takahama U. Inactivation of *Neurospora crassa* conidia by singlet molecular oxygen generated by a photosensitized reaction. J Bacteriol. 1979;138:293–6.
267. Ramadan-Talib Z, Prebble J. Photosensitivity of respiration in *Neurospora* mitochondria. A protective role for carotenoid. Biochem J. 1978;176:767–75.
268. Luque EM, Gutiérrez G, Navarro-Sampedro L, Olmedo M, Rodríguez-Romero J, Ruger-Herreros C, et al. A relationship between carotenoid accumulation and the distribution of species of the fungus *Neurospora* in Spain. PLoS One. 2012;7:e33658.
269. Morris SA, Subden RE. Effects of ultraviolet radiation on carotenoid-containing and albino strains of *Neurospora crassa*. Mut Res. 1974;22:105–9.
270. Blanc PL, Tuveson RW, Sargent ML. Inactivation of carotenoid-producing and albino strains of *Neurospora crassa* by visible light, blacklight, and ultraviolet radiation. J Bacteriol. 1976;125:616–25.
271. Chen SJ, Wang Q, Han JR. Influence of oxidative stress and grains on sclerotial biomass and carotenoid yield of *Penicillium sp*. PT95. J Basic Microbiol. 2010;50:388–91.

Chapter 9
Astaxanthin and Related Xanthophylls

Jennifer Alcaino, Marcelo Baeza, and Victor Cifuentes

Introduction

In 1837, the Swedish chemist Jöns Jacob Berzelius described the yellow pigments extracted from autumn leaves, which he named xanthophylls (from the Greek *xanthos*: yellow and *phyllon*: leaf). Later, the Russian-Italian botanist M.S. Tswett found that these pigments were a complex mixture of "polychromes" and, using adsorption chromatography, isolated and purified xanthophylls and carotenes, which he named carotenoids in 1911. These yellow, orange, or red pigments play important physiological roles in all living organisms, but their synthesis is circumscribed to photosynthetic organisms, some fungi and bacteria. Animals must obtain these essential molecules from food, as they are not able to synthesize carotenoids de novo [1]. Since H.W.F. Wackenroder isolated the first carotenoid from the cells of carrot roots in 1831 [2], more than 750 different chemical structures of natural carotenoids have been described [3]. The annual production of carotenoids is estimated to be more than 100 million tons [4].

The molecular structure of carotenoids consists of a hydrocarbon backbone of forty carbon atoms (C40) usually composed of eight isoprene units joined such that the two methyl groups nearest the center of the molecule are in a 1,6-positional relationship and the remaining nonterminal methyl groups are in a 1,5-positional relationship (Nomenclature of Carotenoids, IUPAC and IUPAC-IUB, rules approved in 1974). All carotenoids derived from the acyclic $C_{40}H_{56}$ structure have a long central chain of conjugated double bonds (that constitutes the chromophoric system of the carotenoids) that may have some chemical modifications such as hydrogenation, the incorporation of oxygen-containing functional groups and the cyclization of one or both ends, resulting in monocyclic or bicyclic carotenoids [5]. The oxygenated carotenoids

J. Alcaino, Sc.D. • M. Baeza, Sc.D. • V. Cifuentes, Sc.D. (✉)
Departamento de Ciencias Ecológicas, Facultad de Ciencias,
Universidad de Chile, Santiago, Chile
e-mail: vcifuentes@uchile.cl

J.-F. Martín et al. (eds.), *Biosynthesis and Molecular Genetics of Fungal Secondary Metabolites*, Fungal Biology, DOI 10.1007/978-1-4939-1191-2_9,

with a hydroxy, epoxy, and/or oxo group form a separate subclass named the xanthophylls [6], while the non-oxygenated carotenoids are named carotenes. The oxygenation in these molecules contributes to the enhanced solubility of xanthophylls and is the reason why they are more polar than the purely hydrocarbon carotenes, thus allowing the separation of xanthophylls from carotenes by chromatography.

Xanthophylls: Functions and Applications

Xanthophylls are synthesized by several organisms in which they fulfill important biological roles. For example, in photosynthetic organisms they work as accessory light harvesting pigments and are involved in the protection against photo-oxidative damage, such as the peroxidation of lipid membranes by reactive radicals [7]. Their photoprotective properties are attributed to their strong light absorption in the 400–500 nm range of the visible spectrum. In recent decades, there have been an increasing number of reports confirming the beneficial effects of xanthophylls to animal and human health, which has positioned these metabolites as a very promising group of phytonutrients. In this sense, the first functional role recognized for carotenoid pigments was as a vitamin A precursor in animals. Because of the coloring properties of xanthophylls, there is interest in their economic impact on the production of animal feed; for example, the poultry industry uses xanthophylls to contribute to chicken and egg yolk pigmentation [8, 9]. For consumers, appearance is one of the most important factors affecting the decision to purchase a food product, and the color together with the freshness are ranked as the main criteria for selection [10]. Furthermore, the outstanding antioxidant properties of xanthophylls have been linked to their capacity to protect animal cells from free radicals. Cumulative reports refer to their positive influence on human health and their anti-disease effects in cancer and obesity. For these reasons, the use of xanthophylls has been explored in several industries over the past 30 years, and they have been used as active ingredients in medicinal pharmaceuticals, as cosmetics ingredients and as colorants and additives in the food industry [11]. In particular, there has been a significant increase in their use in the *Functional Food* field. Currently, the global market for astaxanthin (see below) is similar to that for β(beta)-carotene and is followed by lutein and canthaxanthin in economic importance [12, 13].

Even though efforts to develop commercial methods for the extraction and purification of carotenes and xanthophylls date back to the middle of the twentieth century [14], chemical synthesis remains the main method for the production of xanthophylls and remains in high demand. However, the modern world's penchant for natural products has increased the search of naturally occurring xanthophylls. In this regard, advances in fermentation processes have been driven by a strong demand for microbial sources of carotenoids for the food industry. Moreover, the development of recombinant DNA technologies and metabolic engineering protocols has contributed to advances in the microbial production of some xanthophylls, even in non-carotenogenic organisms [15, 16], which can provide competitive alternatives to chemical synthesis (Table 9.1).

Some commercially relevant xanthophylls are described as follows:

Table 9.1 Xanthophylls with economic relevance

			Producing microorganisms		
Xanthophyll	Applications	Main commercial production	Name	Reported amount	References
Astaxanthin	Fish farming, nutraceutical industries	Chemical	*Xanthophyllomyces dendrorhous*	9.7 mg/g dry weight	[130]
			Haematococcus pluvialis	5.7 mg/g dry weight	[140]
β(beta)-cryptoxanthin	Food color products in Australia and New Zealand	Plant extracts	*Brevibacterium linens*	0.3 mg/mL	[30]
			Flavobacterium lutescens	770 mg/kg dry weight	[31]
Canthaxanthin	Food fortification and coloration	Chemical	*Dietzia natronolimnaea*	7.1 mg/L	[141]
			Mucor circinelloide	200 μg/g dry weight	[142]
			Dietzia natronolimnaea	8923 μg/L	[37]
Capsanthin	Several feed areas: baked goods, instant noodles, biscuits, canned food, meat sauces. In aquatic and cosmetic products	Plant extracts (pepper)	–	–	–
Fucoxanthin	Dietary supplement, mainly weight loss ones	Algae extracts	–	–	–
Lutein	Food additive, health products, cosmetics and pharmaceutical	Plant extracts (marigold)	*Chlorella sorokiniana*	7.0 mg/g dry weight	[143]
			Scenedesmus almeriensis	3.8 mg/L day	[144]
			Muriellopsis sp.	100 mg/m^2 day	[145]
Violaxanthin	–	Not commercially available	*Chlamydomonas reinhardtii*	81.8 Mmol/mol Chla	[146]
Zeaxanthin	Food fortification and coloration. Eye health supplements	Plant extracts (marigold)	*Chlorella saccharophila*	11.32 mg/g	[77]
			Algibacter	3.47 mg/g dry weight	[74]
			Flavobacterium sp.	3.8 g/L	[75]

*Astaxanthin [*3,3′-Dihydroxy-β(Beta),β(Beta)-Carotene-4, 4′-Dione*]*

Astaxanthin is a red-orange pigment naturally synthesized by a number of bacteria, microalgae, and yeasts. The commercial production of this pigment has traditionally been performed by chemical synthesis, but the yeast *Xanthophyllomyces dendrorhous* (e.g., *Phaffia rhodozyma*) and the microalga *Haematococcus pluvialis* appear to be the most promising sources for its industrial biological production. Astaxanthin has strong antioxidant properties, shown to be better than those of β(beta)-carotene or even α(alpha)-tocopherol [17]. There are an increasing number of reports on the potential benefits of astaxanthin on human health, including benefits on cardiovascular diseases [18], the prevention of *Helicobacter pylori* infection in mice [19], the enhancement of the immune response in humans [20], and the inhibition of carcinogenesis in mice [21]. Furthermore, astaxanthin is a very important pigment worldwide. It is used in aquaculture for the pigmentation of salmonid flesh, which is desired by the consumers, thus having considerable economic impact on this industry. In the same way, astaxanthin is used to enrich the nutritional value of egg yolks and to enhance the health and fertility of layer hens [17]. Consequently, the global market for astaxanthin was US$234 million in 2004 [22], which was approximately a quarter of the total global market for carotenoids.

*β(Beta)-Cryptoxanthin [*β(Beta),β(Beta)-Caroten-3-ol*]*

This xanthophyll is mainly found in fruits such as papaya, tangerine, orange, and watermelon, and has the potential to act as a provitamin A [23, 24]. The main medical application reported for β(beta)-cryptoxanthin is in bone homeostasis. It has a stimulatory effect on bone calcification (demonstrated in vitro) and in periodontitis, preventing bone resorption most likely by inhibiting the interleukin production induced by bacterial pathogens and mechanical stress [23, 25]. Cancer-preventative effects of β(beta)-cryptoxanthin have been reported as well. For example, β(beta)-cryptoxanthin protects HeLa and Caco-2 cells from H_2O_2 and visible light damage and induces DNA repair [26]. Moreover, in combination with hesperidin, β(beta)-cryptoxanthin has inhibitory effects on chemically induced tumorigenesis in several rat and mouse tissues [27]. The commercial production of β(beta)-cryptoxanthin is based on natural sources such as citrus and capsicums, but new methods have been developed for its production by the conversion of lutein or lutein esters [28]. There are very few reports on microbial sources of β(beta)-cryptoxanthin. Bacteria transformed with the β(beta)-carotene hydroxylase gene from *Arabidopsis thaliana* were able to produce β(beta)-cryptoxanthin as the principal carotenoid [29]. In addition, although produced in low yields, β(beta)-cryptoxanthin production has been described in *Brevibacterium linens* [30] and in *Flavobacterium lutescens* [31].

Canthaxanthin [β(Beta),β(Beta)-Carotene-4,4′-Dione]

Canthaxanthin has an orange-red color and is naturally produced by some plants, fungi, microalgae, Archaea, and bacteria [32]. Together with other carotenoids, it was reported that canthaxanthin induces gap junction communication in murine fibroblasts and, therefore, intercellular communication. Moreover, it has important effects on the immune response. Due to its antioxidant properties, it has been noted that canthaxanthin is the most potent methyl linoleate inhibitor, providing a model for lipid peroxidation in vivo [33]. Together with astaxanthin, canthaxanthin is the most important pigment used in aquaculture for salmonid flesh coloration and is also used for chicken skin and egg yolk coloring [8, 34]. The main microbial source of commercial canthaxanthin is the alga *Haematococcus lacustris* [35]. Nevertheless, there are laboratory-scale reports of canthaxanthin production by several microorganisms with the potential for use at larger industrial scales for organisms such as *Aspergillus carbonarius* [36], *Dietzia natronolimnaea* [37], and the microalga *Chlorella zofingiensis* [38, 39], to name a few.

Capsanthin [3,3′-Dihydroxy-β(Beta),κ(Kappa)-Caroten-6′-One]

Capsanthin is the major xanthophyll in peppers and in *Lolium lancifolium* "Splendens" flowers (tiger lily) [40]. This pigment is not produced by chemical synthesis and is mainly extracted from red peppers to be used for pigmentation of poultry feed [34]. Recently it was described that paprika pigments contain large amounts of capsanthin and capsorubin and reduce adipocyte chronic inflammation caused by obesity [41]. In addition, epidemiological studies suggested that capsanthin has a strong inhibitory effect on colon carcinogenesis [42].

Fucoxanthin [5,6-Epoxy-3′-Ethanoyloxy-3,5′-Dihydroxy-6′,7′-Didehydro-5,6,7,8,5′,6′-Hexahydro-β(Beta), β(Beta)-Caroten-8-One]

The fucoxanthin pigment is found in different types of comestible seaweeds and is responsible for their brown or olive-green color [43]. Seaweeds are the main sources of this pigment [44] because chemical synthesis is still very expensive. Fucoxanthin is considered to be an anticarcinogenic compound and was recently demonstrated to have apoptosis-inducing effects, most likely through the down-regulation of STAT3/EGFR signaling [45]. Furthermore, anti-obesity and antidiabetic roles have been described for fucoxanthin [43].

Lutein [β(Beta),ε(Epsilon)-Carotene-3,3′-Diol]

Together with zeaxanthin, lutein forms the macular pigment, which is the yellow spot at the center of the human retina. The adequate intake of lutein might help to prevent or ameliorate age-related macular degeneration and other degenerative human diseases [46–48]. Studies on the effects of lutein on the immune response have been performed in several animal species, and its immune-modulatory effect on macrophages was recently reported in both murine and primary-cultured peritoneal macrophages [49]. Lutein is the major xanthophyll present in green leafy vegetables. Currently, lutein is extracted from marigold petals [50], mainly in the esterified form. The market for lutein in the USA is estimated at $150 million [51]. Because of the assumption that esterification diminishes the bioavailability of lutein, a preceding saponification step is performed in commercial formulations to remove esters; however, it has been demonstrated that this modification does not significantly affect lutein bioavailability, which mainly depends on its solubilization [52, 53]. Several studies have been performed to develop carriers to enhance lutein bioavailability, for example, by using solubilized lutein in mixed micelles containing lysophosphatidylcholine [54] and water-soluble, low molecular weight chitosan [55]. There is a constant search for alternative sources of lutein besides plants, and it has been mainly reported that algae and microalgae might become real economic alternatives for the production of lutein. This is the case for *Chlorococcum humicola* [56] and *Coccomyxa acidophila* (which also accumulate significant amounts of β-carotene) [57], *C. zofingiensis* (which also accumulates astaxanthin) [58], and *Dunaliella salina* [59] and *Chlorella protothecoides* (which also contain significant amounts of canthaxanthin, echinenone, and astaxanthin) [60].

Neoxanthin [5′,6′-Epoxy,6,7-Didehydro-5,6,5′,6′-Tetrahydro-β(Beta), β(Beta)-Carotene-3,5,3′-Triol]

Neoxanthin is a precursor of the plant growth regulator abscisic acid [61] in green leafy vegetables, including common edible vegetables [62]. It has been demonstrated that neoxanthin affects the proliferation of human prostate cancer cells, most likely by caspase induction [63, 64].

Violaxanthin [5,6:5′,6′-Diepoxy-5,6,5′,6′-Tetrahydro-β(Beta), β(Beta)-Carotene-3,3′-Diol]

Violaxanthin is a xanthophyll of orange color synthesized by a variety of plants, including the well-known pansies. Significant amounts of violaxanthin have also been reported in orange juices and peels. Together with neoxanthin, violaxanthin is commercially attractive because it is also one of the abscisic acid precursors—the plant hormone indispensable for plant adaptation with important roles in dormancy and embryo development [65]. Recently, low amounts of violaxanthin were reported in intracellular extracts from microalga *Scenedesmus obliquus* strain M2-1 [66].

Zeaxanthin [β(Beta),β(Beta)-Carotene-3,3′-Diol]

Zeaxanthin is a yellow pigment found in vegetables and fruits. By far the main reported role for zeaxanthin in human health is in ocular health, where, together with lutein, zeaxanthin provides protection against age-related macular degeneration [67, 68]. In addition, potential antitumor properties have been described for zeaxanthin [69]. Furthermore, it was recently found that meso-zeaxanthin has an inhibitory effect on the mutagenicity of five mutagenic agents, including nitro-o-phenylenediamine and *N*-methyl- *N′*-nitro-*N*-nitrosoguanidine [70]. At present, the commercial production of zeaxanthin is mainly based on the extraction from plant tissues such as marigold flowers [71] and its use in the generation of new functional foods has been successfully explored [72]. A recent study indicated that spirulina is a rich dietary source of zeaxanthin, as the administration of spirulina increased the zeaxanthin concentration in human serum [73]. No commercial microbial sources for the production of zeaxanthin have been established yet; however, microorganisms that produce zeaxanthin are continuously being described and some of them are promising sources for satisfying the zeaxanthin demands of the future market. Examples include a marine bacterium belonging to the genus *Algibacter* [74], *Flavobacterium* sp. [75], novel bacterial species belonging to the *Sphingobacteriaceae* and *Sphingomonadaceae* families [76] and a new *Chlorella saccharophila* strain that has the potential to be used for biofuel and carotenoid co-production [77].

Biosynthesis of Xanthophylls

The biosynthesis of xanthophylls derives from the synthesis of carotenoids (Fig. 9.1). Although the carotenoid compounds found in nature are enormous in structural diversity, all of them are synthesized through the universally conserved isoprenoid biosynthesis pathway. The biosynthesis of isoprenoids originates from a basic C_5 isoprene unit to which prenyl transferase enzymes sequentially add three other isoprenic units [78] resulting in the formation of C_{20} geranylgeranyl-pyrophosphate (GGPP). The active forms of the isoprene unit are the isopentenyl-pyrophosphate (IPP) and its allylic isomer dimethylallyl-pyrophosphate (DMAPP). In most eukaryotes, IPP derives from the mevalonate pathway [79], while in prokaryotes and in plant plastids, it is synthesized via the 2-C-*m*ethyl-D-*e*rythritol-4-*p*hosphate (MEP) pathway, which is also known as the non-mevalonate pathway [80]. In the first step of isoprenoid biosynthesis, one IPP molecule is isomerized to

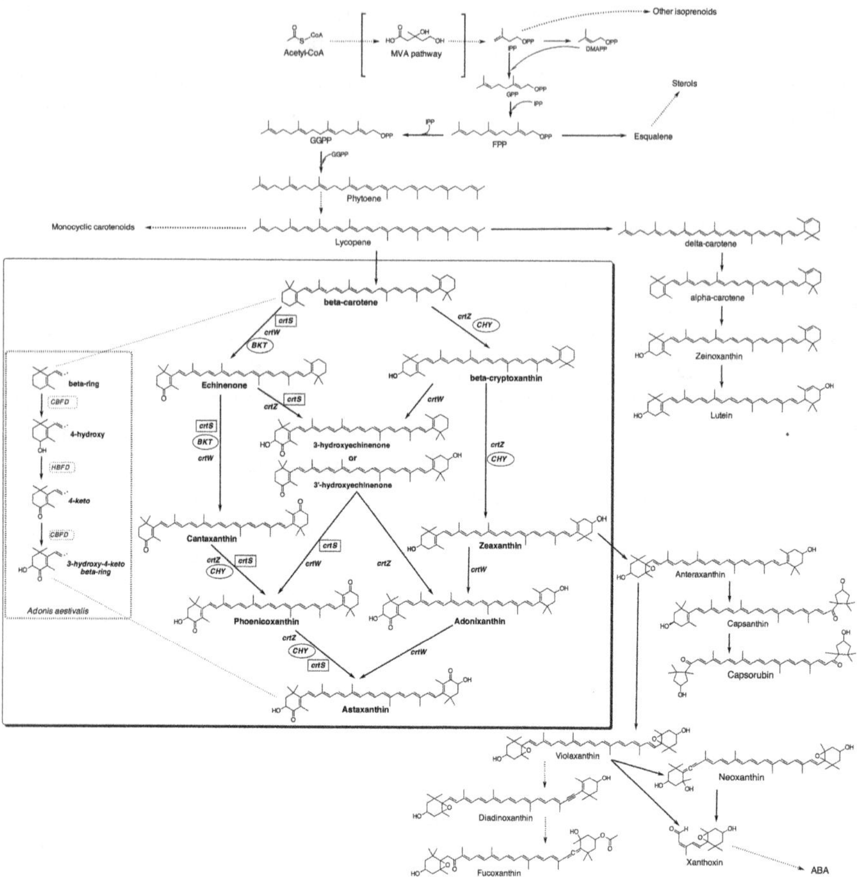

Fig. 9.1 Biosynthetic pathway of xanthophylls. Systematic illustration of the metabolic pathways leading to the synthesis of the xanthophylls described in the text (adapted from [147, 148]). The biosynthesis of astaxanthin is enclosed in a *box* indicating the proposed genes that control each step in *X. dendrorhous* (in *squares*, adapted from [114]), *H. pluvialis* (in *circles*, adapted from [149]), bacteria (no *special mark*, adapted from [150]), and in *A. aestivalis* (in *dotted box*, adapted from [99]). *Abbreviations*: *MVA* mevalonate, *IPP* isopentenyl-pyrophosphate, *DMAPP* dimethylallyl-pyrophosphate, *GGPP* geranylgeranyl-pyrophosphate. Metabolite structures were confirmed according to [151]

DMAPP by the isopentenyl-pyrophosphate isomerase and then both molecules are joined together generating C_{10}-geranyl pyrophosphate (GPP), the precursor of monoterpenes [81]. The addition of a second molecule of IPP to GPP by prenyl transferases gives the precursor of sesquiterpenes, C_{15}-farnesyl pyrophosphate (FPP), which is converted to GGPP (the precursor of diterpenes) by the further addition of IPP by the GGPP synthase enzyme. Next, phytoene synthase condenses two molecules of GGPP in a tail-to-tail manner, yielding phytoene [79]. This is the first carotenoid synthesized in the pathway, which is colorless as it has a symmetrical carotenoid skeleton with only three conjugated double bonds. The huge structural diversity of carotenoids is generated by further modifications such as desaturations, cyclizations, isomerizations, and oxygenations [82].

The phytoene synthase enzyme is well conserved among carotenogenic organisms. It is encoded by the *crtB* gene in bacteria and by the *PSY* gene in plants, algae, and cyanobacteria [11]. Fungi have a bifunctional enzyme, named phytoene β(beta)-carotene synthase (PBS) because it has both phytoene synthase and lycopene cyclase activities, which gives rise to β(beta)-carotene. In this particular enzyme, encoded by the *crtYB* gene, the phytoene synthase and lycopene cyclase activities are restricted to the C-terminal and the N-terminal functional domains, respectively. It is likely that such a bifunctional enzyme was acquired early in the evolution of fungi [83] because genes encoding this unique enzyme have been reported in ascomycetes, zygomycetes, and basidiomycetes such as *Neurospora crassa* [84], *Mucor circinelloides* [85], *Phycomyces blakesleeanus* [86], and *Xanthophyllomyces dendrorhous* [87].

Next, phytoene is desaturated by the incorporation of two, three, four, or five double bonds producing the colored carotenoids ζ(zeta)-carotene (yellow, synthesized by some plants, cyanobacteria, and algae), neurosporene (yellow, accumulates in *Rhodobacter capsulatus* and *R. sphaeroides*), lycopene (red, found in most eubacteria and fungi), or 3,4-didehydrolycopene (found in *N. crassa*) [79], respectively. In photosynthetic organisms, the formation of ζ(zeta)-carotene by the sequential insertion of two double bonds in phytoene is generally performed by a phytoene desaturase, encoded by *PDS* in plants and algae or by *crtP* in cyanobacteria [11]. Next, the ζ(zeta)-carotene is converted into lycopene by the introduction of two additional double bonds, which is catalyzed by a ζ(zeta)-carotene desaturase, encoded by *ZDS* in plants and algae or by *crtQ* in cyanobacteria [11]. In non-photosynthetic carotenogenic organisms, such as fungi and eubacteria, the desaturation of phytoene leading to lycopene is controlled by only one gene, *crtI* [11].

Although there are acyclic carotenoids, the cyclization of lycopene is a frequent step in the biosynthesis of carotenoids, forming three types of ionone rings: β(beta)-, ε(epsilon)-, and γ(gamma)-rings [81]. The β(beta)-ring is the most common form; the ε(epsilon)-type is found in plants and in some algae, and the γ(gamma)-ring is the rarest. Several non-phylogenetically related lycopene β(beta)-cyclases have been described, which are encoded by the *crtL* gene (also known as *LCY*) in plants, cyanobacteria, and algae, and by *crtY* in eubacteria, which produces β(beta)-carotene when a β(beta)-ring is introduced at both ends of lycopene [88]. Another type of lycopene cyclase was described in the actinomycete bacterium *B. linens*, in which a heterodimeric enzyme formed by polypeptides encoded by the *crtYc* and *crtYd* genes (unrelated to *crtY* or *crtL*) is responsible for the conversion of lycopene into β(beta)-carotene [83]. In fungi, the domain of the bifunctional enzyme PBS that exhibits lycopene cyclase activity seems to be related to the *crtYc* and *crtYd* genes of *B. linens*, which led to the hypothesis that PBS developed from a recombination of these two genes and a phytoene synthase gene [83]. Nevertheless, the existence of other types of lycopene cyclases is still expected because no lycopene cyclase genes have been found in the completely sequenced and available genomes of the cyanobacteria *Synechocystis* sp. and *Anabaena* sp., both of which are β(beta)-carotene-producers [89].

The synthesis of xanthophylls involves the oxidation of post-phytoene carotenoid molecules, mainly from α(alpha)- and β(beta)-carotenes, resulting in oxygenated products with hydroxyl-, epoxy-, and oxo-functional groups.

Astaxanthin Biosynthesis in *X. dendrorhous* and in Other Organisms

In the late nineteenth century, Ludwig described a red yeast-like organism responsible for the color of the sap of deciduous trees and named it *Rhodomyces dendrorhous* [90]. In the late 1960s, Herman Phaff and coworkers isolated a red fermenting yeast from natural slime fluxes and exudates on wounded trees from mountainous regions of Japan and Alaska. It was originally designated as *Rhodozyma montanae* nov. gen. et sp, but in 1976 it was renamed *Phaffia rhodozyma* because it has a basidiomycetous origin [91]. Subsequently, many strains were isolated from the European part of Russia, where it was noticed that *P. rhodozyma* was the predominant yeast in the red exudates of trees, suggesting that it corresponded to *Rhodomyces dendrorhous*, as originally described by Ludwig [90]. In 1995, Golubev described the life cycle of this yeast, which was unknown in the basidiosporogenous yeasts, indicating that it was a new teleomorphic genus, and the name *Xanthophyllomyces* was proposed [90]. Currently, the anamorphic strains are designated as *P. rhodozyma* and the teleomorphic strains as *X. dendrorhous*. In 1976, Andrewes and coworkers reported that astaxanthin was the principal carotenoid pigment produced by this yeast and one of the first models for the biosynthesis of astaxanthin in *X. dendrorhous* was suggested [92].

The biosynthesis of astaxanthin is limited to a few organisms such as the microalgae *H. pluvialis* [93], some marine bacteria such as *Paracoccus haeundaensis* [94] and *Brevudimonas* sp. [95], the basidiomycete yeast *X. dendrorhous* [92] and the plant *Adonis*, which accumulates astaxanthin in the petals of the flower [96].

The formation of astaxanthin from β(beta)-carotene involves the introduction of a hydroxyl and a keto group at carbons 3 and 4, respectively, for each of the β(beta)-ionone rings via eight possible intermediate xanthophylls, depending on the producing organism. In the bacterial, plant, and algal systems, these reactions are catalyzed by hydroxylases and ketolases. Ketolases have been described in several organisms that do not produce astaxanthin, but produce other keto-xanthophylls such as echinenone. The bacterial ketolases, encoded by the *crtW* gene, can use a non-substituted β(beta)-ionone ring as well as 3-hydroxylated β(beta)-ionone rings as a substrate [97]. Two paralogous genes with significant identity to *crtW*, *bkt1* and *bkt2*, encode β(beta)-carotene ketolases (BKT) and were described in *H. pluvialis*. The microalga BKTs can only accept a non-substituted β(beta)-ionone ring as a substrate, so it is unlikely that the astaxanthin synthesis from β(beta)-carotene begins with a hydroxylation step in *H. pluvialis* [93]. During the green vegetative phase of *H. pluvialis*, a hydroxylase (CHY) incorporates a hydroxyl group onto the carbon at position 3 of both β(beta)-ionone rings of the β(beta)-carotene substrate, producing zeaxanthin. However, under stress conditions, β(beta)-carotene is converted into astaxanthin primarily via echinenone, canthaxanthin, and phoenicoxanthin, being β(beta)-carotene and echinenone substrates of BKT, while canthaxanthin and phoenicoxanthin are the substrates of CHY. In this way, astaxanthin accumulates as a secondary carotenoid under stress conditions [98]. Further introduction of fatty acids to the hydroxyl groups by esterification leads to the production of mono- and di-esterified astaxanthin in *H. pluvialis* [93].

It has been suggested that in *Adonis* plants, the synthesis of a 3-hydroxy-4-keto-β(beta)-ionone ring from the β(beta)-ionone ring substrate is controlled by two genes and occurs in three steps [99]. First, a 4-hydroxy-β(beta)-ring is formed by carotenoid-β(beta)-ring-4-dehydrogenase (CBFD); second, 4-hydroxy-β(beta)-ring-4-dehydrogenase (HBFD) continues with the further dehydrogenation of carbon 4 giving a keto group at this position; and third, CBFD introduces a hydroxyl group at carbon 3 of the 4-keto-β(beta)-ring to form the 3-hydroxy-4-keto-β(beta)-ring.

There are two major groups of β(beta)-carotene hydroxylases: the non-heme di-iron (NH-di-iron) hydroxylases and the cytochrome P450 monooxygenases (reviewed in [100]). The NH-di-iron hydroxylases are related to fatty acid desaturases, and based on their primary structure, are classified into three groups corresponding to (1) non-photosynthetic eubacteria, (2) plants and green algae, and (3) cyanobacteria. These enzymes require molecular oxygen, iron, ferredoxin, and ferredoxin oxido-reductase for their function; and even though they share low protein identity, carry out the same reaction and conserve iron-coordinating histidines essential for enzyme activity [101]. The bacterial β(beta)-carotene hydroxylases, encoded by *crtZ*, can convert non-substituted and 4-ketolated β(beta)-ionone rings into the respective 3-hydroxylated forms [97]. A cytochrome P450 monooxygenase involved in the β(beta)-carotene hydroxylation was first described in the thermophilic bacterium *Thermus thermophilus* [102]. By functional complementation in an *Escherichia coli* strain carrying the *Erwinia uredovora* carotenoid biosynthetic genes [103], it was demonstrated that this enzyme could introduce hydroxyl groups to both β(beta)-rings of β(beta)-carotene producing zeaxanthin [102].

Cytochrome P450s (P450s) are a large superfamily of heme-containing monooxygenases that have been described in organisms from all domains of life [104, 105], playing significant roles in the oxidative metabolism of a wide range of exogenous and endogenous substrates [106]. They are involved in the metabolism of many physiologically important compounds such as sterols, fatty acids, and vitamins [107]; secondary metabolites [108]; and in the activation and detoxification of many xenobiotics, such as drugs, carcinogens, and environment-polluting chemicals [109]. These enzymes act as a terminal electron acceptor in multicomponent P450-dependent monooxygenation systems (P450 systems) that lead to the reductive activation of molecular oxygen followed by the insertion of one oxygen atom into the substrate molecule and the reduction of the other to water [110]. The two electrons required for cytochrome P450 activity are transferred primarily from NADPH via a redox partner [111], but the specificity of a particular reaction is given by the P450 enzymatic properties and substrate specificity. In the eukaryotic microsomal P450 system, the general P450 redox partner is a cytochrome P450 reductase, CPR [104, 110, 112]. Although in most organisms there are several genes encoding different P450 enzymes, in most species there is only one CPR-encoding gene, with few exceptions [113].

In this regard, *X. dendrorhous* has a single astaxanthin synthase (CrtS, encoded by the *crtS* gene), belonging to the cytochrome P450 protein family, which catalyzes the hydroxylation and ketolation of β(beta)-carotene to produce astaxanthin [114, 115]. To the best of our knowledge, the synthesis of astaxanthin from β(beta)-carotene through a P450 system has only been reported in *X. dendrorhous*, demonstrating that only in this yeast a unique P450 system evolved and specialized for the synthesis of astaxanthin.

An *X. dendrorhous* mutant strain missing the *crtR* wild-type gene, which encodes a CPR-type enzyme (CrtR), accumulates β(beta)-carotene and is unable to synthesize astaxanthin, demonstrating that CrtR is essential for the synthesis of astaxanthin [116]. It is important to highlight the fact that the disruption of *crtR* was not lethal because the mutant strain was able to grow normally under the studied conditions. Additionally, Ukibe and coworkers [15] stated that CrtS has a high specificity for its own CPR. As in metabolically engineered *S. cerevisiae* strains with the *X. dendrorhous* carotenogenic genes, astaxanthin production was only achieved when CrtS was co-expressed with CrtR. This result indicates that the *S. cerevisiae* endogenous CPR was not able to reduce the *X. dendrorhous* CrtS, even though the heterologous expression of several cytochrome P450s in this yeast has been functionally successful [110]. The introduction of *crtR* was crucial for the functional expression of CrtS and astaxanthin production in *S. cerevisiae*, suggesting "that the *X. dendrorhous* CrtS is a unique cytochrome P450 protein that has high specificity for its own P450 reductase" [15].

Genetic Improvement of Astaxanthin Production in *X. dendrorhous*

The specific production of astaxanthin in natural *X. dendrorhous* isolates is too low (200–400 μg/g of dry weight of yeast, ppm) to provide an economically competitive natural source of this xanthophyll [117]. Therefore, several efforts have been made to improve the production of astaxanthin in this yeast (reviewed in [118]). There is a complex interaction between nutritional factors, such as carbon or nitrogen sources and vitamins [119, 120], and physical factors such as oxygen levels [119, 121], pH [122], and light intensities [123], that influence the cell growth and carotenogenesis in *X. dendrorhous*. Moreover, different natural isolates and astaxanthin-hyperproducing mutant *X. dendrorhous* strains may respond differently in their carotenoid production when cultivated under the same conditions, hindering the process of optimization of culture parameters.

In contrast, classical random mutagenesis methods have been applied to generate mutants with increased astaxanthin production [124–126]. *N*-Methyl-*N′*-nitro-N-nitrosoguanidine (NTG) has proven to be an effective chemical mutagen for *X. dendrorhous*, although the achieved astaxanthin levels are still not very attractive from an industrial point of view.

A promising strategy to increase the astaxanthin yield in *X. dendrorhous* is metabolic engineering, and several attempts have been made, including the overexpression of genes involved in carotenoid synthesis. Although the overexpression of the *crtYB* gene (which encodes the phytoene-lycopene synthase) led to an increase in the overall carotenoid synthesis, it was mainly due to higher amounts of β(beta)-carotene and echinenone with an unaffected (or even lower) astaxanthin content observed [127]. In contrast, the overexpression of the *crtI* gene (phytoene desaturase encoding gene) decreased the total carotenoid production and varied its composition. The carbon flux was diverted to the synthesis of monocyclic carotenoids such as 3-hydroxy-3′,4′-didehydro-β(beta)-φ(phi)-caroten-4-one (HDCO), torulene and hydroxy-ketotorulene, while the astaxanthin proportion was reduced to half [127].

An interesting alternative genetic modification is to increase the metabolic flow towards the synthesis of the precursor molecules of a specific pathway. However, when the isopentenyl-isomerase encoding gene (*idi*) was overexpressed in *X. dendrorhous*, the amount of total carotenoids decreased [128]. In contrast, the overexpression of the geranylgeranyl-pyrophosphate synthase encoding gene (*crtE*) resulted in a strain with slightly higher carotenoid levels, which was improved when the strain was cultured under additional air supply and permanent illumination [129]. More recently, a combinatorial approach fused conventional mutagenesis, metabolic pathway engineering (including the simultaneous overexpression of *crtYB* and *crtS*) and different culture medium analysis, and resulted in the highest astaxanthin content for *X. dendrorhous* reported to this day (9,700 ppm) [130]. Although these studies represent a great contribution, the achieved astaxanthin levels are still not industrially sufficient.

Improved astaxanthin yields have been achieved by the addition of the carotenogenesis precursors such as MVA to the culture medium [117]. Many of the regulatory aspects of isoprenoid biosynthesis involve elements of the MVA pathway, which are well conserved throughout evolution. The limiting step in the metabolic flux control of the MVA pathway is catalyzed by the enzyme 3-hydroxy-3-methylglutaryl-CoA (HMG-CoA) reductase [131]. For example, the overexpression of the catalytic domain of a HMGR from *S. cerevisiae* (*HMG1* gene) increased the heterologous production of carotenoids in *Candida utilis* [132]. In the same way, additional *HMGR* gene copies in *N. crassa* also resulted in an increase in the amount of carotenoids produced [133]. In *X. dendrorhous*, only one *HMGR* gene has been detected and results by Miao and coworkers [134] indicated that its transcript levels were increased in an astaxanthin overproducing *X. dendrorhous* strain obtained by random mutagenesis. This mutant also showed lower ergosterol content, suggesting that ergosterol might regulate the *HMGR* gene expression in *X. dendrorhous* and, in turn, affect the carotenoid biosynthesis. Recently, the *X. dendrorhous CYP61* gene involved in the ergosterol biosynthesis was identified and characterized [135]. The disruption of this gene abolished ergosterol production and the carotenoid content, including astaxanthin, was almost doubled relative to the parental strain. Moreover, it was shown that the transcript levels of *HMGR* were significantly increased in the *cyp61*⁻ mutant strains. This background suggests that engineering the steps involved in the MVA pathway could be a good approach for the improvement of astaxanthin production in *X. dendrorhous*.

Conclusion

Future Perspectives

As described in this chapter, the biosynthesis of astaxanthin in *X. dendrorhous* is a complex process. Although the structural genes that control this metabolic pathway are known, knowledge of the mechanisms regulating the synthesis of carotenoids in this yeast is still rather limited. Currently, it is known that carotenogenesis is affected by numerous internal and external factors. Among the external factors, the carbon source plays a fundamental role in the regulation of the synthesis of astaxanthin in this yeast. Specifically, when *X. dendrorhous* is cultivated with glucose as the sole carbon source, carotenogenesis is only induced after this sugar is completely depleted. This point (the late exponential - early stationary phase of growth) also coincides with the maximum ethanol concentration reached, which is produced by the fermentative metabolism of glucose. Along the same lines, when the yeast is cultivated with a non-fermentable carbon source, such as succinate, the synthesis of astaxanthin starts at the beginning of the growth and the achieved carotenoid levels are higher than those reached when the cells are grown in glucose [136]. Furthermore, there are astaxanthin-overproducing strains, obtained by random mutagenesis, in which carotenogenesis starts at the beginning of the culture even in the presence of glucose, indicating that they might be deregulated [137]. Moreover, glucose reduces the mRNA levels of at least three carotenogenic genes (*crtYB*, *crtI* and *crtS*) [138]. All these findings suggest that carotenogenesis in *X. dendrorhous* is regulated by catabolite repression. Similarly, it is known that the carbon to nitrogen (C/N) ratio and the levels of oxidative stress are also relevant factors that can alter the production of carotenoids in this yeast, and may also participate in the mechanisms that govern the carotenogenesis. Among the internal factors that might regulate carotenogenesis in *X. dendrorhous*, we highlight the fact that alternative carotenogenic mRNAs are produced for at least two genes: *crtYB* and *crtI* [139].

Based on the aforementioned observations, understanding the regulatory mechanisms of the astaxanthin biosynthetic pathway in *X. dendrorhous* will have a key impact on the comprehension of the complexity of this biological process. Similarly, the knowledge of the interactions between carotenogenesis and other metabolic processes, whether at the genetic level of the structural and regulatory genes or at the level of their gene products, will be useful in developing new astaxanthin-overproducing strains important for the industrial production of this xanthophyll. Thus, molecular genetic studies leading to the identification of regulatory genes, their targets, and the genetic interactions involved in carotenogenesis will aid in the design of new genetic improvement strategies for the production of astaxanthin in *X. dendrorhous*. Similarly, this knowledge could be applied to develop strains in which the metabolic flux towards intermediary metabolites of carotenogenesis could be diverted to favor the production of new biotechnological products, such as drugs, vitamins, nutrients, or other carotenoids, to name a few.

Finally, the use of metabolic engineering—for example, by modifying the metabolic flux of pathways that interact with carotenogenesis at different levels—will not only generate a greater knowledge of the process itself, but will also create new tools that can be applied in the construction of attractive strains for carotenoid production processes.

Acknowledgements This work was supported by projects: Fondecyt 11121200 and INACH RG_07-12 to JA, Fondecyt 1130333 to MB and Fondecyt 1100324 to VC.

References

1. Goodwin TW. Carotenoid-protein complexes. The biochemistry of the carotenoids, vol. 2. Dordrecht: Springer; 1984. p. 1–21.
2. Sourkes TL. The discovery and early history of carotene. Bull Hist Chem. 2009;34:33.
3. Takaichi S. Carotenoids in algae: distributions, biosynthesis and functions. Mar Drugs. 2011;9:1101–18.
4. Fraser PD, Bramley PM. The biosynthesis and nutritional uses of carotenoids. Prog Lipid Res. 2004;43:228–65.
5. Britton G. Structure and properties of carotenoids in relation to function. FASEB J. 1995;9: 1551–8.
6. Bhosale P, Bernstein PS. Microbial xanthophylls. Appl Microbiol Biotechnol. 2005;68: 445–55.
7. Strzałka K, Kostecka-Gugała A, Latowski D. Carotenoids and environmental stress in plants: significance of carotenoid-mediated modulation of membrane physical properties. Russ J Plant Physiol. 2003;50:168–73.
8. Zhang W, Zhang KY, Ding XM, Bai SP, Hernandez JM, Yao B, Zhu Q. Influence of canthaxanthin on broiler breeder reproduction, chick quality, and performance. Poult Sci. 2011;90:1516–22.
9. Perez-Vendrell AM, Hernandez JM, Llaurado L, Schierle J, Brufau J. Influence of source and ratio of xanthophyll pigments on broiler chicken pigmentation and performance. Poult Sci. 2001;80:320–6.
10. Baker R, Günther C. The role of carotenoids in consumer choice and the likely benefits from their inclusion into products for human consumption. Trends Food Sci Tech. 2004;15:484–8.
11. Sieiro C, Poza M, de Miguel T, Villa TG. Genetic basis of microbial carotenogenesis. Int Microbiol. 2003;6:11–6.
12. Vergari F, Tibuzzi A, Basile G. An overview of the functional food market: from marketing issues and commercial players to future demand from life in space. In: Maria Teresa Giardi, Giuseppina Rea and Bruno Berra (eds.). Bio-farms for nutraceuticals. New York: Springer; 2010:308–21.

13. Breithaupt DE. Modern application of xanthophylls in animal feeding—a review. Trends Food Sci Tech. 2007;18:501–6.
14. Burdick EM. Extraction and utilization of carotenes and xanthophylls. Econ Bot. 1956;10: 267–79.
15. Ukibe K, Hashida K, Yoshida N, Takagi H. Metabolic engineering of *Saccharomyces cerevisiae* for astaxanthin production and oxidative stress tolerance. Appl Environ Microbiol. 2009;75: 7205–11.
16. Shao Z, Zhao H, Zhao H. DNA assembler, an in vivo genetic method for rapid construction of biochemical pathways. Nucleic Acids Res. 2009;37:e16.
17. Higuera-Ciapara I, Felix-Valenzuela L, Goycoolea FM. Astaxanthin: a review of its chemistry and applications. Crit Rev Food Sci Nutr. 2006;46:185–96.
18. Fassett RG, Coombes JS. Astaxanthin in cardiovascular health and disease. Molecules. 2012;17:2030–48.
19. Wang X, Willen R, Wadstrom T. Astaxanthin-rich algal meal and vitamin C inhibit *Helicobacter pylori* infection in BALB/cA mice. Antimicrob Agents Chemother. 2000;44: 2452–7.
20. Park JS, Chyun JH, Kim YK, Line LL, Chew BP. Astaxanthin decreased oxidative stress and inflammation and enhanced immune response in humans. Nutr Metab. 2010;7:18.
21. Yasui Y, Hosokawa M, Mikami N, Miyashita K, Tanaka T. Dietary astaxanthin inhibits colitis and colitis-associated colon carcinogenesis in mice via modulation of the inflammatory cytokines. Chem Biol Interact. 2011;193:79–87.
22. Mortensen A. Supplements. In: Britton G, Liaaen-Jensen S and Pfander H (eds.). Carotenoids. New York: Springer; 2009:67–82.
23. Nishigaki M, Yamamoto T, Ichioka H, Honjo K, Yamamoto K, Oseko F, Kita M, Mazda O, Kanamura N. Beta-cryptoxanthin regulates bone resorption related-cytokine production in human periodontal ligament cells. Arch Oral Biol. 2013;58:880–6.
24. Irwig MS, El-Sohemy A, Baylin A, Rifai N, Campos H. Frequent intake of tropical fruits that are rich in β-cryptoxanthin is associated with higher plasma β-cryptoxanthin concentrations in Costa Rican adolescents. J Nutr. 2002;132:3161–7.
25. Yamaguchi M. Role of carotenoid β-cryptoxanthin in bone homeostasis. J Biomed Sci. 2012;19:1–13.
26. Lorenzo Y, Azqueta A, Luna L, Bonilla F, Dominguez G, Collins AR. The carotenoid beta-cryptoxanthin stimulates the repair of DNA oxidation damage in addition to acting as an antioxidant in human cells. Carcinogenesis. 2009;30:308–14.
27. Tanaka T, Tanaka T, Tanaka M, Kuno T. Cancer chemoprevention by citrus pulp and juices containing high amounts of beta-cryptoxanthin and hesperidin. J Biomed Biotechnol. 2012;2012:516981.
28. Khachik F, Liu Y, Showalter H. Process for the preparation of β-and α-cryptoxanthin. European Patent Application EP1678111 A1. 2012.
29. Louie MTM, Fuerst EJ. Biosynthesis of beta-cryptoxanthin in microbial hosts using an *Arabidopsis thaliana* beta-carotene hydroxylase gene. US Patent Application US20080124755 A1. 2006.
30. Guyomarc'h F, Binet A, Dufosse L. Production of carotenoids by *Brevibacterium linens*: variation among strains, kinetic aspects and HPLC profiles. J Ind Microbiol Biotechnol. 2000;24:64–70.
31. Serrato-Joya O, Jiménez-Islas H, Botello-Álvarez E, Rico-Martínez R, Navarrete-Bolaños JL. Production of β-cryptoxanthin, a provitamin-A precursor, by *Flavobacterium lutescens*. J Food Sci. 2006;71:E314–9.
32. Gharibzahedi SMT, Razavi SH. Microbial canthaxanthin: perspectives on biochemistry and biotechnological production. Eng Life Sci. 2013;00:1–10.
33. Kadian SS, Garg M. Pharmacological effects of carotenoids: a review. Int J Pharm Sci Res. 2012;3:42–8.
34. Breithaupt DR. Xanthophylls in poultry feeding. In: Carotenoids. New York: Springer; 2008, pp. 255–64.

35. Aberoumand A. A review article on edible pigments properties and sources as natural biocolorants in foodstuff and food industry. World J Dairy Food Sci. 2011;6(1):71–8.
36. Krupa D, Nakkeeran E, Kumaresan N, Vijayalakshmi G, Subramanian R. Extraction, purification and concentration of partially saturated canthaxanthin from *Aspergillus carbonarius*. Bioresour Technol. 2010;101:7598–604.
37. Nasri Nasrabadi MR, Razavi SH. Use of response surface methodology in a fed-batch process for optimization of tricarboxylic acid cycle intermediates to achieve high levels of canthaxanthin from *Dietzia natronolimnaea* HS-1. J Biosci Bioeng. 2010;109:361–8.
38. Li H, Fan K, Chen F. Isolation and purification of canthaxanthin from the microalga *Chlorella zofingiensis* by high-speed counter-current chromatography. J Sep Sci. 2006;29(5):699–703.
39. Pelah D, Sintov A, Cohen E. The effect of salt stress on the production of canthaxanthin and astaxanthin by *Chlorella zofingiensis* grown under limited light intensity. World J Microbiol Biotechnol. 2004;20:483–6.
40. Jeknić Z, Morré JT, Jeknić S, Jevremović S, Subotić A, Chen THH. Cloning and functional characterization of a gene for capsanthin-capsorubin synthase from tiger lily (*Lilium lancifolium thunb.* 'Splendens'). Plant Cell Physiol. 2012;53:1899–912.
41. Maeda H, Saito S, Nakamura N, Maoka T. Paprika pigments attenuate obesity-induced inflammation in 3T3-L1 adipocytes. ISRN Inflamm. 2013;2013:1–9.
42. Kim S, Ha TY, Hwang IK. Analysis, bioavailability, and potential healthy effects of capsanthin, natural red pigment from Capsicum spp. Food Rev Int. 2009;25:198–213.
43. Miyashita K, Nishikawa S, Beppu F, Tsukui T, Abe M, Hosokawa M. The allenic carotenoid fucoxanthin, a novel marine nutraceutical from brown seaweeds. J Sci Food Agric. 2011;91:1166–74.
44. Li Y, Li L. Method for producing fucoxanthin. US Patent Application US20100152286 A1. 2009.
45. Wang J, Chen S, Xu S, Yu X, Ma D, Hu X, Cao X. In vivo induction of apoptosis by fucoxanthin, a marine carotenoid, associated with down-regulating STAT3/EGFR signaling in sarcoma 180 (S180) xenografts-bearing mice. Mar Drugs. 2012;10:2055–68.
46. Kijlstra A, Tian Y, Kelly ER, Berendschot TT. Lutein: more than just a filter for blue light. Prog Retin Eye Res. 2012;31:303–15.
47. Calvo MM. Lutein: a valuable ingredient of fruit and vegetables. Crit Rev Food Sci Nutr. 2005;45:671–96.
48. Shegokar R, Mitri K. Carotenoid lutein: a promising candidate for pharmaceutical and nutraceutical applications. J Diet Suppl. 2012;9:183–210.
49. Lo HM, Chen CL, Yang CM, Wu PH, Tsou CJ, Chiang KW, Wu WB. The carotenoid lutein enhances matrix metalloproteinase-9 production and phagocytosis through intracellular ROS generation and ERK1/2, p38 MAPK, and RAR beta activation in murine macrophages. J Leukoc Biol. 2013;93:723–35.
50. Puzio P, Blau A, Plesch G, Kamlage B, Looser R, Schmitz O, Wendel B. Process for the production of lutein. European Patent Application EP2096177 A2. 2009.
51. Fernandez-Sevilla JM, Acien Fernandez FG, Molina GE. Biotechnological production of lutein and its applications. Appl Microbiol Biotechnol. 2010;86:27–40.
52. Bowen PE, Herbst-Espinosa SM, Hussain EA, Stacewicz-Sapuntzakis M. Esterification does not impair lutein bioavailability in humans. J Nutr. 2002;132:3668–73.
53. Yonekura L, Nagao A. Intestinal absorption of dietary carotenoids. Mol Nutr Food Res. 2007;51:107–15.
54. Lakshminarayana R, Raju M, Krishnakantha TP, Baskaran V. Enhanced lutein bioavailability by lyso-phosphatidylcholine in rats. Mol Cell Biochem. 2006;281:103–10.
55. Arunkumar R, Harish P, Veerappa K, Baskaran V. Promising interaction between nanoencapsulated lutein with low molecular weight chitosan: characterization and bioavailability of lutein in vitro and in vivo. Food Chem. 2013;141:327–37.
56. Sivathanu B, Palaniswamy S. Purification and characterization of carotenoids from green algae *Chlorococcum humicola* by HPLC-NMR and LC-MS-APCI. Biomed Prev Nutr. 2012;2:276–82.

57. Casal C, Cuaresma M, Vega JM, Vilchez C. Enhanced productivity of a lutein-enriched novel acidophile microalga grown on urea. Mar Drugs. 2011;9:29–42.
58. Del Campo JA, Rodriguez H, Moreno J, Vargas MA, Rivas J, Guerrero MG. Accumulation of astaxanthin and lutein in *Chlorella zofingiensis* (Chlorophyta). Appl Microbiol Biotechnol. 2004;64:848–54.
59. Garcia-Gonzalez M, Moreno J, Manzano JC, Florencio FJ, Guerrero MG. Production of Dunaliella salina biomass rich in 9-cis-beta-carotene and lutein in a closed tubular photobioreactor. J Biotechnol. 2005;115:81–90.
60. Campenni L, Nobre BP, Santos CA, Oliveira AC, Aires-Barros MR, Palavra AMF, Gouveia L. Carotenoid and lipid production by the autotrophic microalga *Chlorella protothecoides* under nutritional, salinity, and luminosity stress conditions. Appl Microbiol Biotechnol. 2013;97:1383–93.
61. Hartung W. The evolution of abscisic acid (ABA) and ABA function in lower plants, fungi and lichen. Funct Plant Biol. 2010;37:806–12.
62. Garcia-Herrera P, Sanchez-Mata MC, Camara M, Tardio J, Olmedilla-Alonso B. Carotenoid content of wild edible young shoots traditionally consumed in Spain (*Asparagus acutifolius L. Humulus lupulus L. Bryonia dioica Jacq.* and *Tamus communis L.*). J Sci Food Agric. 2013;93:1692–8.
63. Kotake-Nara E, Kushiro M, Zhang H, Sugawara T, Miyashita K, Nagao A. Carotenoids affect proliferation of human prostate cancer cells. J Nutr. 2001;131:3303–6.
64. Kotake-Nara E, Asai A, Nagao A. Neoxanthin and fucoxanthin induce apoptosis in PC-3 human prostate cancer cells. Cancer Lett. 2005;220:75–84.
65. Meléndez-Martínez AJ, Vicario IM, Heredia FJ. Geometrical isomers of violaxanthin in orange juice. Food Chem. 2007;104:169–75.
66. Guedes A, Gião MS, Seabra R, Ferreira AC. Evaluation of the antioxidant activity of cell extracts from microalgae. Mar Drugs. 2013;11(4):1256–70.
67. Abdel-Aal E-SM, Akhtar H, Zaheer K, Ali R. Dietary sources of lutein and zeaxanthin carotenoids and their role in eye health. Nutrients. 2013;5:1169–85.
68. SanGiovanni JP, Neuringer M. The putative role of lutein and zeaxanthin as protective agents against age-related macular degeneration: promise of molecular genetics for guiding mechanistic and translational research in the field. Am J Clin Nutr. 2012;96:1223S–33.
69. Wu NL, Chiang YC, Huang CC, Fang JY, Chen DF, Hung CF. Zeaxanthin inhibits PDGF-BB-induced migration in human dermal fibroblasts. Exp Dermatol. 2010;19:e173–81.
70. Firdous AP, Sindhu ER, Ramnath V, Kuttan R. Anti-mutagenic and anti-carcinogenic potential of the carotenoid meso-zeaxanthin. Asian Pac J Cancer Prev. 2010;11:1795–800.
71. Khachik F. Process for extraction and purification of lutein, zeaxanthin and rare carotenoids from marigold flowers and plants. US Patent Application US6262284 B1. 2007.
72. Zuorro A, Lavecchia R. New functional food products containing lutein and zeaxanthin from marigold (*Tagetes erecta* L.) flowers. J Biotechnol. 2010;150:296.
73. Yu B, Wang J, Suter PM, Russell RM, Grusak MA, Wang Y, Wang Z, Yin S, Tang G. Spirulina is an effective dietary source of zeaxanthin to humans. Br J Nutr. 2012;108:611.
74. Issouf M, Mearns SA, Fraser K, Hodgson R. Biological production of zeaxanthin and carotenoid biosynthesis control. World Intellectual Property Organization (WIPO) Patent Application WO2006120400 A1. 2012.
75. Chavez-Parga M, Munguia-Franco A, Aguilar-Torres M, Escamilla-Silva EM. Optimization of Zeaxanthin production by immobilized *Flavobacterium* sp. cells in fluidized bed bioreactor. Adv Microbiol. 2012;2:598–604.
76. Asker D, Awad TS, Beppu T, Ueda K. Novel zeaxanthin-producing bacteria isolated from a radioactive hot spring water. Methods Mol Biol. 2012;892:99–131.
77. Singh D, Puri M, Wilkins S, Mathur AS, Tuli DK, Barrow CJ. Characterization of a new zeaxanthin producing strain of *Chlorella saccharophila* isolated from New Zealand marine waters. Bioresour Technol. 2013;143:308–14.
78. Liang P-H, Ko T-P, Wang AH-J. Structure, mechanism and function of prenyltransferases. Eur J Biochem. 2002;269:3339–54.

79. Lee PC, Schmidt-Dannert C. Metabolic engineering towards biotechnological production of carotenoids in microorganisms. Appl Microbiol Biotechnol. 2002;60:1–11.
80. Misawa N. Pathway engineering for functional isoprenoids. Curr Opin Biotechnol. 2011;22: 627–33.
81. Britton G. Overview of carotenoid biosynthesis. Carotenoids. 1998;3:13–147.
82. Ajikumar PK, Tyo K, Carlsen S, Mucha O, Phon TH, Stephanopoulos G. Terpenoids: opportunities for biosynthesis of natural product drugs using engineered microorganisms. Mol Pharm. 2008;5:167–90.
83. Krubasik P, Sandmann G. Molecular evolution of lycopene cyclases involved in the formation of carotenoids with ionone end groups. Biochem Soc Trans. 2000;28:806–9.
84. Schmidhauser TJ, Lauter F-R, Schumacher M, Zhou W, Russo VE, Yanofsky C. Characterization of *al-2*, the phytoene synthase gene of *Neurospora crassa*. Cloning, sequence analysis, and photoregulation. J Biol Chem. 1994;269:12060–6.
85. Velayos A, Eslava AP, Iturriaga EA. A bifunctional enzyme with lycopene cyclase and phytoene synthase activities is encoded by the *carRP* gene of *Mucor circinelloides*. Eur J Biochem. 2000;267:5509–19.
86. Arrach N, Fernández-Martín R, Cerdá-Olmedo E, Avalos J. A single gene for lycopene cyclase, phytoene synthase, and regulation of carotene biosynthesis in *Phycomyces*. Proc Natl Acad Sci. 2001;98:1687–92.
87. Verdoes JC, Krubasik KP, Sandmann G, van Ooyen AJ. Isolation and functional characterisation of a novel type of carotenoid biosynthetic gene from *Xanthophyllomyces dendrorhous*. Mol Gen Genet. 1999;262:453–61.
88. Cheng Q. Structural diversity and functional novelty of new carotenoid biosynthesis genes. J Ind Microbiol Biotechnol. 2006;33:552–9.
89. Mochimaru M, Masukawa H, Maoka T, Mohamed HE, Vermaas WF, Takaichi S. Substrate specificities and availability of fucosyltransferase and beta-carotene hydroxylase for myxol 2′-fucoside synthesis in *Anabaena* sp. strain PCC 7120 compared with *Synechocystis* sp. strain PCC 6803. J Bacteriol. 2008;190:6726–33.
90. Golubev WI. Perfect state of *Rhodomyces dendrorhous* (*Phaffia rhodozyma*). Yeast. 1995;11:101–10.
91. Johnson EA. *Phaffia rhodozyma*: colorful odyssey. Int Microbiol. 2003;6:169–74.
92. Andrewes AG, Starr MP. (3R, 3′R)-astaxanthin from the yeast *Phaffia rhodozyma*. Phytochemistry. 1976;15:1009–11.
93. Lemoine Y, Schoefs B. Secondary ketocarotenoid astaxanthin biosynthesis in algae: a multifunctional response to stress. Photosynth Res. 2010;106:155–77.
94. Lee JH, Kim YT. Cloning and characterization of the astaxanthin biosynthesis gene cluster from the marine bacterium *Paracoccus haeundaensis*. Gene. 2006;370:86–95.
95. Yokoyama A, Miki W. Composition and presumed biosynthetic pathway of carotenoids in the astaxanthin-producing bacterium *Agrobacterium aurantiacum*. FEMS Microbiol Lett. 1995;128:139–44.
96. Renstrøm B, Berger H, Liaaen-Jensen S. Esterified, optical pure (3S, 3′S)-astaxanthin from flowers of *Adonis annua*. Biochem Syst Ecol. 1981;9:249–50.
97. Misawa N, Satomi Y, Kondo K, Yokoyama A, Kajiwara S, Saito T, Ohtani T, Miki W. Structure and functional analysis of a marine bacterial carotenoid biosynthesis gene cluster and astaxanthin biosynthetic pathway proposed at the gene level. J Bacteriol. 1995;177:6575–84.
98. Vidhyavathi R, Venkatachalam L, Sarada R, Ravishankar GA. Regulation of carotenoid biosynthetic genes expression and carotenoid accumulation in the green alga *Haematococcus pluvialis* under nutrient stress conditions. J Exp Bot. 2008;59:1409–18.
99. Cunningham FXJ, Gantt E. Elucidation of the pathway to astaxanthin in the flowers of *Adonis aestivalis*. Plant Cell. 2011;23:3055–69.
100. Martín J, Gudiña E, Barredo J. Conversion of β-carotene into astaxanthin: two separate enzymes or a bifunctional hydroxylase-ketolase protein? Microb Cell Fact. 2008;7:3.
101. Tian L, DellaPenna D. Progress in understanding the origin and functions of carotenoid hydroxylases in plants. Arch Biochem Biophys. 2004;430:22–9.

102. Blasco F, Kauffmann I, Schmid RD. CYP175A1 from *Thermus thermophilus* HB27, the first beta-carotene hydroxylase of the P450 superfamily. Appl Microbiol Biotechnol. 2004;64: 671–4.
103. Misawa N, Nakagawa M, Kobayashi K, Yamano S, Izawa Y, Nakamura K, Harashima K. Elucidation of the *Erwinia uredovora* carotenoid biosynthetic pathway by functional analysis of gene products expressed in *Escherichia coli*. J Bacteriol. 1990;172:6704–12.
104. McLean KJ, Sabri M, Marshall KR, Lawson RJ, Lewis DG, Clift D, Balding PR, Dunford AJ, Warman AJ, McVey JP. Biodiversity of cytochrome P450 redox systems. Biochem Soc Trans. 2005;33:796.
105. Zhang H, Im SC, Waskell L. Cytochrome b5 increases the rate of product formation by cytochrome P450 2B4 and competes with cytochrome P450 reductase for a binding site on cytochrome P450 2B4. J Biol Chem. 2007;282:29766–76.
106. Degtyarenko KN, Archakov AI. Molecular evolution of P450 superfamily and P450-containing monooxygenase systems. FEBS Lett. 1993;332:1–8.
107. Bernhardt R. Cytochromes P450 as versatile biocatalysts. J Biotechnol. 2006;124:128–45.
108. Estabrook RW. A passion for P450s (remembrances of the early history of research on cytochrome P450). Drug Metab Dispos. 2003;31:1461–73.
109. Porter TD, Coon MJ. Cytochrome P-450. Multiplicity of isoforms, substrates, and catalytic and regulatory mechanisms. J Biol Chem. 1991;266:13469–72.
110. van den Brink H, van Gorcom RFM, van den Hondel CAMJJ, Punt PJ. Cytochrome P450 enzyme systems in fungi. Fungal Genet Biol. 1998;23:1–17.
111. Sevrioukova IF, Li H, Zhang H, Peterson JA, Poulos TL. Structure of a cytochrome P450—redox partner electron-transfer complex. Proc Natl Acad Sci U S A. 1999;96:1863–8.
112. Munro AW, Girvan HM, McLean KJ. Cytochrome P450—redox partner fusion enzymes. Biochim Biophys Acta. 2007;1770:345–59.
113. Lah L, Krasevec N, Trontelj P, Komel R. High diversity and complex evolution of fungal cytochrome P450 reductase: cytochrome P450 systems. Fungal Genet Biol. 2008;45: 446–58.
114. Ojima K, Breitenbach J, Visser H, Setoguchi Y, Tabata K, Hoshino T, van den Berg J, Sandmann G. Cloning of the astaxanthin synthase gene from *Xanthophyllomyces dendrorhous* (*Phaffia rhodozyma*) and its assignment as a beta-carotene 3-hydroxylase/4-ketolase. Mol Genet Genomics. 2006;275:148–58.
115. Alvarez V, Rodriguez-Saiz M, de la Fuente JL, Gudina EJ, Godio RP, Martin JF, Barredo JL. The *crtS* gene of *Xanthophyllomyces dendrorhous* encodes a novel cytochrome-P450 hydroxylase involved in the conversion of beta-carotene into astaxanthin and other xanthophylls. Fungal Genet Biol. 2006;43:261–72.
116. Alcaino J, Barahona S, Carmona M, Lozano C, Marcoleta A, Niklitschek M, Sepulveda D, Baeza M, Cifuentes V. Cloning of the cytochrome p450 reductase (*crtR*) gene and its involvement in the astaxanthin biosynthesis of *Xanthophyllomyces dendrorhous*. BMC Microbiol. 2008;8:169.
117. Calo P, Gonzalez T. The yeast *Phaffia rhodozyma* as an industrial source of astaxanthin. Microbiologia. 1995;11:386–8.
118. Rodriguez-Saiz M, de la Fuente JL, Barredo JL. *Xanthophyllomyces dendrorhous* for the industrial production of astaxanthin. Appl Microbiol Biotechnol. 2010;88:645–58.
119. Yamane Y, Higashida K, Nakashimada Y, Kakizono T, Nishio N. Influence of oxygen and glucose on primary metabolism and astaxanthin production by *Phaffia rhodozyma* in batch and fed-batch cultures: kinetic and stoichiometric analysis. Appl Environ Microbiol. 1997;63: 4471–8.
120. Flores-Cotera LB, Martín R, Sánchez S. Citrate, a possible precursor of astaxanthin in *Phaffia rhodozyma*: influence of varying levels of ammonium, phosphate and citrate in a chemically defined medium. Appl Microbiol Biotechnol. 2001;55:341–7.
121. Liu YS, Wu JY. Hydrogen peroxide-induced astaxanthin biosynthesis and catalase activity in *Xanthophyllomyces dendrorhous*. Appl Microbiol Biotechnol. 2006;73:663–8.

122. Hu Z-C, Zheng Y-G, Wang Z, Shen Y-C. pH control strategy in astaxanthin fermentation bioprocess by *Xanthophyllomyces dendrorhous*. Enzyme Microb Technol. 2006;39:586–90.
123. An G-H, Johnson EA. Influence of light on growth and pigmentation of the yeast *Phaffia rhodozyma*. Antonie Van Leeuwenhoek. 1990;57:191–203.
124. An G-H, Schuman DB, Johnson EA. Isolation of Phaffia rhodozyma mutants with increased astaxanthin content. Appl Environ Microbiol. 1989;55:116–24.
125. Retamales P, León RUBEN, Martinez C, Hermosilla G, Pincheira G, Cifuentes V. Complementation analysis with new genetic markers in *Phaffia rhodozyma*. Antonie Van Leeuwenhoek. 1998;73:229–36.
126. Ukibe K, Katsuragi T, Tani Y, Takagi H. Efficient screening for astaxanthin-overproducing mutants of the yeast *Xanthophyllomyces dendrorhous* by flow cytometry. FEMS Microbiol Lett. 2008;286:241–8.
127. Verdoes JC, Sandmann G, Visser H, Diaz M, van Mossel M, van Ooyen AJJ. Metabolic engineering of the carotenoid biosynthetic pathway in the yeast *Xanthophyllomyces dendrorhous* (*Phaffia rhodozyma*). Appl Environ Microbiol. 2003;69:3728–38.
128. Visser H, Sandmann G, Verdoes JC. Xanthophylls in fungi. In: Microbial processes and products. Edited by José Luis Barredo. New York: Springer; 2005, pp. 257–72.
129. Breitenbach J, Visser H, Verdoes JC, van Ooyen AJ, Sandmann G. Engineering of geranylgeranyl pyrophosphate synthase levels and physiological conditions for enhanced carotenoid and astaxanthin synthesis in *Xanthophyllomyces dendrorhous*. Biotechnol Lett. 2011;33:755–61.
130. Gassel S, Schewe H, Schmidt I, Schrader J, Sandmann G. Multiple improvement of astaxanthin biosynthesis in *Xanthophyllomyces dendrorhous* by a combination of conventional mutagenesis and metabolic pathway engineering. Biotechnol Lett. 2013;35:565–9.
131. Miziorko HM. Enzymes of the mevalonate pathway of isoprenoid biosynthesis. Arch Biochem Biophys. 2011;505:131–43.
132. Shimada H, Kondo K, Fraser PD, Miura Y, Saito T, Misawa N. Increased carotenoid production by the food yeast *Candida utilis* through metabolic engineering of the isoprenoid pathway. Appl Environ Microbiol. 1998;64:2676–80.
133. Wang G. Amplification of HMG-CoA reductase production enhances carotenoid accumulation in *Neurospora crassa*. Metab Eng. 2002;4:193–201.
134. Miao L, Chi S, Tang Y, Su Z, Yin T, Guan G, Li Y. Astaxanthin biosynthesis is enhanced by high carotenogenic gene expression and decrease of fatty acids and ergosterol in a *Phaffia rhodozyma* mutant strain. FEMS Yeast Res. 2011;11:192–201.
135. Loto I, Gutiérrez MS, Barahona S, Sepúlveda D, Martínez-Moya P, Baeza M, Cifuentes V, Alcaíno J. Enhancement of carotenoid production by disrupting the C22-sterol desaturase gene (*CYP61*) in *Xanthophyllomyces dendrorhous*. BMC Microbiol. 2012;12:235.
136. Lodato P, Alcaino J, Barahona S, Niklitschek M, Carmona M, Wozniak A, Baeza M, Jimenez A, Cifuentes V. Expression of the carotenoid biosynthesis genes in *Xanthophyllomyces dendrorhous*. Biol Res. 2007;40:73–84.
137. Wozniak A, Lozano C, Barahona S, Niklitschek M, Marcoleta A, Alcaino J, Sepulveda D, Baeza M, Cifuentes V. Differential carotenoid production and gene expression in Xanthophyllomyces dendrorhous grown in a nonfermentable carbon source. FEMS Yeast Res. 2011;11:252–62.
138. Marcoleta A, Niklitschek M, Wozniak A, Lozano C, Alcaíno J, Baeza M, Cifuentes V. Glucose and ethanol-dependent transcriptional regulation of the astaxanthin biosynthesis pathway in *Xanthophyllomyces dendrorhous*. BMC Microbiol. 2011;11:1–11.
139. Lodato P, Alcaino J, Barahona S, Retamales P, Cifuentes V. Alternative splicing of transcripts from *crt*I and *crt*YB genes of *Xanthophyllomyces dendrorhous*. Appl Environ Microbiol. 2003;69:4676–82.
140. Dragoş N, Bercea V, Bica A, Drugă B, Nicoară A, Coman C. Astaxanthin production from a new strain of *Haematococcus pluvialis* grown in batch culture. Ann Rom Soc Cell Biol. 2010;15:353–61.
141. Gharibzahedi SMT, Razavi SH, Mousavi SM, Moayedi V. High efficiency canthaxanthin production by a novel mutant isolated from *Dietzia natronolimnaea* HS-1 using central composite design analysis. Ind Crop Prod. 2012;40:345–54.

142. Papp T, Csernetics Á, Nagy G, Bencsik O, Iturriaga EA, Eslava AP, Vágvölgyi C. Canthaxanthin production with modified *Mucor circinelloides* strains. Appl Microbiol Biotechnol. 2013;97:4937–50.
143. Cordero BF, Obraztsova I, Couso I, Leon R, Vargas MA, Rodriguez H. Enhancement of lutein production in *Chlorella sorokiniana* (*Chorophyta*) by improvement of culture conditions and random mutagenesis. Mar Drugs. 2011;9:1607–24.
144. Sánchez JF, Fernández JM, Acién FG, Rueda A, Pérez-Parra J, Molina E. Influence of culture conditions on the productivity and lutein content of the new strain *Scenedesmus almeriensis*. Process Biochem. 2008;43:398–405.
145. Blanco AM, Moreno J, Del Campo JA, Rivas J, Guerrero MG. Outdoor cultivation of lutein-rich cells of *Muriellopsis* sp. in open ponds. Appl Microbiol Biotechnol. 2007;73:1259–66.
146. Scoma A, Krawietz D, Faraloni C, Giannelli L, Happe T, Torzillo G. Sustained H2 production in a *Chlamydomonas reinhardtii* D1 protein mutant. J Biotechnol. 2012;157:613–9.
147. Lohr M, Wilhelm C. Algae displaying the diadinoxanthin cycle also possess the violaxanthin cycle. Proc Natl Acad Sci U S A. 1999;96:8784–9.
148. Dharmapuri S, Giuliano G. Plant metabolic engineering: requirements for success. Br Food J. 2001;103:764–9.
149. Huang J, Zhong Y, Sandmann G, Liu J, Chen F. Cloning and selection of carotenoid ketolase genes for the engineering of high-yield astaxanthin in plants. Planta. 2012;236:691–9.
150. Tao L, Wilczek J, Odom JM, Cheng Q. Engineering a β-carotene ketolase for astaxanthin production. Metab Eng. 2006;8:523–31.
151. Britton G, Liaaen-Jensen S, Pfander H (ed.). Carotenoids. Handbook. – Birkhäuser Verlag, Basel – Boston – Berlin; 2004.

Chapter 10
Gibberellins and the Red Pigments Bikaverin and Fusarubin

Lena Studt and Bettina Tudzynski

Introduction

The genus *Fusarium* includes a number of widely distributed plant-pathogenic species that infect various economically important host-plants thereby causing enormous crop losses (e.g., Wulff et al. [1]). In addition, fusaria produce a highly diverse spectrum of secondary metabolites, including harmful mycotoxins that frequently contaminate food and feed, thus comprising a health risk to animals and humans when consumed. On the contrary, *Fusarium* spp. also produce a number of economically interesting metabolites.

Among the fusaria, *Fusarium fujikuroi* (formerly *Gibberella fujikuroi*) is one of the first described plant-pathogenic fungi. Research on this fungus can be traced back to Japanese plant pathologists who investigated the causes of the *bakanae* (foolish seedling) disease. The causal agents of this disease are gibberellic acids (GAs), a family of plant hormones that are produced and secreted by the fungus [2]. Today, the fungus is used worldwide for the commercial production of GAs, which are applied as a spray or dip to manage fruit crops (e.g., grapes and oranges), to malt barley for beer production, to increase flower size (e.g., gardenia or geranium flowers), and to increase sugar yield in sugarcane [3].

As other fusaria, *F. fujikuroi* is known for the production of a broad spectrum of secondary metabolites, particularly pigments and mycotoxins. Thus, *F. fujikuroi* produces, for example, the mycotoxins moniliformin, fusarins, fumonisins, fusaric acid, and beauvericin [4–10] as well as the pigments neurosporaxanthin, bikaverin, and fusarubins [11–15]. In the recently sequenced genome, 45 potential gene

L. Studt, Ph.D., Dr. rer. nat.
Institute for Biology and Biotechnology of Plants, University of Münster, Münster, Germany

B. Tudzynski, Ph.D., Dr. habil. rer. nat. (✉)
University of Münster, Institute of Biology und Biotechnology of Plants,
Schlossplatz 8, D-48143 Münster, Germany
e-mail: Bettina.Tudzynski@uni-muenster.de; tudzynsb@uni-muenster.de

J.-F. Martín et al. (eds.), *Biosynthesis and Molecular Genetics of Fungal Secondary Metabolites*, Fungal Biology, DOI 10.1007/978-1-4939-1191-2_10,

clusters responsible for the biosynthesis of polyketide synthase (PKS)-, non-ribosomal peptide synthetase (NRPS)-, or terpene cyclase (TC)-derived secondary metabolites were identified [16]. Ongoing work regarding the known secondary metabolites and also the additional putative secondary metabolite gene clusters of *F. fujikuroi* that encode yet unknown metabolites revealed a correlation between nitrogen availability, gene expression, and production capability for most of them [16]. While GAs, bikaverin, and fusarubins were all shown to be repressed by high nitrogen conditions, the two mycotoxins, fusarins and fusaric acid (see Chap. 11), accumulate only under nitrogen-excessive conditions in *F. fujikuroi* [12, 13, 16–18].

In this chapter we will focus on the three nitrogen-repressed metabolites: GAs, bikaverin, and fusarubins. Current knowledge on the distribution of the respective gene clusters, the biosynthetic pathways, and regulatory circuits will be summarized.

Biological Activity of Gibberellins, Bikaverin, and Fusarubins

Gibberellic Acids

The *bakanae* disease led to serious crop losses in rice growing countries such as Japan, Taiwan, and China. Symptoms of this disease are hyperelongation of internodes, yellowish green leaves, stunted roots, and only partially filled, sterile, or empty grains. Affected plants are infertile and hence do not produce edible grains. Already in 1912, the Japanese scientist Sawada published a paper entitled "Diseases of agricultural products in Japan" in which he suggested that the symptoms of the bakanae disease might be due to a toxic compound secreted by a fungus. Some years later, he described the causal agent as *Lisea fujikuroi* [19].

Later, Kurosawa [20] found that culture filtrates from dried rice seedlings caused the same marked elongation in rice as the fungus itself, thereby confirming Sawada's suggestion that the bakanae fungus secretes a "toxin" that stimulates shoot elongation, inhibits chlorophyll formation, and suppresses root growth.

In the 1930s, the toxic principle was isolated from the culture fluid as a noncrystalline solid compound called gibberellin A [21] in allusion to the fungus, which had been renamed *G. fujikuroi* [22]. This "toxin" was then further purified, and the isolated, structurally similar compounds were named accordingly gibberellic acids (GAs) [23]. Due to language barriers and the times of war it took until the mid-1950s before the structure of the first GAs (i.e., GA_1, GA_2, and GA_3) could finally be elucidated [24–26]. At this time, British scientists discovered that GAs are natural regulators of growth and development in all higher plants [27–29]. This discovery stimulated biotechnological production of GAs worldwide, mainly of GA_3, using high-GA titer mutants of the fungus *G. fujikuroi*. The commercial interest led to enhanced genetic, biochemical, and physiological studies on the fungus to optimize the GA production process [30–33].

Currently, there are 136 GAs known from plants, fungi, and even bacteria, which are termed GA_1 to GA_{136} in order of discovery [34–36]. However, only some of them—e.g., GA_1, GA_3, GA_4, and GA_7—have bioactive properties and function as endogenous growth regulators, promoting organ expansion and developmental changes [2, 37]. Due to their function as plant hormones, the concentration of biologically active GAs at their sites of action is tightly regulated and is moderated by numerous developmental and environmental cues. Recent research has focused on regulatory mechanisms, acting primarily on expression of the genes that encode the dioxygenases involved in GA biosynthesis and deactivation. The current state of knowledge on GA metabolism in higher plants with particular emphasis on regulation and maintenance of GA homoeostasis is summarized in several recent reviews [37–39].

Bikaverin

For a long time the fungus was shown to produce the red polyketide bikaverin and its *O*-demethyl derivative norbikaverin that often accompany GA production [12, 40–43]. Bikaverin, formerly known as lycopersin, was first isolated from *F. oxysporum* f. sp. *vasinfectum* and *F. oxysporum* f. sp. *lycopersici* [44, 45]. Later on, Balan et al. [40] independently isolated a compound designated as bikaverin from cultures of *G. fujikuroi*. This compound was found to be structurally identical with lycopersin [42, 46, 47] that was thereafter re-named bikaverin to avoid confusion with lycopersene or lycopene, which are two structurally unrelated compounds.

Bikaverin accumulates during fungal cultivation in the mycelium and is released into the culture filtrate as mostly unsoluble particles upon autolysis. Aqueous solutions prepared from crystalline bikaverin induced vacuolation in hyphal tips of *Aspergillus niger* and 30 other fungal species, affecting fungal growth already at low concentrations [46]. Cell line studies revealed its cytotoxic potential probably due to an inhibition of ATP biosynthesis by uncoupling phosphorylation [48, 49]. In addition, bikaverin has antibiotic activity against the protozoon *L. brasiliensis*, the oomycete Phytophthora infestans as well as the nematode *Bursaphelenchus xylophilus* [40, 50, 51].

Fusarubins

Only recently, a second group of red polyketides structurally related to the heptaketide fusarubin, more precisely 8-*O*-methylfusarubin, 8-*O*-methylnectriafurone, 8-*O*-methyl-13-hydroxynorjavanicin, and 8-*O*-methylanhydrofusarubinlactol, have been identified in liquid cultures of *F. fujikuroi* [13]. In 1970, Cross et al. already identified the first fusarubin-type metabolite in *F. fujikuroi*, *O*-demethylanhydrofusarubin, which co-accumulated with bikaverin and norbikaverin in the liquid culture [42, 52]. Fusarubin itself was first isolated by Ruelius and Gauhe [53] from cultures of *F. solani* (sexual stage: *Nectria haematococca*) and later on was shown to be identical with oxyjavanicin that was described together with its *O*-demethyl derivative javanicin, also called solanione [54], as an antibacterial pigment from *F. javanicum* [55]. Later on, several structural derivatives of fusarubin have been identified (for review, see

Medentsev and Akimenko [56] and references therein). Contrary to bikaverin that mainly remains in the mycelium, these pigments almost completely diffuse into the surrounding medium upon biosynthesis [55, 57]. These naphthazarinoid compounds produced by *Fusarium* spp. show antibiotic, insecticidal, phytotoxic, and antitumor properties [55, 58–63] and have been implicated in disease symptoms in pea, citrus, and cotton [64–66]. 8-*O*-methylfusarubin, that was identified as the predominant end-product in *F. fujikuroi* [13], caused wilting of chickpea cuttings and showed toxicity to cells isolated from chickpea leaflets with an LD_{50} value of 327 ng/mL [67]. These effects are probably due to the production of superoxide radicals (O_2^-), H_2O_2, and semiquinone radicals of the respective compound [68]. In the presence of those naphthoquinone pigments elevated superoxide dismutase and catalase activities were observed, thereby reflecting the increased formation of the superoxide radical and hydrogen peroxide by the fungus. However, the *O*-demethyl derivatives are not able to accept reducing equivalents, thus *O*-methylation of C8 is likely the last biosynthetic step before the pigments are released into the surrounding medium [56].

Distribution of GA, Bikaverin, and Fusarubin Gene Clusters Among Fungi

Gibberellic Acid Gene Cluster

In *F. fujikuroi*, seven genes are involved in GA biosynthesis, all being located adjacent to one another—a common feature of genes involved in secondary metabolite biosynthesis [69] (reviewed in Bömke and Tudzynski [2]). The genes encode a pathway-specific geranylgeranyl diphosphate synthase (Ggs2), the bifunctional *ent-copalyl* diphosphate synthase/ent-kaurene diphosphate synthase (Cps/Ks), four P450 monooxygenases (P450-1 to P450-4), and a desaturase (Des) [70, 71]. While GA biosynthesis is the "trademark" of *F. fujikuroi* and GA is probably one of its best investigated secondary metabolites to date, other species within the *G. fujikuroi* species complex (GFC) as well as closely related species outside of the GFC also contain the GA biosynthetic gene cluster or at least parts of it, e.g., *F. proliferatum*, *F. konzum*, *F. foetens*, *F. napiforme*, *F. miscanthi*, *F. sacchari*, and some *F. oxysporum* strains [2, 16, 72–75]. In contrast, the distantly related species *F. graminearum* and *F. solani* lack the entire GA gene cluster [76]. Recent genome-wide sequence comparison of the *F. fujikuroi* wild-type strain IMI58289 with 15 related *Fusarium* species (i.e., *F. verticillioides*, *F. mangiferae*, *F. circinatum*, and 12 *F. oxysporum* isolates) revealed the presence of the entire GA gene cluster in the genome of *F. mangiferae*, *F. circinatum*, and 5 *F. oxysporum* isolates, while only remnants of the cluster were identified in *F. verticillioides* and the remaining 7 *F. oxysporum* isolates [16]. Intact gene clusters share the same gene order and orientation as described for *F. fujikuroi* [2] and in addition, all genes therein seem to encode functional proteins, except for *P450-2* and *P450-3* in *F. oxysporum* isolates PHW815 and FOSC 3-a, respectively, which are interrupted by premature stop codons [16] (Fig. 10.1a).

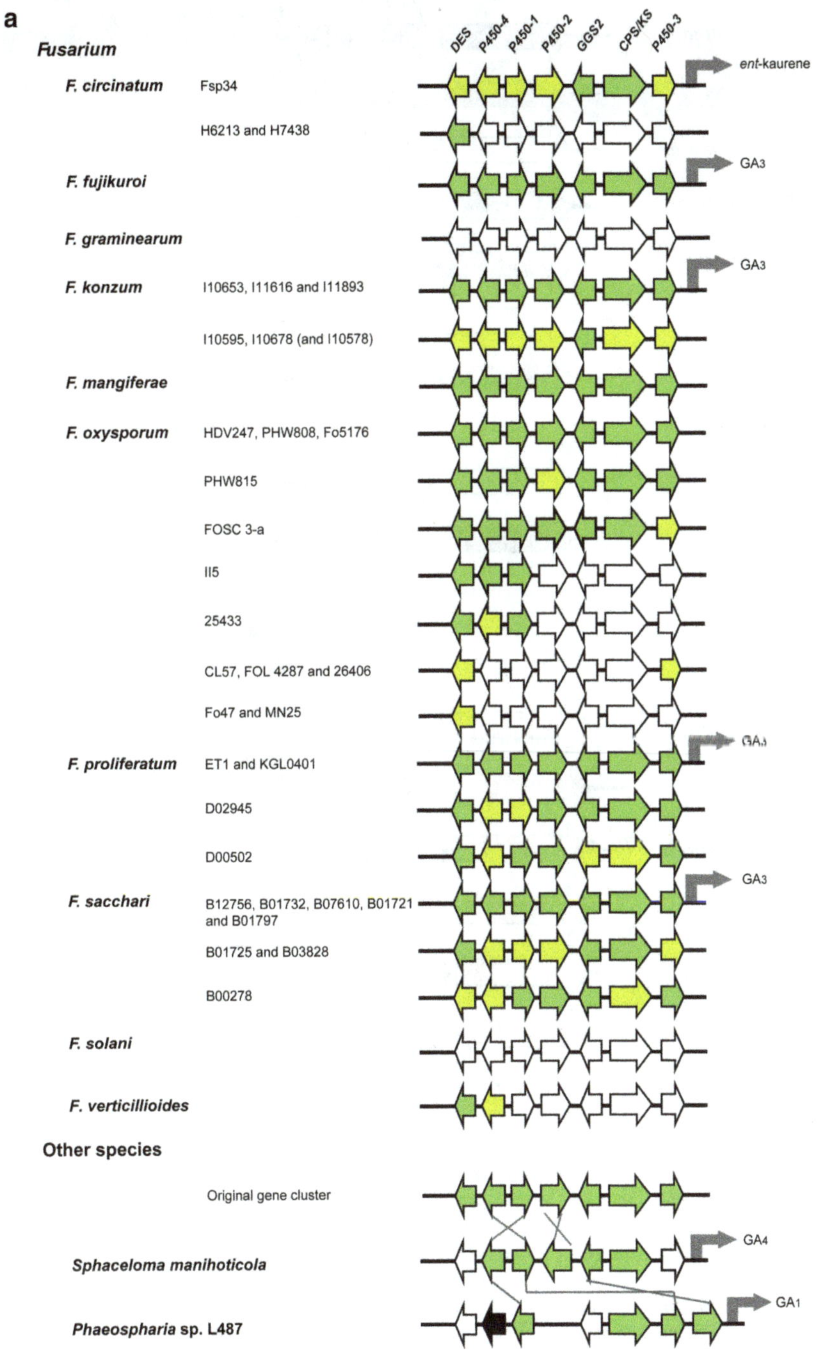

Fig. 10.1 Distribution of secondary metabolite gene clusters in fusaria and related species. Genes within a cluster are depicted as *arrows*; direction indicates direction of translation. (**a**) Gibberellin gene cluster, i.e., *DES*, *P450-4*, *P450-1*, *P450-2*, *GGS2*, *CPS/KS*, and *P450-2*.

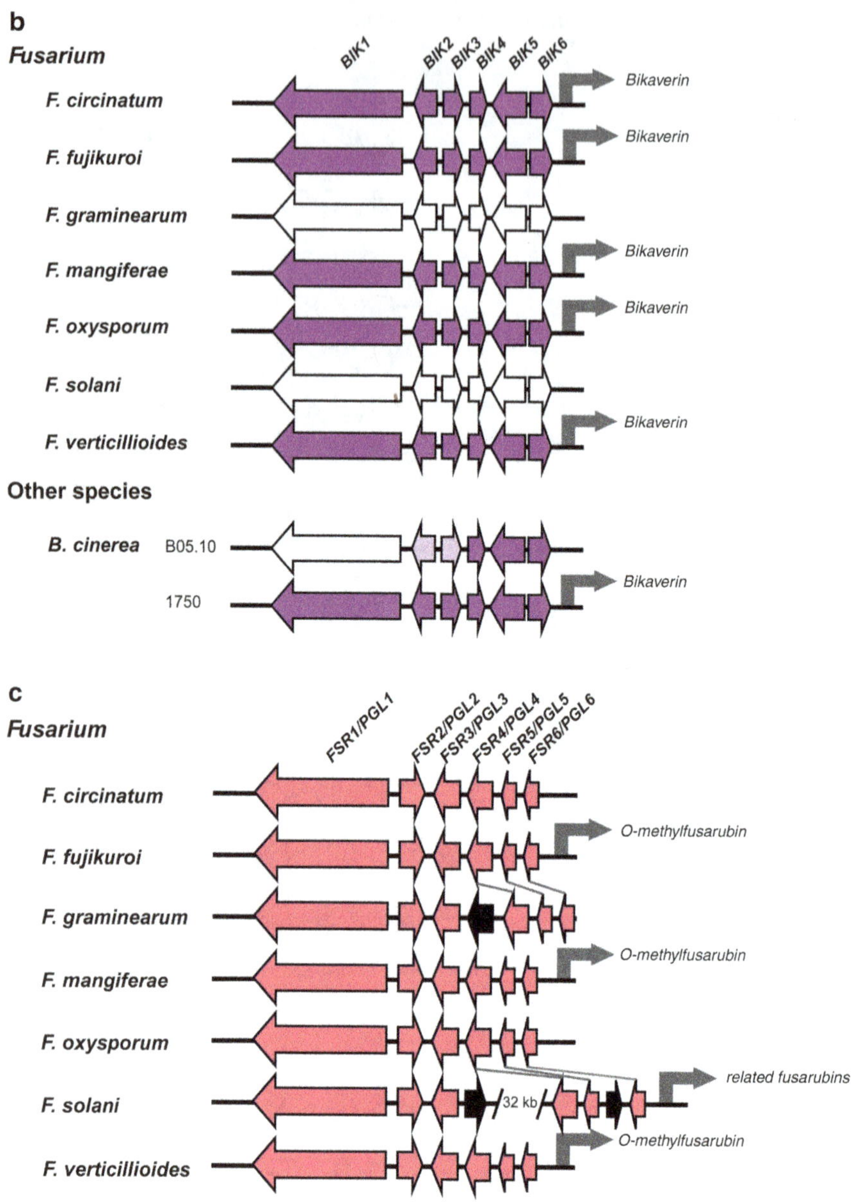

Fig. 10.1 (continued) *Green colored arrows* indicate the presence of the respective gene—*dark green* when active and *light green* when inactive. In case genes are marked *darkly green*, but no gibberellin (GA) production was detected no detailed characterization of the genes has been done yet. *White arrows* indicate the absence of the corresponding genes. GA or GA-intermediate biosynthesis is indicated as *gray arrows. Black arrows* stand for genes that do not belong to the original GA gene cluster. (**b**) Bikaverin gene cluster, i.e., *BIK1*, *BIK2*, *BIK3*, *BIK4*, *BIK5*, and *BIK6*. The presence of the gene cluster is depicted as *purple arrows—light purple* and *white arrows* indicate nonfunctional and missing genes, respectively. Biosynthesis of bikaverin by the respective fungal strain is indicated as a *gray arrow*. (**c**) Fusarubin gene cluster, i.e., *FSR1/PGL1*, *FSR2/PGL2*, *FSR3/PGL3*, *FSR4/PGL4*, *FSR5/PGL5*, and *FSR6/PGL6*. The presence of the gene cluster is indicated by *pink arrows. Black arrows* depict genes missing in the original fusarubin gene cluster. Actual production of either *O*-methylfusarubin or a related fusarubin derivative is indicated as a *gray arrow*

Despite the presence of the complete gene cluster, only a few other *Fusarium* spp., besides *F. fujikuroi*, share the ability to actually synthesize the phytohormones, including some but not all strains of *F. konzum*, *F. proliferatum*, *F. sacchari*, and *F. mangiferae* [75, 77, 78]. A strain of *F. circinatum*, a member of the American clade of the GFC, has the entire GA gene cluster, but produces only the first committed intermediate in the GA pathway, *ent*-kaurene, and accordingly transcripts are detectable for *CPS/KS* in northern blot analysis [16]. Interestingly, two fungi outside the genus *Fusarium*, *Sphaceloma manihoticola*, and *Phaeosphaeria* sp. L487, both belonging to the class of Dothideomycetes, were also shown to produce GAs, but the final products are GA_4 and GA_1, respectively, due to the lack of some GA biosynthetic genes [2, 79–82] (Fig. 10.1a).

The evolutionary mechanism by which the two latter fungal species acquired the GA gene cluster is not yet understood. An increasing number of reports suggests that the presence of homologous secondary metabolite gene clusters in distantly related fungal species can result from horizontal gene transfer (HGT) [83, 84]; one example being the presence of the bikaverin gene cluster in the distantly related fungus *Botrytis cinerea* [85, 86].

Bikaverin Gene Cluster

Sequencing of the *F. fujikuroi* genome revealed the presence of 17 PKSs, of which only three (PKS3, PKS4, and PKS18) are non-reducing PKSs and thus most likely involved in the biosynthesis of colored secondary metabolites [16]. While the final product of PKS18 is still unknown, we recently identified and characterized the corresponding metabolites of PKS3 and PKS4. PKS3 (Fsr1 in *F. fujikuroi*) is a homologue of PGL1 from *F. graminearum* [87] and *F. verticillioides* [88], and is responsible for precursor biosynthesis of the fusarubin-derived perithecial pigments in *F. fujikuroi* and likely also in other fusaria harboring this gene cluster [13]. PKS4 (re-named Bik1) releases pre-bikaverin, the first intermediate of the bikaverin pathway, that is further converted to the final products norbikaverin and bikaverin by a monooxygenase (BIK2) and a methyltransferase (BIK3) [12, 16, 89, 90].

The presence of the bikaverin biosynthetic gene cluster (*BIK1* to *BIK6*) as well as production of this red polyketide is conserved in the GFC species—i.e., in *F. fujikuroi*, *F. mangiferae*, *F. verticillioides*, *F. circinatum*, and *F. proliferatum* and also in *F. oxysporum* [16]; while distantly related *Fusarium* species, such as *F. graminearum*, lack the entire gene cluster and produce a structurally different red pigment—i.e., aurofusarin [91, 92] (Fig. 10.1b). The cluster organization is maintained throughout the different genomes including at least one side of the flanking genes, while synteny is abrogated at the other flanking region [16, 93]. Although bikaverin biosynthesis is restricted to some *Fusarium* spp., a few strains of *B. cinerea* contain either the entire (*B. cinerea* 1750) or at least remnants (*B. cinerea* B05.10) of the bikaverin biosynthetic gene cluster (Fig. 10.1b) with an astonishing level of nucleotide sequence identity, suggesting that an ancestor of this fungus gained the intact gene cluster through HGT from a *Fusarium* species [85, 86]. In addition, *B. cinerea* strain 1750 is able to synthesize this polyketide resulting in a pink variant of the gray mold fungus [86].

Fusarubin Gene Cluster

While the bikaverin biosynthetic gene cluster and the ability to produce bikaverin are conserved mainly in GFC species, the presence of the fusarubin biosynthetic gene cluster is a common feature for several *Fusarium* spp. [16]. Interestingly, organization of the fusarubin biosynthetic gene cluster (*FSR1*/*PGL1* to *FSR6*/*PGL6*) is completely maintained within the GFC, but reorganization occurs outside of the GFC (Fig. 10.1c). While a core gene cluster consisting of the first three genes (*FSR1*/*PGL1*, *FSR2*/*Omt1*, and *FSR3*/*Fdm1*) is conserved also in the distantly related species *F. graminearum* and *F. solani*, the synteny is abrogated between *FSR3*/*Fdm1* and *FSR4* compared to *F. fujikuroi* [13, 16, 88]. Despite the presence of an intact gene cluster only the GFC members *F. fujikuroi*, *F. verticillioides*, *F. mangiferae*, and *F. proliferatum* actually produced the final product *O*-methylfusarubin under the used conditions [13, 14, 16]. Remarkably, variations regarding the optimal production conditions may already occur in closely related species. Thus, *F. proliferatum* wild-type strain D4854 produces fusarubins under acidic culture conditions that were shown to inhibit fusarubin biosynthesis in *F. fujikuroi* [14]. Therefore, different regulatory mechanism may apply also for the *FSR*/*PGL* biosynthesis gene cluster in *F. circinatum* and *F. oxysporum* that did not produce detectable *O*-methylfusarubin concentrations in our hands albeit the presence of an intact gene cluster [16]. Several fusaria outside the GFC also produce fusarubins including *F. solani*, *F. decemcellulare*, *Neocosmospora vasinfecta*, and *N. africana* [94–97]. Noteworthy, no fusarubin-like metabolites have been identified in *F. graminearum* so far, albeit the presence of the core gene cluster, *FSR1* to *FSR3* (Fig. 10.1c), suggests distinct yet unknown regulatory mechanisms in this fungus [16, 88, 94].

Secondary Metabolite Biosynthesis

Despite the great diversity of secondary metabolites in fungi, they are all synthesized from only a few precursors, which are derived from primary metabolism. They include ubiquitous small molecules such as amino acids, sugars, or Krebs cycle intermediates such as acetyl coenzyme A (acetyl-CoA), which are all used as building blocks for the different classes of secondary metabolites. Fungal polyketides, such as bikaverin and fusarubin, are synthesized by type I PKSs, which generate their products from short-chain carboxylic acids, usually acetyl-CoA and malonyl-CoA, via the polyketide pathway [98] (Fig. 10.2). Similarly, diterpenes like gibberellins are also synthesized from the same building block, acetyl CoA, via the mevalonic acid pathway (Fig. 10.2). The mevalonic acid or HMG-CoA reduc-

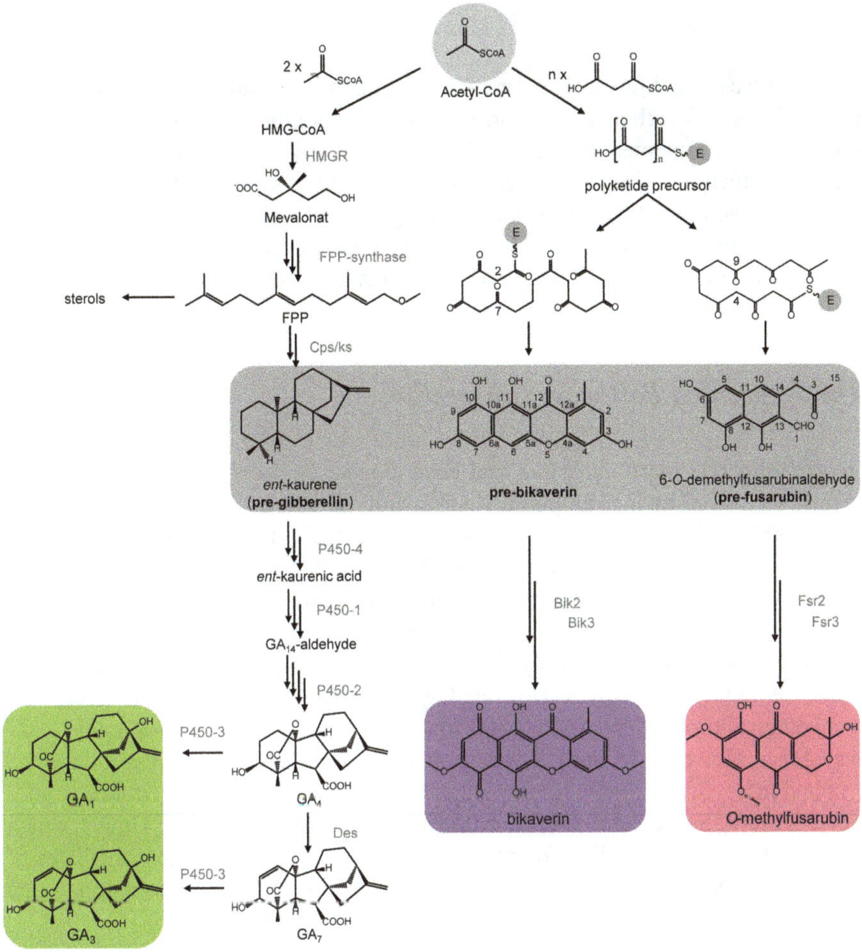

Fig. 10.2 Biosynthetic pathways leading to the formation of gibberellin, bikaverin, and fusarubins. Biosynthesis of all three secondary metabolites starts with acetyl-CoA. In case of gibberellin (GA) three acetyl-CoA subunits result in the formation of β(beta)-Hydroxyl-β(beta)-Methylglutaryl-Coenzyme-A (HMG-CoA) that is further reduced by HMG reductase (HMGR) leading to mevalonate. Mevalonat is further converted into farnesylpyrophosphate by the FPP-synthase. Then the geranylgeranyl pyrophosphate synthase (Ggs2) generates geranylgeranyl diphosphate further modified into *ent*-copalyl diphosphate and *ent*-kaurene, the first committed pathway intermediate, driven by the copalyl phosphate synthase/*ent*-kaurene synthase (Cps/Ks). Then 7β(beta)-hydroxylation leads to formation of *ent*-7α(alpha)-hydroxykaurenoic acid, oxidation at C6 results in contraction of the B-ring, and formation of GA12-aldehyde, which is then 3β(beta)-hydroxylated to GA14-aldehyde and oxidized at C7 to form GA14. Subsequently GA14 is converted to GA4 by C20-oxidation and desaturation of the latter at C1,2 results in formation of GA7, which is finally converted to GA3 by 13-hydroxylation. Bikaverin and fusarubins are generated from one acetyl-CoA subunit and eight or six malonyl-CoA subunits, respectively. In case of bikaverin the product template domain of Bik1catalyzes C2/C7 cyclization yielding pre-bikaverin. Pre-bikaverin is then further modified by Bik2 (FAD-dependent monooxygenase) and Bik3 (*O*-methyltransferase). Bik2 incorporates hydroxyl groups at position C5 and C7, and Bik3 is responsible for *O*-methylation of C3 and C8, finally resulting in the formation of bikaverin. In case of fusarubin biosynthesis, the PKS Fsr1 catalyzes C4/C9 aldol condensation leading to 6-*O*-demethylfusarubinaldehyde (pre-fusarubin). Then Fsr2 (*O*-methyltransferase) methylates hydroxyl groups at C6 and C8 and Fsr3 (FAD-dependent monooxygenase) incorporates hydroxyl groups at positions C5 and C10. Enzymes involved in the biosynthesis are highlighted in grey. The enzyme-bound state of the polyketide precursors is depicted as E

tase pathway is highly conserved and present in all higher eukaryotes. It provides the 5-carbon isoprenoid building blocks dimethylallyl pyrophosphate (DMAPP) and isopentenyl pyrophosphate (IPP), which serve not only as precursors for the production of isoprenoid secondary metabolites, such as gibberellins, carotenoids, and acorenol [16, 99], but also for ergosterol biosynthesis. In the following section, the biosynthetic pathways resulting in the formation of either GAs, bikaverin, or fusarubins, all starting from acetyl-CoA, are presented in more detail.

The Isoprenoid Pathway: Gibberellins

Before gibberellin-specific genes had been cloned, the biosynthetic pathway was mainly recovered by incorporation studies with ^{14}C-labeled acetate and mevalonic acid [100, 101] as well as by analysis of different UV mutants blocked at certain steps of the biosynthetic pathway [102–105]. Later, the genes of the early isoprenoid pathway encoding the three key enzymes of the mevalonate pathway, the hydroxymethylglutaryl coenzyme A reductase (HmgR), the farnesyl diphosphate synthase (FppS), and the geranyl diphosphate synthase (Ggs1) have been cloned [106–108]. One possibility to generate high-producing strains for the efficient biotechnological production of GAs is the manipulation of these early isoprenoid pathway genes. However, recently it was shown that overexpression of *hmgR* and *fppS* resulted in a reduced production level, probably due to a negative feedback regulation of HmgR. Subsequent deletion of the transmembrane domains of HmgR, which are known to be responsible for negative feedback regulation by intermediates of sterol biosynthesis, and overexpression of the remaining catalytic domain led to a 2.5-fold increased GA content [109].

A highlight in gibberellin research was the identification of biosynthetic genes in higher plants [110–113]. Some years later, the genes encoding the key enzyme of fungal gibberellin biosynthesis, *ent*-copalyl diphosphate/*ent*-kaurene synthase (*CPS*/*KS*) were isolated from *F. fujikuroi* and *Phaeosphaeria* spp. [114, 115], and recently from a third GA-producing fungus, *Sphaceloma manihoticola* [79]. After discovery of GA biosynthetic genes from plants on the one hand and from *F. fujikuroi* (and the other two fungi) on the other hand it became particularly apparent that GA biosynthesis in fungi differs fundamentally from that in higher plants on chemical, enzymatic, and genetic levels [39, 116]. For example, the first pathway-specific reactions, the cyclization of geranylgeranyl diphosphate to form *ent*-kaurene via *ent*-copalyl diphosphate is catalyzed by two enzymes (Cps and Ks) in plants, but only one single bifunctional enzyme (Cps/Ks) in fungi. Furthermore, while only P450 monooxygenases catalyze all oxidation steps in fungi, both 2-oxoglutarate-dependent dioxygenases and cytochrome P450s are involved in the plant pathway. These profound differences indicate that higher plants and fungi have evolved their complex biosynthetic pathways to generate GAs independently and not by HGT from plants to fungi as it had been proposed previously [117].

In the following years, seven GA biosynthetic genes, all located in a gene cluster, have been cloned and functionally characterized [70, 71, 118–121] (Fig. 10.1). Starting with the general mevalonic acid pathway, GA biosynthesis surprisingly branches off already at the stage of farnesyl diphosphate (FDP), although the formation of geranylgeranyl diphosphate (GGDP) was expected to be catalyzed by Ggs1 [107] in the general mevalonic acid pathway. The reason for this is the presence of a second GGDP synthase-encoding gene, *GGS2* that belongs to the GA gene cluster [71]. Thus, GGDP for GA biosynthesis is exclusively synthesized by the GA pathway-specific GGDP synthase, Ggs2 [71], while Ggs1 catalyzes the first step of carotenoid biosynthesis in fungi [122].

After formation of GGDP by Ggs2, the first GA-specific intermediate *ent-kaurene* is synthesized in a two-step cyclization via *ent*-copalyl diphosphate (CPP) by the bifunctional terpene cyclase Cps/Ks [71]. *Ent*-kaurene is then converted by sequential oxidation to *ent*-kaurenoic acid by the kaurene oxidase P450-4 [71], which is further oxidized to 7α(alpha)-hydroxy-kaurenoic acid by an unusual multifunctional enzyme, P450-1. The different functions of P450-1 were identified by a combination of two approaches: the deletion of the *P450-1* gene and its insertion into the GA-deficient *F. fujikuroi* mutant SG139 that lacks the entire GA gene cluster. While the deletion mutant accumulated *ent*-kaurenoic acid, cultures of complemented transformants converted *ent*-[^{14}C]kaurenoic acid, *ent*-7α(alpha)-hydroxy[^{14}C]kaurenoic acid, [^{14}C]GA_{12}-aldehyde, and [^{14}C]GA_{12} acid efficiently into [^{14}C]GA_{14} via [^{14}C]GA_{14}-aldehyde. These data clearly indicate that P450-1 catalyzes four sequential steps in the GA biosynthetic pathway: 7β(beta)-hydroxylation, contraction of ring B by oxidation at C-6, 3β(beta)-hydroxylation, and oxidation at C-7 [118]. Besides, the [^{14}C]-labeled GA precursors were also converted to kaurenolides and fujenoic acids, which are by-products of GA biosynthesis in *F. fujikuroi* [118, 123]. Thus, P450-1 displays remarkable multifunctionality and may be responsible for the formation of 12 products altogether.

The subsequent conversion of GA_{14} to GA_4 by the C20 oxidase P450-2 results in loss of carbon C20 and the formation of the first biologically active C19 gibberellin GA_4 [120] (Fig. 10.2). Interestingly, in plants the removal of C20 by progressive oxidation of the C20 methyl is catalyzed by multifunctional 2-oxoglutarate-dependent dioxygenases instead of a cytochrome P450 monooxygenase, which is a clear evidence for independent evolution of the GA biosynthetic pathways in plants and fungi. Desaturation of GA_4 at C1,2 by the GA_4 1,2-desaturase Des results in the formation of GA_7 [121]. For a long time the nature of the desaturase that converts GA_4 to GA_7 was unknown. When first described in 2003, only one potential homologue had been found by BLAST search with very low sequence similarity, a 7α(alpha)-cephem-methoxylase from *Nocardia lactamdurans* [121]. Only recently it has been shown by heterologous gene expression in *Escherichia coli* that Des has the characteristics of a 2-oxoglutarate-dependent dioxygenase, however with very low amino acid sequence similarity with known 2-oxoglutarate-dependent dioxygenases [124]. Expression of the fungal *DES* gene from the cauliflower mosaic virus 35S promoter in the plant species *Solanum nigrum*, *S. dulcamara*, and

Nicotiana sylvestris resulted in substantial growth stimulation, with a threefold increase in height compared with controls. Thus, expression of the *F. fujikuroi DES* in plants has the potential to enable substantial growth increases, with practical implications, for example, in biomass production [124]. In a final step, GA_7 is converted by the C13 oxidase P450-3 to the main product in *F. fujikuroi*, gibberellic acid (GA_3) [121]. The same enzyme also catalyzes 13-hydroxylation of GA_4 to the minor product GA_1. However, when the desaturase-encoding gene *DES* is deleted, GA_1 accumulates as the main product in *F. fujikuroi* instead of GA_4 (Fig. 10.2). To generate a GA_4-producing strain, both the desaturase-encoding gene *DES* and the 13-hydroxylase-encoding gene *P450-3* had to be deleted [121, 125].

Summarizing, the gibberellin biosynthetic genes in *F. fujikuroi* are organized in a gene cluster that contains seven genes encoding a pathway-specific geranylgeranyl diphosphate synthase (Ggs2), the bifunctional *ent*-copalyl diphosphate synthase/*ent*-kaurene diphosphate synthase (Cps/Ks), four P450 monooxygenases (P450-1 to P450-4), and an unusual desaturase (Des) [70, 71]. In contrast, *S. manihoticola* has only five GA biosynthetic genes missing *DES* and *P450-3* at the left and right cluster borders, respectively (Fig. 10.1a). Subsequently, this fungus produces GA_4 as final product [79].

The Polyketide Route: Pigment Biosynthesis

Biosynthesis of naphthoquinones is widespread in nature, arousing interest regarding their formation. Gatenbeck and Bentley [126] already suggested that biosynthesis of naphthoquinones proceeds via the polyketide route by formation of a common precursor—a product of the acetate malonate pathway, resulting from the condensation of an acetyl-CoA unit with a defined number of malonyl-CoA units.

Bikaverin Biosynthesis

The structures of bikaverin and the co-existing *O*-demethyl derivative norbikaverin were elucidated by X-ray crystallography, UV and IR spectra, and nuclear magnetic resonance measurements as well as by retrosynthetic analysis [12, 42, 46, 47, 127]. Both compounds contain a PKS-derived benzoxanthone ring system [42]. Given the unusual chain length of C20 atoms, it was suggested that orsellinic acid might function as a starter unit [128] instead of the canonical acetyl-CoA unit [42]. A biogenetic approach to proof the assumed polyketide origin was initiated only a few years later. Feeding experiments with isotopically labeled acetic acid ([1,2-^{13}C]acetic acid) using ^{13}C single-frequency homonuclear decoupling yielded satellites at every resonance associated with the bikaverin carbon skeleton. The sample of [1,2-^{13}C] acetic acid-enriched bikaverin yielded spectra that unequivocally indicated that the entire carbon skeleton of bikaverin is assembled from intact pairs of carbon atoms derived from nine acetate units thereby proofing the polyketide origin [129]. However, obtained data did not allow distinction whether the molecule was formed by folding of a single polyketide chain or if indeed building blocks, such as the proposed orsellinic acid [42], occurred during biosynthesis as suggested earlier.

Later, the gene encoding the bikaverin-specific PKS *BIK1*, formerly *PKS4*, was identified [12, 89]. Bik1 resembles the typical domains of a non-reducing PKS: a starter unit: ACP transacylase (SAT) domain that selects the starter unit [128], a ketoacyl synthase (KS) domain responsible for repeated decarboxylative condensation, an acyl-carrier protein (ACP) that tethers the polyketide chain via a phosphopantetheinyl residue [130], a malonyl-CoA:acyl carrier protein (ACP) transacylase (MAT) domain responsible for exception of the extender unit, a product template (PT) domain that controls regioselective cyclization of the polyketide chain [131], and a canonical thioesterase/claisen cyclase (TE/CLC) domain [89, 90, 132]. Sequence analysis of the coding regions surrounding *BIK1* revealed five additional genes, designated as *BIK2* through *BIK6*, that were co-regulated with *BIK1* under bikaverin-inducing conditions (nitrogen starvation and acidic pH). The five genes encode for a putative FAD-dependent monooxygenase (Bik2), an *O*-methyltransferase (Bik3), an NMR-like protein (Bik4), a $Zn_2(II)Cys_6$ transcription factor (Bik5), and an efflux pump of the major facilitator superfamily (MFS) (Bik6). Targeted deletion of these genes verified their participation in bikaverin biosynthesis. Only Bik1, Bik2, and Bik3 are necessary for bikaverin and norbikaverin formation, while Bik4 and Bik5 function as positive regulators for bikaverin biosynthesis. Bik6 is most likely involved in the transport of bikaverin outside of the cell, although low bikaverin concentrations are still detectable in the Δ(Delta)*bik6* mutant [12].

To identify the first pathway-specific intermediate Pks4/Bik1 was heterologously expressed in *E. coli*. Interestingly, the enzyme was functional despite the absence of the fungal-specific phosphopantetheinyl transferase (PPTase) NpgA/Ppt1 responsible for posttranslational modification and thereby activation of the ACP domain [133, 134]. Hence Bik1 is most likely activated by the *E. coli holo*-ACP synthase. By that, one predominant compound accumulated in the culture, which was subsequently purified and structurally elucidated as the first pathway-specific intermediate named SMA76a [90]. Only a few years later the identical compound, referred to as pre-bikaverin, was purified from the liquid culture of a *F. fujikuroi* Δ(Delta)Δ(Delta)*bik2*/*bik3* double mutant strain expressing only the first biosynthetic gene *BIK1* thereby confirming the assumption that pre-bikaverin (SMA76a) is a true intermediate in the bikaverin biosynthetic pathway [12] (Fig. 10.2).

Interestingly, the heterologously expressed Pks4 was found to utilize malonyl-CoA as starter unit instead of the canonical acetyl-CoA utilized by the vast majority of fungi. This was explained by the lack of the GXCXG motif required for starter unit attachment [90]. However, re-characterization of the Pks4 SAT domain showed that the initial sequence was incorrect. The revised SAT domain of Pks4 preferentially accepted acetyl-CoA, while malonyl-CoA or other longer-chain fatty acids were not or only rarely accepted thereby disproving the previous assumption by Ma et al. [90] and also that of Kjaer et al. who suggested the acceptance of orsellinic acid as a starter unit [131]. Characterization of the PT domain established C2-C7-specific cyclization of Bik1 and, furthermore, removal of the TE/CLC domain proved its involvement in the C1-C10 Claisen condensation during release of SMA76a/pre-bikaverin [90, 135] (Fig. 10.2).

The proposed enzymatic functions of Bik2 and Bik3 suggest that both modify the PKS-derived pre-bikaverin by incorporation of hydroxyl groups (C6 and C7) and *O*-methylation of already existing hydroxyl groups (C3 and C8), respectively. However, the precise biosynthetic steps and the chronological sequence, if they exist, of both enzymes have yet to be verified. Coexistence of di-methylated bikaverin and the mono-methylated derivative norbikaverin in the liquid culture suggests that methylation of both methyl groups is not a premise for product transport.

In summary, the six bikaverin biosynthetic genes are clustered and encode a non-reducing PKS (Bik1), an FAD-dependent monooxygenase (Bik2), an *O*-methyltransferase (Bik3), an NMR-like protein (Bik4), a $Zn_2(II)Cys_6$ transcription factor (Bik5), and an efflux pump of the major facilitator superfamily (MFS) (Bik6). Only Bik1-Bik3 are involved in biosynthetic steps while the other three act as regulators (Bik4, Bik5) and transport proteins (Bik6) [12, 89].

Fusarubin Biosynthesis

Early experiments using ^{14}C-labeled acetate revealed that the first identified fusarubin-type pigment javanicin also proceeds via the polyketide route [126, 136], and incorporation studies using ^{13}C-acetate suggested a single heptaketide-chain origin for the fusarubins [137]. The presence of several fusarubins under some and the total lack of all fusarins under different conditions further indicated their origin from a common precursor [64]. In *F. fujikuroi* four structurally related compounds—i.e., 8-*O*-methylfusarubin, 8-*O*-methylnectriafurone, 8-*O*-methyl-13-hydroxynorjavanicin, and 8-*O*-methylanhydrofusarubinlactol—have been isolated and characterized from the liquid culture under low nitrogen, alkaline conditions. All were structurally similar to fusarubin, differing from the latter only in constitution of the C-ring. 8-*O*-methylfusarubin was found to be the major metabolite under the investigated condition in *F. fujikuroi* [13].

Gatenbeck and Bentley [126] and later also Arsenault [138] proposed an aromatic acid as the first metabolite, which is then further methylated resulting in a carboxylic acid. Successive reduction of this acid should then result via an aldehyde in the formation of an alcohol resulting subsequently in javanicin, solaniol, and bostrycoidin—all compounds that are structurally related to fusarubin [126, 138]. However, neither the aldehyde nor the alcohol had been identified by then. Later, isolation of the proposed carboxylic acid, referred to as fusarubinoic acid, from a pigment-overproducing mutant strain led to the assumption that this heptaketide is indeed the first pathway-specific intermediate in the fusarubin biosynthesis [139, 140]. Kurobane et al. [137, 141] proposed that the aromatic acid (i.e., fusarubinoic acid) is directly converted into the diastereomeric 13,14-dihydrofusarubin (according to our numbering). These compounds are thought to be the true end-products in the fusarubin biosynthetic pathway, while fusarubin accumulation was described as nonenzymatic oxidation of the diastereomers under alkaline conditions in *F. solani* [141]. However, dihydrofusarubin is not detectable at any time-point under pigment-inducing conditions in *F. fujikuroi*, thus indicating that 8-*O*-methylfusarubin and the three other identified fusarubins are the true metabolites in the fusarubin biosynthetic pathway in this fungus [13]. Nevertheless, the metabolite spectrum seems to be strictly dependent on culture conditions as well as the producing *Fusarium* species. Several other fusarubin-type pigments have been identified in different fusaria (Medentsev et al. [94] and references therein). Furthermore, optimal production conditions varied from species to species, e.g., production of fusarubins was induced under acidic conditions in *F. decemcellulare* [56], but not in *F. oxysporum* f. sp. *vasinfectum*, *F. verticillioides*, and *F. fujikuroi*. In the latter two, a pH above 5 is a prerequisite for the formation of the fusarubin-like pigments [13, 64, 142].

In *F. fujikuroi* the responsible non-reducing PKS (i.e., Pks3/Fsr1) was identified by phylogenetic analyses. Five additional genes downstream of *FSR1* were co-regulated under pigment-inducing conditions, referred to as *FSR1* to *FSR6*, encoding a non-reducing PKS (Fsr1), an *O*-methyltransferase (Fsr2), an FAD-dependent monooxygenase (Fsr3), a Zn2(II)Cys6 fungal-specific transcription factor (Fsr6), and two further unknown proteins (Fsr4 and Fsr5). Single targeted deletion of each gene indicated that all six are involved in the regulation of fusarubin biosynthesis in *F. fujikuroi*. However, only Fsr1, Fsr2, and Fsr3 are directly involved in formation of the final products, whereas deletions of *FSR4*, *FSR5*, and *FSR6* only exhibited altered metabolite quantities [13]. Deletion of *FSR1*, responsible for synthesis of the first pathway-specific intermediate, completely abolished pigment biosynthesis. Structural analysis of Fsr1 revealed the typical domain structure of a non-reducing PKS with some exceptions: Fsr1 has an additional ACP domain and instead of the canonical CLC/TE domain this protein harbors a reductive release (R) release domain, suggesting that the first pathway-specific intermediate is not released as a carboxylic acid as previously proposed [126, 138, 140, 143], but rather as an aldehyde [13]. This hypothesis was later verified by deletion of *FSR2-FSR5* and simultaneous overexpression of *FSR1*, which led to identification of 6-*O*-demethylfusarubinaldehyde. This heptaketide lacks the methyl groups at C6 and C8 as well as the hydroxyl groups at C5 and C10 present in the final product 8-*O*-methylfusarubin and is therefore the earliest pathway-specific intermediate [13] (Fig. 10.2). A similar compound, referred to as 6-*O*-demethylnectriachrysone, was described by Parisot et al. [143] from *yelY* mutants of *F. solani*, most likely resulting from rearrangement of the released aldehyde to a more stable intermediate during product isolation. Thus, 6-*O*-demethylfusarubinaldehyde is probably also the first fusarubin pathway-specific intermediate in *F. solani*. 6-*O*-demethylfusarubinaldehyde was later on also identified as the first pathway-specific intermediate in *F. solani* by heterologous expression of Fsr1/Pgl1 in *E. coli* [144] thereby proving our assumption. Ring closure of the first two rings most likely results from C4/C9-type aldol cyclization and subsequent aromatization driven by the identified PT domain of Fsr1, similar to PksA from *Aspergillus fumigatus* during aflatoxin formation as Fsr1 shows high similarity to PksA [13, 128, 131, 145] (Fig. 10.2).

Deletion of *FSR2* resulted in production of 6-*O*-demethyl-10-deoxyfusarubin and 6-*O*-demethylfusarubinaldehyde, and deletion of *FSR3* led to the production of fusarubinaldehyde, thereby confirming the hypothesized functions of those two enzymes. Fsr3 incorporates hydroxyl groups at position C5 and C10 and Fsr2 is responsible for *O*-methylation of C6 and C8 [13]. Thus, *FSR2* and *FSR3* are most likely identical to the previously characterized genes, *yelY* and *yelJ*, from *F. solani*, respectively [143, 146]. However, methylation of C8 seems not to be a prerequisite for product release as fusarubin, and not 8-*O*-methylfusarubin, is also found as a major component in other fusaria and 13-hydroxynorjavanicin is also observed in *F. fujikuroi* wild-type strains. Interestingly, only the aldehyde is detectable in the *FSR3* deletion mutant, compared to Δ(Delta)*fsr2*, suggesting that Fsr3 is necessary for the reduction as well as oxidation of the respective aldehyde as well as for the oxidation and 13C-hydroxylation after cleavage of CO_2 [13]. This would require multiple oxidation reactions by a single enzyme, similarly to the GA and also the trichothecene biosynthesis in *F. fujikuroi* and *F. graminearum*, respectively [118–120, 147].

These results led us to propose the following biosynthetic pathway for the fusarubins identified in *F. fujikuroi*: Fsr1 is responsible for the condensation of seven acetyl-CoA units, most likely from one acetyl-CoA and six malonyl-CoA units, to form a heptaketide that is subsequently released as 6-*O*-demethylfusarubin aldehyde. Then Fsr2 and Fsr3 are responsible for methylation of hydroxyl groups at C6 and C8 and incorporation of hydroxyl groups at C5 and C10. Fsr3 furthermore reduces the aldehyde to an alcohol that equilibrates mainly with 8-*O*-methylfusarubin and, in small amounts, also with 8-*O*-methylnectriafurone on the one hand, or, on the other hand, oxidizes the aldehyde to a carboxylic acid. The carboxylic acid either equilibrates with a lactone that after further loss of water and reduction of the lactone results in 8-*O*-methylanhydrofusarubinlactol or can undergo decarboxylation and subsequent hydroxylation of C13, most likely also catalyzed by Fsr3 to give 8-*O*-methyl-13-hydroxynorjavanicin (Fig. 10.2).

Parisot et al. [148] proposed that 8-*O*-methylanhydrofusarubin lactol directly results from cyclization of the fusarubinaldehyde rather than via prior oxidation yielding the lactone. However, the absence of anhydrofusarubin lactol in the culture of Δ(Delta)*fsr3* mutants in *F. fujikuroi* suggests that the latter version is more likely [13].

Regulatory Components That Affect Secondary Metabolite Regulation

Nitrogen Regulation

Our recent genome-wide expression analysis of all 45 potential secondary metabolite gene clusters revealed a strong dependency on nitrogen availability for many of them. The GA, bikaverin, and fusarubin biosynthesis genes have in common that they are repressed under high nitrogen conditions [16].

It has been well known for a long time that fermentation of *F. fujikuroi* for gibberellin production delivers highest yields under nitrogen limiting conditions, and that bikaverin is a by-product that is induced under the same conditions [30, 149]. However, the molecular mechanism of nitrogen regulation of gibberellin and bikaverin biosynthesis differs. Gibberellins were the first secondary metabolites for which a strict dependency on the GATA-type transcription factor AreA was shown [150, 151]. These findings were unexpected as AreA is known to activate the expression of genes involved in utilization of alternative nitrogen sources when preferred nitrogen sources such as glutamine and ammonium are not available [152–155]. Although gibberellins have no nitrogen in the structure and therefore cannot serve as nitrogen source for the fungus, a direct binding of AreA to double GATA motifs in promoters of gibberellin biosynthetic genes was unequivocally proven [150]. Recently we have shown that also a second GATA transcription factor, AreB, is essential for expression of gibberellin genes [156]. Surprisingly, though co-regulated with the gibberellin biosynthetic genes, expression of bikaverin genes (*BIK1-BIK6*) does not depend on the presence of AreA. Therefore, a second noncanonical, AreA-independent mechanism of nitrogen metabolite repression must exist [12, 157]. The different mechanisms of nitrogen regulation are also obvious from the clear upregulation of bikaverin genes under repressing (nitrogen sufficient) conditions in a double *areA/meaB* knock-out mutant. MeaB is a nitrogen-responsive bZIP transcription factor, which negatively affects both gibberellin and bikaverin genes [158]. While the single deletion of *MEAB* resulted in slightly increased expression of both AreA-dependent gibberellin and AreA-independent bikaverin genes under inducing (nitrogen limiting) conditions, the double deletion mutant revealed a strong deregulation of bikaverin, but not of gibberellin genes in high nitrogen conditions [158]. Further experiments will allow an unambiguous integration of MeaB into the established regulation network (Fig. 10.3a).

Another highly unexpected regulator of secondary metabolite production in *F. fujikuroi* is the glutamine synthetase (GS), the only enzyme that synthesizes glutamine. Since glutamine is probably the most favorite nitrogen source leading to strong downregulation of all nitrogen-repressed genes, its deletion was thought to result in lower intracellular levels of glutamine and higher expression of gibberellin and bikaverin genes. However, the opposite was the case: Deletion of the GS-encoding gene *GLN1* led to the total downregulation of biosynthetic genes and subsequent loss of gibberellin and bikaverin production [159] (Fig. 10.3a).

Besides GAs and bikaverin, fusarubins also are repressed under nitrogen-sufficient conditions, and several putative double GATA motifs are present in the promoter sequences of fusarubin biosynthetic genes. However, in contrast to bikaverin, the fusarubin biosynthetic genes need alkaline conditions in addition to nitrogen limitation [13, 15]. To maintain alkaline pH for optimal fusarubin production, sodium nitrate instead of glutamine is used [13]. However, nitrate can neither be used by *AREA* nor by *GLN1* deletion mutants. Therefore, the question if AreA and GS are essential for fusarubin biosynthesis remains to be elucidated.

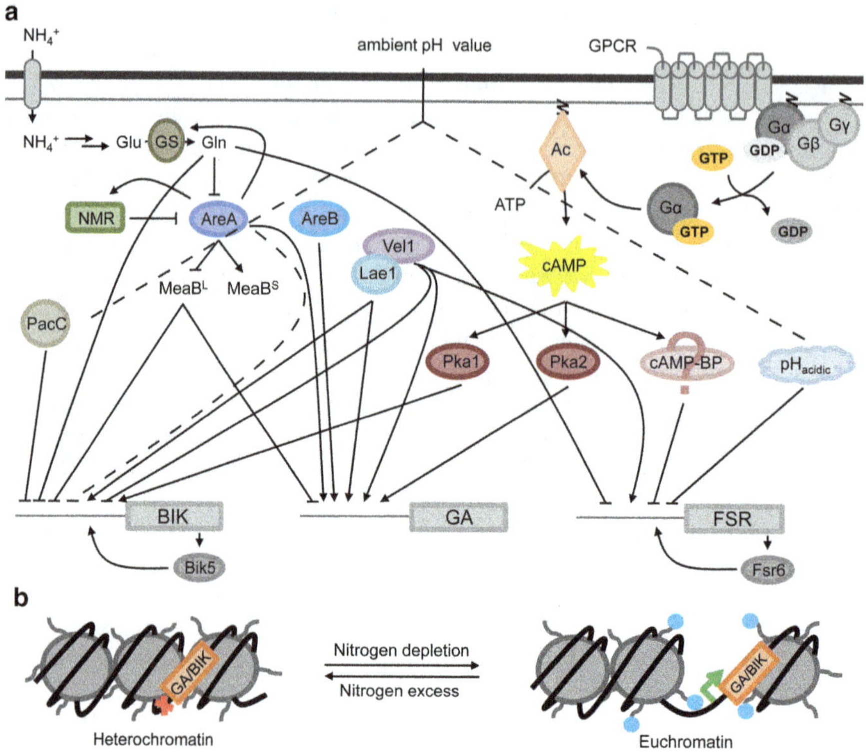

Fig. 10.3 Regulatory mechanisms that affect gibberellin, bikaverin, and fusarubin biosynthesis. (**a**) All three secondary metabolites are strictly repressed by nitrogen. While biosynthesis of bikaverin (BIK) and fusarubins (FSR) strictly depends on the ambient pH values in a PacC-dependent and PacC-independent manner, respectively, GAs are synthesized in either pH value. In the case of GA biosynthesis, nitrogen metabolite repression is regulated via the GATA transcription factors AreA and AreB, while these GATA transcription factors are not essential for bikaverin biosynthesis. AreA, furthermore, represses the bZIP transcription factor $MeaB^L$, which itself represses both bikaverin and GA biosynthesis. The presence of glutamine strongly represses transcription of all three nitrogen-repressed secondary metabolites but, surprisingly, deletion of the glutamine synthase-encoding gene, *GLN1*, results in completely abolished biosynthesis of GAs and bikaverin; nothing is known regarding the molecular mechanism of fusarubin N-regulation yet. The global regulators FfVel1 and FfLae1 both affect secondary metabolite biosynthesis in *F. fujikuroi*. While FfVel1 functions as positive regulator for GA and fusarubin biosynthesis, bikaverin is negatively regulated by FfVel1. Deletion of *FfLAE1* led to a downregulation of both GA and bikaverin biosynthesis. Its involvement in fusarubin regulation still needs to be investigated. In addition, heterotrimeric G protein-mediated signaling affects all three secondary metabolites in a distinct way. While the adenylyl cyclase FfAc functions as positive regulator for both GA and bikaverin biosynthesis, regulation of bikaverin biosynthesis is further mediated through FfPka1, and GA biosynthesis through FfPka2. The upstream target involved in the regulation of both has yet to be determined. Fusarubin biosynthesis is negatively regulated by FfAc, which itself is stimulated by FfG1 and FfG3. However, neither FfPka1 nor FfPka2 are involved in fusarubin regulation, suggesting another yet unidentified cAMP-binding protein (cAMP-BP) beside the regulatory subunit of the Pka. (**b**) The influence of chromatin on secondary metabolite biosynthesis in *F. fujikuroi*. The chromatin may be present in one of the two states: silent heterochromatin (on the *left*) or active euchromatin (on the *right*). Acetylation of lysine 9 at histone 3 (H3K9Ac) results in the formation of euchromatin, here depicted as blue balls. H3K9Ac is enriched in the chromatin landscape of both GA and bikaverin gene clusters present under biosynthesis-inducing conditions thus enabling biosynthesis (*green arrow*), but absent under normally repressing conditions thus prohibiting sufficient gene transcription (*red cross*)

Taken together, these findings substantiate the hypothesis of additional nitrogen response regulators besides AreA, AreB, and MeaB that are involved in nitrogen regulation of secondary metabolite biosynthesis in *F. fujikuroi*.

Regulation Via the Velvet Complex

Proteins of the velvet family are highly conserved and restricted to fungi [160]. In *A. nidulans*, the velvet complex has at least four members—i.e., VeA, VelB, LaeA, and VosA—and is known to have major impact on secondary metabolism and differentiation in several fungi [161]. VeA functions in most cases as an activating protein for secondary metabolite biosynthesis including sterigmatocystin in *A. nidulans* [162], aflatoxin, cyclopiazonic acid and aflatrem in *A. flavus* [163, 164], beauvericin and fusaric acid in *F. oxysporum* [165], deoxynivalenol in *F. graminearum* [166], or fumonisin and fusarins in *F. verticllioides* [167]. In *F. fujikuroi* three velvet proteins have been characterized so far: FfVel1 (homologous to AnVeA), FfVelB (homologous to AnVeB), and FfLae1 (homologous to AnLaeA) [168]. Similar to the observation in other filamentous fungi, gibberellin biosynthesis is also positively regulated by FfVel1 as deletion results in downregulation of the respective biosynthetic genes, and biosynthesis of GAs is almost completely abolished under production conditions, a phenotype that is reversed by overexpression of the respective *FfVEL1* gene. Interestingly, bikaverin biosynthesis is controversely regulated by FfVel1 in *F. fujikuroi*. Deletion of *FfVEL1* results in upregulation of bikaverin gene expression, and the pigment is overproduced under inducing and also under normally repressing conditions [168] (Fig. 10.3a). Although bikaverin is produced by several fusaria, it is not yet known if this deregulation is specific to *F. fujikuroi* or can also be observed in other fusaria. Recently, a functional bikaverin gene cluster was also detected in some isolates of the distantly related *B. cinerea* that was most likely acquired by HGT from a *Fusarium* sp. However, in *B. cinerea* bikaverin biosynthesis is positively regulated by BcVel1 as deletion of *bcvel1* in the bikaverin-producing strain 1750 resulted in decreased bikaverin gene expression. Heterologous expression of BcVel1 in the Δ(Delta)*FfVEL1* mutant restored bikaverin biosynthesis to wild-type level, suggesting that this opposing regulation is rather due to downstream signaling targets and not the velvet protein itself [86, 169]. Interestingly, fusarubins are opposingly regulated by FfVel1. Fusarubin expression is totally abolished, and no fusarubin formation is detectable, neither under inducing nor repressing conditions. Under fusarubin-inducing conditions, bikaverin was produced in the Δ(Delta)*FfVEL1* mutant instead of fusarubins that accumulated in the wild type [15, 168] (Fig. 10.3a).

Characterization of FfVel2 revealed a similar regulation with regard to gibberellin and bikaverin biosynthesis compared to FfVel1 in *F. fujikuroi*, while FfLaeA functions as a positive regulator for both secondary metabolites [168]. Influence of both velvet proteins regarding fusarubin biosynthesis awaits elucidation.

Influence of Signaling Components

Secondary metabolites are synthesized in response to certain environmental stimuli (reviewed in Brakhage [170]). In order to trigger biosynthesis of a secondary metabolite, the external signals have to be sensed and transduced to downstream targets that in turn activate expression of the respective genes. Heterotrimeric G proteins illustrate such signaling components. Once activated they stimulate downstream targets such as the adenylyl cyclase (AC) resulting in formation of the second messenger cyclic AMP (cAMP) and subsequent activation of the protein kinase A (Pka), which in turn phosphorylates target proteins, including transcription factors, yielding altered protein activities [171–173]. The influence of this signaling cascade with regard to secondary metabolism has been studied in various fungi, including *F. fujikuroi*. Interestingly, each of the three secondary metabolites discussed here is affected by cAMP signaling, however, each in a distinct way. While neither of the studied Gα(alpha) subunits, FfG1 or FfG3, affects GA biosynthesis, deletion of the adenylyl cyclase-encoding gene *FfAC* resulted in a dramatic reduction of GA accumulation in liquid cultures [15]. This is in accordance with a recent publication that describes decreased GA_3 amounts upon deletion of *acyA* (*FfAC*) in the *F. fujikuroi* wild-type strain MRC-1995 [174]. Furthermore, deletion of one of the two catalytic subunits of the Pka, *FfPKA2*, but not of *FfPKA1*, led to a similar phenotype, indicating that both FfAc and FfPka2 positively regulate GA biosynthesis in *F. fujikuroi* [15]. Whether FfG2 or other upstream effectors that stimulate the adenylyl cyclase—e.g., ambient pH values (reviewed in Vandamme et al. [175])—CO_2 state of the cell [176–178], or the small GTPase Ras [179, 180] are involved in GA biosynthesis still awaits proof.

Similar to GAs, bikaverin biosynthesis is also positively regulated by FfAc, but in this case the signal is transduced via FfPka1 and not FfPka2 (Fig. 10.3a). Here deletion of *FfG1* as well as *FfG3* resulted in decreased bikaverin gene expression and pigment biosynthesis. However, constitutive expression of *FfG1* did not lead to enhanced bikaverin biosynthesis, suggesting that bikaverin is only indirectly regulated through FfG1 and possibly also FfG3. Contrary to this, fusarubin biosynthesis is negatively regulated by FfAc, which is itself activated by FfG1 and FfG3. Interestingly, neither FfPka1 nor FfPka2 are involved in FfG1/FfG3/FfAc-mediated regulation of fusarubin biosynthesis, suggesting that an additional yet unknown cAMP-binding protein must exist that represses fusarubin biosynthesis in the wild type [15] (Fig. 10.3a). Not much is known regarding the influence of signaling components on pigment biosynthesis in other fusaria. Deletion of the adenylyl cyclase in *F. proliferatum* and *F. verticillioides* resulted in overproduction of a reddish pigment that diffused into the surrounding medium, which the authors interpreted as bikaverin [181, 182]. However, this was never proven unequivocally and, furthermore, contradicts the findings regarding the regulation of bikaverin in *F. fujikuroi*. The presence of the fusarubin gene cluster in both fungi and the results obtained during studies with *F. fujikuroi* [15, 174] suggest that in the case of both *F. proliferatum* and *F. verticillioides* the red pigmentation is due to fusarubins rather than bikaverin, an assumption that needs clarification in the future.

Epigenetic Regulation by Histone Modifications

Recent studies in different fungi provide strong evidence that secondary metabolite biosynthesis depends on the chromatin state, which differs between active (euchromatic) and silent (heterochromatic) regions. Histone-modifying enzymes that add or remove certain histone marks decide on the fate of the chromatin and thus may serve as markers for either gene transcription or silencing. Prominent histone marks are, for example, acetylation of lysine 9 or 14 at histone 3 (H3K9Ac/H3K14Ac) or methylation of lysine 4 (H3K4me2/3) all associated with gene activation, while trimethylation of lysine 9 at histone 3 (H3K9me3) is associated with gene silencing (for recent reviews, see [170, 183–185]). ChIP-sequencing of *F. fujikuroi* under acidic, low and high nitrogen conditions using antibodies against H3K9Ac, H3K4me2 as well as H3K9me3 revealed the presence of H3K9me3 mainly in centromeric and telomeric regions, while H3K9Ac and H3K4me2 were dispersed along the chromosomal arms [16]. Furthermore, the presence of the active histone mark H3K9Ac correlated with gene expression at the GA and bikaverin gene cluster under low nitrogen conditions, while this mark was almost abolished under repressing high nitrogen conditions in both cases (Fig. 10.3b). Contrary to this, the second activating histone mark H3K4me2 accumulated at the GA gene cluster, but only at two of the seven genes under biosynthesis-inducing conditions but not under biosynthesis-repressing conditions. H3K4me2 level was not elevated for the bikaverin genes under any condition tested [16]. Fusarubin biosynthetic genes are not expressed under these acidic conditions and were therefore not investigated. Hence future experiments under alkaline conditions will show if the same correlation can be seen also for fusarubin biosynthesis. The presence of activating histone marks under biosynthesis inducing and absence under repressing conditions in some, but not all, cases indicates that secondary metabolites are regulated by a complex network of histone-modifying enzymes. Targeted deletion of the respective genes will shed light on this aspect in the future.

Conclusion

Summarizing the different regulation levels (Fig. 10.3), the biosynthesis of all three secondary metabolites is strictly repressed by nitrogen. While bikaverin and fusarubin biosynthesis strictly depends on the ambient pH values in a PacC-dependent and PacC-independent manner, respectively, GAs are synthesized in either pH value. In the case of GA biosynthesis, nitrogen metabolite repression is regulated via the GATA transcription factors AreA and AreB. Deletion of either of these genes led to a lack of GA biosynthesis, while these GATA transcription factors are not essential for bikaverin biosynthesis. AreA, furthermore, activates the small construct of the bZIP transcription factor MeaB, $MeaB^S$, but simultaneously represses the large construct of MeaB, $MeaB^L$, which itself represses both bikaverin and GA biosynthesis.

The presence of glutamine strongly represses transcription of all three nitrogen-repressed secondary metabolites but, surprisingly, deletion of the glutamine synthase-encoding gene, *GLN1*, results in completely abolished biosynthesis of GAs and bikaverin. Although fusarubin biosynthesis is repressed by the presence of glutamine, nothing is known regarding the molecular mechanism of its N-regulation yet. The global regulators FfVel1 and FfLae1 both affect secondary metabolite biosynthesis in *F. fujikuroi*. While FfVel1 functions as positive regulator for GA and fusarubin biosynthesis, bikaverin is negatively regulated by FfVel1. Deletion of *FfLAE1* led to a down-regulation of both GA and bikaverin biosynthesis. Its involvement in fusarubin regulation still needs to be investigated. In addition, heterotrimeric G protein-mediated signaling affects all three secondary metabolites in a distinct way. While the adenylyl cyclase FfAc functions as positive regulator for both GA and bikaverin biosynthesis, regulation of bikaverin biosynthesis is further mediated through FfPka1, and GA biosynthesis through FfPka2. The upstream target involved in the regulation of both has yet to be determined. Fusarubin biosynthesis is negatively regulated by FfAc, which itself is stimulated by FfG1 and FfG3. However, neither FfPka1 nor FfPka2 are involved in fusarubin regulation, suggesting another yet unidentified cAMP-binding protein beside the regulatory subunit of the Pka. (Fig. 10.3a).

References

1. Wulff EG, Sørensen JL, Lübeck M, Nielsen KF, Thrane U, Torp J. *Fusarium* spp. associated with rice Bakanae: ecology, genetic diversity, pathogenicity and toxigenicity. Environ Microbiol. 2010;12:649–57.
2. Bömke C, Tudzynski B. Diversity, regulation, and evolution of the gibberellin biosynthetic pathway in fungi compared to plants and bacteria. Phytochemistry. 2009;70:1876–93.
3. Sponsel VM. Commercial uses of gibberellins. In: Taiz L, Zeiger E, editors. Plant physiology online. 5th ed. Sunderland: Sinauer Associates; 2010.
4. Barrero AF, Sánchez JF, Oltra JE, Tamayo N, Cerdá-Olmedo E, et al. Fusarin C and 8Z-fusarin C from *Gibberella fujikuroi*. Phytochemistry. 1991;30:2259–63.
5. Bacon CW, Porter JK, Norred WP, Leslie JF. Production of fusaric acid by *Fusarium* species. Appl Environ Microbiol. 1996;62:4039–43.
6. Desjardins AE, Plattner RD, Nelson PE. Production of fumonisin B_1 and moniliformin by *Gibberella fujikuroi* from rice from various geographic areas. Appl Environ Microbiol. 1997;63:1838–42.
7. Fotso J, Leslie JF, Smith JS. Production of beauvericin, moniliformin, fusa-proliferin, and fumonisins b(1), b(2), and b(3) by fifteen ex-type strains of *Fusarium* species. Appl Environ Microbiol. 2002;68:5195–7.
8. Kvas M, Marasas WFO, Wingfield BD, Wingfield MJ, Steenkamp ET. Diversity and evolution of *Fusarium* species in the *Gibberella fujikuroi* complex. Fungal Divers. 2009;34:1–21.
9. Moretti A, Mulè G, Ritieni A, Logrieco A. Further data on the production of beauvericin, enniatins and fusaproliferin and toxicity to *Artemia salina* by *Fusarium* species of *Gibberella fujikuroi* species complex. Int J Food Microbiol. 2007;118:158–63.
10. Nelson PE. Taxonomy and biology of *Fusarium moniliforme*. Mycopathologia. 1992;117: 29–36.
11. Prado-Cabrero A, Schaub P, Díaz-Sánchez V, Estrada AF, Al-Babili S, et al. Deviation of the neurosporaxanthin pathway towards beta-carotene biosynthesis in *Fusarium fujikuroi* by a point mutation in the phytoene desaturase gene. FEBS J. 2009;276:4582–97.

12. Wiemann P, Willmann A, Straeten M, Kleigrewe K, Beyer M, et al. Biosynthesis oft he red pigments bikaverin in Fusarium fujikuroi: genes, their function and regulation. Mol Microbiol. 2009;72:931–46.
13. Studt L, Wiemann P, Kleigrewe K, Humpf HU, Tudzynski B. Biosynthesis of fusarubins accounts for pigmentation of *Fusarium fujikuroi* perithecia. Appl Environ Microbiol. 2012;78:4468–80.
14. Studt L, Troncoso C, Gong F, Hedden P, Toomajian C, et al. Segregation of secondary metabolite biosynthesis in hybrids of *Fusarium fujikuroi* and *Fusarium proliferatum*. Fungal Genet Biol. 2012;49:567–77.
15. Studt L, Humpf HU, Tudzynski B. Signaling governed by G proteins and cAMP is crucial for growth, secondary metabolism and sexual development in *Fusarium fujikuroi*. PLoS One. 2013;8(2):e58185.
16. Wiemann P, Sieber CMK, von Bargen KW, Studt L, Niehaus EM, et al. Deciphering the cryptic genome: genome-wide analyses of the rice pathogen *Fusarium fujikuroi* reveal complex regulation of secondary metabolism and novel metabolites. PLoS Pathog. 2013;9(6): e1003475.
17. Díaz-Sánchez V, Avalos J, Limón MC. Identification and regulation of fusA, the polyketide synthase gene responsible for fusarin production in *Fusarium fujikuroi*. Appl Environ Microbiol. 2012;78:7258–66.
18. Niehaus EM, Kleigrewe K, Wiemann P, Studt L, Sieber CMK, et al. Multifactorial manipulation of the *Fusarium fujikuroi* fusarin gene cluster yields insight into the complex regulation and fusarin biosynthetic pathway. Chem Biol. 2013;20:1055–66.
19. Sawada K. Contributions on Formosan fungi. Trans Nat Hist Soc Formosa. 1917;7: 131–3.
20. Kurosawa E. Experimental studies on the nature of the substance secreted by the "bakanae" fungus. Nat Hist Soc Formosa. 1926;16:213–27.
21. Yabuta T. Biochemistry of the bakanae fungus of rice. Agr Hort. 1935;10:17–22.
22. Ito S, Kimura J. Studies on the 'bakanae' disease of rice plant. Rep Hokkaido Natl Agric Exp Stn. 1931;27:1–99.
23. Yabuta T, Hayashi T. Biochemical studies on the bakanae fungus of rice. II. Isolation of "Gibberellin", the active principle which makes the rice seedlings grow slenderly. J Agric Chem Soc Jpn. 1939;15:257–66.
24. Cross BE, Grove JF, MacMillan J, Moffatt JS, Mulholland TPC, et al. A revised structure for gibberellic acid. Proc Chem Soc. 1959;30:302–3.
25. Curtis PJ, Cross BE. Gibberellic acid. A new metabolite from the culture filtrates of *Gibberella fujikuroi*. Chem Ind. 1954;1066.
26. Takahashi N, Kitamura H, Kawarada A, Seta Y, Takai M, et al. Isolation of gibberellins and their properties. Bull Agric Chem Soc Jpn. 1955;19:267–77.
27. Kato J, Purves WK, Phinney BO. Gibberellin-like substances in plants. Nature. 1962;196:687–8.
28. Lang A. Bolting and flowering in biennial *Hyoscyamus niger*, induced by gibberellic acid. Plant Physiol. 1956;31(suppl):35.
29. Radley M. Occurrence of substances similar to gibberellic acid in higher plants. Nature. 1956;178:1070–1.
30. Bu'Lock J, Detroy R, Hostalek Z, Munin-al-Shakardu M. Regulation of secondary biosynthesis in *Gibberella fujikuroi*. Trans Br Mycol Soc. 1974;62:377–89.
31. Jefferys EG. The gibberellin fermentation. Adv Appl Microbiol. 1973;13:283–315.
32. Kumar PK, Lonsane BK. Gibberellic acid by solid state fermentation: consistent and improved yields. Biotechnol Bioeng. 1987;30:267–71.
33. Shukla R, Srivastava AK, Chand S. Bioprocess strategies and recovery processes in gibberellic acid fermentation. Biotechnol Bioproc Eng. 2003;8:269–78.
34. http://www.plant-hormones.info/gibberellin_nomenclature.htm. Accessed 21 Aug 2013.
35. MacMillan J. Occurrence of gibberellins in vascular plants, fungi, and bacteria. J Plant Growth Regul. 2002;20:387–442.

36. Sponsel VM, Hedden P. Gibberellin biosynthesis and inactivation. In: Davies PJ, editor. Plant hormones: biosynthesis, signal transduction, action! Dordrecht: Kluwer Academic Publishers; 2004. p. 63–94.
37. Hedden P, Thomas SG. Gibberellin biosynthesis and its regulation. Biochem J. 2012;444: 11–25.
38. Hauvermale AL, Ariizumi T, Steber CM. Gibberellin signaling: a theme and variations on DELLA repression. Plant Physiol. 2012;160:83–92.
39. Yamaguchi S. Gibberellin metabolism and its regulation. Annu Rev Plant Biol. 2008;59:225–51.
40. Balan J, Fuska J, Kuhr I, Kuhrová V. Bikaverin, an antibiotic from *Gibberella fujikuroi*, effective against *Leishmania brasiliensis*. Folia Microbiol (Praha). 1970;15:479–84.
41. Giordano W, Avalos J, Fernández-Martín R, Cerdá-Olmedo E, Domenech CE. Lovastatin inhibits the production of gibberellins but not sterol or carotenoid biosynthesis in *Gibberella fujikuroi*. Microbiology. 1999;145(Pt 10):2997–3002.
42. Kjaer D, Kjaer A, Pedersen C, Bu'lock JD, Smith JR. Bikaverin and norbikaverin, benzoxanthentrione pigments of *Gibberella fujikuroi*. J Chem Soc Perkin 1. 1971;16:2792–7.
43. Rodríguez-Ortiz R, Mehta BJ, Avalos J, Limón MC. Stimulation of bikaverin production by sucrose and by salt starvation in *Fusarium fujikuroi*. Appl Microbiol Biotechnol. 2010;85: 1991–2000.
44. Kreitman G, Nord FF. Lycopersin, a pigment from *Fusarium lycopersici*. Arch Biochem. 1949;21:457.
45. Kreitman G, Sebek OK, Nord FF. On the mechanism of enzyme action. XLIII. Chemistry and interaction of lycopersin in the carbohydrate fat conversion by Fusarium vasinfectum. Arch Biochem. 1950;28:77–93.
46. Cornforth JW, Ryback G, Robinson PM, Park D. Isolation and characterization of a fungal vacuolation factor (bikaverin). J Chem Soc Perkin 1. 1971;16:2786–8.
47. De Boer JJ, Bright D, Dallinga G, Hewitt TG. Crystal and molecular structure of the chloroform solvate of bikaverin. J Chem Soc Perkin 1. 1971;16:2788–91.
48. Henderson JF, Battell ML, Zombor G, Fuska J, Nemec P. Effects of bikaverin on purine nucleotide synthesis and catabolism in Ehrlich ascites tumor cells in vitro. Biochem Pharmacol. 1977;26:1973–7.
49. Kovac L, Bohmerova E, Fuska J. Inhibition of mitochondrial functions by the antibiotics bikaverin and duclauxine. J Antibiot. 1978;31:616–20.
50. Kwon HR, Son SW, Han HR, Choi GJ, Jang KS, et al. Nematicidal activity of bikaverin and Fusaric acid isolated from *Fusarium oxysporum* against pine wood nematode, *Bursaphelenchus xylophilus*. Plant Pathol J. 2007;23:318–21.
51. Son SW, Kim HY, Choi GJ, Lim HK, Jang KS, et al. Bikaverin and fusaric acid from *Fusarium oxysporum* show antioomycete activity against *Phytophtora infestans*. J Appl Microbiol. 2008;104:692–8.
52. Cross BE, Myers PL, Webster GRB. New metabolites of *Gibberella fujikuroi*. XIV. *O*-Demethyl-anhydrofusarubin. J Chem Soc. 1970;7:930.
53. Ruelius HW, Gauhe A. Über Fusarubin, einen Naphthochinonfarbstoff aus Fusarien. Justus Liebigs Ann Chem. 1950;569:38–59.
54. Weiss S, Nord FF. On the mechanism of enzyme action; solanione a pigment from *Fusarium solani* D2 purple. Arch Biochem. 1949;22:288–313.
55. Arnstein HRV, Cook AH, Lacey MS. An antibacterial pigment from *Fusarium javanicum*. Nature. 1946;157:333–4.
56. Medentsev AG, Akimenko VK. Naphthoquinone metabolites of the fungi. Phytochemistry. 1998;47:935–59.
57. Arnstein HRV, Cook AH. Production of antibiotics by fungi. Part III. Javanicin. An antibacterial pigment from *Fusarium javanicum*. J Chem Soc. 1947:1021–8.
58. Baker RA, Tatum JH, Nemec Jr S. Toxin production by *Fusarium solani* from fibrous roots of blight-diseased citrus. Physiol Biochem. 1981;71:951–4.

59. Brimble MA, Duncalf LJ, Nairn MR. Pyranonaphthoquinone antibiotics—isolation, structure and biological activity. Nat Prod Rep. 2006;16:267–81.
60. Issaq HJ, Barr EW, Wei TT, Meyers C, Aszalos AA. Thin-layer chromatographic classification of antibiotics exhibiting antitumor properties. J Chromatogr. 1977;133:291–301.
61. Kern H. Phytotoxins produced by *Fusaria*. In: Wood RKS, Ballio A, Graniti A, editors. Phytotoxins in plant diseases. New York: Academic; 1972. p. 35–48.
62. Kern H, Naef-Roth S. The formation of phytotoxins by *Neocosmospora vasinfecta* and *Neocosmospora africana*. Phytopathol Z. 1965;53:45–64.
63. McCulloch AW, McInnes AG, Smith DG, Kurobane I, Vining LC. Alkaline oxidation of diastereoisomeric 4a, 10a-dihydrofusarubins to norjavanicin, fusarubin, and a new antibiotic isofusarubin: nonenzymic formation of products in *Fusarium solani* cultures. Can J Chem. 1982;60:2943–9.
64. Bell AA, Wheeler MH, Liu J, Stipanovic RD, Puckhaber LS, Orta H. United States Department of Agriculture-Agricultural Research Service studies on polyketide toxins of *Fusarium oxysporum* f sp *vasinfectum*: potential targets for disease control. Pest Manag Sci. 2003;59:736–47.
65. Holenstein JE, Défago G. Inheritance of naphthazarin production and pathogenicity to pea in *Nectria haematococca*. J Exp Bot. 1983;34:927–35.
66. Van Rensburg JCJ, Labuschagne N, Nemec S. Occurrence of *Fusarium*-produced naphthazarins in citrus trees and sensitivity of rootstocks to isomarticin in relation to citrus blight. Plant Pathol. 2001;50:258–65.
67. Gopalakrishnan S, Beale MH, Ward JL, Strange RN. Chickpea wilt: identification and toxicity of 8-*O*-methyl-fusarubin from *Fusarium acutatum*. Phytochemistry. 2005;66:1536–9.
68. Medentsev AG, Akimenko VK. Mechanism of phytotoxic action of naphthoquinone pigments of the fungus Fusarium decemcellulare. Phytochemistry. 1992;31:77–9.
69. Keller NP, Hohn TM. Metabolic pathway gene clusters in filamentous fungi. Fungal Genet Biol. 1997;21:17–29.
70. Linnemannstöns P, Voß T, Hedden P, Gaskin P, Tudzynski B. Deletions in the gibberellin biosynthesis gene cluster of *Gibberella fujikuroi* by restriction enzyme-mediated integration and conventional transformation-mediated mutagenesis. Appl Environ Microbiol. 1999;65: 2558–64.
71. Tudzynski B, Hölter K. Gibberellin biosynthetic pathway in *Gibberella fujikuroi*: evidence for a gene cluster. Fungal Genet Biol. 1998;25:157–70.
72. Malonek S, Bömke C, Bornberg-Bauer E, Rojas MC, Hedden P, et al. Distribution of gibberellin biosynthetic genes and gibberellin production in the *Gibberella fujikuroi* species complex. Phytochemistry. 2005;66:1296–311.
73. Malonek S, Rojas MC, Hedden P, Gaskin P, Hopkins P, et al. Functional characterization of two cytochrome P450 monooxygenase genes, P450-1 and p 450-4, of the gibberellic acid gene cluster in Fusarium proliferatum (Gibberella fujikuroi MP-D). Appl Environ Microbiol. 2005;71:1462–72.
74. Malonek S, Rojas MC, Hedden P, Hopkins P, Tudzynski B. Restoration of gibberellin production in *Fusarium proliferatum* by functional complementation of enzymatic blocks. Appl Environ Microbiol. 2005;71:6014–25.
75. Troncoso C, González X, Bömke C, Tudzynski B, Gong F, et al. Gibberellin biosynthesis and gibberellin oxidase activities in *Fusarium sacchari*, *Fusarium konzum* and *Fusarium subglutinans* strains. Phytochemistry. 2010;71:1322–31.
76. Bömke C, Rojas MC, Hedden P, Tudzynski B. Loss of gibberellin production is due to a deletion in the gibberellic acid gene cluster in *Fusarium verticillioides* (*G. fujikuroi* MP-A). Appl Environ Microbiol. 2008;74:7790–801.
77. Rim SO, Lee JH, Choi WY, Hwang SK, Suh SJ, et al. *Fusarium prolifertum* KGL0401 as a new gibberellin-producing fungus. J Microbiol Biotechnol. 2005;15:809–14.
78. Tsavkelova EA, Bömke C, Netrusov AI, Weiner J, Tudzynski B. Production of gibberellic acids by an orchid-associated *Fusarium proliferatum* strain. Fungal Genet Biol. 2008;45:1393–403.

79. Bömke C, Rojas MC, Gong F, Hedden P, Tudzynski B. Isolation and characterization of the gibberellin biosynthetic gene cluster in *Sphaceloma manihoticola*. Appl Environ Microbiol. 2008;74:5325–39.
80. Kawaide H. Biochemical and molecular analyses of gibberellin biosynthesis in fungi. Biosci Biotechnol Biochem. 2006;70:583–90.
81. Rademacher W, Graebe JE. Gibberellin A4 produced by *Sphaceloma manihoticola*, the cause of the superelongation disease of cassava (*Manihot esculenta*). Biochem Biophys Res Commun. 1979;91:35–40.
82. Zeigler RS, Powell LE, Thurston HD. Gibberellin A4 production by *Sphaceloma manihoticola*, causal agent of cassava superelongation disease. Phytopathology. 1980;70:589–93.
83. Khaldi N, Collemare J, Lebrun MH, Wolfe KH. Evidence for horizontal transfer of a secondary metabolite gene cluster between fungi. Genome Biol. 2008;9:R18.
84. Slot JC, Rokas A. Horizontal transfer of a large and highly toxic secondary metabolic gene cluster between fungi. Curr Biol. 2011;21:134–9.
85. Campbell MA, Rokas A, Slot JC. Horizontal transfer and death of a fungal secondary metabolic gene cluster. Genome Biol Evol. 2012;4:289–93.
86. Schumacher J, Gautier A, Morgant G, Studt L, Ducrot PH, et al. A functional bikaverin biosynthesis gene cluster in rare strains of *Botrytis cinerea* is positively controlled by VELVET. PLoS One. 2013;8(1):e53729.
87. Gaffoor I, Brown DW, Plattner RD, Proctor RH, Qi W, et al. Functional analysis of the polyketide synthase genes in the filamentous fungus *Gibberella zeae* (anamorph *Fusarium graminearum*). Eukaryot Cell. 2005;4:1926–33.
88. Proctor RH, Butchko RAE, Brown DW, Moretti A. Functional characterization, sequence comparisons and distribution of a polyketide synthase gene required for perithecial pigmentation in some *Fusarium* species. Food Addit Contam. 2007;24:1076–87.
89. Linnemannstöns P, Schulte J, Prado MM, Proctor RH, Avalos J, et al. The polyketide synthase gene *pks4* from *Gibberella fujikuroi* encodes a key enzyme in the biosynthesis of the red pigment bikaverin. Fungal Genet Biol. 2002;37:134–48.
90. Ma SM, Zhan J, Watanabe K, Xie X, Zhang W, et al. Enzymatic synthesis of aromatic polyketides using PKS4 from *Gibberella fujikuroi*. J Am Chem Soc. 2007;129:10642–3.
91. Kim JE, Han KH, Jin J, Kim H, Kim JC, et al. Putative polyketide synthase and laccase genes for biosynthesis of aurofusarin in *Gibberella zeae*. Appl Environ Microbiol. 2005;71:1701–8.
92. Malz S, Grell MN, Thrane C, Maier FJ, Rosager P, et al. Identification of a gene cluster responsible for the biosynthesis of aurofusarin in the *Fusarium graminearum* species complex. Fungal Genet Biol. 2005;42:420–33.
93. Brown DW, Butchko RAE, Proctor RH. Genomic analysis of Fusarium verticillioides. Food Addit Contam Part A Chem Anal Control Expo Risk Assess. 2008;25:1158–65.
94. Medentsev AG, Arinbasarova AI, Akimenko VK. Biosynthesis of naphthoquinone pigments by fungi of the genus *Fusarium*. Prikl Biokhim Mikrobiol. 2005;41:573–7.
95. Parisot D, Maugin M, Gerlinger C. Genetic and epigenetic factors involved in the excretion of naphthoquinone pigments into the culture medium by *Nectria haematococca*. J Gen Microbiol. 1981;126:443–57.
96. Roos A. Zur Physiologie und Pathologie von *Neoeosmoiporu vasinfecta* E. F. Smith. Phytopathol Z. 1977;88:238–71.
97. Stoessl A. Structure and biogenetic relations: fungal non-host specific toxins. In: Durbin RD, editor. Toxins in plant diseases. New York: Academic; 1981. p. 144–7.
98. Keller NP, Turner G, Bennett JW. Fungal secondary metabolism—from biochemistry to genomics. Nat Rev Microbiol. 2005;3:937–47.
99. Brock NL, Huss K, Tudzynski B, Dickschat JS. Genetic dissection of sesquiterpene biosynthesis by *Fusarium fujikuroi*. Chembiochem. 2013;14:311–5.
100. Birch AJ, Richards RW, Smith H. The biosynthesis of gibberellic acid. Proc Chem Soc. 1958:192–3.
101. Birch AJ, Richards RW, Smith H. The biosynthesis of terpenoid compounds in fungi. In: Wolstenholme GEW, editor. Biosynthesis of terpines and sterols. Boston: Little, Brown & Co; 1959. p. 245–66.

102. Avalos J, Casadesús J, Cerdá-Olmedo E. *Gibberella fujikuroi* mutants obtained with UV radiation and N-methyl-N′-nitro-N-nitrosoguanidine. Appl Environ Microbiol. 1985;49: 187–91.
103. Bearder JR, MacMillan J, Wels CM, Chaffey MB, Phinney BO. Position of the metabolic block for gibberellin biosynthesis in mutant B1-41a of *Gibberella fujikuroi*. Phytochemistry. 1974;13:911–7.
104. Bearder JR. In vivo diterpenoid biosynthesis in *Gibberella fujikuroi*: the pathway after *ent-kaurene*. In: Crozier A, editor. The biochemistry and physiology of gibberellins. New York: Praeger; 1983. p. 251–387.
105. Fernández-Martín R, Reyes F, Domenech CE, Cabrera E, Bramley PM, et al. Gibberellin biosynthesis in gib mutants of *Gibberella fujikuroi*. J Biol Chem. 1995;270:14970–4.
106. Homann V, Mende K, Arntz C, Ilardi V, Macino G, et al. The isoprenois pathway: cloning and characterization of fungal FPPS genes. Curr Genet. 1996;30:232–9.
107. Mende K, Homann V, Tudzynski B. Molecular characterization of the geranylgeranyl diphosphate synthase gene of *Gibberella fujikuroi*. Mol Gen Genet. 1997;255:96–105.
108. Woitek S, Unkles SE, Kinghorn JR, Tudzynski B. 3-Hydroxy-3-methylglutaryl-CoA reductase gene of *Gibberella fujikuroi*: isolation and characterization. Curr Genet. 1997;31:38–47.
109. Albermann S, Elter T, Teubner A, Krischke W, Tudzynski B. Characterization of novel mutants with an altered gibberellin spectrum in comparison to different wild-type strains of *Fusarium fujikuroi*. Appl Microbiol Biotechnol. 2013;97(17):7779–90.
110. Hedden P, Phillips AL. Gibberellin metabolism: new insights revealed by the genes. Trends Plant Sci. 2000;5:523–30.
111. Lange T, Hedden P, Graebe JE. Expression cloning of a gibberellin 20-oxidase, a multifunctional enzyme involved in gibberellin biosynthesis. Proc Natl Acad Sci U S A. 1994;91:8552–6.
112. Phillips AL, Ward DA, Unkles S, Appleford NE, Lange T, et al. Isolation and expression of three gibberellin 20-oxidase cDNA clones from *Arabidopsis*. Plant Physiol. 1995;108:1049–57.
113. Sun TP, Kamiya Y. The Arabidopsis GA1 locus encodes the cyclase *ent*-kaurene synthetase A of gibberellin biosynthesis. Plant Cell. 1994;6:1509–18.
114. Kawaide H, Imai R, Sassa T, Kamiya Y. *Ent*-kaurene synthase from the fungus *Phaeosphaeria* sp. L487. cDNA isolation, characterization, and bacterial expression of a bifunctional diterpene cyclase in fungal gibberellin biosynthesis. J Biol Chem. 1997;272:21706–12.
115. Tudzynski B, Kawaide H, Kamiya Y. Gibberellin biosynthesis in *Gibberella fujikuroi*: cloning and characterization of the copalyl diphosphate synthase gene. Curr Genet. 1998;34:234–40.
116. Hedden P, Phillips AL, Rojas MC, Carrera E, Tudzynski B. Gibberellin biosynthesis in plants and fungi: a case of convergent evolution? J Plant Growth Regul. 2002;20:319–31.
117. Chapman DJ, Regan MA. Evolution of a biochemical pathway: evidence from comparative biochemistry. Annu Rev Plant Physiol. 1980;31:639–45.
118. Rojas MC, Urrutia O, Cruz C, Gaskin P, Tudzynski B, et al. Kaurenolides and fujenoic acids are side products of the gibberellin P450-1 monooxygenase in *Gibberella fujikuroi*. Phytochemistry. 2001;65:821–30.
119. Tudzynski B, Hedden P, Carrera E, Gaskin P. The *P450-4* gene of *Gibberella fujikuroi* encodes *ent*-kaurene oxidase in the gibberellin biosynthetic pathway. Appl Environ Microbiol. 2001;67:3514–22.
120. Tudzynski B, Rojas MC, Gaskin P, Hedden P. The *Gibberella fujikuroi* gibberellin 20-oxidase is a multifunctional monooxygenase. J Biol Chem. 2002;277:21246–53.
121. Tudzynski B, Mihlan M, Rojas MC, Linnemannstöns P, Gaskin P, et al. Characterization of the final two genes of the gibberellin biosynthesis gene cluster of *Gibberella fujikuroi*. J Biol Chem. 2003;278:28635–43.
122. Avalos J, Prado-Cabrero A, Estrada AF. Neurosporaxanthin production by *Neurospora* and *Fusarium*. Methods Mol Biol. 2012;898:263–74.
123. Urrutia O, Hedden P, Rojas MC. Monooxygenases involved in GA12 and GA14 synthesis in *Gibberella fujikuroi*. Phytochemistry. 2001;56:505–11.
124. Bhattacharya A, Kourmpetli S, Ward DA, Thomas SG, Gong F, et al. Characterization of the fungal gibberellin desaturase as a 2-oxoglutarate-dependent dioxygenase and its utilization for enhancing plant growth. Plant Physiol. 2012;160:837–45.

125. Albermann S, Linnemannstöns P, Tudzynski B. Strategies for strain improvement in *Fusarium fujikuroi*: overexpression and localization of key enzymes of the isoprenoid pathway and their impact on gibberellin biosynthesis. Appl Microbiol Biotechnol. 2013;97:2979–95.
126. Gatenbeck S, Bentley R. Naphthoquinone biosynthesis in molds: the mechanism for formation of javanicin. Biochem J. 1965;94:478–81.
127. Barton DH, Cottier L, Freund K, Luini F, Magnus PD, et al. Total synthesis of bikaverin (6,11-dihydroxy-3,8-dimethoxy-1-methylbenzo(b)xanthem-7,10,12-trione). J Chem Soc Perkin 1. 1976;5:499–503.
128. Crawford JM, Dancy BC, Hill EA, Udwary DW, Townsend CA. Identification of a starter unit acyl-carrier protein transacylase domain in an iterative type I polyketide synthase. Proc Natl Acad Sci U S A. 2006;103:16728–33.
129. McInnes AG, Walter JA, Smith DG, Wright JL, Vining LC. Biosynthesis of bikaverin in *Fusarium oxysporum*. Use of 13C nuclear magnetic resonance with homonuclear 13C decoupling to locate adjacent 13C labels. J Antibiot (Tokyo). 1976;29:1050–7.
130. Wattana-amorn P, Williams C, Płoskoń E, Cox RJ, Simpson TJ, et al. Solution structure of an acyl carrier protein domain from a fungal type I polyketide synthase. Biochemistry. 2010;49: 2186–93.
131. Crawford JM, Korman TP, Labonte JW, Vagstad AL, Hill EA, et al. Structural basis for biosynthetic programming of fungal aromatic polyketide cyclization. Nature. 2009;461:1139–43.
132. Fujii I, Watanabe A, Sankawa U, Ebizuka Y. Identification of Claisen cyclase domain in fungal polyketide synthase WA, a naphthopyrone synthase of *Aspergillus nidulans*. Chem Biol. 2001;8:189–97.
133. Crawford JM, Vagstad AL, Ehrlich KC, Townsend CA. Starter unit specificity directs genome mining of polyketide synthase pathways in fungi. Bioorg Chem. 2008;36:16–22.
134. Wiemann P, Albermann S, Niehaus EM, Studt L, von Bargen K, et al. The Sfp-type 4′-phosphopantetheinyl transferase Ppt1 of *Fusarium fujikuroi* controls development, secondary metabolism and pathogenicity. PLoS One. 2012;7(5):e37519.
135. Zhang W, Li Y, Tang Y. Engineered biosynthesis of bacterial polyketides in *Escherichia coli*. Proc Natl Acad Sci U S A. 2008;105:20683–8.
136. Holenstein JE, Stoessl A, Kern H, Stothers JB. The marticins: confirmation of structure, elucidation of biosynthetic origin by ^{13}Cmr studies, and revision of stereochemical assignments. Can J Chem. 1984;62:1971–6.
137. Kurobane I, Vining LC, McInnes AG, Walter JA. Use of 13C in biosynthetic studies. The labeling pattern in dihydrofusarubin enriched from [13C]- and [13C,2H]acetate in cultures of *Fusarium solani*. Can J Chem. 1980;58:1380–5.
138. Arsenault GP. Fungal metabolites III. Quinones from *Fusarium solani* D 2 purple and structure of (+)-solaniol. Tetrahedron. 1968;24:4745–9.
139. Parisot D, Devys M, Barbier M. Structure and biosynthesis of 5-deoxyfusarubin and anhydro-5-deoxyfusarubin, naphthaquinone pigments from *Nectria haematococca*. Phytochemistry. 1985;24:1977–9.
140. Parisot D, Devys M, Barbier M. Fusarubinic acid, a new naphthoquinone from the fungus *Nectria haematococca*. Phytochemistry. 1988;27:3002–4.
141. Kurobane I, Vining LC, McInnes AG, Smith DG. Diastereoisomeric 4a, 10a-dihydrofusarubins: true metabolites of *Fusarium solani*. Can J Chem. 1978;56:1593–4.
142. Boonyapranai K, Tungpradit R, Lhieochaiphant S, Phutrakul S. Optimization of submerged culture for the production of naphthoquinone pigments by *Fusarium verticillioides*. Chiang Mai J Sci. 2008;35:457–66.
143. Parisot D, Devys M, Barbier M. Heptaketide-derived polyenes from the fungus *Nectria haematococca*. Phytochemistry. 1992;31:4357–8.
144. Awakawa T, Kyji T, Wakimoto T, Abe I. A heptaketide naphthaldehyde produced by a polyketide synthase from *Nectria haematococca*. Bioorg Med Chem Lett. 2012;22: 4338–40.
145. Li Y, Xu W, Tang Y. Classification, prediction, and verification of the regioselectivity of fungal polyketide synthase product template domains. J Biol Chem. 2010;285:22764–73.

146. Parisot D, Devys M, Barbier M. Nectriachrysone, a new metabolite related to fusarubin produced by the fungus *Nectria haematococca*. J Chem Soc Perkin Trans 1. 1991:2280–1.
147. Tokai T, Koshino H, Takahashi-Ando N, Sato M, Fujimura M, et al. *Fusarium* Tri4 encodes a key multifunctional cytochrome P450 monooxygenase for four consecutive oxygenation steps in trichothecene biosynthesis. Biochem Biophys Res Commun. 2007;353:412–7.
148. Parisot D, Devys M, Barbier M. Anhydrofusarubinlactol from *Nectria haematococca*. Phytochemistry. 1989;28:3240–1.
149. Borrow A, Brown S, Jefferys EG, Kessell RH, Lloyd EC, et al. The kinetics of metabolism of *Gibberella fujikuroi* in stirred culture. Can J Microbiol. 1964;10:407–44.
150. Mihlan M, Homann V, Liu TW, Tudzynski B. AREA directly mediates nitrogen regulation of gibberellin biosynthesis in *Gibberella fujikuroi*, but its activity is not affected by NMR. Mol Microbiol. 2003;47:975–91.
151. Tudzynski B, Homann V, Feng B, Marzluf GA. Isolation, characterization and disruption of the *areA* nitrogen regulatory gene of *Gibberella fujikuroi*. Mol Gen Genet. 1999;261: 106–14.
152. Arst Jr HN, Cove DJ. Nitrogen metabolite repression in *Aspergillus nidulans*. Mol Gen Genet. 1973;126:111–41.
153. Caddick MX, Peters D, Platt A. Nitrogen regulation in fungi. Antonie Van Leeuwenhoek. 1994;65:169–77.
154. Marzluf GA. Genetic regulation of nitrogen metabolism in the fungi. Microbiol Mol Biol Rev. 1997;61:17–32.
155. Schönig B, Brown DW, Oeser B, Tudzynski B. Cross-species hybridization with *Fusarium verticillioides* microarrays reveals new insights in *Fusarium fujikuroi* nitrogen regulation and the role of AreA and NMR. Eukaryot Cell. 2008;7:1831–46.
156. Michielse CB, Pfannmueller A, Macios M, Rengers P, Dzikowska A, et al. The interplay between the GATA transcription factors AreA, the global nitrogen regulator and AreB in *Fusarium fujikuroi*. Mol Microbiol. 2014;91(3):472–93.
157. Wiemann P, Tudzynski B. The nitrogen regulation network and its impact on secondary metabolism and pathogenicity. In: Brown DW, Proctor RH, editors. *Fusarium*: genomics, molecular and cellular biology. Norwich: Caister Academic Press; 2013. p. 111–42.
158. Wagner D, Schmeinck A, Mos M, Morozov IY, Caddick MX, et al. The bZIP transcription factor MeaB mediates nitrogen metabolite repression at specific loci. Eukaryot Cell. 2010;9:1588–601.
159. Teichert S, Schönig B, Richter S, Tudzynski B. Deletion of the *Gibberella fujikuroi* glutamine synthetase gene has significant impact on transcriptional control of primary and secondary metabolism. Mol Microbiol. 2004;53:1661–75.
160. Ni M, Yu JH. A novel regulator couples sporogenesis and trehalose biogenesis in *Aspergillus nidulans*. PLoS One. 2007;2(10):e970.
161. Bayram O, Braus GH. Coordination of secondary metabolism and development in fungi: the velvet family of regulatory proteins. FEMS Microbiol Rev. 2012;36:1–24.
162. Kato N, Brooks W, Calvo AM. The expression of sterigmatocystin and penicillin genes in *Aspergillus nidulans* is controlled by *veA*, a gene required for sexual development. Eukaryot Cell. 2003;2:1178–86.
163. Amaike S, Keller NP. Distinct roles for VeA and LaeA in development and pathogenesis of *Aspergillus flavus*. Eukaryot Cell. 2009;8:1051–60.
164. Duran RM, Cary JW, Calvo AM. Production of cyclopiazonic acid, aflatrem, and aflatoxin by Aspergillus flavus is regulated by veA, a gene necessary for sclerotial formation. Appl Microbiol Biotechnol. 2007;3:1158–68.
165. López-Berges MS, Hera C, Sulyok M, Schäfer K, Capilla J, et al. The velvet complex governs mycotoxin production and virulence of *Fusarium oxysporum* on plant and mammalian hosts. Mol Microbiol. 2013;87:49–65.
166. Jiang J, Liu X, Yin Y, Ma Z. Involvement of a velvet protein FgVeA in the regulation of asexual development. Lipid and secondary metabolisms and virulence in *Fusarium graminearum*. PLoS One. 2011;6:e28291.

167. Myung K, Zitomer NC, Duvall M, Glenn AE, Riley RT, Calvo AM. The conserved global regulator VeA is necessary for symptom production and mycotoxin synthesis in maize seedlings by *Fusarium verticillioides*. Plant Pathol. 2012;61:152–60.
168. Wiemann P, Willmann A, Brown DW, Kleigrewe K, Bok JW, et al. FfVel1 and FfLae1, components of a Velvet-like complex in *Fusarium fujikuroi*, affect differentiation, secondary metabolism and virulence. Mol Microbiol. 2010;77:972–94.
169. Schumacher J, Pradier JM, Simon A, Traeger S, Moraga J, et al. Natural variation in the VELVET Gene *bcvel1* affects virulence and light-dependent differentiation in *Botrytis cinerea*. PLoS One. 2012;7(10):e47840.
170. Brakhage AA. Regulation of fungal secondary metabolism. Nat Rev Microbiol. 2013;11: 21–32.
171. Bölker M. Sex and crime: heterotrimeric G proteins in fungal mating and pathogenesis. Fungal Genet Biol. 1998;25:143–56.
172. Lengeler KB, Davidson RC, D'Souza C, Harashima T, Shen WC, et al. Signal transduction cascades regulating fungal development and virulence. Microbiol Mol Biol Rev. 2000;64:746–85.
173. Li L, Wright SJ, Krystofova S, Park G, Borkovich KA. Heterotrimeric G protein signaling in filamentous fungi. Annu Rev Microbiol. 2007;61:423–52.
174. García-Martínez J, Adám AL, Avalos J. Adenylyl cyclase plays a regulatory role in development, stress resistance and secondary metabolism in Fusarium fujikuroi. PLoS One. 2012;7(1):e28849.
175. Vandamme J, Castermans D, Thevelein JM. Molecular mechanisms of feedback inhibition of protein kinase A on intracellular cAMP accumulation. Cell Signal. 2012;24:1610–8.
176. Bahn YS, Mühlschlegel FA. CO_2 sensing in fungi and beyond. Curr Opin Microbiol. 2006;9:572–8.
177. Klengel T, Liang WJ, Chaloupka J, Ruoff C, Schröpper K, et al. Fungal adenylyl cyclase integrates CO_2 sensing with cAMP signaling and virulence. Curr Biol. 2005;15:2021–6.
178. Mogensen EG, Janbon G, Chaloupka J, Steegborn C, Fu AS, et al. *Cryptococcus neoformans* Senses CO_2 through the carbonic anhydrase Can2 and the adenylyl cyclase Cac1. Eukaryot Cell. 2006;5:103–11.
179. Santagelo GM. Glucose signaling in *Saccharomyces cerevisiae*. Microbiol Mol Biol Rev. 2006;70:253–82.
180. Schumacher J, Kokkelink L, Huesmann C, Jimenez-Teja D, Collado IG, et al. The cAMP-dependent signaling pathway and its role in conidial germination, growth and virulence of the gray mold *Botrytis cinerea*. Mol Plant Microbe Interact. 2008;21:1443–59.
181. Choi YE, Brown JA, Williams CB, Canales LL, Shim WB. GAC1, a gene encoding a putative GTPase-activating protein regulates bikaverin biosynthesis in *Fusarium verticillioides*. Mycologia. 2008;100:701–9.
182. Kohut G, Oláh B, Adám AL, García-Martínez J, Hornok L. Adenylyl cyclase regulates heavy metal sensitivity, bikaverin production and plant tissue colonization in *Fusarium proliferatum*. J Basic Microbiol. 2010;50:59–71.
183. Bannister AJ, Kouzarides T. Regulation of chromatin by histone modifications. Cell Res. 2011;21:381–95.
184. Gacek A, Strauss J. The chromatin code of fungal secondary metabolite gene clusters. Appl Microbiol Biotechnol. 2012;95:1389–404.
185. Strauss J, Reyes-Dominguez Y. Regulation of secondary metabolism by chromatin structure and epigenetic codes. Fungal Genet Biol. 2011;48:62–9.

Chapter 11
Fusarins and Fusaric Acid in *Fusaria*

Eva-Maria Niehaus, Violeta Díaz-Sánchez, Katharina Walburga von Bargen, Karin Kleigrewe, Hans-Ulrich Humpf, M. Carmen Limón, and Bettina Tudzynski

Introduction

Fungi of the genus *Fusarium* produce a great variety of secondary metabolites that are diverse in structure and biological activity [1]. Some species produce mycotoxins that threaten human and animal health after consumption of contaminated grain or grain-derivatives [2, 3]. *Fusarium* produces polyketide metabolites such as fusaric acid (FA), fusarins, bikaverin, and fusarubin. The pigments bikaverin and fusarubin have been reviewed in Chap. 10 of this book while FA and fusarins are the objects of this chapter.

Fusarin C

Fusarins are polyketides produced by many *Fusarium* species and by the entomopathogenic fungus *Metarhizium anisopliae* [4] (Table 11.1). Fusarin A, B, C, and D were first described in 1981 by Wiebe and Bjeldanes in Berkeley, California [5]. The same authors already had isolated and partially characterized fusarin C [5], and its chemical structure was completed in 1984 [6]. The structure of fusarin C consists of a polyenic chromophore with a substituted 2-pyrrolidone (Fig. 11.1a) [6].

E.-M. Niehaus, Dr. rer. nat. • B. Tudzynski, Dr. rer. nat. (✉)
University of Münster, Institute of Biology und Biotechnology of Plants, Schlossplatz 8, D-48143 Münster, Germany
e-mail: tudzynsb@uni-muenster.de

V. Díaz-Sánchez, Ph.D. • M.C. Limón, Ph.D. (✉)
Departamento de Genética, Universidad de Sevilla, E-41012 Sevilla, Spain
e-mail: carmenlimon@us.es

K.W. von Bargen, Dr. rer. nat. • K. Kleigrewe, Dr. rer. nat. • H.-U. Humpf, Dr. rer. nat.
University of Münster, Institute of Food Chemistry, Corrensstr. 45, D-48149 Münster, Germany

J.-F. Martín et al. (eds.), *Biosynthesis and Molecular Genetics of Fungal Secondary Metabolites*, Fungal Biology, DOI 10.1007/978-1-4939-1191-2_11,

Table 11.1 Fungal species that produce fusarin and fusaric acid [1, 8–10, 13, 19, 131, 132]

Fusarin	Fusaric acid
F. avenaceum	*F. crookwellense*
F. crookwellense	*F. fujikuroi*
	F. heterosporum
F. culmorum	*F. moniliforme*
F. fujikuroi	*F. napiforme*
F. graminearum	*F. nigamai*
F. oxysporum	*F. oxysporum*
F. poae	*F. proliferatum*
F. sambucinum	*F. ramigenum*
F. sporotrichioides	*F. sacchari*
F. venenatum	*F. sambucinum*
	F. solani
Metarhizium anisopliae	*F. subglutinans*
	F. thapsinum
	F. verticilloides

During the isolation of fusarin C, it was found that it spontaneously converted to two different compounds whereas exposure to long-wave UV light gave rise to three other compounds (8Z, 6Z, and 10Z stereoisomers). Two other related compounds, fusarin A and D, with similar UV absorption properties were produced by *Fusarium*. A third kind of fusarin-like molecule was isolated from *F. moniliforme*, which was called fusarin F [7]. Kleigrewe et al. found out that this published structural assignment of fusarin F is incorrect, since fusarin C epimerizes to epi-fusarin C via the open-chain fusarin C structure [84]. Furthermore they discovered a new fusarin C-like structure called dihydrofusarin C. The structures are depicted in Fig. 11.1a.

Fusaric Acid

In 1934, Yabuta, a Japanese agricultural chemist, was trying to isolate gibberellins (see Chap. 10) when he separated a crystalline compound from the culture filtrate of *Gibberella fujikuroi* [8]. At that time the fungus was classified as *F. heterosporum* Nees. This compound was named FA and its chemical structure was 5-butylpicolic acid. The production of FA is broadly distributed among the entire genus *Fusarium*. Not only members of the *Gibberella fujikuroi* species complex (GFC)—e.g., *F. fujikuroi*, *F. proliferatum*, *F. circinatum*, *F. mangiferae*, *F. verticillioides*, and *F. subglutinans*—but also more distantly related *Fusarium* species, such as *F. crookwellense*, *F. heterosporum*, *F. napiforme*, *F. oxysporum*, and *F. solani* are producers of FA (Table 11.1) [1, 8–10].

FA analogs isolated from culture filtrates include 9,10-dehydrofusaric acid, diacid analog of fusaric acid, 10,11-dihydroxy fusaric acid, and methyl esters of fusaric and

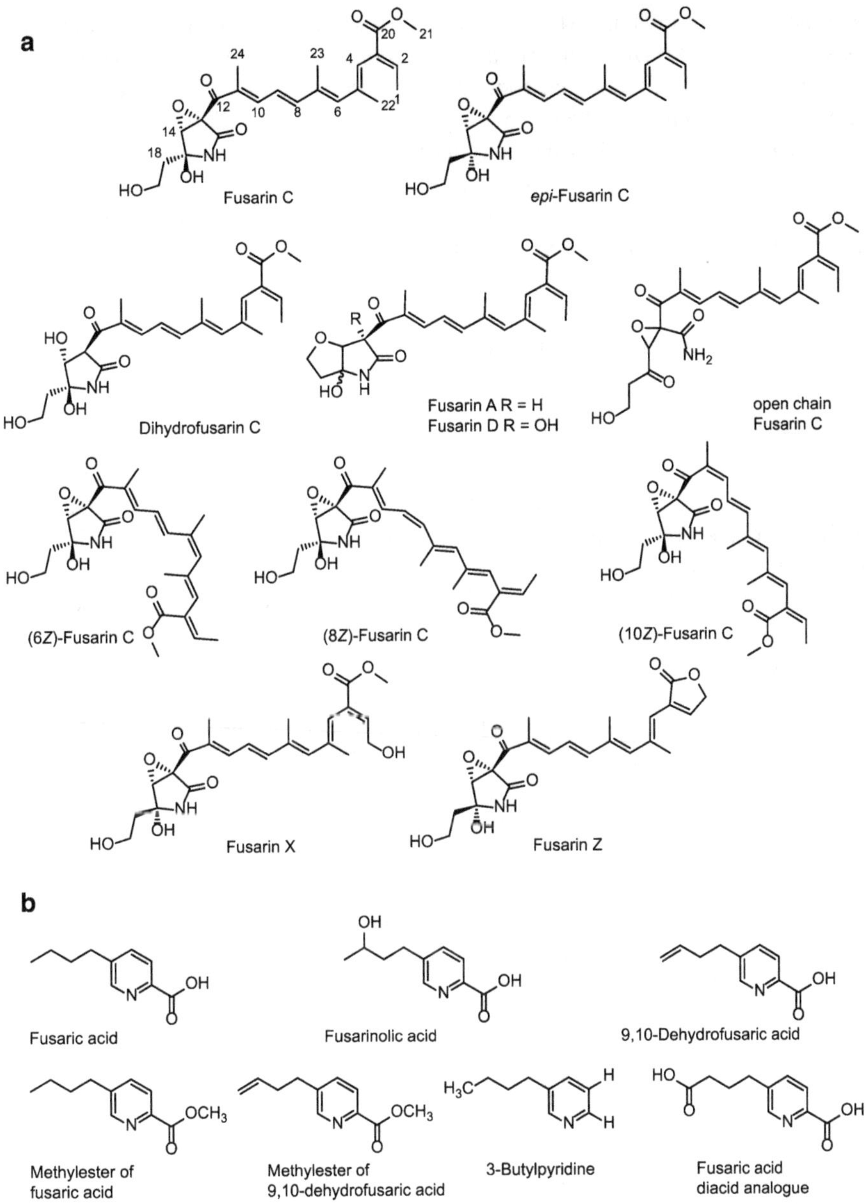

Fig. 11.1 Structures of fusarins and fusaric acid-related compounds. (**a**) Fusarin C, *epi*-fusarin C, dihydrofusarin C, fusarin D and A, open-chain fusarin C, (6*Z*)-fusarin C, (8*Z*)-fusarin C, (10*Z*)-fusarin C, fusarin X, and fusarin Z. (**b**) Fusaric acid, two derivatives fusarinolic acid and 9,10-dihydrofusaric acid, methylester of fusaric acid, methylester of 9,10-dehydrofusaric acid, 3-butylpyridine and fusaric acid diacid analog

9,10-dehydrofusaric acids (Fig. 11.1b) [11–14]. The 9,10-dehydrofusaric acid was reported in *Fusarium nygamai* and *Fusarium oxysporum* at high concentration. The methyl esters of both compounds were produced at very low amounts in *F. nygamai* and were not detected in *F. oxysporum* [11]. *F. moniliforme* NRRL 13,163 produced the analogs, 10,11-dihydroxyfusaric acid and a diacid of fusaric acid ([12]).

FA is a substituted pyridine. Fusaric acid methyl ester is degraded to 3-butylpyridine (Fig. 11.1b) in the soil [15] and FA is also converted into the same compound, vivotoxin, by cotton plants [1].

Biological Activity of Fusarins and Fusaric Acid

Fusarins

Fusarin C was considered a mutagen according to the Ames test, which detects the ability to reverse mutations in different *Salmonella typhimurium* strains [5, 16, 17]. However, fusarin C reversed mutations only if it was previously activated by a liver homogenate from rats induced with the barbiturate phenobarbital (S-9 mix) [17]. These authors discovered that the C13-14 epoxide of fusarin C was responsible for mutagenicity [17]. Fusarins A and D lacking the C13-14 epoxide are not mutagenic [17, 18]. Fusarin C is one-half as potent as the mycotoxin aflatoxin B1 and one-fourth as potent as sterigmatocystin [19]. The three stereoisomers of fusarin C, fusarin 8Z, 6Z, and 10Z conserve their mutagenicity [20].

Sister chromatid exchanges, chromosome aberrations, and cells with micronuclei in V79 cells induced with S-9 mix are significantly increased with fusarin C extracted from *F. moniliforme* [16]. The same authors also reported the induction of 6-thioguanine-resistant mutants in V79 cells after incubation with fusarin C and S-9 mix [16].

Fusarin C was metabolized to fusarin X and fusarin Z in microsomal mixtures from phenobarbital-induced rat livers [21]. Fusarin Z and fusarin X were 60 and 500 times more mutagenic than non-activated fusarin C in the Ames test, respectively. The non-activated fusarin C shows little mutagenic activity in comparison to fusarin X and Z. However, if fusarin C is activated it is almost as mutagenic as fusarin Z. Fusarin X has been isolated from cultures of *F. verticilloides* but fusarin Z has not been detected in cultures [21].

Different assays have been carried out to determine if fusarin C is responsible for carcinogenesis in rodents. Two groups of rats were fed with culture material from two *F. moniliforme* strains: MRC 826 and MRC 1069. The first was a highly toxic strain, whereas the latter was a nontoxic one. However, MRC 1069 produces a higher amount of fusarin C than MRC 826. Rats fed with 5 % of *F. moniliforme* MRC 1069 culture did not develop tumors. However, rats fed with only 0.5 % of *F. moniliforme* MRC 826 culture developed carcinomas such as cholangiocarcinomas, hepatocellular carcinomas, carcinomas of the forestomach epithelium, and esophageal papilloma [22]. Hence, fusaric C was not the metabolite responsible for carcinogenesis. Other trials done on mouse skin or rat liver agreed that fusarin C was not involved in cancer initiation [23]. It was proposed that fusarin C could be inactivated by glutathione [24].

Fusarin C caused a dose-dependent decrease in viability of human cells such as colorectal cancer Caco 2, prostate cancer PC3, and multiple myeloma U266 cells. The estimated IC_{50} for these cells varied from 5.6 to 42.8 μM and all of them were inhibited at 100 μM [25]. However, fusarin C stimulates growth of the human breast adenocarcinoma cell line MCF-7. This opposite effect could be explained by α(alpha)-and β(beta)-estrogen receptors present in this cell. The induction concentration ranged from 100 nM to 20 μM and the EC_{50}=890 nM. But above 20 μM these cells were also inhibited with an IC_{50}=46.8 μM [25]. Direct binding of fusarin C to estrogen receptors was demonstrated using an assay with chemical activated luciferase in recombinant cells containing an estrogen response element-luciferase reporter vector. Fusarin C can act as an estrogenic agonist and should be classified as a mycoestrogen as the mycotoxin zearalone from *F. graminearum*. Furthermore, fusarin C inhibited the activation of macrophages by macrophage activation factor and the cytotoxic activity of activated macrophages [26].

Fusaric Acid

FA is considered as a plant toxin. In fact, FA was one of the first fungal metabolites implicated in plant pathogenesis, in concrete in the tomato wilt symptoms caused by *F. oxysporum* f. sp. *lycopersici* Schlecht. emend. Snyd. and Hans [27]. Phytotoxicity assays with FA and picolinic acid analogs revealed that the addition of alkyl groups to the 5-position of picolinic acid increased their phytotoxicity [28]. The FA analog 3-butylpyridine (Fig. 11.1b) was described to be 100-fold more toxic to cotton than FA [29–31], but Stipanovic et al. found that this analog did not provoke necrosis on cotton cotyledons [28]. However, methylester of FA was the most toxic analog at any concentration tested [28].

The toxic effects of FA on plants include alteration of membrane permeability (modification of cell membrane potential), decrease of mitochondrial activity and oxygen uptake, inhibition of ATP synthesis and inhibition of root growth [32–35]. These effects are observed at toxic concentrations ($>10^{-5}$ M). Nontoxic concentrations ($<10^{-6}$ M) of FA induce synthesis of the phytoalexin camalexin, which produces reactive oxygen species (ROS) and increases cytosolic Ca^{2+}. It has been suggested that FA could act as an elicitor of plant responses to pathogen attack [34, 35].

The concentration of FA positively correlated with *Fusarium* wilt index. *Fusarium*-infected banana seedlings showed a higher leaf temperature than that of non-infected plants as determined by thermal imaging. The same results were obtained treating plants with purified FA. Infected plants had reduced stomata conductance and transpiration rate, which resulted in lower levels of water loss than in control plants [36]. The main cause of the disease is the damage of the membrane system caused by FA, the toxin produced by the pathogen.

FA has antiviral activities against DNA and RNA viruses that are important for humans and animals, such as cytomegalovirus (CMV), varicella-zoster virus (VZV), and human herpes simplex virus (HSV) [37]. FA interacts with some metalloproteins and zinc finger peptides (ZFPs), both of them involved in different functions. For example, metallopanstimullin-1 (MPS-1) is a ribosomal protein involved in DNA repair, transcription, and biogenesis of ribosomes. MPS contains a zinc finger and it is overexpressed in cancer tissues.

FA is a potent chelator of divalent cations such as Zn^{2+}, Ca^{2+}, Cu^{2+}, Se^{2+}, and Fe^{2+} that inhibit conserved zinc finger protein [38]. FA inhibits viruses by targeting ZFPs. The mode of action of FA is by titration of both zinc and Zn-peptides. An example is the inhibition of a HIV indispensable regulatory protein called Tat [39]. Future therapies could be based on FA that would release the Zn^{2+} from ZFPs and metalloproteins.

FA is also a bactericide [40, 41] that inhibits the growth of *Bacillus subtillis*. FA represses the synthesis of the antimicrobial polyketide 2,4-diacetylphloroglucinol (2,4-DAPG) in *Pseudomonas fluorescens* [42]. Some strains of *P. fluorescens* are insensitive to FA because they are able to deacetylate 2,4-DAPG [43]. FA is detoxified by bacterial species such as *Klebsiella oxytoca*, *Pseudomonas cepacia*, and *Pseudomonas aeruginosa* [44, 45]. In the case of *K. oxytoca* FA was used as a carbon source. Genes responsible for detoxification have been cloned from those species. *Stenotrophomonas maltophilia* displays an intrinsic resistance to fusaric acid that is due to a tripartite fusaric acid efflux pump, FuaABC. A *fuaABC* knockout mutant lost their fusaric acid resistance [46].

FA and dehydrofusaric acid showed activity against clinical trophozoites of the genus *Acanthamoeba*. This protozoon can cause keratitis due to contact lenses [47, 48], encephalitis in immunocompromised patients, and diseases caused by their endosymbiotic pathogens [49]. Some authors have suggested that FA could be used as an acanthamoebicide because of its low toxicity to humans [50, 51].

The best-known effects of FA on animals are vomit and hypotension. Feeding young dogs with FA caused low appetite, vomiting, hypotension, and suppressed weight gain. Moreover, at high doses FA was lethal, causing significant hypotension and gastrointestinal, hepatic and pneumonic bleeding [52, 53]. FA also caused vomiting, lethargy, and neurochemical changes in swine [54]. Feeding pigs with grains contaminated with both deoxynivalenol (DON, vomitoxin) and different FA concentrations provoked a reduction in weight gain, demonstrating a toxicological synergism between both toxins [55]. Such toxicological potentiation between DON and FA has been demonstrated in piglets where DON toxicity was augmented when FA was added in diet [55].

In addition to the hypotensive and vomiting effects, FA also increases brain serotonin concentrations in a manner similar to other *Fusarium* mycotoxins, such as DON and T-2 toxin [56]. As a consequence of serotonin increase in the brain, FA and DON administration provoked appetite loss, lethargy, and loss of muscle coordination in pigs [57]. Tryptophan is carried in blood bound to albumin but enters the brain in a free form. Because FA is derived from tryptophan, both compounds compete to bind to albumin, increasing the amount of free tryptophan in blood and increasing the synthesis of brain serotonin [58]. The effect of both toxins is the same although the biochemical mechanism is different for each toxin.

FA has many different pharmacological effects [59], as demonstrated on neurotransmitters in the brain. Its presence in the diet elevated the level of serum melatonin in pineal cell monolayer cultures [60]. When FA was administrated intraperitoneally, it provoked an increase in the levels of serotonin, tyrosine, and dopamine in the brain, and it decreased the level of norepinephrine [59]. The effect of FA on neurotransmitters may contribute to its toxicity.

FA depresses blood pressure, especially in stressed animals. FA calcium salt was assayed in elderly hypertensive Japanese with satisfactory results [51]. Trials carried out with FA derivatives also lowered blood pressure but provoked side effects [59]. FA inhibits the luteinizing hormone [59] and inhibits both in vitro and in vivo dopamine β(beta)-monooxygenase (DBM), the enzyme that converts dopamine into (R)-noradrenaline [61]. It has been demonstrated that the hypotensive effect of FA is explained by the inhibition of DBM [62].

FA has been shown to be cytotoxic to different cell lines including dog kidney fibroblast, rat hepatoma, and Chinese hamster ovary [63]. On human cells the cytotoxic effect of FA was more evident in cell lines from colon and adenocarcinoma than on epidermoid carcinoma cells [64]. Cells derived from a squamous carcinoma were reduced in number when treated with FA or in combination with paclitaxel or carboplatin [65]. Paclitaxel, isolated from *Taxus brevifolia*, prevents microtubule de-polymerization arresting cells in the G2-M phase, and carboplatin intercalates into DNA and interferes with DNA synthesis during S phase.

Saffron root cells treated with 50–100 μM FA suffered DNA fragmentation. Increases in FA concentration provoked chromatin condensation, nucleus budding, and destruction. Most of the cells treated with 100 μM FA were apoptotic. However, However, high doses of FA stimulated necrosis [66]. FA provoked release of cytochrome *c* to the cytosol and production of H_2O_2, but these effects were not observed in the presence of caspase inhibitors. Additionally, FA induces programmed cell death (PCD) in tobacco suspension cells [67]. The cell death increased with both FA concentration and incubation time. FA-treated cells have dilated endoplasmic reticulum cisternae, ruptures of mitochondrial membranes, condensate heterochromatin, and fragmentized DNA.

Most fungi, including *Penicillium* spp., *Aspergillus fumigatus*, *Cladosporium warneckii* [68], and *Fusarium* producers, transform FA into a compound with reduced phytotoxicity, 5-*N*-(3-hydroxybutyl)-pyridine-2-carboxylic acid. However, FA is not degraded in old cultures of *F. oxysporum* [45]. Plants are able to modify the chemical structure of several mycotoxins to defend themselves against xenobiotics [69]. Thus, FA is metabolized into less toxic compounds such as *N*-methyl fusaric acid amide (fusaric acid methylamide). A correlation between FA detoxification ability and the resistance to *F. oxysporum* f. sp. *lycopersici* has been described [70].

Distribution of Gene Clusters

Most of the fungal secondary metabolite (SM) genes are organized in gene clusters [71]. Only recently the fusarin and FA gene clusters have been identified by microarray and gene knockout approaches [1, 14, 72, 73]. With the increasing number of available fungal genome sequences, the distribution of these gene clusters among closely related *Fusaria* on one hand, and distantly related fungal species on the other hand, can now be studied by comparative genomics.

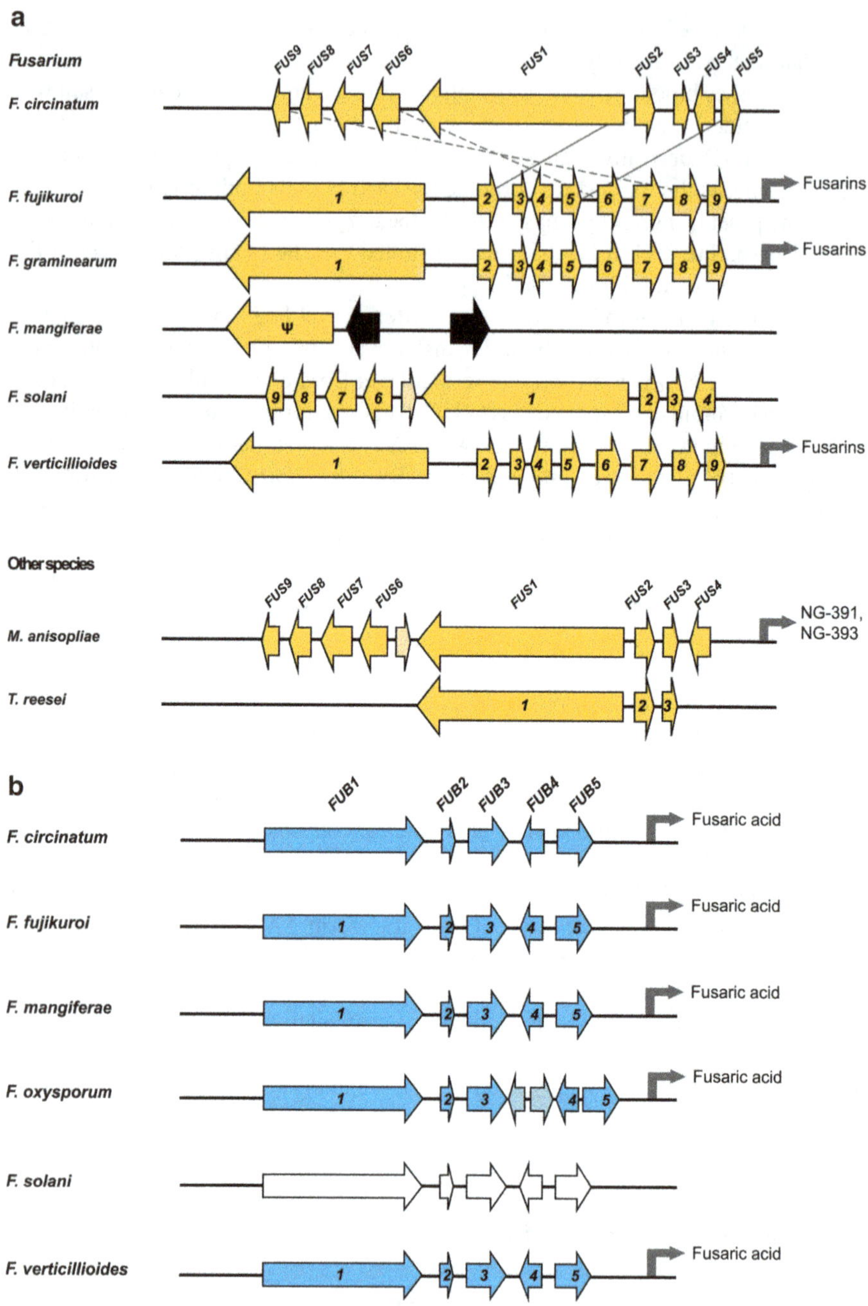

Fig. 11.2 Distribution of fusarin and fusaric acid gene clusters in *Fusaria* and other ascomycetes. (**a**) The fusarin gene cluster consisting of nine genes, *FUS1-FUS9*. Genes within a cluster are depicted as *arrows*; *direction* indicates direction of transcription. The numbers within these arrows correspond to *FUS* gene numbers. *Yellow arrows* indicate presence of the respective gene. *Black arrows* stand for genes that do not belong to the original fusarin gene cluster. There are two types of organization of the *FUS* cluster: *F. circinatum*, *F. solani*, and *M. anisopliae* have the same order and

Fusarin C Gene Cluster

The first described fusarin biosynthetic gene, *FUSS*, was identified in *F. moniliforme* and *F. venenatum* and encodes an unusual hybrid of a type I polyketide synthase (PKS) fused to a nonribosomal peptide synthetase (NRPS) module [74]. Further studies led to the discovery of *FUSS* homologs in *F. graminearum* (*GzFUS1*), *F. verticillioides* (*fus1*) and *F. fujikuroi* (*fusA*), respectively [72, 75, 76]. Recently, the entire fusarin gene cluster consisting of nine co-regulated genes (*FUS1-FUS9*) has been characterized in *F. fujikuroi*. However, only four of them (*FUS1*, *FUS2*, *FUS8*, and *FUS9*) are involved in the biosynthesis of the fusarins [73]. *FUS1* encodes the key enzyme, the hybrid PKS/NRPS. The other genes encode enzymes with similarities to an α(alpha)/β(beta) hydrolase with a predicted peptidase domain (*FUS2*), a glutathione *S*-transferase (*FUS3*), a peptidase A1 (*FUS4*), a serine hydrolase (*FUS5*), a major facilitator superfamily (MFS) transporter (*FUS6*), an aldehyde dehydrogenase (*FUS7*), a cytochrome P450 monooxygenase (*FUS8*), and the recently characterized methyltransferase (*FUS9*) [73, 77]. Besides *F. fujikuroi* other *Fusarium* species inside and outside of the GFC contain the whole fusarin cluster or parts of it; e.g., *F. circinatum*, *F. graminearum*, *F. solani*, and *F. verticillioides* ([1, 10, 76, 78] (Fig. 11.2a), and also the recently sequenced genome of *F. proliferatum* (Tudzynski, Güldener and coworkers, unpublished data).

Beside the members of the genus *Fusarium*, a similar gene cluster or parts of it have been identified in *Metarhizium anisopliae* and *Trichoderma reesei* [4, 10] (Fig. 11.2a). In the distantly related fungus *M. anisopliae* this gene cluster is responsible for the production of NG-391 and NG-393, 7-desmethyl analogs of fusarin C and (8*Z*)-fusarin C, respectively ([4]). *T. reesei* has only three genes left: *FUS1*, *FUS2*, and *FUS3* (Fig. 11.2a), and a fusarin-like product has not been identified in this fungus.

During its history of evolution, the fusarin cluster has undertaken one obvious rearrangement resulting in two different arrangements of cluster genes: first as in *F. circinatum*, *F. solani*, and *M. anisopliae*, and second, as in *F. fujikuroi*, *F. graminearum*, and *F. verticillioides* (Fig. 11.2a). Phylogenetic analyses of *FUS1* indicate that the second rearrangement is descended from the first rearrangement after the divergence of the genus *Fusarium* from the ancestor strain of *Metarhizium* [10]. Interestingly, both *F. solani* and *M. anisopliae* have an additional, hypothetical gene with unknown function in their clusters (Fig. 11.2a). However the fusarin-like gene cluster of *F. solani* is still uncharacterized.

Fig. 11.2 (continued) the probably later rearrangement in *F. fujikuroi*, *F. graminearum*, and *F. verticillioides*. Other species like *M. anisopliae* have a similar gene cluster and this fungus produces NG-391 and NG-393 (7-desmethyl analogs of fusarin C and (8*Z*) fusarin C). *T. reesei* has only *FUS1*, *FUS2*, and *FUS3* left, since now no fusarin-like product has been characterized. (**b**) The fusaric acid biosynthetic gene cluster consists of five genes, *FUB1-FUB5*. *Horizontal arrows* and their direction represent the direction of transcription. The numbers within these arrows correspond to *FUB* gene numbers. In *F. oxysporum*, the *FUB* cluster is interrupted by two genes (*light blue*). White arrows indicate that no fusaric acid production was detected, or no detailed characterization of the genes has been done yet

The differences in gene cluster organization between two closely related members of the GFC (*F. circinatum* and *F. fujikuroi*) are surprising and untypical. Furthermore, although *F. fujikuroi* and *F. circinatum* have both the whole gene cluster, only *F. fujikuroi* is able to produce fusarins [10]. On the other hand, *F. mangiferae*, belonging to the Asian clade of the GFC together with *F. fujikuroi*, has only remnants of the fusarin cluster that is not functional (Fig. 11.2a).

Fusaric Acid Gene Cluster

Recently, the first FA biosynthetic gene, *FUB1*, has been identified in *F. verticillioides* by gene replacement, and the putative FA gene cluster was assumed to consist of five genes based on their co-regulation in a microarray experiment (*FUB1-FUB5*) [1]. The *FUB* genes encode a PKS (Fub1), an uncharacterized protein (Fub2), a putative aspartate kinase (Fub3), a putative serine hydrolase (Fub4), and a putative homoserine *O*-acyltransferase (Fub5). In *F. fujikuroi*, *FUB1* has also been deleted resulting in total loss of FA production. However, deletion of the other four assumed biosynthetic genes, *FUB2-FUB5*, revealed that all mutants except for Δ(Delta)*FUB4* still produced FA suggesting that only two of the co-regulated genes, *FUB1* and *FUB4*, are responsible for FA biosynthesis [14]. To confirm this unexpected result, a multiple Δ(Delta)*FUB*2-5 deletion mutant expressing only the PKS-encoding gene *FUB1*, was transformed with *FUB4*. The generated strains containing only the two cluster genes *FUB1* and *FUB4* were able to produce FA—though in lower amounts than the wild type [14].

The production of FA is broadly distributed among the entire genus *Fusarium* (Table 11.1).

The organization of the FA gene cluster is highly conserved in the species belonging to the GFC (*F. fujikuroi*, *F. mangiferae*, and *F. circinatum*, *F. verticillioides*) (Fig. 11.2b). The FA gene cluster in *F. oxysporum* contains two additional genes (Fig. 11.2b) [10]. The function of these genes is not yet clear.

F. graminearum lacks the entire FA cluster and has never been described as FA producer. On the other hand, some *F. solani* strains were found to produce FA [8], but the sequenced *F. solani* isolate does not contain a FA gene cluster (Fig. 11.2b). However, it is possible that other isolates of the same species contain the *FUB* genes.

Methods for Detection and Structural Analysis

The risk of food and feed contamination by mycotoxins such as fusarins and FA is an important food safety concern for many field crops. Therefore, the development of rapid and reliable determination methods for each known mycotoxin is important to provide exposure data for risk assessments.

Fusarin C

Up to now, only limited data about the occurrence of fusarin C and its analogs in food and feed samples is available. A recent study by Kleigrewe et al. showed that the fusarin C levels varied in food samples from not detectable to 28 μg/kg, and in kernels of corn ears from not detectable to 83 mg/kg [79]. In the 1980s fusarin C levels of 0.39 mg/kg were found in corn samples, which caused equine leucoencephalomalacia in the USA [80]. Furthermore, fusarin C was detected in corn samples of the Transkei region, South Africa, and in Linxian County, China [18, 81].

All fusarins have a 2-pyrrolidone ring and a polyene chain as typical structural elements in common. Due to UV light the pentaene chain can easily rearrange to form (Z)-isomers [23, 82, 83]. *Epi*-fusarin C is formed chemically by epimerization via an open-chain form [84]. Fusarin A is formed in the presence of glutathione either chemically or enzymatically [24] (Fig. 11.1a). Fusarin D was recently shown to be formed by chemical rearrangement. When fusarin C and *epi*-fusarin C stock solutions were heated to 100 °C for 1 h, HPLC-UV-FTMS/MS analysis revealed the formation of fusarin D under these conditions (K. Kleigrewe, unpublished).

Fusarin C is a yellow oil and has an optical rotation of [α(alpha)]^{23} +47.04 (2.0 % in methanol) [85]. The reported extinction coefficients vary from 25,655 at 365 nm, 32,000 at 358 nm to 33,000 at 360 nm depending on the solvent and the age of the stock solution [5, 6, 85, 86]. Fusarin A has an extinction coefficient from 21,300 at 352 nm [5, 87]. Fusarin C can be detected with normal phase or reversed phase by UV-detection at 350–360 nm. As column material C18-, silica- or cyanophase can be used [5, 7, 79, 80, 84, 86, 88–91]. Thin-layer chromatography methods use solvent mixtures of chloroform/methanol (19/1, v/v), methylene chloride/methanol (95/5, v/v), or diethyl ether/ethyl acetate/methanol (25/25/1, v/v/v) [5, 18, 20, 82, 92, 93]. After silylation fusarin C is detectable with gas chromatography (GC) coupled to a flame ionization detector (FID) [91]. Mass spectrometry (MS) with electron ionization (EI) was used to confirm the presence of fusarin C in fungal extracts [82, 91]. Fusarin C was one of the 474 mycotoxins that could be detected by a HPLC-UV-electrospray (ESI)-MS multi method [94].

To detect fusarin C in food and feed, a new HPLC-MS/MS method in the multiple reaction monitoring mode (MRM) has been recently developed. For quantitation either a matrix-matched calibration or a stable isotope dilution assay was used [77, 79]. Due to different fragmentation mechanisms, fusarin C and its analogs can be differentiated by its characteristic MS^2 fragmentation patterns of the sodium-adduct [79, 84].

In the last years, monoclonal antibodies were developed to detect fusarin C and A. However, this enzyme-linked immunosorbent assay was never used to analyze food and feed samples so far [95].

Sample preparation to detect fusarin C usually involves the extraction of the toxin with an organic solvent and a purification step, which either consists of a liquid–liquid extraction or a solid phase extraction. To detect fusarin C in food and feed the sample preparation is performed with a dispersive solid phase extraction, also known as QuEChERS [77, 79]. The limit of detection is 1 µg/kg and the limit of quantitation is 4 µg/kg [77]. Former methods suspended ground corn with 50 mL water, which were extracted with methylene chloride/2-propanol (1/1, v/v) [80]. The extract was concentrated and alternately extracted with petroleum ether and chloroform. The petroleum ether fraction was afterwards extracted with acetonitrile, which was combined with the chloroform fraction. The extract was fractionated with a silica column and methanol/methylene chloride (1:19) as solvent. Afterwards fusarin C was detected by normal phase HPLC at 360 nm [80].

As internal standard for the detection of fusarin C phenothiazine, phenazine-1-carboxylic acid and d_3-fusarin C are reported in literature [77, 90]. The synthesis of d_3-fusarin C was achieved by isolating carboxyfusarin C from a *F. fujikuroi* mutant, deficient of an *O*-methyltransferase (Δ[Delta]*FUS9*), which was than derivatized by a simple chemical reaction with d_3-labeled diazomethane [77].

Since fusarin C is not commercially available, the method of choice for large-scale isolation of fusarin C and its analogs is the combination of solid phase extraction of culture filtrates and a final normal-phase HPLC separation step [84].

Fusarin C is not stable when exposed to UV light, high temperatures, and extreme pH values. Under UV light fusarin C rearranges to form (6Z)-, (8Z)-, and (10Z)-fusarin C and over a longer time period of exposure it degrades completely [17, 20, 86, 88, 90, 96, 97]. A stock solution of fusarin C is stable when stored at −80 °C in the dark in methanol/water (50/50, v/v) [79]. Jackson et al. reported a degradation of fusarin C when stored at −20 °C within 10 days [90].

Fusarin C rearranges under reversed phase chromatographic conditions. The involvement of the 2-pyrrolidone ring in the rearrangement of fusarin C was demonstrated by the synthesis of 15-methoxy-fusarin C and the isolation of an open-chain fusarin C. On the basis of these data and due to detailed NMR measurements and density functional theory calculations (DFT), the rearrangement product of fusarin C was identified as *epi*-fusarin C [84]. A summary of all fusarin structures can be found in the publication of Kleigrewe et al. [84].

Fusaric Acid

The analysis of FA emerged in the 1950s and was based on bioassays or paper disk chromatography [98]. A first quantitative approach used spectro-photometric detection of the paper extract [98]. In the following years, for instance, paper chromatography with detection of copper complexes of FA [99, 133] and optical density measurements after extraction of the culture filtrate were used to estimate the FA amounts [100, 110, 136–137]. Later screening used thin-layer chromatography with UV-detection or bio assays [101, 134, 135].

The first method for determination in biological fluids used GC coupled to a flame ionization detector (FID). FA was methylated on-column and could be detected down to a concentration of 1 µg/mL [102].

Present instrumental analytical methods are either GC- or HPLC-based. Prior to GC analysis FA is derivatized and then detected using a mass spectrometer [8, 103–105] or FID [104]. Without prior derivatization, HPLC coupled to UV is useful [8, 11, 106, 138]. There are also several methods for multi-mycotoxin analaysis using HPLC-MS/MS reported that analyse, amongst others, FA are also reported [107–109, 139, 140]. A recent publication describes HPLC-HRMS for an estimation of the There are also several methods for multi-mycotoxin analaysis using HPLC-MS/MS reported that analyse, amongst others, FA content besides other mycotoxins in fungal culture media by using the culture filtrate directly for injection [10].

The bioassay using the bioluminescent *Vibrio qinghaiensis* sp. Nov. Q67 was presented as an alternative for the quantitation of FA production of fungal cultures by instrumental analysis. This method is applicable down to a concentration of 5 μg/mL [110]. In summary, there are only few methods for the analysis of FA, which might be due to missing legal limits for this mycotoxin in food and feed.

Besides FA, some analogs have been identified in culture fluids of producing fungi. Thus, fusarinolic acid [111, 112] and dehydrofusaric acid [113] are found together with FA in cultures of *Gibberella fujikuroi* [112], *F. fujikuroi* [14] or *F. nygamai* where they occur together with their methyl esters [13]. 10,11-dihydroxyfusaric acid and diacid of FA have been identified in a *Fusarium moniliforme* strain [12].

Biosynthetic Pathways

Fusarin C

The study of the biosynthesis of fusarins has its beginning years ago. Feeding experiments with ^{13}C-labeled acetate could demonstrate that the methyl groups of C-20, C-22, C-23, and C-24 are resulting from *S*-adenosylmethionine (*S*AM) using the *C*-methyltransferase (*C*MeT) domain of the PKS (Steyn and Vleggaar 1985) (Fig. 11.1a). The methyl group of C-21 derived from the methyltransferase Fus9 [73, 77].

By combining single and multiple gene deletions, overexpression of cluster genes and co-cultivation of knockout mutants with different genetic blocks, Niehaus et al. recently provided evidence that only four of the nine co-regulated cluster genes (*FUS1*, *FUS2*, *FUS8*, and *FUS9*) are essential for the production of fusarins [73]. The PKS part of the hybrid PKS/NRPS enzyme is responsible for the condensation of one acetyl CoA and six malonyl CoA units [87]. Feeding studies with [1,2-$^{13}C_2$,^{15}N]-L-homoserine revealed the incorporation of this amino acid into the structure of fusarin C [114]. The NRPS part of the hybrid PKS/NRPS is able to activate this amino acid. After the activation, homoserine is attached to the peptidyl carrier domain. The PKS/NRPS forms the amide linkage between the heptaketide part of the PKS and the activated homoserine of the NRPS part [114]. The PKS/NRPS product is released as an alcohol, prefusarin, which is then hydroxylated by the P450 monooxygenase Fus8 at carbon C-20 to 20-hydroxy-fusarin ([73]). Only after this oxidation step, Fus2 catalyzes the formation of the 2-pyrrolidone ring what enables the P450 monooxygenase Fus8 to catalyze additional oxidation steps at the C-20 atom. At last, the methyltransferase Fus9 methylates the hydroxy group at C-21 and forms fusarin C [73] (Fig. 11.3).

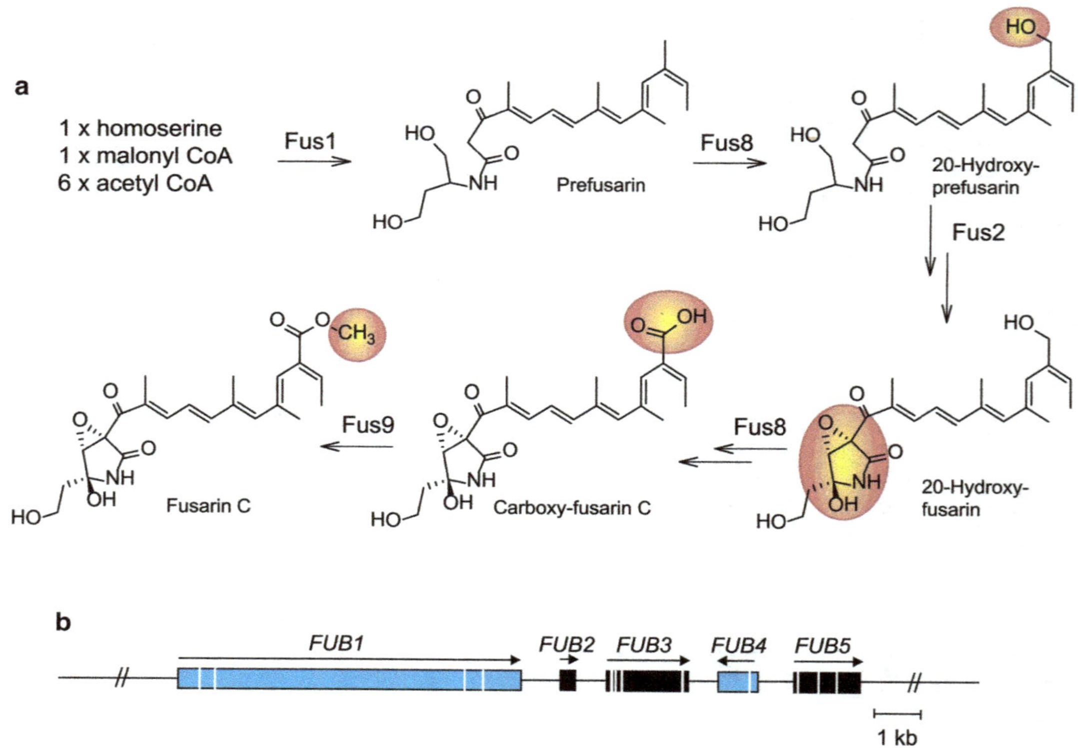

Fig. 11.3 Fusarin biosynthetic pathway and the fusaric acid gene cluster. (**a**) After formation of prefusarin by the key enzyme, the PKS/NRPS Fus1, the P450 monooxygenase Fus8 oxidizes this first intermediate at C-20 to form 20-hydroxy-prefusarin. After that, the putative multifunctional enzyme Fus2 is involved in the formation of the 2-pyrrolidone ring, the oxidation at C-15, and the epoxidation at C-13 and C-14 (20-hydroxy-fusarin). Then, Fus8 oxidizes C-20 to form the carboxylic acid group (carboxyfusarin C). The last step is the methylation by the methyltransferase Fus9 of the hydroxyl group of C-21 to form fusarin C. (after Niehaus et al. [73]). (**b**) The fusaric acid cluster contains five co-regulated genes, but only two of them (blue color) are responsible for the production of fusaric acid

Fusaric Acid

The biosynthesis of FA and its analogs is not yet elucidated although it has been investigated for several years. Feeding studies with FA-producing fungi *Gibberella* (*Fusarium*) *fujikuroi* and *F. oxysporum* using radioactively labeled substrates showed that acetate or related metabolites are precursor-molecules in FA biosynthesis [100, 115, 116]. Further experiments using ^{14}C and ^{15}N labeled aspartate suggested that the nitrogen is transferred to an oxalacetate or oxalacetate-like compound by transamination and then incorporated into fusaric acid [115]. These early findings were confirmed by recent experiments. $^{13}C_2$-acetate, ^{13}C and ^{15}N labeled aspartate, and ^{15}N glutamine were fed to *F. oxysporum* f. sp. *vasinfectum* and yielded to the hypothesis that three acetate units and one tricarboxylic acid cycle (TCA)-derivative are important precursors in the FA biosynthesis with glutamine as a more likely nitrogen source than aspartate [28]. Pitel and Vining proposed that FA might only be an intermediate of the biosynthesis as they observed an enrichment of 9,10-dehydrofusaric acid compared over the time [112].

As the polyketide gene cluster responsible for FA biosynthesis has been recently identified in *F. verticillioides* (FVEG_12523—FVEG_12519) ([1]) and *F. fujikuroi* [14] the biosynthetic pathway can be studied now in more detail by using *FUB* gene deletion and overexpression mutants. Based on the data known so far, Niehaus et al. [14] showed that only two of the five co-regulated genes in the FA cluster are sufficient for FA production. Beside the PKS-encoding gene *FUB1* only *FUB4* encoding a putative hydrolase is required for the FA biosynthesis. Previously, the FA biosynthetic pathway has been postulated [1]. According to this hypothetical pathway, the PKS Fub1 condensates three acetate units. This triketide together with glutamine and oxaloacetate are putative building blocks that are possibly transformed to FA by the aspartate kinase Fub3 and the hypothetical acetyltransferase Fub5. The methylester might be cleaved by the hydrolase Fub4 while Fub2 is supposed to have no role in the biosynthesis [1]. However, after the finding that two genes are sufficient for FA biosynthesis [14], the real role of *FUB2*, *FUB3*, and *FUB5* remains mysterious.

Until now, analytical data for structures of the intermediates are still missing. As the expected intermediates of the pathway are probably quite usual biomolecules from the primary metabolism (e.g., an amino acid and acetate) [14, 1], elucidation of the biosynthetic pathway appears to be quite challenging.

Regulation of the Cluster Gene Expression

Nitrogen Regulation

Many SM gene clusters were shown to be regulated by nitrogen availability in *F. fujikuroi* [10]. Thus, the two PKS-derived pigments, bikaverin and fusarubins, and the diterpenoid gibberellins are repressed by nitrogen [117–119]. In contrast, the two mycotoxins fusarin C and FA are activated under nitrogen excess conditions

[14, 72, 73, 112, 120]. While fusarins are preferentially produced at acidic pH conditions (60 mM glutamine) [73], FA and its derivatives are produced under both alkaline and acidic pH values (60 mM glutamine, acidic; 120 mM $NaNO_3$, alkaline) [14]. Indeed, the composition of these mycotoxins in cultures of *F. fujikuroi* differs significantly in both conditions: FA is almost exclusively produced in media with alkaline pH, while in media with glutamine (acidic pH) a mixture of FA and two derivatives, fusarinolic acid and 9,10-dehydrofusaric acid, accumulates [14].

Although both mycotoxins are induced by nitrogen, the mechanism of nitrogen regulation seems to differ: The nitrogen responsive GATA transcription factors AreA and AreB have only a slight impact on the expression of fusarin genes [73]. However, expression studies showed that the FA cluster genes are positively regulated by AreB [14]. This is the first example that a gene cluster is activated by nitrogen excess in an AreB-dependent manner.

Additionally, expression of fusarin genes depends on the presence of an active glutamine synthetase (GS) [73]. Although the GS is not a transcription factor, it seems to have regulatory functions in addition to its enzymatic activity as the only glutamine-forming enzyme. Besides fusarins, also the expression of nitrogen-repressed gibberellin-and bikaverin genes depends on the GS [121, 122].

Regulation Via the Fungal-Specific Velvet Complex

The fungal-specific *velvet* complex is well known as global regulator of secondary metabolism and differentiation in fungi [123]. The core components of this complex are two members of the velvet protein family, VeA and VelB, and the putative histone methyltransferase LaeA. In some fungi, additional proteins such VosA and VelC, were shown to belong to the complex [123].

Deletion of the FvVe1 homologue in *F. verticillioides* resulted in reduced production of fusarins [124]. The same effect was shown for the homologue FfVel1 in *F. fujikuroi* [73, 125]. Also expression of FA biosynthetic genes and biosynthesis of FA depends on the *velvet* complex. The *F. oxysporum* Δ(Delta)*veA* mutant produces less FA in human blood [126], and the *F. fujikuroi* Δ(Delta)*vel1* mutant produces significantly reduced amounts of FA in axenic cultures [14]. Compared to VeA (Vel1), the VelB (Vel2) protein has only a minor effect on many SMs. Deletion of *vel2* in *F. fujikuroi* resulted in slightly reduced production of fusarins [73] but had almost no effect on FA formation [14]. In contrast to the Velvet proteins, which regulate both secondary metabolism and differentiation, LaeA acts as a global transcriptional regulator mainly of SM clusters in several fungi. Similarly to what was observed in *F. verticillioides*, the putative methyltransferase FfLae1 is an activator of the fusarin genes in *F. fujikuroi* [73, 125, 127]. Also the FA genes are downregulated in the Δ(Delta)*laeA* mutant in *F. verticillioides*, *F. oxysporum*, and *F. fujikuroi* [14, 126, 127].

Epigenetic Control: The Role of Histone Modifications

SM clusters in fungi tend to be located in subtelomeric regions of chromosomes. These regions often show an increased efficiency of epigenetic regulation such as histone acetylation and methylation [128]. Genome-wide studies in several fungi revealed that chromatin modifications differ in regions with active (euchromatin) and silent (heterochromatin) gene transcription [10, 129]. A large number of enzymes can modify histones at multiple lysine residues resulting in activation or inactivation of specific regions. Thus, active gene expression has been associated with acetylation of histone H3 lysine 9 (H3K9ac) and dimethylation of histone H3 lysine 4 (H3K4me2), whereas gene silencing has been associated with trimethylation of histone H3 lysine 9 (H3K9me3). Several silent SM gene clusters in fungi were shown to be activated by deletion or overexpression of genes encoding chromatin-modifying enzymes, such as histone acetylases, deacetylases, and methylases [129, 130].

In order to show whether the contrasting expression of SM genes under high and low nitrogen conditions is linked with the chromatin landscape, a genome-wide chromatin immunoprecipitation (ChIP) experiment followed by high-throughput sequencing ("ChIP-seq") was recently performed in *F. fujikuroi* [10]. For several gene clusters (e.g., those for gibberellins and bikaverin) a correlation was found between gene expression under low nitrogen conditions and acetylation of histone 3 at lysine 9 (H3K9ac). In the case of the fusarin genes, acetylation of H3K9ac was enriched across the gene cluster under activating high nitrogen conditions [73]. The FA cluster showed a little enrichment for H3K9Ac under inducing (high levels of nitrogen) conditions [10].

Conclusion

Future Perspectives

Entire fusarin C and FA gene clusters have been identified in *F. fujikuroi*. The first one has suffered different rearrangements during evolution in many *Fusarium* and other fungal species, while the FA cluster has been found to be very homogeneous in *Fusarium* spp. Although many years have been needed to identify and characterize the gene clusters for fusarin and FA biosynthesis, there are still some questions open. Thus, the role of hydrolase-like Fus2 in the formation of the 2-pyrrolidone ring in fusarin C biosynthesis is still not well understood. Furthermore, protein Fub2, coded by the FA cluster, does not have a homologue in any fungal genome that has been characterized yet. Two other enzymes, the putative aspartate kinase Fub3 and the putative homoserine *O*-acyltransferase, Fub5, do not seem to have catalytic functions in FA biosynthesis. Understanding the role of these enzymes would help to decipher the fusarin and FA biosynthetic pathways in more detail.

Although regulation of *FUS* and *FUB* gene expression has been intensively studied since the identification of the two gene clusters, these investigations are not finished as well as the discovery of new regulators and regulatory networks is an ongoing process.

References

1. Brown DW, Butchko RAE, Busman M, Proctor RH. Identification of gene clusters associated with fusaric acid, fusarin, and perithecial pigment production in *Fusarium verticillioides*. Fungal Genet Biol. 2012;49:521–32.
2. Glen AE. Mycotoxigenic *Fusarium* species in animal feed. Animal Feed Sci Technol. 2007;137:213–40.
3. Morgavi DP, Riley RT. An historical overview of field disease outbreaks known or suspected to be caused by consumption of feeds contaminated with *Fusarium* toxins. Animal Feed Sci Technol. 2007;137:201–12.
4. Krasnoff SB, Sommers CH, Moon Y-S, Donzelli BGG, Vandenberg JD, Churchill ACL, et al. Production of mutagenic metabolites by *Metarhizium anisopliae*. J Agric Food Chem. 2006;54:7083–8.
5. Wiebe LA, Bjeldanes LF. Fusarin C, a mutagen from *Fusarium moniliforme* grown on corn. J Food Sci. 1981;46:1424–6.
6. Gelderblom WCA, Marasas WFO, Steyn PS, Thiel PG, Van Der Merwe KJ, Van Rooyen PH, et al. Structure elucidation of fusarin C, a mutagen produced by *Fusarium moniliforme*. J Chem Soc Chem Commun. 1984;7:122–4.
7. Savard ME, Miller DJ. Characterization of fusarin F, a new fusarin from *Fusarium moniliforme*. J Nat Prod. 1992;55:64–70.
8. Bacon CW, Porter JK, Norred WP, Leslie JF. Production of fusaric acid by *Fusarium* species. Appl Environ Microbiol. 1996;62:4039–43.
9. Abbas HK, Mirocha CJ, Kommedahl T, Vesonder RF, Golinski P. Production of trichothecene and non-trichothecene mycotoxins by *Fusarium* species isolated from maize in Minnesota. Mycopathologia. 1989;108:55–8.
10. Sieber CMK, von Bargen KW, Studt L, Niehaus E-M, Espino JJ, et al. Deciphering the cryptic genome: Genome-wide analyses of the rice pathogen *Fusarium fujikuroi* reveal complex regulation of secondary metabolism and novel metabolites. PLoS Pathog. 2013;9: e1003475.
11. Amalfitano C, Pengue R, Andolfi A, Vurro M, Zonno MC, Evidente A. HPLC analysis of fusaric acid, 9,10-dehydrofusaric acid and their methyl esters, toxic metabolites from weed pathogenic *Fusarium* species. Phytochem Anal. 2002;13:277–82.
12. Burmeister HR, Grove MD, Peterson RE, Weisleder D, Plattner RD. Isolation and characterization of two new fusaric acid analogs from *Fusarium moniliforme* NRRL 13,163. Appl Environ Microbiol. 1985;50:311–4.
13. Capasso R, Evidente A, Cutignano A, Vurro M, Zonno MC, Bottalico A. Fusaric and 9,10-dehydrofusaric acids and their methyl esters from *Fusarium nygamai*. Phytochemistry. 1996;41:1035–9.
14. Niehaus E-M, von Bargen KW, Espino JJ, Pfannmüller A, Humpf H-U, Tudzynski B. Characterization of the fusaric acid gene cluster in *Fusarium fujikuroi*. Appl Microbiol Biotechnol. 2014;98(4):1749–62.
15. Vischetti C, Esposito A. Degradation and transformation of a potential natural herbicide in three soils. J Agric Food Chem. 1999;47:3901–4.
16. Cheng SJ, Jiang YZ, Li MH, Lo HZ. A mutagenic metabolite produced by *Fusarium moniliforme* isolated from Linxian county, China. Carcinogenesis. 1985;6:903–5.

17. Gelderblom WCA, Thiel PG, van der Merwe KJ. Metabolic activation and deactivation of fusarin C, a mutagen produced by *Fusarium moniliforme*. Biochem Pharmacol. 1984;33: 1601–3.
18. Gelderblom WCA, Thiel PG, Marasas WFO, van Der Merwe KJ. Natural occurrence of fusarin C, a mutagen produced by *Fusarium moniliforme*, in corn. J Agric Food Chem. 1984;32:1064–7.
19. Desjardins AE. Other selected mycotoxins. In: Desjardins AE, editor. *Fusarium* mycotoxins. St. Paul: The American Phytopathological Society; 2006, pp. 109–29
20. Gelderblom WCA, Thiel PG, van der Merwe KJ, Marasas WFO, Spies HSC. A mutagen produced by *Fusarium moniliforme*. Toxicon. 1983;21:467–73.
21. Zhu B, Jeffrey AM, Fusarin C. Isolation and identification of two microsomal metabolites. Chem Res Toxicol. 1993;6:97–101.
22. Jaskiewicz K, van Rensburg SJ, Marasas WF, Gelderblom WC. Carcinogenicity of *Fusarium moniliforme* culture material in rats. J Natl Cancer Inst. 1987;78:321–5.
23. Gelderblom WCA, Thiel PG, Jaskiewicz K, Marasas WFO. Investigations on the carcinogenicity of fusarin C—a mutagenic metabolite of *Fusarium moniliforme*. Carcinogenesis. 1986;7:1899–901.
24. Gelderblom WCA, Thiel PG, van Der Merwe KJ. The role of rat liver microsomal enzymes in the metabolism of the fungal metabolite fusarin C. Food Chem Toxicol. 1988;26:31–6.
25. Sondergaard TE, Hansen FT, Purup S, Nielsen AK, Bonefeld-Jorgensen EC, Giese H, Sorensen JL. Fusarin C acts like an estrogenic agonist and stimulates breast cancer cells *in vitro*. Toxicol Lett. 2011;205:116–21.
26. Dong ZY, Zhan YH. Inhibitory effect of a mycotoxin, fusarin C, on macrophage activation and macrophage mediated cytotoxicity to tumor cells in mice. J Exp Clin Cancer Res. 1987;6:31–8.
27. Gäumann E. Fusaric acid as a wilt toxin. Phytopathology. 1957;47:342–57.
28. Stipanovic RD, Wheeler MH, Puckhaber LS, Liu J, Bell AA, Williams HJ. Nuclear magnetic resonance (NMR) studies on the biosynthesis of fusaric acid from *Fusarium oxysporum* f. sp. *vasinfectum*. J Agric Food Chem. 2011;59:5351–6.
29. Bekker EE, Dovletmuradov KD, Pushkareva ID, Poletaeva VF, Shilina SG, Yasakova EI. Nature and biosynthesis of the toxin of the causative agents of fusariosis wilt, the mechanism of its actions, and its possible transformation within the cotton plant. Izv Akad Nauk SSSR Ser Biol. 1971;5:749–54.
30. Becker SE, Pushkareva ID, Poletaeva VF, Shilina SG, Yasakova EI. Nature and biosynthesis of Fusarium wilt toxin, its mechanism of action, and its transformation in the cotton plant. Bodenkultur. 1972;23:256–71.
31. Shilina, SG, Bekker ZE, Goshaev MG. Isolation and comparative characterization of vivotoxin from wilt-infected cotton plants and of fusaric acid. Ekologo-Fiziologlcheskle Metody Bor'be Fuzarioznym Viltom Khlop. 1973;2:219–30. [*Chem. Abstracts, 84, 39502 (1976)*]
32. D'Alton A, Etherton B. Effects of fusaric acid on tomato root hair membrane potentials and ATP levels. Plant Physiol. 1984;74:39–42.
33. Marré MT, Vergani P, Albergoni FG. Relationship between fusaric acid uptake and its binding to cell structures by leaves of *Egeria densa* and its toxic effects on membrane permeability and respiration. Physiol Mol Plant Pathol. 1993;42:141–57.
34. Bouizgarne B, El-Maarouf-Bouteau H, Frankart C, Reboutier D, Madiona K, Pennarun AM, et al. Early physiological responses of *Arabidopsis thaliana* cells to fusaric acid: toxic and signalling effects. New Phytol. 2006;169:209–18.
35. Bouizgarne B, El-Maarouf-Bouteau H, Madiona K, Biligui B, Monestiez M, Pennarun AM, et al. A putative role for fusaric acid in biocontrol of the parasitic angiosperm *Orobanche ramosa*. Mol Plant Microbe Interact. 2006;19:550–6.
36. Dong X, Ling N, Wang M, Shen Q, Guo S. Fusaric acid is a crucial factor in the disturbance of leaf water imbalance in *Fusarium*-infected banana plants. Plant Physiol Biochem. 2012;60:171–9.

37. Fernández-Pol JA. Conservation of multifunctional ribosomal protein metallopanstimulin-1 (RPS27) through complex evolution demonstrates its key role in growth regulation in Archaea, eukaryotic cells, DNA repair, translation and viral replication. Cancer Genomics Proteomics. 2011;8:105–26.
38. Stack Jr BC, Hansen JP, Ruda JM, Jaglowski J, Shvidler J, Hollenbeak CS. Fusaric acid: a novel agent and mechanism to treat HNSCC. Otolaryngol Head Neck Surg. 2004;131:54–60.
39. Ramautar A, Mabandla M, Blackburn J, Daniels WM. Inhibition of HIV-1 tat-induced trans-activation and apoptosis by the divalent metal chelators, fusaric acid and picolinic acid—implications for HIV-1 dementia. Neurosci Res. 2012;74:59–63.
40. Schnider-Keel U, Seematter A, Maurhofer M, Blumer C, Duffy B, Gigot-Bonnefoy C, et al. Autoinduction of 2,4-diacetylphloroglucinol biosynthesis in the biocontrol agent *Pseudomonas fluorescens* CHA0 and repression by the bacterial metabolites salicylate and pyoluteorin. J Bacteriol. 2000;182:1215–25.
41. Bacon CW, Hinton DM, Porter JK, Glenn AE, Kuldau G. Fusaric acid, a *Fusarium verticillioides* metabolite, antagonistic to the endophytic biocontrol bacterium *Bacillus mojavensis*. Can J Bot. 2004;82:878–85.
42. Notz R, Maurhofer M, Dubach H, Haas D, Defago G. Fusaric acid-producing strains of *Fusarium oxysporum* alter 2,4-diacetylphloroglucinol biosynthetic gene expression in *Pseudomonas fluorescens* CHA0 *in vitro* and in the rhizosphere of wheat. Appl Environ Microbiol. 2002;68:2229–35.
43. Schouten A, van den Berg G, Edel-Hermann V, Steinberg C, Gautheron N, Alabouvette C, et al. Defense responses of *Fusarium oxysporum* to 2,4-diacetylphloroglucinol, a broad-spectrum antibiotic produced by *Pseudomonas fluorescens*. Mol Plant Microbe Interact. 2004;17:1201–11.
44. Toyoda H, Katsuragi K, Tamai T, Ouchi S. DNA sequence of genes for detoxification of fusaric acid, a wilt-inducing agent produced by *Fusarium* species. J Phytopathol. 1991;133:265–77.
45. Karlovsky P. Biological detoxification of fungal toxins and its use in plant breeding, feed and food production. Nat Toxins. 1999;7:1–23.
46. Hu RM, Liao ST, Huang CC, Huang YW, Yang TC. An inducible fusaric acid tripartite efflux pump contributes to the fusaric acid resistance in *Stenotrophomonas maltophilia*. PLoS One. 2012;7:e51053.
47. Booton GC, Visvesvara GS, Byers TJ, Kelly DJ, Fuerst PA. Identification and distribution of *Acanthamoeba* species genotypes associated with nonkeratitis infections. J Clin Microbiol. 2005;43:1689–93.
48. Boonman N, Prachya S, Boonmee A, Kittakoop P, Wiyakrutta S, Sriubolmas N, et al. *In vitro* acanthamoebicidal activity of fusaric acid and dehydrofusaric acid from an endophytic fungus *Fusarium* sp. Tlau3. Planta Med. 2012;78:1562–7.
49. Iovieno A, Ledee DR, Miller D, Alfonso EC. Detection of bacterial endosymbionts in clinical acanthamoeba isolates. Ophthalmology. 2010;117:445–52, 452.e1–3
50. Hidaka H, Nagatsu T, Takeya K, Takeuchi T, Suda H. Fusaric acid, a hypotensive agent produced by fungi. J Antibiot. 1969;22:228–30.
51. Terasawa F, Ying LH, Kameyama M. The hypotensive effect of fusaric acid: the results of long-term administration of fusaric acid in elderly hypertensive patients. Jpn Circ J. 1976;40: 1017–23.
52. Matsuzaki M, Yoshida A, Akutsu S, Tsuchida M, Okuyama D. Studies on toxicity of fusaric acid-Ca. IV. Chronic toxicity in dogs. Jpn J Antibiot. 1976;29:518–42.
53. Matsuzaki M, Yoshida A, Tsuchida M, Sekino M, Asano M. Studies on toxicity of fusaric acid-Ca. III. Subacute toxicity. Jpn J Antibiot. 1976;29:491–517.
54. Smith TK, MacDonald EJ. Effect of fusaric acid on brain regional neurochemistry and vomiting behavior in swine. J Anim Sci. 1991;69:2044–9.
55. Smith TK, McMillan EG, Castillo JB. Effect of feeding blends of *Fusarium* mycotoxin-contaminated grains containing deoxynivalenol and fusaric acid on growth and feed consumption of immature swine. J Anim Sci. 1997;75:2184–91.

56. Boermans HJ, Leung MC. Mycotoxins and the pet food industry: toxicological evidence and risk assessment. Int J Food Microbiol. 2007;119:95–102.
57. Leathwood PD. Tryptophan availability and serotonin synthesis. Proc Nutr Soc. 1987;46:143–56.
58. Chaouloff F, Laude D, Merino D, Serrurrier B, Elghozi JL. Peripheral and central short-term effects of fusaric acid, a DBH inhibitor, on tryptophan and serotonin metabolism in the rat. J Neural Transmission. 1986;65:219–32.
59. Wang H, Ng TB. Pharmacological activities of fusaric acid (5-butylpicolinic acid). Life Sci. 1999;65:849–56.
60. Rimando AM, Porter JK. Fusaric acid increases melatonin levels in the weanling rat and in pineal cell cultures. J Toxicol Environ Health. 1997;50:275–84.
61. Hidaka H. Fusaric (5-butylpicolinic) acid, an inhibitor of dopamine beta-hydrolase, affects serotonin and noradrenaline. Nature. 1971;231:54–5.
62. Dove S. Picolinic acids as inhibitors of dopamine β-monooxygenase: QSAR and putative binding site. Arch Pharm (Weinheim). 2004;337:645–53.
63. Vesonder RF, Gasdorf IE, Peterson RE. Comparison of the cytotoxicities of *Fusarium* metabolites and *Alternaria* metabolite AAL-toxin to cultured mammalian cell lines. Arch Environ Contam Toxicol. 1993;24:473–7.
64. Fernåndez-Pol JA, Klos DJ, Hamilton PD. A growth factor inducible gene encodes a novel protein with zinc finger structure. J Biol Chem. 1993;268:21198–204.
65. Jaglowski JR, Stack Jr BC. Enhanced growth inhibition of squamous cell carcinoma of the head and neck by combination therapy of fusaric acid and paclitaxel or carboplatin. Cancer Lett. 2006;243:58–63.
66. Samadi L, Shahsavan Behboodi B. Fusaric acid induces apoptosis in saffron root-tip cells: roles of caspase-like activity, cytochrome c, and H_2O_2. Planta. 2006;225:223–34.
67. Jiao J, Zhou B, Zhu X, Gao Z, Liang Y. Fusaric acid induction of programmed cell death modulated through nitric oxide signalling in tobacco suspension cells. Planta. 2013;238: 727–37.
68. Utsumi R, Hadama T, Noda M, Toyoda H, Hashimoto H, Ohuchi S. Cloning of fusaric acid-detoxifying gene from *Cladosporium werneckii*: a new strategy for the prevention of plant diseases. J Biotechnol. 1988;8:311–6.
69. Berthiller F, Crews C, Dall'Asta C, Saeger SD, Haesaert G, Karlovsky P, Oswald IP, Seefelder W, Speijers G, Stroka J. Masked mycotoxins: a review. Mol Nutr Food Res. 2013;57:165–86.
70. Kluepfel D. Über die Biosynthese und die Umwandlung von Fusarinsäure in Tomatenpflanzen. Phytopathol Z. 1957;29:349–79.
71. Keller NP, Hohn TM. Metabolic pathway gene clusters in filamentous fungi. Fungal Genet Biol. 1997;21:17–29.
72. Díaz-Sánchez V, Avalos J, Limón MC. Identification and regulation of *fusA*, The polyketide synthase gene responsible for fusarin production in *Fusarium fujikuroi*. Appl Environ Microbiol. 2012;78:7258–66.
73. Niehaus E-M, Kleigrewe K, Wiemann P, Studt L, Sieber CMK, Connolly LR, et al. Genetic manipulation of the *Fusarium fujikuroi* fusarin gene cluster yields insides into the regulation and fusarin biosynthetic pathway. Chem Biol. 2013;20:1055–66.
74. Song Z, Cox RJ, Lazarus CM, Simpson TJ. Fusarin C biosynthesis in *Fusarium moniliforme* and *Fusarium venenatum*. Chembiochem. 2004;5:1196–203.
75. Brown DW, Butchko RAE, Baker SE, Proctor RH. Phylogenomic and functional domain analysis of polyketide synthases in *Fusarium*. Fungal Biol. 2012;116:318–31.
76. Gaffoor I, Brown DW, Plattner R, Proctor RH, Qi W, Trail F. Functional analysis of the polyketide synthase genes in the filamentous fungus *Gibberella zeae* (anamorph *Fusarium graminearum*). Eukaryot Cell. 2005;4:1926–33.
77. Kleigrewe K, Niehaus E-M, Wiemann P, Tudzynski B, Humpf H-U. New approach via gene knockout and single-step chemical reaction for the synthesis of isotopically labeled fusarin C as an internal standard for the analysis of this *Fusarium* mycotoxin in food and feed samples. J Agric Food Chem. 2012;60:8350–5.

78. Donzelli BGG, Krasnoff SB, Churchill ACL, Vandenberg JD, Gibson DM. Identification of a hybrid PKS-NRPS required for the biosynthesis of NG-391 in *Metarhizium robertsii*. Curr Genet. 2010;56:151–62.
79. Kleigrewe K, Söhnel AC, Humpf H-U. A new high-performance liquid chromatography-tandem mass spectrometry method based on dispersive solid phase extraction for the determination of the mycotoxin fusarin C in corn ears and processed corn samples. J Agric Food Chem. 2011;59:10470–6.
80. Thiel PG, Gelderblom WCA, Marasas WFO, Nelson PE, Wilson TM. Natural occurrence of moniliformin and fusarin C in corn screenings known to be hepatocarcinogenic in rats. J Agric Food Chem. 1986;34:773–5.
81. Li M-X, Cheng S-J. Carcinogenesis of esophageal cancer in Linxian, China. Chin Med J. 1984;97:311–6.
82. Barrero AF, Sánchez JF, Enrique Oltra J, Tamayo N, Cerdá-Olmedo E, Candau R, et al. Fusarin C and 8Z-fusarin C from *Gibberella fujikuroi*. Phytochemistry. 1991;30:2259–63.
83. Eilbert F, Thines E, Arendholz WR, Sterner O, Anke H. Fusarin C, (7*Z*)-fusarin C and (5*Z*)-fusarin C. Inhibitors of dihydroxynaphthalene-melanin biosynthesis from *Nectria coccinea* (*Cylindrocarpon* sp.). J Antibiot. 1997;50:443–5.
84. Kleigrewe K, Aydin F, Hogrefe K, Piecuch P, Bergander K, Würthwein E, et al. Structure elucidation of new fusarins revealing insights in the rearrangement mechanisms of the *Fusarium* mycotoxin fusarin C. J Agric Food Chem. 2012;60:5497–505.
85. Gaddamidi V, Bjeldanes LF, Shoolery JN, Fusarin C. Structure determination by natural abundance ^{13}C-^{13}C coupling and deuterium-induced ^{13}C shifts. J Agric Food Chem. 1985;33:652–4.
86. Scott PM, Lawrence GA, Matula TI. Analysis of toxins of *Fusarium moniliforme*. In: Steyn PS, Vleggaar R, editors. Mycotoxins and phycotoxins. Amsterdam: Elsevier; 1985. p. 306–16.
87. Steyn PS, Vleggaar R. Biosynthetic studies on the fusarins, metabolites of *Fusarium moniliforme*. J Chem Soc Chem Comm. 1985;17:1189–91
88. Bever Jr RJ, Couch LH, Sutherland JB, Williams AJ, Beger RD, Churchwell MI, et al. DNA adduct formation by *Fusarium* culture extracts: lack of role of fusarin C. Chem Biol Interact. 2000;128:141–57.
89. Farber JM, Sanders GW. Production of fusarin C by *Fusarium* spp. J Agric Food Chem. 1986;34:963–6.
90. Jackson MA, Stewart JN, Peterson RE, Slininger PJ. Fusarin C purification and measurement in submerged cultures of *Fusarium moniliforme* by high-performance liquid chromatography. J Agric Food Chem. 1990;38:1511–4.
91. Tseng TC, Chung CS, Li I. Production of fusarin C mycotoxin by *Fusarium moniliforme* isolates of Taiwan. Bot Bul Acad Sinica. 1990;31:169–74.
92. Cantalejo MJ, Torondel P, Amate L, Carrasco JM, Hernández E. Detection of fusarin C and trichothecenes in *Fusarium* strains from Spain. J Basic Microbiol. 1999;39:143–53.
93. Farber JM, Sanders GW. Fusarin C production by North American isolates of *Fusarium moniliforme*. Appl Environ Microbiol. 1986;51:381–4.
94. Nielsen KF, Smedsgaard J. Fungal metabolite screening: database of 474 mycotoxins and fungal metabolites for dereplication by standardised liquid chromatography-UV-mass spectrometry methodology. J Chromatogr A. 2003;1002:111–36.
95. Maragos CM, Busman M, Plattner RD. Development of monoclonal antibodies for the fusarin mycotoxins. Food Addit Contam. 2008;25:105–14.
96. Lu F-X, Jeffrey AM. Isolation, structural identification, and characterization of a mutagen from *Fusarium moniliforme*. Chem Res Toxicol. 1993;6:91–6.
97. Zhu B, Jeffrey AM. Stability of fusarin C: effects of the normal cooking procedure used in China and pH. Nutr Cancer. 1992;18:53–8.
98. Zähner H (1954) Die Bestimmung der Fusarinsäure mit Hilfe der Papierchromatographie. Phytopathol Z 22:227–8.
99. Lakshminarayanan K, Subramanian D. Chromatographic detection and estimation of fusaric acid. Experientia. 1957;13:350–1.

100. Dobson TA, Desaty D, Brewer D, Vining LC. Biosynthesis of fusaric acid in cultures of *Fusarium oxysporum* Schlecht. Can J Biochem. 1967;45:809–23.
101. Paterson RRM, Rutherford MA. A simplified rapid technique for fusaric acid detection in *Fusarium* strains. Mycopathologia. 1991;113:171–3.
102. Fung KK, Koda RT, Maronde RF, Cohen JL. Rapid GLC determination of fusaric acid in biological fluids. J Pharm Sci. 1976;65:596–8.
103. Fogliano V, Monti SM, Ferracane R, Ambrosino P, Grottola A, Ritieni A. Rapid fusaric acid analysis by GC-MS. Annali Della Facolt. Di Agraria Della R. Universit. Di Napoli Federico II—Portici. 1999;33:9–14
104. Meyer K, Wenz S, Bauer J. Vorkommen und Nachweis von Fusarinsäure in Weizen. Mycotox Res. 2002;18:70–3.
105. Porter JK, Bacon CW, Wray EM, Hagler Jr WM. Fusaric acid in *Fusarium moniliforme* cultures, corn, and feeds toxic to livestock and the neurochemical effects in the brain and pineal gland of rats. Nat Toxins. 1995;3:91–100.
106. Smith TK, Sousadias MG. Fusaric acid content of swine feedstuffs. J Agric Food Chem. 1993;41:2296–9.
107. Mogensen JM, Sørensen SM, Sulyok M, van der Westhuizen L, Shephard GS, Frisvad JC, et al. Single-kernel analysis of fumonisins and other fungal metabolites in maize from South African subsistence farmers. Food Addit Contam Part A Chem Anal Control Expo Risk Assess. 2011;28:1724–34.
108. Shimshoni JA, Cuneah O, Sulyok M, Krska R, Galon N, Sharir B, et al. Mycotoxins in corn and wheat silage in Israel. Food Addit Contam Part A Chem Anal Control Expo Risk Assess. 2013;30:1614–25.
109. van Pamel E, Verbeken A, Vlaemynck G, De Boever J, Daeseleire E. Ultrahigh-performance liquid chromatographic-tandem mass spectrometric multimycotoxin method for quantitating 26 mycotoxins in maize silage. J Agric Food Chem. 2011;59:9747–55.
110. Li J, Jiang G, Yang B, Dong X, Feng L, Lin S, Chen F, Ashraf M, Jiang Y. A luminescent bacterium assay of fusaric acid produced by Fusarium proliferatum from banana. Anal Bioanal Chem. 2012;402:1347–54.
111. Braun R. Ueber Wirkungsweise und Umwandlungen der Fusarinsäure. Phytopathol Z. 1960;39:197–241.
112. Pitel DW, Vining LC. Accumulation of dehydrofusaric acid and its conversion to fusaric and 10-hydroxyfusaric acids in cultures of *Gibberella fujikuroi*. Can J Biochem. 1970;48:623–30.
113. Stoll C. Über Stoffwechsel und biologisch wirksame Stoffe von *Gibberella fujikuroi* (Saw.) Woll., dem Erreger der Bakanaekrankheit. Phytopathol Z. 1954;22:233–74.
114. Rees DO, Bushby N, Cox RJ, Harding JR, Simpson TJ, Willis CL. Synthesis of [1,2-$^{13}C_2$, ^{15}N]-L-homoserine and its incorporation by the PKS-NRPS system of *Fusarium moniliforme* into the mycotoxin fusarin C. Chembiochem. 2007;8:46–50.
115. Desaty D, McInnes AG, Smith DG, Vining LC. Use of ^{13}C in biosynthetic studies. Incorporation of isotopically labelled acetate and aspartate in fusaric acid. Can J Biochem. 1968;46:1293–300.
116. Hill RD, Unrau AM, Canvin DT. The biosynthesis of fusaric acid from ^{14}C-labelled acetate in *Gibberella fujikuroi*. Can J Chem. 1966;44:2077–82.
117. Bömke C, Tudzynski B. Diversity, regulation, and evolution of the gibberellin biosynthetic pathway in fungi compared to plants and bacteria. Phytochemistry. 2009;70(15–16):1876–93.
118. Studt L, Wiemann P, Kleigrewe K, Humpf H-U, Tudzynski B. Biosynthesis of fusarubins accounts for pigmentation of *Fusarium fujikuroi* perithecia. Appl Environ Microbiol. 2012;78:4468–80.
119. Wiemann P, Willmann A, Straeten M, Kleigrewe K, Beyer M, Humpf H-U, et al. Biosynthesis of the red pigment bikaverin in *Fusarium fujikuroi*: genes, their function and regulation. Mol Microbiol. 2009;72:931–46.
120. Yabuta T, Sumiki Y, Aso K, Tamura T, Igarashi H, Tamari K. Biochemical studies on the bakanae fungus. IV. The culture conditions for producing gibberellin or fusaric acid. J Agric Chem Soc Jpn. 1939;15:1209–20.

121. Teichert S, Schönig B, Richter S, Tudzynski B. Deletion of the *Gibberella fujikuroi* glutamine synthetase gene has significant impact on transcriptional control of primary and secondary metabolism. Mol Microbiol. 2004;53:1661–75.
122. Wagner D, Wiemann P, Huß K, Brandt U, Fleißner A, Tudzynski B. A sensing role of the glutamine synthetase in the nitrogen regulation network in *Fusarium fujikuroi*. PLoS One. 2013;8(11):e80740.
123. Bayram O, Braus GH. Coordination of secondary metabolism and development in fungi: the velvet family of regulatory proteins. FEMS Microbiol Rev. 2012;36:1–24.
124. Myung K, Li S, Butchko RAE, Busman M, Proctor RH, Abbas HK, et al. FvVE1 regulates biosynthesis of the mycotoxins fumonisins and fusarins in *Fusarium verticillioides*. J Agric Food Chem. 2009;57:5089–94.
125. Wiemann P, Brown DW, Kleigrewe K, Bok JW, Keller NP, Humpf H-U, et al. FfVel1 and FfLae1, components of a *velvet*-like complex in *Fusarium fujikuroi*, affect differentiation, secondary metabolism and virulence. Mol Microbiol. 2010;77:972–94.
126. López-Berges MS, Hera C, Sulyok M, Schäfer K, Capilla J, Guarro J, et al. The velvet complex governs mycotoxin production and virulence of *Fusarium oxysporum* on plant and mammalian hosts. Mol Microbiol. 2013;87:49–65.
127. Butchko RAE, Brown DW, Busman M, Tudzynski B, Wiemann P. Lae1 regulates expression of multiple secondary metabolite gene clusters in *Fusarium verticillioides*. Fungal Genet Biol. 2012;49:602–12.
128. Shwab EK, Jin WB, Tribus M, Galehr J, Graessle S, Keller NP. Histone deacetylase activity regulates chemical diversity in *Aspergillus*. Eukaryot Cell. 2007;6:1656–64.
129. Brakhage AA. Regulation of fungal secondary metabolism. Nat Rev Microbiol. 2013;11: 21–32.
130. Bok JW, Chiang Y, Szewczyk E, Reyes-Domingez Y, Davidson AD, Sanchez JF, et al. Chromatin-level regulation of biosynthetic gene clusters. Nat Chem Biol. 2009;5:462–4.
131. Moretti A, Ferracane L, Somma S, Ricci V, Mulè G, Susca A, Ritieni A, Logrieco AF. Identification, mycotoxin risk and pathogenicity of Fusarium species associated with fig endosepsis in Apulia, Italy. Food Addit Contam Part A Chem Anal Control Expo Risk Assess. 2010;27:718–28.
132. Thrane U, Adler A, Clasen PE, Galvano F, Langseth W, Lew H, Logrieco A, Nielsen KF, Ritieni A. Diversity in metabolite production by *Fusarium langsethiae*, *Fusarium poae*, and *Fusarium sporotrichioides*. Int J Food Microbiol. 2004;95:257–66.
133. Lakshminarayanan K, Subramanian D (1955) Is fusaric acid a vivotoxin? Nature 176:697–698
134. Luz JM, Paterson RRM, Brayford D (1990) Fusaric acid and other metabolite production in Fusarium oxysporum f. sp. vasinfectum. Lett Appl Microbiol 11:141–144
135. Ďuračková Z, Betina V, Nemec P (1976) Systematic analysis of mycotoxins by thin-layer chromatography. J Chromatogr 116:141–154
136. Kalyanasundaram R (1955) Bioassay of Fusarium toxin. Agar-cup method for quantitative evaluation (only abstract in English). J Indian Bot Soc 34:43–46
137. Šrobárová A, Eged S, Teixeira Da Silva J, Ritieni A, Santini A (2009) The use of Bacillus subtilis for screening fusaric acid production by Fusarium spp. Czech J Food Sci 27:203–209
138. Appell M, Jackson MA, Wang LC, Ho C-H, Mueller A (2014) Determination of fusaric acid in maize using molecularly imprinted SPE clean-up. J Sep Sci 37:281–286
139. Abia WA, Warth B, Sulyok M, Krska R, Tchana AN, Njobeh PB, Dutton MF, Moundipa PF (2013) Determination of multi-mycotoxin occurrence in cereals, nuts and their products in Cameroon by liquid chromatography tandem mass spectrometry (LC-MS/MS). Food Control 31:438–453
140. Warth B, Parich A, Atehnkeng J, Bandyopadhyay R, Schuhmacher R, Sulyok M, Krska R (2012) Quantitation of mycotoxins in food and feed from Burkina Faso and Mozambique using a modern LC-MS/MS multitoxin method. J Agric Food Chem 60:9352–9363

Chapter 12
Lovastatin, Compactin, and Related Anticholesterolemic Agents

David Dietrich and John C. Vederas

Introduction

Cardiovascular disease (CVD) is a leading cause of death worldwide, with recent statistics stating that one-third of all deaths reported in the USA were a result of CVD in 2009 [1]. Elevated levels of cholesterol are a key indicator of CVD, and as such, therapies aimed at lowering cholesterol levels are of great significance. Endo and coworkers at Sankyo Co., Ltd. recognized the importance of 3-hydroxy-3-methyl-glutaryl-CoA reductase (HMGR) as a target for pharmaceutical cholesterol management, and screened secondary metabolite extracts of several fungi for inhibitors of this enzyme. Activity-guided fractionation of *Penicillium citrinum* extracts led to the identification of compactin (**1**, synonyms: ML-236B, mevastatin) [2]. Similarly, a research group from Merck Sharp and Dohme identified lovastatin (**2**, synonyms: monacolin K, mevinolin, Mevacor) in *Aspergillus terreus* cultures [3]. Lovastatin has also been identified in other fungal cultures, including, *Monascus ruber* [4] and compactin was found in *P. brevicomaptum* [5]. The discovery of these metabolites, along with their proven ability to reduce cholesterol levels in rats, led to an explosion of molecular and biosynthetic studies of these molecules, as well as the synthesis of unnatural analogs, collectively referred to as statins [2–5]. This class of pharmaceutical includes semisynthetic analog simvastatin (**3**, synonym: Zocor) and the totally synthetic drugs rosuvastatin (**4**, synonym: Crestor) and atorvastatin (**5**, synonym: Lipitor) (Fig. 12.1a). The statins are the best selling class of prescription medication, with annual revenues for certain blockbusters numbering in the

---

D. Dietrich, B.Sc., Ph.D. • J.C. Vederas, B.Sc., Ph.D. (✉)
Department of Chemistry, University of Alberta, Edmonton, AB, Canada
e-mail: ddietrich01@gmail.com; john.vederas@ualberta.ca

J.-F. Martín et al. (eds.), *Biosynthesis and Molecular Genetics of Fungal Secondary Metabolites*, Fungal Biology, DOI 10.1007/978-1-4939-1191-2_12,

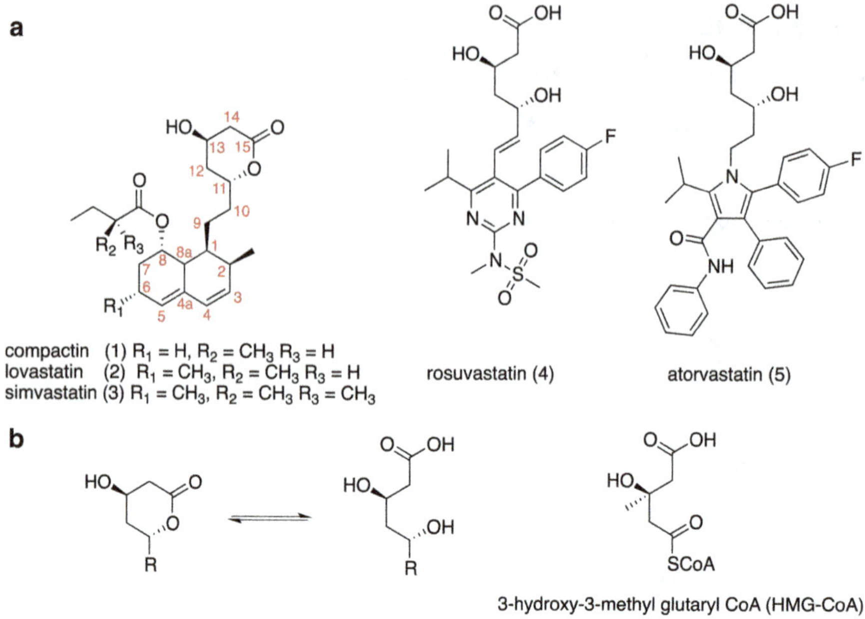

Fig. 12.1 (**a**) The chemical structure of various statins. (**b**) Cyclic and open-chain equilibrium of the molecular warhead of the anticholesterolemic statins, and comparison of the open form to HMG-CoA

billions of dollars (USD). While the statins vary in their overall chemical structure, they all bear the same biologically active warhead—a 3(*R*),5(*R*)-dihydroxypentanoate moiety, which is in equilibrium with its lactone form (Fig. 12.1b).

These pharmaceutical agents share a common biological target, which accounts for their cholesterol-lowering abilities. Specifically, these molecules inhibit the enzyme HMGR [6–8]. The biosynthesis of cholesterol belongs to the mevalonic acid pathway, where HMGR is responsible for the biosynthesis of mevalonate by reduction of 3-hydroxy-3-methyl-glutaryl-CoA (HMG-CoA) (Fig. 12.1b). This biosynthetic transformation has been shown to be the key regulation step in cholesterol biosynthesis, and thus inhibition of this process leads to lowered cholesterol production [9]. Close examination of the warhead of the statins reveals that these inhibitors are mimics of HMG-CoA, the natural substrate for the enzyme. Replacement or modification of the 3(*R*),5(*R*)-dihydroxypentanoate moiety leads to inactive analogs.

This review will focus on biosynthetic and molecular genetic studies of lovastatin (**2**), and will compare this information to the biosynthesis of compactin (**1**) (Table 12.1). Early biosynthetic studies led to detailed analysis of the gene cluster responsible for biosynthesis of these metabolites, and the function of these gene products will be discussed in detail [10–12]. At the end, we will briefly discuss related polyketides that employ similar biosynthetic machinery.

Table 12.1 List of abbreviations

Abbreviation	Full title
6-MSA	6-Methylsalicylic acid
ACP	Acyl carrier protein
AT	Acyl transferase
CoA	Coenzyme A
CON	Condensation domain (NRPS)
CP450	Cytochrome P450
CPR	Cytochrome P450 reductase
CVD	Cardiovascular disease
DH	Dehydratase
DMB-S-MMP	*S*-(methyl 3-mercaptopropionate) dimethylbutyrate
DML	Dihydromonacolin L
ER	Enoyl reductase
FAS	Fatty acid synthase
HMG-CoA	3-Hydroxy-3-methyl-glutaryl-CoA
HMGR	HMG-CoA reductase
K_M	Michaelis constant
KR	Ketoreductase
KS	Ketosynthase
LDKS	Lovastatin diketide synthase
LNKS	Lovastatin nonaketide synthase
MeT	Methyl transferase
MMP	Methyl 3-mercaptopropionate
NADPH	Nicotinamide adenine dinucleotide phosphate
NRPS	Non-ribosomal peptide synthetase
PKS	Polyketide synthase
SAM	*S*-adenosyl methionine
SNAC	*N*-acetyl cysteamine
TE	Thioesterase
USD	US dollars

Early Biosynthetic Studies of Lovastatin and Compactin

Shortly after the structural elucidation of lovastatin and compactin, these metabolites were shown to be generated via a polyketide biosynthetic pathway by a variety of labeling studies [13–16]. For both metabolites, it was shown that the molecule is synthesized by the consecutive incorporation of nine acetate units in the main chain (a nonaketide), plus a diketide that is attached via esterification to a hydroxyl group at the 8-position [17–19]. The oxygen atom at carbon 8 is incorporated through a post-PKS oxidative pathway, which was proposed to involve at least one cytochrome P450 (CP450) enzyme (Fig. 12.2).

The methyl butyrate side chain was shown to be a diketide, with installation of the methyl group coming from *S*-adenosyl methionine (SAM). Surprisingly, this side chain was not derived from isoleucine. What separates compactin (**1**) and

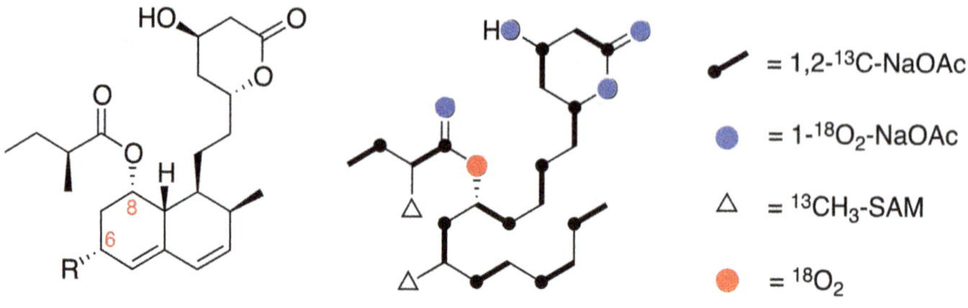

Fig. 12.2 Biosynthetic labeling studies confirmed the polyketide nature of compactin (**1**) and lovastatin (**2**), and accounted for the molecular oxygen incorporated

lovastatin (**2**) is the presence of a methyl group at the 6-position. In lovastatin, this methyl group is incorporated during chain elongation via transfer from SAM, whereas compactin does not introduce this additional functionality. A variety of lovastatin- and compactin-like metabolites were also identified in *Penicillium* and *Aspergillus* cultures, which were proposed to be precursors that were interrupted in the biosynthetic pathway. This information was compiled to propose a biosynthetic pathway for **2** (Fig. 12.3).

The biosynthetic pathway shown in Fig. 12.3 begins with the addition of nine acetate units (in the form of malonyl-CoA, **6**) and the methyl group from one molecule of SAM (**7**) in a typical polyketide synthase (PKS) fashion to generate enzyme-linked intermediate **8**. The first isolable intermediate is dihydromonacolin L (DML, **9**). This intermediate is oxidized to monacolin L (**10**) and again to monacolin J (**11**). Separately, two malonyl-CoA units and one molecule of SAM are responsible for the preparation of the enzyme-linked 2(*S*)-methylbutyrate side chain (**12**), which is transacylated onto **11** to furnish lovastatin (**2**). The order of each step in this pathway was proposed based on the identification of intermediates **9–11**. While this pathway accounts for the structure of the final metabolites, the stepwise programming of what was proposed to be an iterative PKS remained a mystery.

One interesting aspect of the biosynthetic pathway of lovastatin and compactin is the intramolecular Diels-Alder [4+2] cycloaddition reaction, which generates the didehydrodecalin core of DML (**9**) [13]. Specifically, this reaction was thought to be the result of an enzyme-catalyzed process. The Diels-Alder reaction is a well-known stereoselective transformation to synthetic chemists; however, it had only been proposed to exist in nature. The involvement of a Diels-Alder reaction was also suggested to be involved in the biogenesis of other natural products, including cytochalasin E [20] and solanapyrone A [21]. For reviews that discuss the current state of biological Diels-Alder reactions, see [22–25]. Biochemical proof of an enzyme-catalyzed version of this reaction, however, had not been described. Lovastatin provides a particularly intriguing example of an enzymatically catalyzed Diels-Alder reaction,

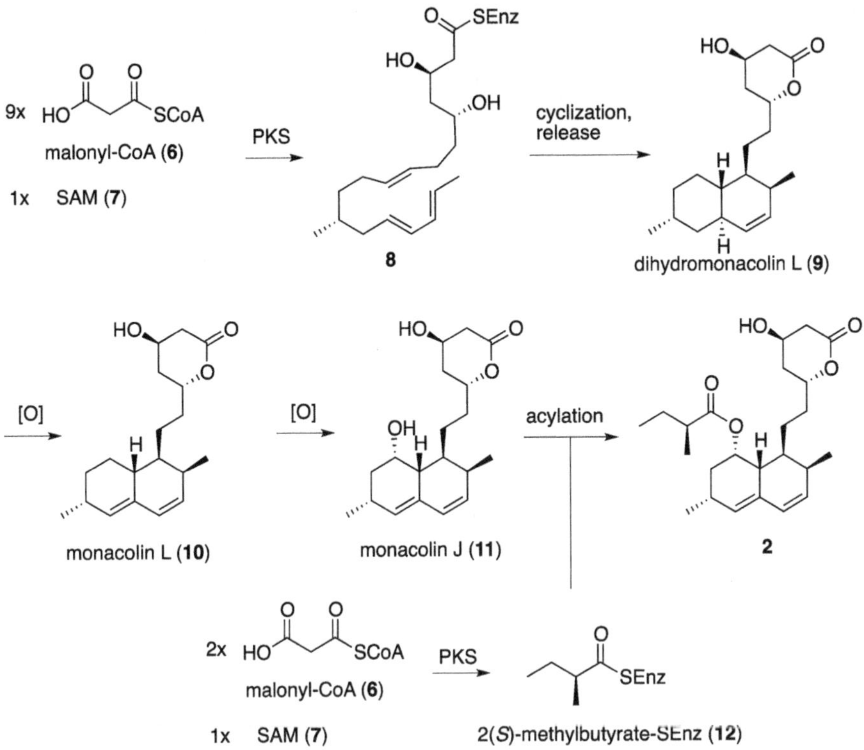

Fig. 12.3 Proposed biosynthetic pathway for compactin and lovastatin. In early studies the exact timing of the cyclization was not confirmed

as the stereochemical outcome of the cyclization reaction produces a diasteromer where the methyl groups are found in an unfavored axial orientation. It is worthwhile noting that in compactin, the 6-methyl group is not present, and therefore does not contribute to the stability of the Diels-Alder cyclization products.

Using a synthetic hexaketide precursor bearing an *N*-acetylcysteamine (SNAC) thioester (**13**), it was shown that in solution, this triene substrate produces a 1:1 mixture of **14c** and **14d**, and that the rate of this reaction is accelerated in aqueous solution ($t_{½}=60$ h in water) (Fig. 12.4) [26]. The other diastereomers (**14a** and **14b**) were not observed in this control reaction. When the same triene thioester was incubated with a homogenously purified PKS enzyme involved in lovastain biosynthesis (LovB), a change in the ratio of the diastereomeric products was observed, and the natural isomer (**14a**) was identified [27]. This represents the first evidence than an enzyme can catalyze the Diels-Alder reaction. Similarly, Katayama et al. [28, 29] showed that enzymes can catalyze Diels-Alder cyclizations, using purified enzymes involved in solanapyrone biosynthesis. These results point toward the fact that enzymes do catalyze the Diels-Alder reaction.

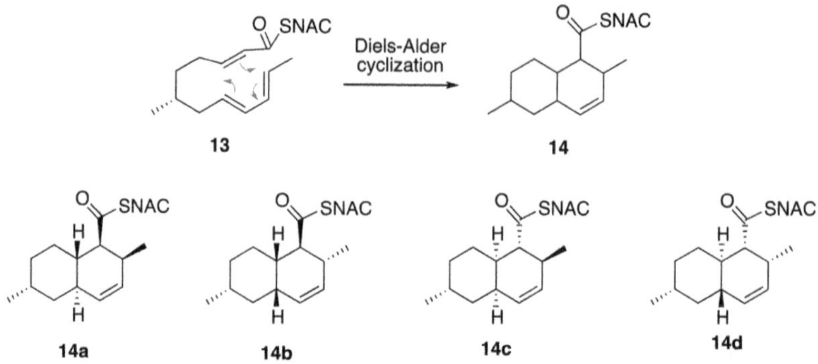

	14a	14b	14c	14d
in H_2O:	0%	0%	50%	50%
in H_2O + LovB	3%	0%	48.5%	48.5%

Fig. 12.4 Stereochemical result of Diels-Alder reaction on triene **13** in the absence and presence of LovB

Molecular Genetics of Lovastatin Biosynthesis

Identification of the Lovastatin Biosynthetic Gene Cluster

As previously mentioned, lovastatin and compactin are polyketide in origin. These fungal secondary metabolites represent highly reduced polyketides. Highly reduced polyketides produced by bacterial biosynthetic machinery operate in a modular fashion using each catalytic domain one single time [30]. The highly reduced polyketides lovastatin and compactin were of particular interest since fungal PKSs are iterative in nature; each catalytic domain of the megasynthase is used repetitively during chain elongation. The existence of highly reduced fungal polyketide metabolites implied that a unique subset of these domains is used during any given round of chain elongation. To provide the tools to study this hypothesis, the biosynthetic gene cluster responsible for lovastatin production in *A. terreus* was determined. When *A. terreus* mutants that lacked the ability to produce lovastatin were identified, the β-ketoacyl synthase (KS) domain of the 6-methylsalicylic acid (6-MSA) PKS system was used as a probe to discover a new putative PKS that contributed to lovastatin biosynthesis [31].

One new protein estimated to be ~250 kDa was identified, and was termed the lovastatin nonaketide synthase (LNKS). Furthermore, Hutchinson, Vederas and coworkers used this information to identify the full gene cluster surrounding LNKS (Note: LNKS is synonymous with LovB) [32]. The expected activity of this gene cluster was confirmed through knockout studies. The cluster was shown to contain 18 protein-encoding genes, of which nine were proposed to be directly involved in lovastatin synthesis, namely *lovA* through *lovI* and *orf5, 8,* and *10* (Fig. 12.5).

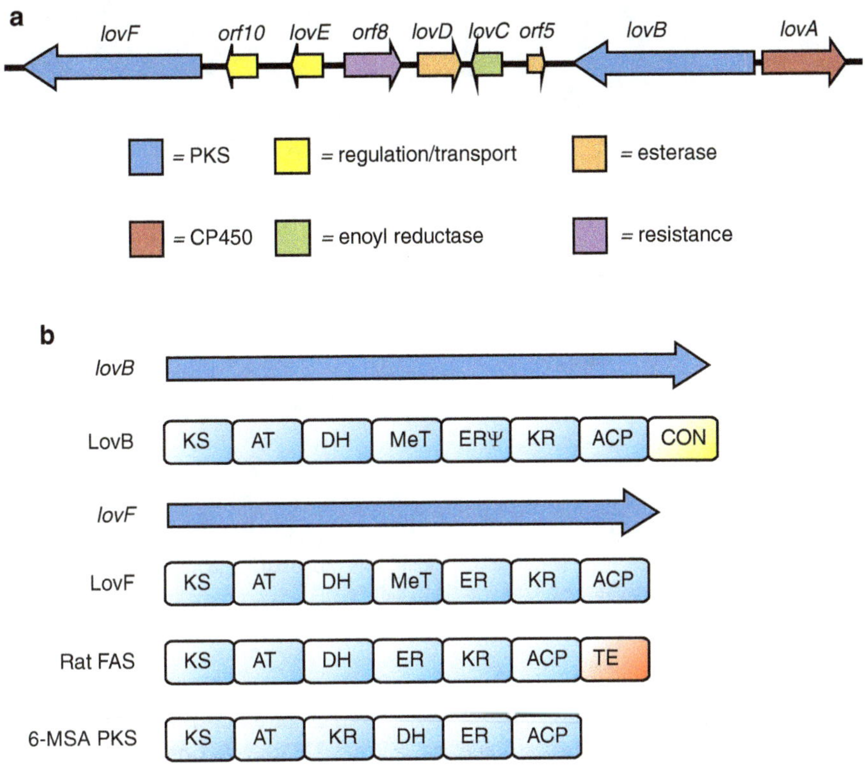

Fig. 12.5 (**a**) Annotated gene cluster highlighting the genes responsible for lovastatin biosynthesis. (**b**) Comparison of LovB and LovF to 6-MSA PKS and a mammalian fatty acid synthase. The Ψ symbol indicates an inactive domain

Two prominent PKS genes were identified in this gene cluster: the previously reported LNKS (LovB) and LovF. When the organization of the catalytic domains within LovB and LovF was compared to that of 6-MSA PKS and a mammalian fatty acid synthase (FAS), some interesting comparisons were drawn [33]. Strikingly, with few exceptions, the LovB and LovF domain architecture resembled that of a mammalian FAS. LovB and LovF showed seven putative functional domains: β-ketoacyl synthase (KS), malonyl-CoA acyl transferase (AT), dehydratase (DH), methyl transferase (MeT), enoyl reductase (ER), ketoreductase (KR), and acyl carrier protein (ACP). Six of these domains are also present in FAS, with the MeT domain being the only exception. One interesting function absent in both LovB and LovF relative to FAS is a thioesterase (TE) domain; this domain is thought to be responsible for the release of the completed product.

Two other interesting features of LovB were also observed: the proposed ER domain did not correspond to a functional ER domain, based on the poor homology in the nucleotide-binding domain responsible for nicotinamide adenine dinucleotide phosphate (NADPH) recognition. There was also a fragment at the C-terminus of

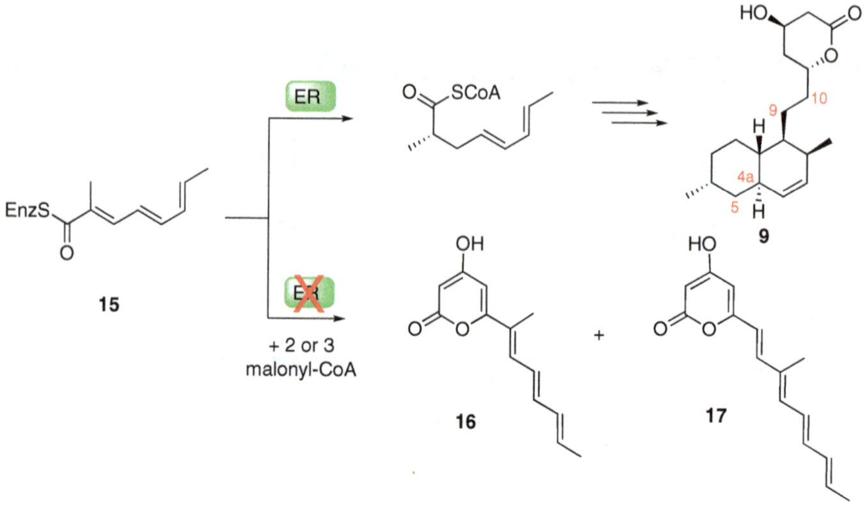

Fig. 12.6 The expected product **9** was not observed when LovB was heterologously expressed in *A. nidulas*; a functional enoyl reductase (ER) domain was proposed to be required. In this case, pyrones **16** and **17** were generated as part of a shunt pathway

LovB that resembled the condensation (CON) domain of a non-ribosomal peptide synthetase (NRPS). The lack of a functional ER, no TE (or other product releasing domain), and the partial NRPS module led to the need for further biosynthetic studies of this enzyme.

Functional Characterization of the LovB-LovC PKS System

To probe the function of LovB, the megasynthase responsible for nonaketide synthase, it was heterologously expressed in the non-lovastatin producing organism *Aspergillus nidulans* under the control of the *alcA* promoter [32]. Transformants that successfully over-expressed the 335-kDa LovB protein were fermented, and the culture supernatant was extracted. Dihydromonacolin L (DML, **9**), an expected product of a fully functioning LovB enzyme was not detected, but two new metabolites were identified. These were purified and characterized by various spectroscopic techniques, and further shown through synthetic standards to be pyrones **16** and **17** (Fig. 12.6) [32, 34]. The formation of these pyrones implied that LovB was not functioning properly. Since the ER domain of LovB appeared to be nonfunctional, these pyrones were predicted to be shunt products. In the absence of a functional enoyl reductase, enzyme-linked tetraketide intermediate **15** cannot be reduced, and LovB recognizes this as the wrong product, and induces the shunt pathway. The shunt pathway repetitively adds two or three malonyl-CoA units, to offload as pyrones **16** and **17** respectively. It was proposed that a second protein in the gene cluster must interact with LovB to provide the enoyl reductase activity.

Silencing and complementation studies showed that LovC was needed for generation of DML (**9**). It was also shown that LovC, a 363 amino acid protein, has sequence homology to the ER domain of other PKSs. Cell-free extracts of *A. nidulans* that co-expressed LovB and LovC generated the expected DML product. These results imply that these two proteins act in concert, and that there must be critical programming steps that ensure the formation of the correct product. This represented one of the first examples where a nonfunctioning domain of a protein is complemented by an endogenous enzyme [35, 36]. When one (or several) of the expected transformations is missing, the PKS LovB will follow a shunt pathway to offload undesired products, typically as pyrones. When expression levels of LovB and LovC are mismatched, monacolin N (9,10-dehydro,4a,5-dihydromonacolin L) is observed. This is a byproduct that results from no ER activity at the heptaketide stage [37].

Despite the significant improvements that were made with respect to understanding the biosynthesis of lovastatin, all of these early studies of LovB were performed in cell culture. This limitation does not allow for the controlled addition of substrates, which can be used to gain further understanding of the programming of the highly reducing iterative PKS. A pictorial view of each round of chain extension, with focus on domain use, is shown in Fig. 12.7.

This pathway implies that four substrates/cofactors are required for DML production: malonyl-CoA, SAM, NADPH (for KR and ER), and LovC. The ability to study the effect of these cofactors was hampered by the ability to obtain significant purified quantities of the enzyme. This problem was overcome when LovB was heterologously expressed in an engineered *Saccharomyces cerevisiae* host and purified to homogeneity with yields of ~4.5 mg/L [38]. With pure LovB in hand, Tang, Vederas, and coworkers were in a position to reconstitute the PKS activity with careful control of the substrates supplied to the megasynthase. When LovB was incubated with malonyl-CoA (malonyl-CoA can serve as both the starter and extender unit in lovastatin biosynthesis), the major product identified was triketide pyrone **18** (Fig. 12.8a). This system was lacking NADPH, so the KR of LovB cannot function, and the enzyme goes into shunt mode to produce **18**.

When NADPH was added into the system, the KR domain was active, and new shunt products **19–21** were observed (Fig. 12.8b). These products imply that LovB was able to successfully extend three malonate units, until the MeT and endogenous ER domain from LovC are needed. The absence of LovC and SAM renders these functions absent, and pyrones **19–21** are produced. When LovB was incubated with malonyl-CoA, NADPH, and SAM, pyrones **15** and **16** were the main products identified (Fig. 12.8c). These pyrones are similar to **20** and **21** with the exception that the methyl functionality was incorporated. Since the ER function is still absent, the same shunt pathway that produced **20** and **21** is followed. Interestingly, when LovB was reconstituted with malonyl-CoA, NADPH, and LovC, pyrones **19–21** were again identified (not shown). This implies that methylation plays a critical role in the function of the endogenous ER, and that the absence of this methyl group leads

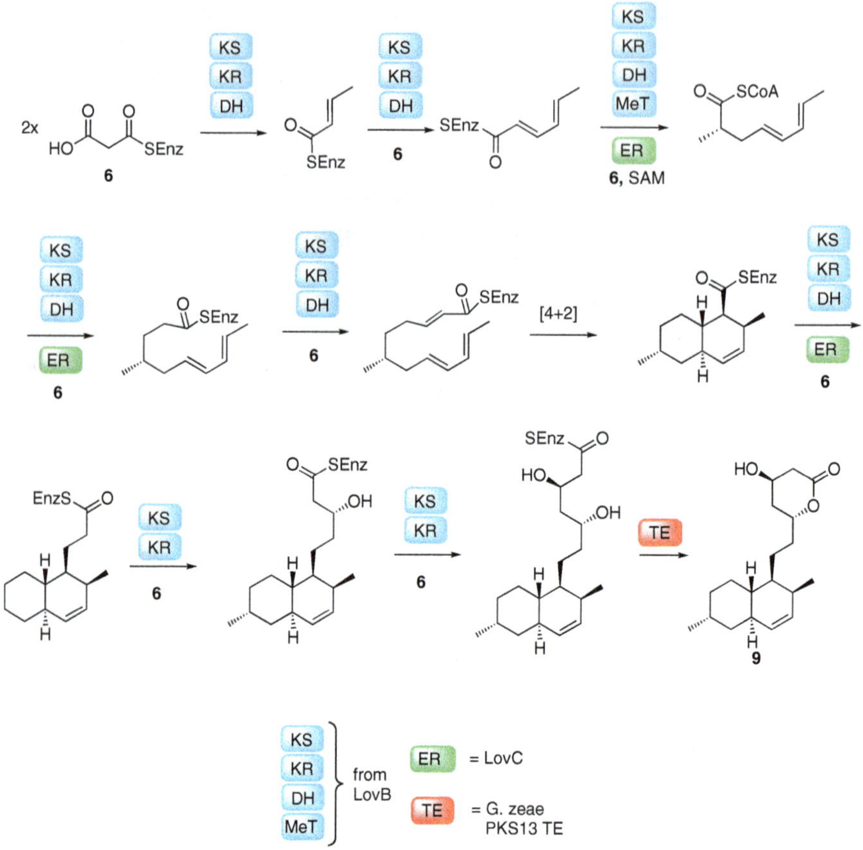

Fig. 12.7 Complete iterative PKS steps employed in the formation of DML (**9**). Each step of chain elongation uses a unique set of PKS domains, but these domains are on the same protein

to shunt products. Finally, LovB was incubated with all substrates and cofactors (malonyl-CoA, NADPH, SAM, and LovC). Surprisingly, no significant products were extracted from this system. Due to the lack of a TE domain in LovB, which is present in typical FAS systems, it was proposed that the fully elaborated nonaketide remained attached to the PKS. When LovB incubated in these conditions was treated with base, the thioester intermediate was released from the enzyme (Fig. 12.8d). These conditions rendered LovB a stoichiometric reagent. Since heterologous host *A. nidulans* expressing LovB and LovC was able to produce significant amounts of DML, it was expected that exogenous TE domains could facilitate the release of the elaborated thioester intermediate. When heterologously expressed TE domain from *Gibberella zeae* PKS13 [39] was added to the in vitro reaction of LovB treated with LovC, NADPH, SAM, and malonyl-CoA, production of DML (**9**) was observed.

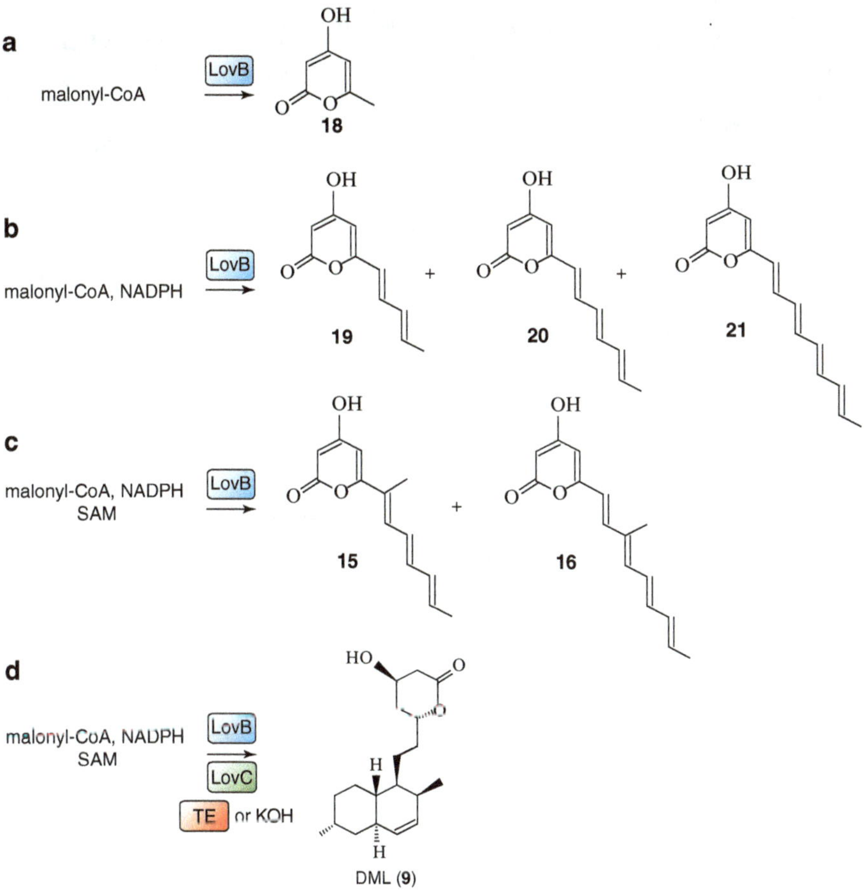

Fig. 12.8 (**a–d**) Summary of the results obtained following the reconstitution of LovB in the presence of various combinations of cofactors (see text)

At this point, a fully functional LovB-LovC system was successfully reconstituted, and important conclusions about the biosynthetic programming of lovastatin biosynthesis could be made. Importantly, all of the required cofactors must be present for the PKS to elaborate malonyl-CoA units to the expected DML product—the absence of any single component will lead to improperly tailored intermediates and induce a shunt process that typically offloads as pyrones. The exogenous ER enzyme LovC is not only responsible for the reduction of enoyl functionalities, but is also responsible for recognizing the correct installation of the methyl group at C6. Further information about this specificity may be gleaned from the crystal structure of LovC [40]. Fully reconstituted LovB is not capable of off-loading the correct product, but requires an exogenous TE domain. It was unclear at the time if this off-loading was catalyzed by a specific enzyme in the lovastatin gene cluster, or if this was performed by a nonspecific enzyme from the producing fungus.

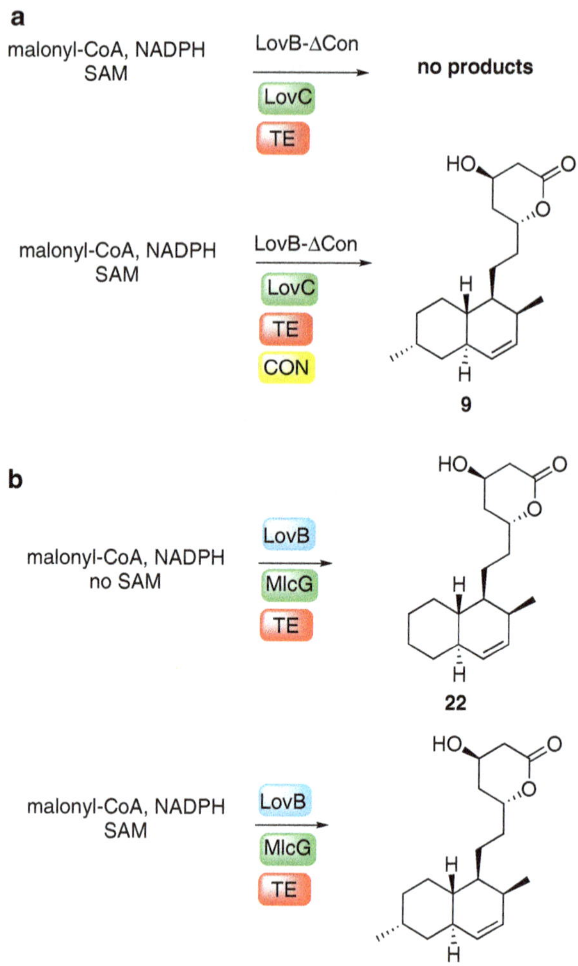

Fig. 12.9 (**a**) Effect of the CON domain of LovB in the biosynthesis of DML (**9**). (**b**) Effect of the exogenous enoyl reductase domain from compactin biosynthesis (MlcG) on LovB in the biosynthesis of DML and its desmethyl analog (**22**)

With a fully operational in vitro PKS system in hand, the function of the NRPS CON domain that exists at the C-terminus of LovB was examined. When LovB variants that were truncated in the CON domain (ΔCon), were incubated with malonyl-CoA, NADPH, SAM, LovC, and *G. zeae* PKS13 TE, no products were detected. Similar to previous experiments where one component was removed, the LovB-ΔCon construct was able to produce pyrones **15**, **16,** and **19–21**. This result implies that the CON domain is responsible for stabilization of the later stage steps of DML production, and could even be involved in the Diels-Alder cyclization. Interestingly, when the fully reconstituted LovB-ΔCon system was supplemented with heterologously expressed CON domain, functional activity was recovered (Fig. 12.9a) albeit, with lower yields. This shows that the function of the CON domain can work in trans with LovB and LovC.

This in vitro expression system also allowed for a detailed analysis of the function of the endogenous ER protein. Complementation experiments showed that LovC is required for the function of LovB and that it requires correct installation of the methyl group at C6. Since compactin does not contain this methyl group, the effect of the LovC analog involved in compactin biosynthesis (MlcG) was cloned, over-expressed and used in similar reconstitution assays with LovB.

When LovB was incubated with malonyl-CoA, NADPH, *G. zeae* PKS13 thioesterase and the exogenous ER from compactin (MlcG), the 6-desmethyl analog **22** was generated (Fig. 12.9b). Alternatively, when SAM was added to the mixture, fully elaborated DML (**9**) was produced. This result implies that the exogenous ER MlcG is not affected by the presence of the methyl group, whereas LovC requires methylation at the triketide stage; LovC will not produce DML in the absence of SAM. Since either DML or its desmethyl analog can be produced when LovB is coupled with MlcG, the downstream events catalyzed by this PKS must not be affected the presence of the 6-methyl functionality. This valuable collection of experiments provides detailed information regarding the remarkable programming that is used by the iterative highly reducing PKS LovB in the biosynthesis of lovastatin.

Functional Characterization of LovF

Analysis of the lovastatin biosynthetic gene cluster showed a second PKS, known as LovF (synonymous with lovastatin diketide synthase or LDKS), which was proposed to be responsible for the generation of the 2(*S*)-methylbutyrate side chain. In *lovF* gene disrupted mutants of *A. terreus*, loss of lovastatin production was accompanied by increased production of monacolin J (**11**), supporting this hypothesis [32]. The deduced sequence of this 2,532 amino acid PKS implied the same catalytic domains as in LovB (KS, AT, DH, MeT, ER, KR, and ACP), but does not have the NRPS CON domain at the C-terminus. One important difference in this PKS is that the ER domain retains a functional NADPH binding region, which implies LovF does not require the action of an endogenous enoyl reductase.

The function and independent nature of LovF was confirmed in mutant cultures of *A. terreus* that contain a disrupted LovC (which cannot produce lovastatin). These mutants did not produce any detectable amount of the expected LovF product 2(*S*)-methylbutyrate or any shunt products that could be the result of an unfunctional ER in LovF. When exogenous monacolin J (**11**) or its desmethyl analog **23** was introduced into the LovC disruptant cultures, conversion of these intermediates to lovastatin and compactin (respectively) was restored (Fig. 12.10) [32, 41, 42]. This implies that these LovC mutants are able to successfully generate the diketide moiety, and transfer it onto fully elaborated intermediates without the need for an endogenous enoyl reductase. The inability to detect the 2(*S*)-methylbutyrate diketide or its shunt product in cultures is not surprising. This is likely due to the fact that LovF does not contain a TE domain. This means that the diketide product generated by LovF remains attached to the ACP domain, until it is directly transesterified

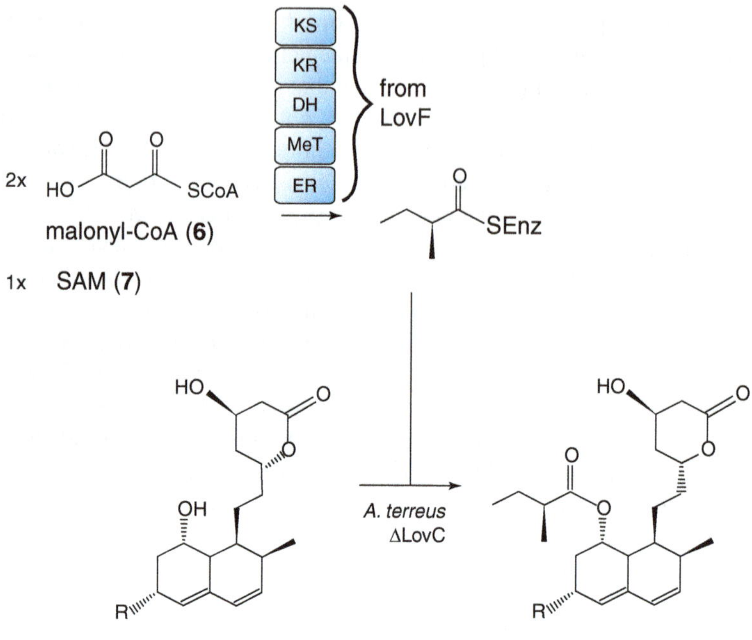

Fig. 12.10 When *A. terreus* cultures deficient in LovC production are treated with monacolin J (**11**), or its desmethyl analog **23**, production of lovastatin (**2**) and compactin (**1**) is observed. This implies that LovF is an independent PKS involved in production of the 2(*S*)-methylbutyrate side chain

onto monacolin J or else onto an accessory protein. Since the diketide could also be transferred onto the desmethyl analog **23** to produce compactin, the enzyme(s) responsible for the transesterification do not have a structural requirement at the 6-position of the substrate.

Functional Characterization of LovD

Since *A. terreus* mutants deficient in LovC production did not lead to an increased concentration of 2(*S*)-methylbutyrate in the culture supernatant, it was proposed that the diketide intermediate remains attached to the PKS LovF until it is enzymatically transferred to its substrate [32]. Analysis of the lovastatin gene cluster implied that LovD could be responsible for this transfer [43]. The *lovD* gene encodes for a 46 kDa protein that shows homology to other esterases, including β-lactamases, carboxypeptidases, and lipases. To determine whether the transfer of the diketide from LovF to monacolin J (**11**) happens directly, or with the aid of LovD, mutant strains of *A. terreus* were used. Cultures that produced functional LovB, LovC, and LovF, but were deficient in LovD showed an accumulation of monacolin J. This

24

11

2

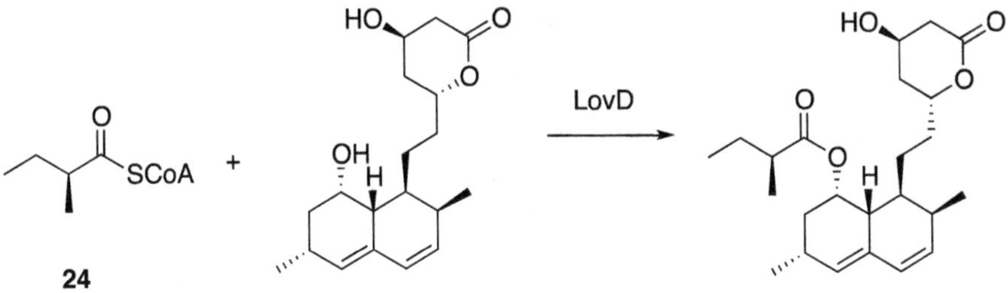

Fig. 12.11 The enzyme LovD is responsible for the transacylation of diketides onto monacolin J (**11**) in lovastatin (**2**) biosynthesis. This is transferred directly from LovF in the host, but CoA thioesters, such as **24**, can replace LovF in vitro

supports the idea that LovD is involved in the esterification reaction. This activity was further confirmed when LovD was heterologously expressed in *Escherichia coli* and purified by Ni^{2+}-affinity chromatography. When LovD was incubated in buffer containing monacolin J (**11**) and the CoA-thioester of 2(*S*)-methylbutyrate (**24**), lovastatin (**2**) was generated (Fig. 12.11).

The CoA-thioester **24** was used in place of the LovF-tethered substrate for ease in preparation; based on this replacement, slower rates were obtained. This is likely due to protein–protein interactions between LovD and the ACP domain of LovF. It was later shown that indeed, interactions between LovF and LovD are involved in the transacylation reaction [44]. These results not only confirm the activity of LovD, but also that it can function in the absence of other biosynthetic enzymes. This result has important implications in engineering novel biosynthetic pathways.

Simvastatin (**3**) is a semisynthetic analog of lovastatin (**2**) that is marketed by Merck under the trade name Zocor. The structural difference between lovastatin and simvastatin lies in the diketide side chain. In lovastatin, the 8-hydroxyl group of monacolin J is esterified with 2(*S*)-methylbutyrate. The methyl group is prone to epimerization, and therefore its replacement with an achiral analog was developed. Specifically, the α-dimethylbutyrate analog was prepared by semisynthetic means. This derivative (simvastatin) earned >$4 billion USD annually, before it lost patent protection. In Merck's semisynthetic approach, lovastatin is isolated from *A. terreus* culture and hydrolyzed to give monacolin J (**11**). A series of protecting group manipulations, followed by esterification with α-dimethylbutyric acid, then deprotection and lactonization is used to produce simvastatin [45].

In an attempt to avoid the laborious synthetic steps that are involved in transforming lovastatin into simvastatin, the Tang group sought to develop a method of directly preparing this material chemoenzymatically using LovD. The acyltransferase was shown to transform a wide variety of CoA- and SNAC-thioesters, including the natural substrate 2(*S*)-methylbutyrate, as well as α-dimethylbutyrate. These reactions were successful both in in vitro conditions and in LovD-expressing *E. coli* supplemented with monacolin J. In whole-cell biocatalysis, the yields suffered from low-density fermentation conditions, slow acylation, and premature nonspecific hydrolysis [46]. Acyl donor analogs were studied in detail, including the identity of

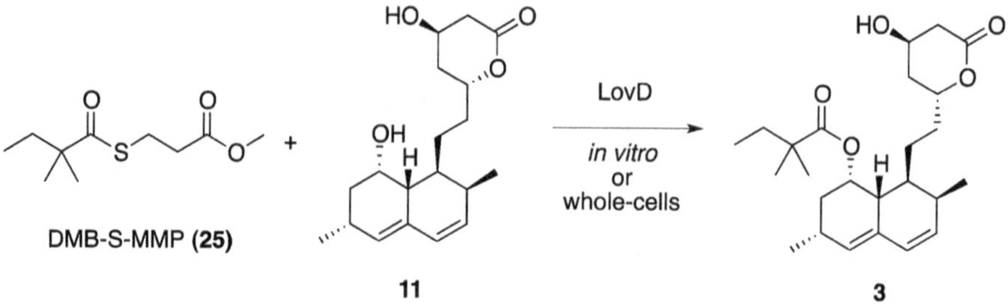

Fig. 12.12 Transesterase LovD can catalyze the formation of simvastatin (**3**) both in vitro and in whole-cell biocatalysis using unnatural substrate **25**

the thiol leaving group. It was found that the transesterification rate could be increased more than 30-fold using methyl 3-mercaptopropionate (MMP) thioesters of α-dimethylbutyric acid (**25**, DMB-S-MMP) (Fig. 12.12).

The use of DMB-S-MMP (**25**) as a substrate not only improved the rate of the acylation reaction, but also overcame a kinetic issue that was encountered with some thioester donors. Since LovD operates via a ping-pong mechanism, the K_M for both the thioester and monacolin J substrates is important. Poor thioester recognition by LovD results in inhibition by monacolin J. Since the K_M for DMB-S-MMP was improved relative to the SNAC-thioester, this inhibition is averted.

Another problem encountered in the whole-cell biocatalytic approach used to prepare simvastatin was the nonspecific hydrolysis of the thioester donors. Through an intense screening process, it was found that DMB-S-MMP is inactivated in *E. coli* cultures by one single native lipase: BioH. Deletion of the gene responsible for the production of BioH in *E. coli* and over-expression of LovD in the same strain significantly improved the stability of the DMB-S-MMP donor in cell cultures [47]. The whole-cell bioconversion of monacolin J to simvastatin was even further improved through both a rational design [48] and directed evolution approach [49]. In these experiments, the activity of LovD was improved such that LovD mutants increased the overall rate of transesterification of DMB-S-MMP onto monacolin J. These experiments also improved the solubility and thermal stability of LovD. Crystal structures of the parent enzyme and various mutants pointed to a more compact conformation contributing to this improved catalysis.

Functional Characterization of LovA

The nonaketide synthesized by LovB (acting in concert with LovC) is dihydromonacolin L (DML, **9**), which must be further oxidized to generate monacolin J (**11**), before LovD can add the diketide product of LovF. Generation of the double bond likely goes through hydroxylation at the 3-position (Fig. 12.13). Isolation of the lactone form of interrupted metabolite **27** (which can spontaneously undergo dehydration to yield the conjugated diene) supports this theory [50]. The hydroxyl group at the 8-position is likely inserted directly. Sequencing of the lovastatin gene cluster indicated that *lovA* and *orf17* encode CP450 oxidases. It was proposed that at least

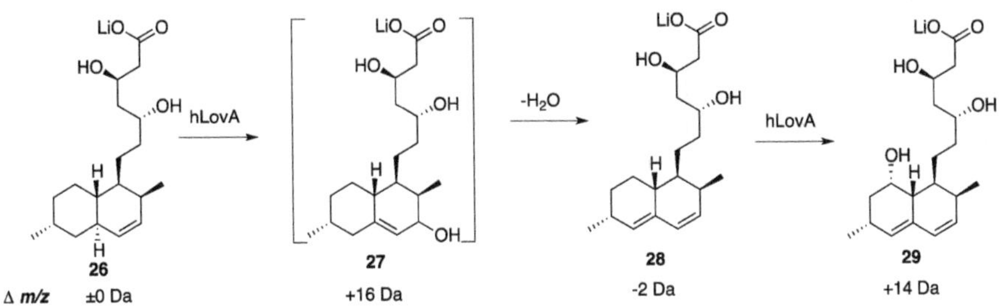

Fig. 12.13 A hybrid cytochrome P450 oxidase hLovA was shown to be responsible for both oxidative processing steps that convert dihydromonacolin L (Li⁺ salt, **26**) to monacolin J (Li⁺ salt, **29**). The changes in *m/z* are all relative to starting material **26**

one of these gene products was responsible for the conversion of DML into monacolin J. Mutational studies showed that when LovA deficient *A. terreus* was grown, DML was the only product isolated—no intermediate oxidized products were identified [51]. This result implies that LovA is involved in at least one oxidative processing step, and likely acts first. Heterologous expression of LovA proved to be challenging. Successful expression was achieved when an N-terminally engineered LovA with yeast codon optimization was used. The hybrid enzyme (hLovA) contained the N-terminal fragment of a lettuce P450 oxidase (LsGAO), which was designed to improve endoplasmic reticulum localization [52].

Co-expression of hLovA in a yeast clone producing *A. terreus* CP450 oxidoreductase (CPR, to regenerate CP450 activity) led to an oxidatively functional system. It was shown that the open-chain form of DML was a much better substrate for the engineered LovA than the lactone form. When yeast microsomes from the hLovA-CPR co-expressing system were incubated with the lithium salt of DML acid (**26**), two new major products were identified, with mass differences relative to the DML substrate of −2 and +14 Da (Fig. 12.13). Furthermore, when this system was incubated with the lithium salt of monacolin L acid (**28**), a +16 Da product was observed. Taken together, these results indicate the hLovA is responsible for the oxidation of DML first to the 3α-hydroxy intermediate **27**, which spontaneously dehydrates to produce monacolin L derivative **28** (−2 Da). Furthermore, hLovA inserts the hydroxyl group at the 8-position to generate **29** (+14 Da). This evidence points toward LovA being the sole CP450 responsible for the oxidative processing that is required for the biosynthetic conversion of the LovB-LovC product DML (**9**) to LovD substrate monacolin J (**11**). This bifunctional enzyme was shown to be ordered; hydroxylation at the 8-position cannot occur unless the diene system is present. The function of the hypothetical CP450 encoded for by *orf17* is unknown.

Functional Characterization of LovG (Orf5)

At this point, most of the biosynthetic pathway that accounts for the production of lovastatin from the simple biogenic precursors malonyl-CoA, *S*-adenosyl methionine and atmospheric oxygen were well understood. Two megasynthase PKSs (LovB and LovF) are responsible for the head-to-tail connection of the

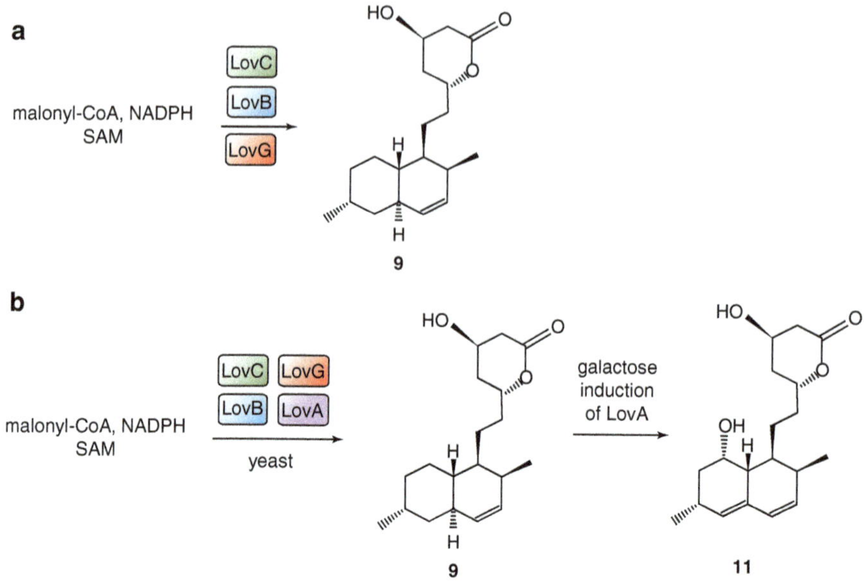

Fig. 12.14 (**a**) Heterologously expressed LovG was shown to successfully offload the nonaketide product generated by the concerted action of LovB and LovC. (**b**) A whole-cell yeast system that co-expresses LovA, LovB, LovC, and LovG can produce monacolin L (**9**), or in the presence of galactose, monacolin J (**11**)

malonyl-CoA substrates. LovB requires the action of exogenous enoyl reductase LovC for activity and programming. LovA carries out the post-PKS oxidation steps, and esterase LovD is responsible for transacylation of the LovF diketide onto monacolin J. However, one question still remained: How is the nonaketide product of LovB released in the native host, *A. terreus*? In the in vitro reconstitution system with LovB and LovC, exogenous TE domains were required for release of DML (**9**). Convinced that a dedicated thioesterase must be present in the lovastatin biosynthetic gene cluster, Tang and coworkers carefully examined the previously reported genetic data, with emphasis on *orf*5; this gene was originally assigned to encode for a hypothetical protein or an oxidoreductase [53]. Conserved domain analysis of *orf5* from the *A. terreus* lovastatin biosynthetic gene cluster revealed homology to the esterase family of enzymes, with close similarity to genes involved in compactin and lovastatin production in *Penicillium citrinum* (*mlcF*) and *Monascus pilosus* (*mokD*). In all three organisms, this gene is located between the PKS and its trans-acting ER partner (*lovB* and *lovC* in *A. terreus*). Furthermore, this gene was shown to be co-transcribed with other genes in the biosynthetic cluster. Disruption of the *orf5* (renamed *lovG*) gene in *A. terreus* led to a decrease in production of lovastatin, implying the importance of LovG in lovastatin production.

To confirm the action of LovG, a His-tagged version of the enzyme was expressed heterologously in *E. coli* and purified. This enzyme was added to an assay mixture containing LovB, LovC, malonyl-CoA, SAM, and NADPH (Fig. 12.14a). In these conditions, DML production was observed. These results confirm the function of LovG. With these results in hand, the gene sequence encoding LovG was cloned into yeast that already expressed LovB and LovC. These cultures were able to produce DML without the addition of any exogenous substrates. Furthermore, when the yeast was cloned with the genes to express LovA and its CPR partner under the control of a galactose promoter, it was possible to induce the formation of DML (**9**) (no galactose) as well as oxidation products monacolin L and monacolin J (**11**) (galactose added) (Fig. 12.14b). In this system, the majority of the enzymes responsible for lovastatin biosynthesis were reconstituted in a single heterologous host, which is able to produce monacolin J in whole-cell fermentation assays. The addition of *lovD* and *lovF* to this yeast system could reasonably generate a host capable of producing lovastatin, or alternatively, this could be engineered into a simvastatin-producing organism.

Comparison of Compactin and Lovastatin Biosynthesis

To this point, much of the discussion has centered on the well-studied lovastatin biosynthetic machinery from *A. terreus*. Concurrent programs focused on isolating the gene clusters responsible for the biosynthesis of compactin in *P. citrinum* [54] and also for lovastatin from an alternate lovastatin producing species, *Monascus pilosus* [55]. Not surprisingly, the overall architecture of these two gene clusters (*mlc* and *mok* respectively) has high homology to the *lov* system. A comparison of the gene clusters is shown in Fig. 12.15. In each case, knockout studies were conducted to confirm the action of these clusters in the production of their cognate secondary metabolite. In all three cases, two PKSs were identified (one with a nonfunctional ER), an exogenous

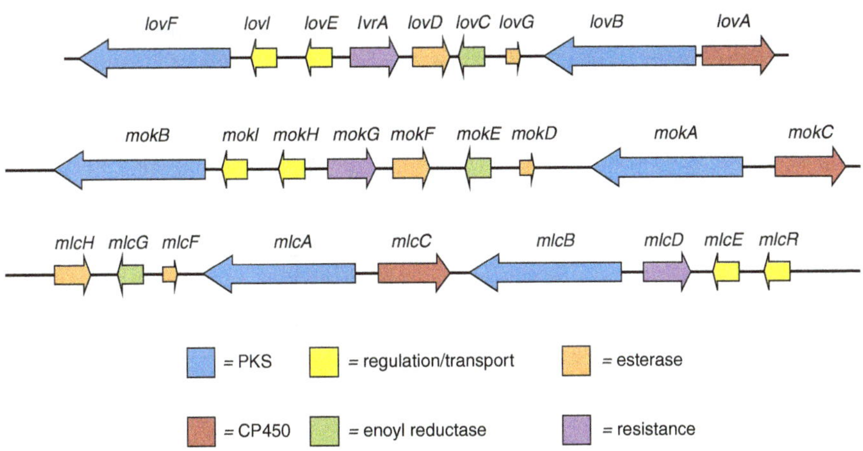

Fig. 12.15 Comparison of two lovastatin (*lov* and *mok*) and a compactin (*mlc)* biosynthetic gene cluster

ER, a CP450 oxidase, a transesterase, and a gene proposed to be an oxidase, which in the case of *A. terreus*, was shown to encode another esterase (*lovG*, homologous to *mokD* and *mlcF*).

Three other genes were identified that are present in all three species: *lovE* (homologous to *mlcR* and *mokH*) that encodes for a transcription factor, *lvrA* (formerly *orf8*, homologous to *mlcD* and *mokG*) that encodes for an HMG-CoA reductase and is involved in resistance in the host species, and *lovI* (formerly *orf10*, homologous to *mlcE, mokI*) that is an efflux pump. The remarkable similarity between these three species in terms of the overall assembly of the gene cluster, as well as protein sequence similarity, points to the fact that lovastatin and compactin are produced following the same biosynthetic steps. One interesting feature was noted when the Yuan group reported the *mok* gene cluster. When considering the *lov* system, it showed higher overall homology to *mok* than *mlc*. This may not be surprising when you consider that *lov* and *mok* produce the same product, however *A. terreus* and *P. citrinum* belong to the family Trichocomaceae, while *M. pilosus* is in the Monascaceae family. If these genes were to follow similarity based on evolution, it is expected that the *lov* system would be closer in homology with *mlc*. Also, the number and location of the introns in the MeT domain of MlcA was different than that in MokA and LovB. The amino acid sequence in the MeT domain of LovB and MokA conforms to the MeT consensus sequence. In MlcA, a number of amino acid mutations were noted that could lead to inactive MeT domain. With the sequences of these species in hand, researchers are currently in a position to further understand the biosynthetic programming steps involved in lovastatin and compactin biosynthesis at the amino acid level.

Fungal Polyketides with Similar PKS Machinery

Other fungal polyketide metabolites have been isolated that share important biosynthetic properties with lovastatin and compactin, but the spectrum of bioactivity varies widely. Here, we chose to briefly introduce three of them: solanapyrone B (**30**), equisetin (**31**), and cytochalasin E (**32**). The biosynthesis of each of these three fungal polyketide metabolites involves important similar features that are present in lovastatin and compactin biosynthesis.

Solanapyrones, including solanapyrone B (**30**), are phytotoxins isolated from *Alternaria solani* [56]. The biosynthesis of solanapyrones also follows a PKS pathway, using eight acetate units and two molecules of SAM. As discussed earlier, an enzyme-catalyzed Diels-Alder reaction is responsible for formation of the didehydrodecalin fused ring system [21]. One difference of note is that in the case of solanapyrones, it was shown that a putative oxidase involved in post-PKS modification is responsible for the cycloaddition reaction. The biosynthetic gene cluster of **30** revealed a PKS (Sol1) that contains a functional ER domain, and does not have a residual C-terminal NRPS fragment [57]. This lack of the NRPS fragment is a point of interest for genetic analysis of novel Diels-Alder catalyzing enzymes. In lovastatin,

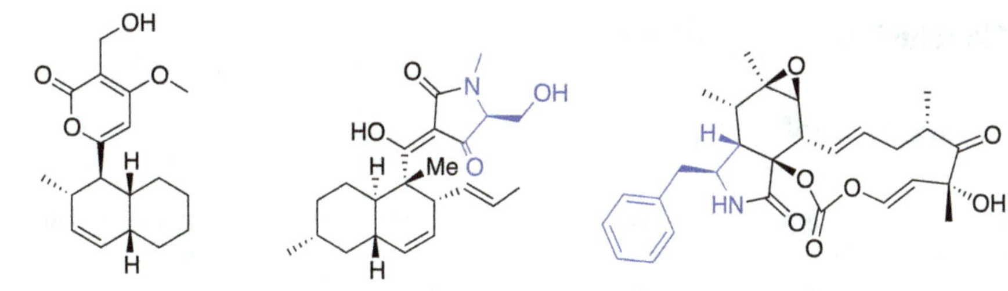

solanapyrone B (**30**)
Alternaria solani

equisetin (**31**)
Fursarium heterosporum

cytochalasin E (**32**)
Aspergillus clavatus

Fig. 12.16 Structurally related fungal polyketides that use similar biosynthetic machinery but do not display anticholesterolemic activity. Amino acid moieties are shown in *blue*

compactin and **31** and **32** (see later), it is proposed that the PKS catalyzes the [4+2] cyclization reaction, and these enzymes all contain an NRPS module. Interestingly, a dedicated enzyme catalyzing the Diels-Alder reaction in the biosynthesis of spinosyn A was recently described [58, 59].

Equisetin (**31**) is a fungal secondary metabolite of *Fusarium heterosporum*, which is an inhibitor of HIV-1 integrase [60]. Through labeling studies, it became clear that this metabolite incorporates a single molecule of serine (blue in Fig. 12.16) into the polyketide skeleton, which is further transformed into the tetramic acid moiety. Analysis and sequence of the biosynthetic gene cluster responsible for the biosynthesis of **31** showed several similarities to lovastatin [61]. This includes a highly reducing iterative PKS (EqiS) that resembles LovB, and requires an endogenous enoyl reductase. The bicyclic core is likely generated by a Diels-Alder cyclization, albeit with different overall stereoselectivity. One intriguing difference was at the C-terminal region of the PKS. While LovB bears only a partial NRPS module, EqiS contains the CON domain as well as adenylation (A) and thiolation (T) domains of a fully functional NRPS module. This result accounts for the addition of the serine moiety.

Cytochalasin E (**32**) is one member of a large family of fungal metabolites known collectively as cytochalasins and chaetoglobosins [62]. These molecules have a wide spectrum of activity. In the case of **32**, it displays anti-angiogenic activity. Similar to equisetin, **32** incorporates an amino acid (phenylalanine, blue in Fig. 12.16) into its PKS skeleton. An intramolecular Diels-Alder reaction is proposed in the biosynthesis of cytochalasin E, but in this case it forms the 5,6-fused ring system, once a tetramic acid moiety has been generated. The biosynthetic gene cluster of **32** was recently disclosed, and a PKS that showed similarity to both LovB and EqiS was described [63]. Similar to EqiS, CcsA is a highly reducing iterative PKS, lacking a functional ER domain, and bearing a full NRPS module at the C-terminus. Current studies are underway to study the programming of the various post-PKS steps that must be involved in the biosynthesis of **32**, including the Diels-Alder reaction and the various oxidations steps, which includes a possible Baeyer-Villiger monoxygenase.

Conclusion

Future Outlook

Compactin (**1**) and lovastatin (**2**) are fungal secondary metabolites with incredible medicinal value. These compounds spawned the statin revolution for pharmaceutical cholesterol management, which generates billions of dollars of revenue annually. While the importance of these compounds for human health cannot be overlooked, they also have interesting biochemistry that has generated some important discoveries in basic science. These were among the first metabolites demonstrated to be synthesized by a highly reducing iterative PKS (LovB or MlcA). The PKS requires the participation of an endogenous enoyl reductase protein (LovC or MlcG) to generate the desired nonaketide intermediates. (For a recent review of fungal secondary metabolites that use similar machinery, see Boettger and Hertweck [64] and for in depth reviews on fungal PKSs, see Cox and Simpson [65] and Fujii [66].) Remarkable programming steps are involved in ensuring correct domain use during each round of chain elongation. The polyketide intermediate undergoes an enzyme-catalyzed stereoselective Diels-Alder cyclization to generate the bicyclic core. It is possible that the CON domain of the partial NRPS at the C-terminus of LovB can contribute to this reaction, but this remains to be seen. Despite the current evidence, controversy in the ability for enzymes to catalyze a Diels-Alder reaction exists [25]. This can be observed in the enzyme macrophomate synthase [67–69]. Two oxidative processing steps that act on the nonaketide intermediate are carried out by a single CP450 oxidase (LovA or MlcC). The transesterase LovD (MlcH) is involved in esterification of the diketide side chain, and shows promise in developing whole-cell biocatalytic processes to manufacture new statins. Metabolic engineering efforts have cloned nearly every enzyme in the lovastatin biosynthetic pathway into a single yeast strain, which can be used for host-free production of this metabolite.

The important genetic and metabolic information that was learned in the process of determining the function of each of the lovastatin biosynthetic gene products can be applied to the production of other fungal secondary metabolites. This is of particular interest in the field of synthetic biology. Lovastatin can serve as a template for the host-free preparation of natural and unnatural metabolites with a wide range of important biological activity.

References

1. Go AS, Mozaffarian D, Roger VL, Benjamin EJ, Berry JD, Borden WB, et al. Heart disease and stroke statistics-2013 update: a report from the American Heart Association. Circulation. 2013;127(1):E6–245.
2. Endo A, Kuroda M, Tsujita Y. ML-236a, ML-236b, and ML-236c, new inhibitors of cholesterogenesis produced by *Penicillium citrinum*. J Antibiot (Tokyo). 1976;29(12): 1346–8.
3. Alberts AW, Chen J, Kuron G, Hunt V, Huff J, Hoffman C, et al. Mevinolin—a highly potent competitive inhibitor of hydroxymethylglutaryl-coenzyme-A reductase and a cholesterol-lowering agent. Proc Natl Acad Sci U S A. 1980;77(7):3957–61.

4. Endo A. Monacolin K, a new hypocholesterolemic agent produced by a *Monascus* species. J Antibiot (Tokyo). 1979;32(8):852–4.
5. Brown AG, Smale TC, King TJ, Hasenkamp R, Thompson RH. Crystal and molecular structure of compactin, a new antifungal metabolite from *Penicillium brevicompactum*. J Chem Soc Perkin 1. 1976(11):1165–73.
6. Endo A. The discovery and development of HMG-CoA reductase inhibitors. J Lipid Res. 1992;33(11):1569–82.
7. Endo A, Hasumi K. HMG-CoA reductase inhibitors. Nat Prod Rep. 1993;10(6):541–50.
8. Kita T, Brown MS, Goldstein JL. Feedback regulation of 3-hydroxy-3-methyglutaryl coenzyme A reductase in livers of mice treated with mevinolin, a competitive inhibitor of the reductase. J Clin Invest. 1980;66(5):1094–100.
9. Goldstein JL, Brown MS. Regulation of the mevalonate pathway. Nature. 1990;343(6257): 425–30.
10. Campbell CD, Vederas JC. Biosynthesis of lovastatin and related metabolites formed by fungal iterative PKS enzymes. Biopolymers. 2010;93(9):755–63.
11. Hutchinson CR, Kennedy J, Park C, Auclair K, Kendrew S, Vederas J. Molecular genetics of lovastatin and compactin biosynthesis. Handbook of industrial mycology. New York: Marcel Dekker; 2005. p. 479–91.
12. Hutchinson CR, Kennedy J, Park C, Kendrew S, Auclair K, Vederas J. Aspects of the biosynthesis of non-aromatic fungal polyketides by iterative polyketide synthases. Antonie Van Leeuwenhoek. 2000;78(3–4):287–95.
13. Moore RN, Bigam G, Chan JK, Hogg AM, Nakashima TT, Vederas JC. Biosynthesis of the hypocholesterolemic agent mevinolin by *Aspergillus terreus*—determination of the origin of carbon, hydrogen, and oxygen atoms by C13-NMR and mass spectrometry. J Am Chem Soc. 1985;107(12):3694–701.
14. Wagschal K, Yoshizawa Y, Witter DJ, Liu YQ, Vederas JC. Biosynthesis of ML-236C and the hypocholesterolemic agents compactin by *Penicillium aurantiogriseum* and lovastatin by *Aspergillus terreus*: determination of the origin of carbon, hydrogen and oxygen atoms by C13-NMR spectrometry and observation of unusual labelling of acetate-derived oxygens by 18O2. J Chem Soc Perkin 1. 1996(19):2357–63.
15. Yoshizawa Y, Witter DJ, Liu YQ, Vederas JC. Revision of the biosynthetic origin of oxygens in mevinolin (lovastatin), a hypocholesterolemic drug from *Aspergillus terreus* MF-4845. J Am Chem Soc. 1994;116(6):2693–4.
16. Chan JK, Moore RN, Nakashima TT, Vederas JC. Biosynthesis of mevinolin—spectral assignment by double-quantum coherence NMR after high C-13 incorporation. J Am Chem Soc. 1983;105(10):3334–6.
17. Endo A, Negishi Y, Iwashita T, Mizukawa K, Hirama M. Biosynthesis of ML-236b (compactin) and monacolin K. J Antibiot (Tokyo). 1985;38(3):444–8.
18. Greenspan MD, Yudkovitz JB. Mevinolinic acid biosynthesis by *Aspergillus terreus* and its relationship to fatty-acid biosynthesis. J Bacteriol. 1985;162(2):704–7.
19. Treiber LR, Reamer RA, Rooney CS, Ramjit HG. Origin of monacolin L from *Aspergillus terreus* cultures. J Antibiot (Tokyo). 1989;42(1):30–6.
20. Tamm C. The biosynthesis of mycotoxins. New York: Academic Press; 1980.
21. Oikawa H, Yokota T, Abe T, Ichihara A, Sakamura S, Yoshizawa Y, et al. Biosynthesis of solanapyrone A, a phytotoxin of *Alternaria solani*. J Chem Soc Chem Commun. 1989;17:1282–4.
22. Kelly WL. Intramolecular cyclizations of polyketide biosynthesis: mining for a "Diels-Alderase"? Org Biomol Chem. 2008;6(24):4483–93.
23. Kim HJ, Ruszczycky MW, Liu HW. Current developments and challenges in the search for a naturally selected Diels-Alderase. Curr Opin Chem Biol. 2012;16(1–2):124–31.
24. Oikawa H, Tokiwano T. Enzymatic catalysis of the Diels-Alder reaction in the biosynthesis of natural products. Nat Prod Rep. 2004;21(3):321–52.
25. Stocking EM, Williams RM. Chemistry and biology of biosynthetic Diels-Alder reactions. Angew Chem Int Ed Engl. 2003;42(27):3078–115.
26. Witter DJ, Vederas JC. Putative Diels-Alder-catalyzed cyclization during the biosynthesis of lovastatin. J Org Chem. 1996;61(8):2613–23.

27. Auclair K, Sutherland A, Kennedy J, Witter DJ, Van den Heever JP, Hutchinson CR, et al. Lovastatin nonaketide synthase catalyzes an intramolecular Diels-Alder reaction of a substrate analogue. J Am Chem Soc. 2000;122(46):11519–20.
28. Katayama K, Kobayashi T, Oikawa H, Honma M, Ichihara A. Enzymatic activity and partial purification of solanapyrone synthase: first enzyme catalyzing Diels-Alder reaction. Biochim Biophys Acta. 1998;1384(2):387–95.
29. Oikawa H, Kobayashi T, Katayama K, Suzuki Y, Ichihara A. Total synthesis of (−)-solanapyrone A via enzymatic Diels-Alder reaction of prosolanapyrone. J Org Chem. 1998;63(24):8748–56.
30. Cane DE, Walsh CT, Khosla C. Harnessing the biosynthetic code: combinations, permutations, and mutations. Science. 1998;282(5386):63–8.
31. Hendrickson L, Davis CR, Roach C, Nguyen DK, Aldrich T, McAda PC, et al. Lovastatin biosynthesis in *Aspergillus terreus*: characterization of blocked mutants, enzyme activities and a multifunctional polyketide synthase gene. Chem Biol. 1999;6(7):429–39.
32. Kennedy J, Auclair K, Kendrew SG, Park C, Vederas JC, Hutchinson CR. Modulation of polyketide synthase activity by accessory proteins during lovastatin biosynthesis. Science. 1999;284(5418):1368–72.
33. Hopwood DA. Molecular genetics of polyketides and its comparison to fatty-acid biosynthesis. Annu Rev Genet. 1990;24:37–66.
34. Burr DA, Chen XB, Vederas JC. Syntheses of conjugated pyrones for the enzymatic assay of lovastatin nonaketide synthase, an iterative polyketide synthase. Org Lett. 2007;9(1):161–4.
35. Guenzi E, Galli G, Grgurina I, Gross DC, Grandi G. Characterization of the syringomycin synthetase gene cluster—a link between prokaryotic and eukaryotic peptide synthetases. J Biol Chem. 1998;273(49):32857–63.
36. Rangaswamy V, Jiralerspong S, Parry R, Bender CL. Biosynthesis of the *Pseudomonas* polyketide coronafacic acid requires monofunctional and multifunctional polyketide synthase proteins. Proc Natl Acad Sci U S A. 1998;95(26):15469–74.
37. Sorensen JL, Vederas JC. Monacolin N, a compound resulting from derailment of type I iterative polyketide synthase function en route to lovastatin. Chem Commun. 2003;13: 1492–3.
38. Ma SM, Li JWH, Choi JW, Zhou H, Lee KKM, Moorthie VA, et al. Complete reconstitution of a highly reducing iterative polyketide synthase. Science. 2009;326(5952):589–92.
39. Kim YT, Lee YR, Jin JM, Han KH, Kim H, Kim JC, et al. Two different polyketide synthase genes are required for synthesis of zearalenone in *Gibberella zeae*. Mol Microbiol. 2005;58(4):1102–13.
40. Ames BD, Nguyen C, Bruegger J, Smith P, Xu W, Ma S, et al. Crystal structure and biochemical studies of the trans-acting polyketide enoyl reductase LovC from lovastatin biosynthesis. Proc Natl Acad Sci U S A. 2012;109(28):11144–9.
41. Kimura K, Komagata D, Murakawa S, Endo A. Biosynthesis of monacolins—conversion of monacolin J to monacolin K (mevinolin). J Antibiot (Tokyo). 1990;43(12):1621–2.
42. Auclair K, Kennedy J, Hutchinson CR, Vederas JC. Conversion of cyclic nonaketides to lovastatin and compactin by a LovC deficient mutant of *Aspergillus terreus*. Bioorg Med Chem Lett. 2001;11(12):1527–31.
43. Xie XK, Watanabe K, Wojcicki WA, Wang CCC, Tang Y. Biosynthesis of lovastatin analogs with a broadly specific acyltransferase. Chem Biol. 2006;13(11):1161–9.
44. Xie XK, Meehan MJ, Xu W, Dorrestein PC, Tang Y. Acyltransferase mediated polyketide release from a fungal megasynthase. J Am Chem Soc. 2009;131(24):8388–9.
45. Hoffman WF, Alberts AW, Anderson PS, Chen JS, Smith RL, Willard AK. 3-Hydroxy-3-methylglutaryl-coenzyme A reductase inhibitors. 4. Side-chain ester derivatives of mevinolin. J Med Chem. 1986;29(5):849–52.
46. Xie XK, Tang Y. Efficient synthesis of simvastatin by use of whole-cell biocatalysis. Appl Environ Microbiol. 2007;73(7):2054–60.
47. Xie XK, Wong WW, Tang Y. Improving simvastatin bioconversion in *Escherichia coli* by deletion of *bioH*. Metab Eng. 2007;9(4):379–86.

48. Xie X, Pashkov I, Gao X, Guerrero JL, Yeates TO, Tang Y. Rational improvement of simvastatin synthase solubility in *Escherichia coli* leads to higher whole-cell biocatalytic activity. Biotechnol Bioeng. 2009;102(1):20–8.
49. Gao X, Xie XK, Pashkov I, Sawaya MR, Laidman J, Zhang WJ, et al. Directed evolution and structural characterization of a simvastatin synthase. Chem Biol. 2009;16(10):1064–74.
50. Nakamura T, Komagata D, Murakawa S, Sakai K, Endo A. Isolation and biosynthesis of 3-alpha-hydroxy-3,5-dihydromonacolin-L. J Antibiot (Tokyo). 1990;43(12):1597–600.
51. Sorensen JL, Auclair K, Kennedy J, Hutchinson CR, Vederas JC. Transformations of cyclic nonaketides by *Aspergillus terreus* mutants blocked for lovastatin biosynthesis at the *lovA* and *lovC* genes. Org Biomol Chem. 2003;1(1):50–9.
52. Barriuso J, Nguyen DT, Li JWH, Roberts JN, MacNevin G, Chaytor JL, et al. Double oxidation of the cyclic nonaketide dihydromonacolin L to monacolin J by a single cytochrome P450 monooxygenase, LovA. J Am Chem Soc. 2011;133(21):8078–81.
53. Xu W, Chooi YH, Choi JW, Li S, Vederas JC, Da Silva NA, et al. LovG: the thioesterase required for dihydromonacolin L release and lovastatin nonaketide synthase turnover in lovastatin biosynthesis. Angew Chem Int Ed Engl. 2013;52(25):6472–5.
54. Abe Y, Suzuki T, Ono C, Iwamoto K, Hosobuchi M, Yoshikawa H. Molecular cloning and characterization of an ML-236B (compactin) biosynthetic gene cluster in *Penicillium citrinum*. Mol Genet Genomics. 2002;267(5):636–46.
55. Chen YP, Tseng CP, Liaw LL, Wang CL, Chen IC, Wu WJ, et al. Cloning and characterization of monacolin K biosynthetic gene cluster from *Monascus pilosus*. J Agric Food Chem. 2008;56(14):5639–46.
56. Ichihara A, Tazaki H, Sakamura S. Solanapyrones A, B and C, phytotoxic metabolites from the fungus *Alternaria solani*. Tetrahedron Lett. 1983;24(48):5373–6.
57. Kasahara K, Miyamoto T, Fujimoto T, Oguri H, Tokiwano T, Oikawa H, et al. Solanapyrone synthase, a possible Diels-Alderase and iterative type I polyketide synthase encoded in a biosynthetic gene cluster from *Alternaria solani*. Chembiochem. 2010;11(9):1245–52.
58. Hess BA, Smentek L. Concerted, highly asynchronous, enzyme-catalyzed [4+2] cycloaddition in the biosynthesis of spinosyn A; computational evidence. Org Biomol Chem. 2012;10(37):7503–9.
59. Kim HJ, Ruszczycky MW, Choi SH, Liu YN, Liu HW. Enzyme-catalysed [4+2] cycloaddition is a key step in the biosynthesis of spinosyn A. Nature. 2011;473(7345):109–12.
60. Singh SB, Zink DL, Goetz MA, Dombrowski AW, Polishook JD, Hazuda DJ. Equisetin and a novel opposite stereochemical homolog phomasetin, two fungal metabolites as inhibitors of HIV-1 integrase. Tetrahedron Lett. 1998;39(16):2243–6.
61. Sims JW, Fillmore JP, Warner DD, Schmidt EW. Equisetin biosynthesis in *Fusarium heterosporum*. Chem Commun. 2005;2:186–8.
62. Scherlach K, Boettger D, Remme N, Hertweck C. The chemistry and biology of cytochalasans. Nat Prod Rep. 2010;27(6):869–86.
63. Qiao KJ, Chooi YH, Tang Y. Identification and engineering of the cytochalasin gene cluster from *Aspergillus clavatus* NRRL 1. Metab Eng. 2011;13(6):723–32.
64. Boettger D, Hertweck C. Molecular diversity sculpted by fungal PKS-NRPS hybrids. Chembiochem. 2013;14(1):28–42.
65. Cox RJ, Simpson TJ. Fungal type I polyketide synthases. Methods Enzymol. 2009;459:49–78.
66. Fujii I. Functional analysis of fungal polyketide biosynthesis genes. J Antibiot (Tokyo). 2010;63(5):207–18.
67. Guimaraes CRW, Udier-Blagovic M, Jorgensen WL. Macrophomate synthase: QM/MM simulations address the Diels-Alder versus Michael-Aldol reaction mechanism. J Am Chem Soc. 2005;127(10):3577–88.
68. Ose T, Watanabe K, Mie T, Honma M, Watanabe H, Yao M, et al. Insight into a natural Diels-Alder reaction from the structure of macrophomate synthase. Nature. 2003;422(6928):185–9.
69. Ose T, Watanabe K, Yao M, Honma M, Oikawa H, Tanaka I. Structure of macrophomate synthase. Acta Crystallogr D Biol Crystallogr. 2004;60:1187–97.

Chapter 13
Meroterpenoids

Yudai Matsuda and Ikuro Abe

Introduction

Meroterpenoids are hybrid natural products that are partially derived from terpenoid origins, and thus they have the prefix "mero-", meaning "part, partial, or segment" [1]. The fusion of the structurally diverse terpenoid and non-terpenoid moieties is a characteristic of the unique and complex molecular structures of this class of natural products. Meroterpenoids are widely distributed among plants, actinomycetes, and fungi, and the fungal meroterpenoids have especially diversified chemical structures with a wide range of biological activities, as exemplified by pyripyropenes, arisugacins, and territrems [2, 3]. The pyripyropenes produced by *Aspergillus fumigatus* have very strong inhibitory activity against acyl-coenzyme A: cholesterol acyltransferase (ACAT). They selectively inhibit the ACAT-2 isoform, with much less inhibitory activity against the ACAT-1 isoform. This feature makes the pyripyropenes promising leads for cholesterol-lowering and anti-atherosclerotic drugs, since ACAT-2 is expressed predominantly in the liver and intestine, while ACAT-1 is expressed ubiquitously [4]. On the other hand, arisugacins and territrems, isolated from *Penicillium* sp. and *Aspergillus terreus*, respectively, show selective acetylcholinesterase inhibitory activities and are expected to be developed as clinical drugs for Alzheimer's disease.

As described above, the molecular structures and the biological activities of fungal meroterpenoids are attracting great interest, although the details of the biosynthetic pathways and the enzymes involved in the production of the compounds have remained elusive for a long time. However, the sequences of many fungal genomes are now available, thus accelerating the discovery of biosynthetic gene clusters for fungal meroterpenoids and biosynthetic studies. Two different approaches have been undertaken primarily to elucidate the functions of the biosynthetic genes: the

Y. Matsuda, M.Sc. • I. Abe, Ph.D. (✉)
Graduate School of Pharmaceutical Sciences, The University of Tokyo, Tokyo, Japan
e-mail: abei@mol.f.u-tokyo.ac.jp

J.-F. Martín et al. (eds.), *Biosynthesis and Molecular Genetics of Fungal Secondary Metabolites*, Fungal Biology, DOI 10.1007/978-1-4939-1191-2_13,

reconstitution of the biosynthetic pathway in a heterologous host, and the disruption of each gene in the cluster. This chapter summarizes the recent studies on several fungal meroterpenoids, such as pyripyropene A, meroterpenoids derived from 3,5-dimethylorsellinic acid (DMOA), and indole-diterpenes.

Pyripyropene A

Pyripyropene A (**1**) has a unique carbon structure composed of a tricyclic C15-terpenoid moiety and a polyketide-derived pyrone with an attached pyridine ring. Previous precursor incorporation studies demonstrated the biosynthetic origin of **1**, and based on the study, the biosynthetic pathway for **1** was predicted as follows [5]: First, nicotinic acid (**2**)-derived coenzyme A (CoA), nicotinyl-CoA (**3**), is introduced into the polyketide pathway and is condensed with two molecules of malonyl-CoA to form the pyridino-α(alpha)-pyrone moiety (**4**). Farnesyl pyrophosphate (FPP, **5**), produced from three mevalonate molecules, is then attached to **4**. The following epoxidation of the terminal double bond and cyclization of the terpenoid moiety yield the basic core structure of the pyripyropenes (**9**). Subsequent hydroxylations and acetylations at three distinct positions finally produce the structure of **1** (Fig. 13.1). Based on the predicted pathway, pyripyropene A biosynthesis should require the involvement of a polyketide synthase (PKS), a prenyltransferase (PT), and a terpene cyclase, as well as several modification enzymes, such as a hydroxylase and an acetyltransferase. The biosynthetic genes for a specific compound are adjacent to each other and clustered on the chromosome in microorganisms. Therefore, the biosynthetic gene cluster for **1** was searched in a publicly available

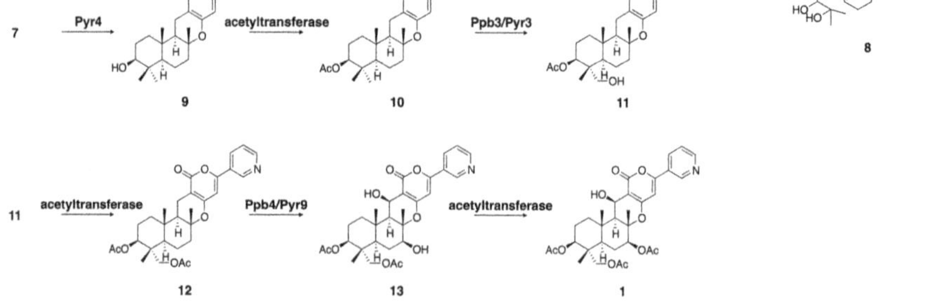

Fig. 13.1 Biosynthetic pathway for pyripyropene A (**1**)

genome database of *A. fumigatus* Af293, and a 23-kb gene cluster with nine genes (the *pyr* cluster) was found [6]. The cluster included PKS, PT, CoA ligase, flavin adenine dinucleotide (FAD)-dependent monooxygenase (FMO), cytochrome P450 monooxygenase (P450), and acetyltransferase genes, which are all presumed to be involved in the biosynthesis of pyripyropene A, although none of the nine genes encodes a protein homologous to the known terpene cyclases.

To characterize the functions of the genes in the cluster, a reconstitution approach was undertaken using *Aspergillus oryzae* as a heterologous expression host [6]. First, the *pyr1* and *pyr2* genes, encoding CoA-ligase and type I PKS, respectively, were coexpressed in *A. oryzae* under the regulation of the amylase promoter, and the transformant was cultured in induction medium supplemented with the starter substrate **2**. As a result, the predicted polyketide intermediate, 4-hydroxy-6-(3-pyridinyl)-2*H*-pyran-2-one (HPPO, **4**), was obtained. Next, the function of the prenyltransferase encoded by *pyr6*, which shares sequence homology with UbiA, an enzyme involved in ubiquinone biosynthesis, was examined. To this end, *pyr6* was coexpressed with the CoA ligase and PKS genes to give the farnesylated compound farnesyl-HPPO (**6**), which was further converted to dihydroxyfarnesyl-HPPO (**8**) by an FMO, Pyr5. The diol product **8** was presumably derived from the hydrolysis of epoxyfarnesyl-HPPO (**7**), and thus Pyr5 should be the epoxidase responsible for the epoxidation of the terminal olefin of **6**.

The subsequent biosynthetic step should be the cyclization of the terpenoid moiety, but as mentioned previously, no gene in the cluster encodes a protein homologous to the known terpene cyclases. Since the P450s and acetyltransferases encoded by the cluster did not seem to be involved in the cyclization reaction, the only possibility appeared to be the gene encoding a putative "integral membrane protein," *pyr4*, a protein that consists of 242 amino acid residues. Importantly, homologues of *pyr4* are widely distributed among the biosynthetic gene clusters for several meroterpenoids in fungi, as described later, as well as in actinomycetes [7]. The wide existence of *pyr4* homologues suggested that they play an important role in the biosynthesis of meroterpenoids, and may be involved in the cyclization reaction. To test whether Pyr4 is responsible for the cyclization, *pyr4* was coexpressed with the PT and FMO genes and incubated with **4**, and the cyclized product deacetyl-pyripyropene E (**9**) was obtained. Furthermore, in an in vitro assay of Pyr4, using the microsome fraction of Pyr4-expressing fungi, chemically synthesized **7** was successfully transformed into **9**, demonstrating that Pyr4 is a novel and independently functioning terpene cyclase.

The biosynthetic gene cluster for **1** was also identified in another filamentous fungus, *Penicillium coprobium* PF1169, and the cluster is quite similar to that of *A. fumigatus*, with all nine genes corresponding to each gene in the *A. fumigatus* cluster [8]. Two P450s encoded by the cluster, Ppb3 (95 % identity with Pyr3) and Ppb4 (74 % identity with Pyr9), were expressed separately in *A. oryzae,* and the transformants were incubated with several substrate candidates. The biotransformation experiment revealed that Ppb3 catalyzes the hydroxylation of pyripyropene E (**10**) to give 11-deacetyl-pyripyropene O (**11**), whereas Ppb4 hydroxylates pyripyropene O (**12**) at two distinct positions to form deacetyl-pyripyropene A (**13**), suggesting that Pyr3 and Pyr9, whose functions have yet to be elucidated, possess the same activities as Ppb3 and Ppb4, respectively. In pyripyropene A biosynthesis, it is still unclear which enzymes are responsible for the acetylation reactions. Two acetyltransferases, Pyr7 and Pyr8, are considered to be involved in the reactions, but their functions have not been characterized.

Meroterpenoids Derived from 3,5-Dimethylorsellinic Acid

3,5-Dimethylorsellinic acid (DMOA, **14**)-derived meroterpenoids comprise an especially large number of structurally diverse compounds among the fungal meroterpenoids, including austinol (**16**), terretonin (**17**), andrastin A (**18**), and anditomin (**19**) [9–12] (Fig. 13.2). Previous precursor incorporation studies indicated that these compounds are all derived from the polyketide intermediate **14** and FPP (**5**), but variations in the terpenoid moiety cyclization and post-cyclization modification reactions contribute to the structural diversity of the compounds [13–16]. According to the biosynthetic origins of the DMOA-derived meroterpenoids, the biosyntheses of these compounds should also require the involvement of PKSs, prenyltransferases, and presumably novel terpene cyclases homologous to Pyr4, as in the biosynthesis of pyripyropene A (**1**).

The first gene involved in the biosynthesis of DMOA-derived meroterpenoids was identified in the genome database of *Aspergillus nidulans* FGSC A4, by the disruption of all of the PKS genes in *A. nidulans* [17]. The gene, designated as *ausA*, encodes a type I PKS that synthesizes **14** from one molecule of acetyl-CoA and three molecules of malonyl-CoA, and is responsible for the production of **16** and dehydroaustinol (**20**). A subsequent gene disruption study was undertaken to identify the whole biosynthetic gene cluster for **16** and **20** (the *aus* cluster). Interestingly, the results revealed that two separately located gene clusters are both responsible for the production of **16** and **20**; the first cluster includes the PKS gene and the second one includes the PT and terpene cyclase genes [18]. To fully understand the biosynthetic pathway for **16** and **20**, all of the genes in the cluster, as well as the genes located nearby, were deleted one by one, thus demonstrating the involvement of 14 genes in the biosyntheses of **16** and **20**.

First, the deletion of *ausN*, encoding a PT that is also homologous to UbiA and Pyr6, caused the accumulation of the polyketide intermediate **14**, suggesting that AusN attaches a farnesyl moiety to **14** to yield the prenylated polyketide (**15**).

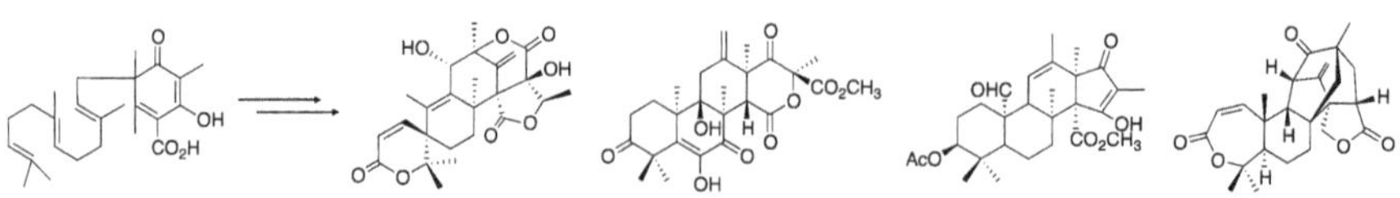

15 16 17 18 19

Fig. 13.2 Representative DMOA-derived meroterpenoids

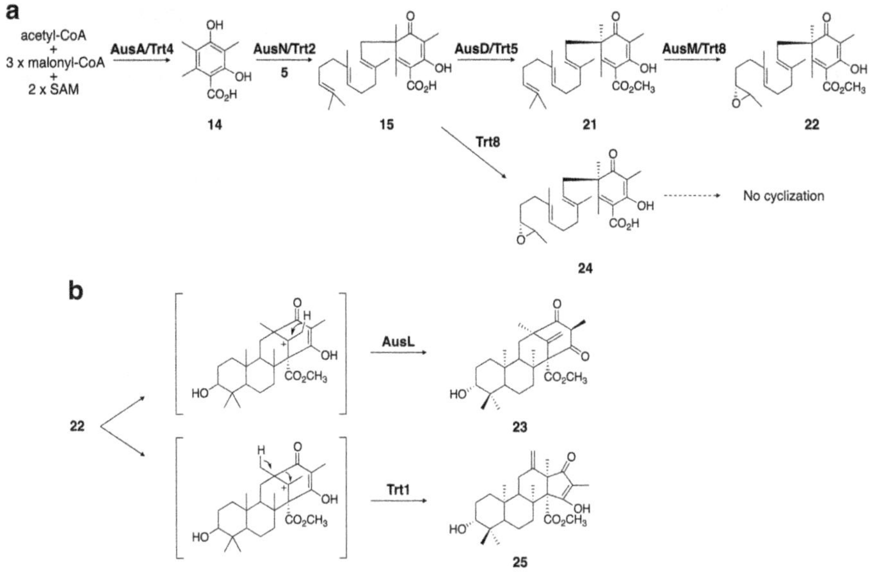

Fig. 13.3 (**a**) Shared biosynthetic pathway in austinol (**16**) and terretonin (**17**) syntheses, and (**b**) differences in the cyclization reactions

The next biosynthetic steps include the epoxidation of the terminal double bond of **15** and the cyclization of the terpenoid moiety, and these reactions are presumably catalyzed by an FMO, AusM, and a terpene cyclase, AusL, respectively. In addition, since the predicted AusL-catalyzed product, protoaustinoid A (**23**), possesses a methyl carboxylate group, the methylation reaction, which is possibly catalyzed by the methyltransferase AusD, must occur before the cyclization reaction by AusL (Fig. 13.3a). However, when either *ausD*, *ausM*, or *ausL* was deleted, no biosynthetic intermediate was observed, and therefore the functions of these genes as well as that of *ausN*, and the biosynthetic scheme to **23** from **14**, were still ambiguous at that point.

The biosynthetic pathway toward **23** was subsequently confirmed through the biosynthetic study on terretonin (**17**), using a reconstitution approach [19] (Fig. 13.3a). The biosynthetic gene cluster for **17**, which consists of 13 genes (the *trt* cluster), was discovered in the genome database of *A. terreus* NIH 2624 [20]. The cluster was revised subsequently, because 4 of the 13 genes (*trt10* to *trt13*) are not actually involved in the biosynthesis of **17** and the requirement of one additional gene (*trt14*) was found [21]. In the reconstitution study using *A. oryzae* as a host, the PKS gene *trt4* (*ausA* homologue) was expressed at first to give **14**, and then the PT gene *trt2* (*ausN* homologue) was coexpressed with the PKS gene, to successfully yield the prenylated polyketide, farnesyl-DMOA (**15**). Next, the FMO gene *trt8* (*ausM* homologue) was coexpressed with the PKS and PT genes to yield the epoxide compound, epoxyfarnesyl-DMOA (**24**), but unexpectedly **24** was not accepted by the terpene cyclase encoded by *trt1* (*ausL* homologue). Given the presence of the methyl carboxylate group in the structure of **23**, it was possible that the methylation is required before the cyclization reaction. Since the methyltransferase gene *trt5* (*ausD* homologue) resides in the cluster, *trt5* was then assessed for the methylation of the carboxyl group. To characterize the function of *trt5*, *trt5* was coexpressed with the PKS gene alone or with both the PKS and PT genes. As a result, the methylated intermediate, farnesyl-DMOA methyl ester (**21**), was obtained from the three-gene expression system, whereas no methylated compound was observed from the two-gene expression system. These results revealed that the methylation occurs after the prenylation, and it was then expected that **21** would be further transformed into a cyclized product by the FMO and terpene cyclase. As expected, the cyclized product, preterretonin A (**25**), was produced by the transformant expressing all five genes, demonstrating that the cyclization reaction by Trt1 requires the methylation of the carboxyl group as an essential factor. Finally, to obtain insight into the biosynthesis of **23**, *trt1* was replaced with *ausL* in the five-gene expression system, and **23** was successfully produced (Fig. 13.3b). As in the case of Trt1, AusL only accepted the methylated substrate, epoxyfarnesyl-DMOA methyl ester (**22**), but not the non-methylated substrate **24**, which confirmed the biosynthetic pathway for **23** and suggested that methylation is widely required in the biosynthesis of DMOA-derived meroterpenoids.

The next biosynthetic pathways, from **23** to austinol (**16**) and dehydroaustinol (**20**) or from **25** to terretonin (**17**), should require many modification reactions, including carbon skeleton rearrangement, which would significantly contribute to the structural differences between **16** and **17**. The modification reactions and pathways have been predicted well by the gene disruption studies of each gene in the biosynthetic gene clusters [18, 21] (Fig. 13.4).

In the biosyntheses of **16** and **20**, **23** is accepted by another FMO, AusB. However, the product from the reaction catalyzed by AusB has not been identified, and the biosynthetic scheme from **23** to preaustinoid A3 (**26**) has also not been determined. Subsequently, **26** undergoes the acid-catalyzed keto-rearrangement and ring contraction of the polyketide moiety to generate the preaustinoid A4 (**27**), in a reaction catalyzed by AusJ, which is simply annotated as a “hypothetical protein” in the database. The C-5′ keto moiety of **27** is then reduced to a hydroxyl group, and the generated hydroxyl oxygen attacks the carbonyl of the methyl ester group to yield isoaustinone (**28**). This reaction is catalyzed by AusK, an enzyme homologous to Nor1, which reduces the 1′ keto group of norsolorinic acid to a hydroxyl group. Interestingly, the reaction catalyzed by AusK requires the support of another protein, AusH; in the absence of AusH, AusK produces a stereoisomer of **28** at the 5′ position, (5′*S*)-isoaustinone (**29**). Although AusH is simply annotated as a “hypothetical protein” and its function has yet to be elucidated, AusH could function as an accessory enzyme that works in tandem with AusK to produce **28** in a stereocontrolled manner. Two P450s, AusI and AusG, then convert **28** into austinol (**16**). In these reactions, AusI, a Baeyer-Villiger monooxygenase, inserts an oxygen atom between the two carbons at C-4′ and C-3′ to generate austinolide (**30**), and AusG catalyzes the C-11 hydroxylation of **30** to form **16**. The final biosynthetic step is the formation of **20** from **16**, but the enzyme(s) responsible for the reaction has not been identified yet. Furthermore, three additional genes, *ausC*, *ausE*, and *ausF*, are involved in the biosyntheses of **16** and **20**, but their functions also have yet to be elucidated.

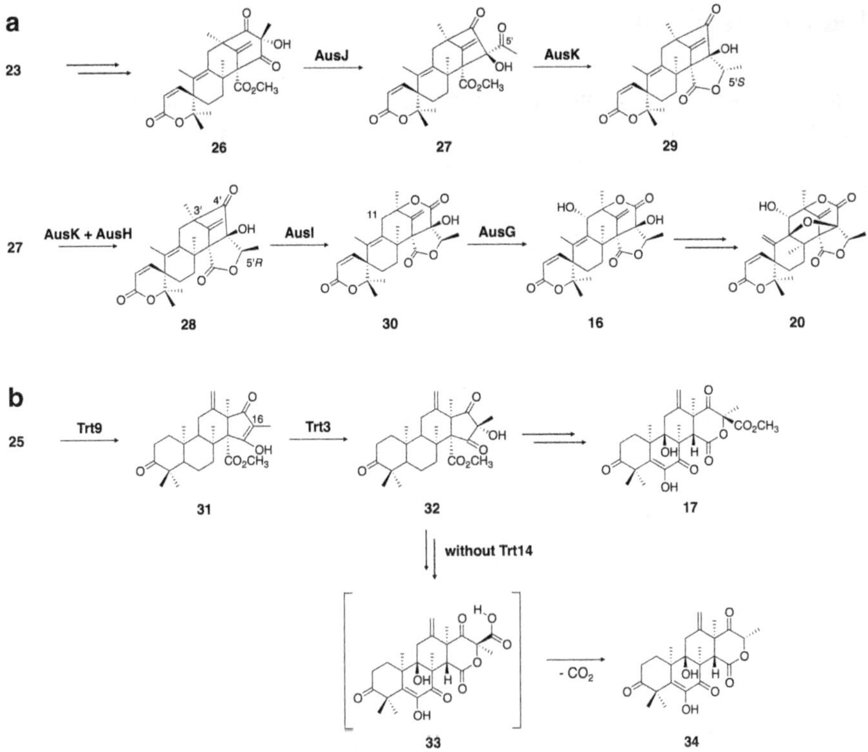

Fig. 13.4 Post-cyclization modification reactions in the biosyntheses of (**a**) austinol (**16**) and dehydroaustinol (**20**), and (**b**) terretonin (**17**)

In the biosynthesis of **17**, the C-3 hydroxyl group of **25** is oxidized to form a preterrenoid (**31**) by Trt9, a short-chain dehydrogenase, and the following hydroxylation at C-16 is catalyzed by another FMO, Trt3, to yield the terrenoid (**32**). The next biosynthetic step involves a P450, Trt6, but the product from the reaction catalyzed by Trt6 has not been identified, and the biosynthetic scheme from **32** toward **17** has yet to be determined. Interestingly, when *trt14*, which encodes a hypothetical protein, was deleted, **17** was not produced but terretonin C (**34**) was accumulated, suggesting that the deletion of *trt14* may allow the accumulation of **33**, which could be easily converted into **34** via spontaneous decarboxylation. In addition, *trt7*, which encodes a protein similar to phytanoyl-CoA dioxygenase [22], was also found to be responsible for the biosynthesis of **17**, but no intermediate was accumulated in the *trt7*-deleted strain, and therefore its function has yet to be elucidated.

The biosynthetic genes for other DMOA-derived meroterpenoids, such as andrastin A (**18**) and anditomin (**19**), have not been identified, but they should also share the same biosynthetic pathways as those for austinol (**16**) and terretonin (**17**), and enzymes homologous to those for **16** and **17** production should also be involved in their biosyntheses.

Indole-Diterpenes

Indole-diterpenes also comprise a structurally diverse group among the fungal meroterpenoids, and many of them are known as potent tremorgenic mammalian mycotoxins [23]. All indole-diterpenes share the same biosynthetic intermediate, 3-geranylgeranyl-indole (**43**), which is presumably derived from indole-3-glycerol phosphate (**41**) and geranylgeranyl pyrophosphate (GGPP, **42**). As seen in the biosynthesis of DMOA-derived meroterpenoids, the variations in the terpenoid moiety cyclization and post-cyclized modification reactions significantly contribute to the structural diversity of this class of compounds. Paxilline (**35**), aflatrem (**36**), and terpendole K (**37**) are all biosynthesized via paspaline (**47**), whereas emindole SA (**38**), radarin A (**39**), and thiersinine A (**40**) are derived from differently cyclized intermediates from **47** [23] (Fig. 13.5). Among them, the biosyntheses of paspaline-derived indole-diterpenes have been extensively studied by the gene disruption and complementation approaches.

The first biosynthetic gene cluster for indole-diterpenes was reported in 2001, for the production of **35** in *Penicillium paxilli* [24]. Subsequent analyses by gene deletion and gene transfer to a paxilline non-producing mutant confirmed that at least seven genes are involved in the biosynthesis of **35** [25]. The biosynthesis of **35** is initiated by the formation of **41** by a geranylgeranyl pyrophosphate synthase (GGPS), PaxG. Three additional genes, *paxB*, *paxC*, and *paxM*, were then found to be involved in the biosynthesis of **47**, but until quite recently, their functions had been unclear.

A reconstitution approach using *A. oryzae* as a host revealed the biosynthetic scheme to **47**, for the first time [26] (Fig. 13.6). First, *paxC*, encoding a prenyltransferase, was coexpressed with the GGPS gene to generate **43**, and an in vitro study of PaxC revealed that it utilizes **41** as its preferred substrate. It should be noted that PaxC

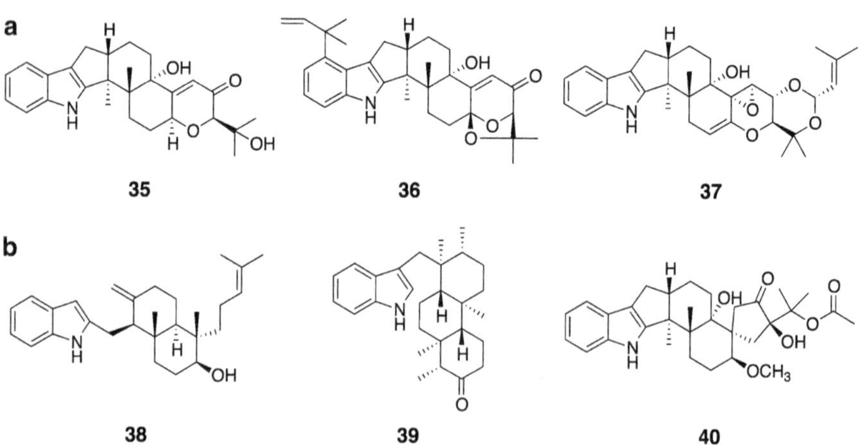

Fig. 13.5 (**a**) Paspaline-derived and (**b**) non-paspaline-derived indole-diterpenes

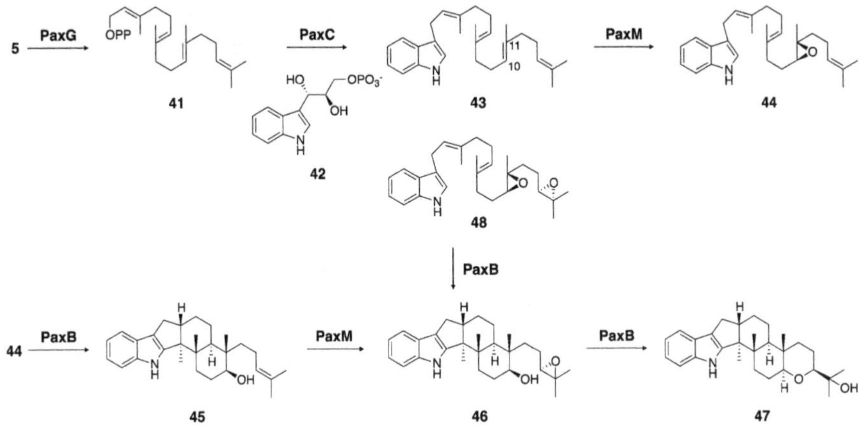

Fig. 13.6 Biosynthetic pathway for paspaline (**47**) in *Penicillium paxilli*

is a soluble protein, and thus is quite different from the UbiA-like prenyltransferases, such as Pyr6, AusN, and Trt2. The next biosynthetic steps involve the epoxidation and cyclization of the terpenoid moiety, which are catalyzed by the FMO PaxM and the membrane-bound terpene cyclase PaxB. Previously, it was predicted that PaxM would generate a bisepoxide compound, but the transformant that expressed *paxG*, *paxC*, and *paxM* produced a monoepoxide compound at the C-10 and C-11 positions (**44**). Since the further expression of *paxB* along with the other three genes successfully produced **47**, a biosynthetic scheme was suggested, in which the first cyclization of **44** by PaxB yields the emindole SB (**45**), and then **45** undergoes further epoxidation and cyclization steps to produce **47**. To clarify the biosynthetic pathway from **44** to **47**, biotransformation experiments were performed. The monoepoxide **44** was transformed into **45** by the transformant that expressed only *paxB*, and the synthetic bisepoxide **48** was transformed into **47** as well as the pentacyclic intermediate **46** by the same transformant, confirming the stepwise epoxidation and cyclization mechanism to produce **47**. The next steps toward **35** were well studied through gene inactivation and biotransformation experiments with the presumed substrates, which revealed the requirement of two P450s, PaxP and PaxQ [27]. PaxP catalyzed multiple oxidation steps to form 13-desoxypaxilline (**50**) from **47** via β(beta)-PC-M6 (**49**), and PaxQ catalyzed the hydroxylation of **50** to yield paxilline (**35**) (Fig. 13.7a).

The biosynthetic gene clusters for other paspaline-derived indole-diterpenes, aflatrem (**36**), terpendoles, and lolitrems were also previously identified, and the functional analyses of the genes in the cluster have been reported [28–30]. The biosynthesis of **36** in *Aspergillus flavus* shares the biosynthetic pathway toward **50** with that of paxilline (**35**), involving five genes, *atmG*, *atmM*, *atmC*, *atmB*, and *atmP*, but the reconstitution study revealed that AtmQ, a P450 homologous to PaxQ, possesses different activities from that of PaxQ [31] (Fig. 13.7b). When *atmQ* was expressed in a *P. paxilli* mutant lacking the *paxQ* gene, paspalicine (**51**) and paspalinine (**52**) were produced instead of **35**, suggesting that AtmQ catalyzes the oxidation at C-7 first to produce **51**, and then oxidizes C-13 to form **52**. The final step involves a

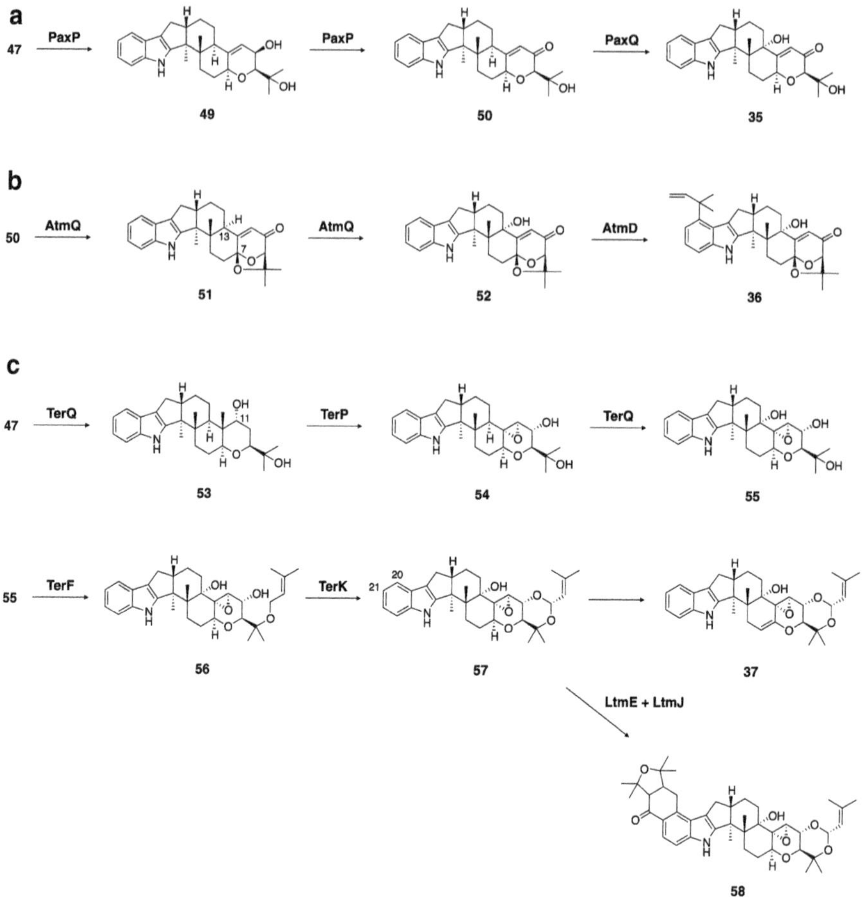

Fig. 13.7 Biosynthetic diversity in the biosyntheses of (**a**) paxilline (**35**), (**b**) aflatrem (**36**), and (**c**) terpendoles and lolitrems

prenyltransferase that transfers dimethylallyl pyrophosphate (DMAPP) to **52** in a reverse manner to yield **36**, in a reaction presumably catalyzed by AtmD, which is homologous to the indole prenyltransferases. On the other hand, the biosynthesis of terpendoles in *Chaunopycnis alba* is more complex than those of **35** and **36** (Fig. 13.7c). The biosynthetic gene cluster (the *ter* cluster) for terpendoles includes seven genes. Five of the seven genes, *terM*, *terC*, *terB*, *terP*, and *terQ*, are homologous to those of **35** and **36**, and the first three are also responsible for the production of **47**, but the cluster also contains two additional, unique genes, *terF* and *terK*. The functions of *terP* and *terQ* were studied by gene disruption and feeding experiments. First, *terP*, which encodes a P450, was deleted and terpendole E (**53**) accumulated. Next, to investigate the involvement of **53** in the biosynthesis of terpendole K (**37**), **53** was incubated with a *C. alba* strain that lacks all of the *ter* genes except for *terP*, and the strain converted **53** into 13-desoxyterpendole I (**54**). In contrast, **53** was not metabolized by a *C. alba* strain that lacks all of the *ter* genes except for *terQ*.

This *terQ*-expressing strain, on the other hand, transformed **54** into terpendole I (**55**). To obtain further insight into the reaction catalyzed by TerQ, a *terQ* deletion mutant was then constructed, and the mutant accumulated **47**, indicating that TerQ catalyzes the hydroxylation of **47** at C-11 to form **53**. Interestingly, **48** and **50**, the biosynthetic intermediates of both **35** and **36**, were not involved in the biosynthesis of terpendoles, although TerP and TerQ actually convert **47** into **50** via **48**. The next biosynthetic step toward terpendole C (**57**) from **55** is presumably catalyzed by TerF and TerK, with TerF catalyzing the O-prenylation of **55** to yield terpendole J (**56**) and the other P450, TerK, converting **56** into **57**. The biosynthetic intermediates of terpendoles are also involved in the biosynthesis of lolitrems. The biosynthetic gene clusters (the *ltm* clusters) for lolitrems include ten genes in three separate regions, and two of the ten genes, *ltmE* and *ltmJ*, are unique to the *ltm* clusters. LtmE and LtmJ are predicted to be involved in the last steps of lolitrem biosynthesis, and are responsible for the transformation of **57** into lolitrem B (**58**) [32]. Here, LtmE catalyzes prenylations at two positions, C-20 and C-21, and the P450 LtmJ catalyzes multiple oxidations and ring closures to yield **58** (Fig. 13.7c).

Conclusion

The fungal meroterpenoids exhibit tremendous structural diversity in nature, and biosynthetic studies of fungal meroterpenoids with unique skeletons can provide great lessons on how nature designs diverse molecules from simple starter units. As seen in the biosynthetic pathways of the DMOA-derived meroterpenoids and the paspaline-derived indole-diterpenes, the cyclization and post-cyclization modification reactions greatly contribute to the molecular diversity of these classes of natural products. Many genes and enzymes involved in these biosynthetic steps have been identified, thus providing important insights into the biosynthetic schemes. However, little is known about the means by which these enzymes produce such different products from the same or similar substrates. For example, we mentioned the terpene cyclases. As described previously, AusL and Trt1 both accept the same substrate to produce differently cyclized products, and the indole-diterpene biosyntheses should also involve terpene cyclases with unique activities. The only part that is known about the catalytic mechanism of the terpene cyclases is that two acidic amino acid residues (E63 and D218) of Pyr4, which are conserved among all of the terpene cyclases described in this chapter, are very important for the catalytic activity [6], but other aspects, such as how AusL and Trt1 produce different compounds, have yet to be elucidated. The differences in the activities of the P450s that oxidize paspaline are also quite interesting. The elucidation of the mechanisms by which the biosynthetic enzymes control the product selectivity may result in the creation of novel unnatural activities by the rational engineering of enzymes of interest.

Further investigations of the biosynthetic genes of other fungal meroterpenoids, elucidations of biosynthetic pathways, and the creation of novel meroterpenoids with useful activities by enzyme engineering or combining the biosynthetic pathways of several molecules are some of the future goals in this field.

References

1. Geris R, Simpson T. Meroterpenoids produced by fungi. Nat Prod Rep. 2009;26(8):1063–94.
2. Tomoda H, Nishida H, Kim YK, Obata R, Sunazuka T, Omura S, et al. Relative and absolute stereochemistry of pyripyropene A, a potent, bioavailable inhibitor of acyl-CoA:Cholesterol acyltransferase (ACAT). J Am Chem Soc. 1994;116:12097–8.
3. Kuno F, Otoguro K, Shiomi K, Iwai Y, Omura S. Arisugacins A and B, novel and selective acetylcholinesterase inhibitors from *penicillium* sp. FO-4259. J Antibiot. 1996;49(8):742–51.
4. Parini P, Davis M, Lada A, Erickson S, Wright T, Gustafsson U, et al. ACAT2 is localized to hepatocytes and is the major cholesterol-esterifying enzyme in human liver. Circulation. 2004;110(14):2017–23.
5. Tomoda H, Tabata N, Nakata Y, Nishida H, Kaneko T, Obata R, et al. Biosynthesis of pyripyropene A. J Org Chem. 1996;61:882–6.
6. Itoh T, Tokunaga K, Matsuda Y, Fujii I, Abe I, Ebizuka Y, et al. Reconstitution of a fungal meroterpenoid biosynthesis reveals the involvement of a novel family of terpene cyclases. Nat Chem. 2010;2(10):858–64.
7. Xu Z, Baunach M, Ding L, Hertweck C. Bacterial synthesis of diverse indole terpene alkaloids by an unparalleled cyclization sequence. Angew Chem Int Ed. 2012;51(41):10293–7.
8. Hu J, Okawa H, Yamamoto K, Oyama K, Mitomi M, Anzai H. Characterization of two cytochrome P450 monooxygenase genes of the pyripyropene biosynthetic gene cluster from *penicillium coprobium*. J Antibiot. 2011;64(3):221–7.
9. Simpson TJ, Stenzel DJ, Bartlett AJ, Obrien E, Holker JSE. Studies on fungal metabolites part 3. 13C NMR spectral and structural studies on austin and new related meroterpenoids from *Aspergillus ustus*, *Aspergillus variecolor*, and *Penicillium diversum*. J Chem Soc Perkin Trans 1. 1982(11):2687–92.
10. Springer JP, Dorner JW, Cole RJ, Cox RH. Terretonin, a toxic compound from *Aspergillus terreus*. J Org Chem. 1979;44(29):4852–4.
11. Omura S, Inokoshi J, Uchida R, Shiomi K, Masuma R, Kawakubo T, et al. Andrastins A similar to C, new protein farnesyltransferase inhibitors produced by *penicillium* sp. FO-3929. J Antibiot (Tokyo). 1996;49(5):414–24.
12. Simpson TJ, Walkinshaw MD. Anditomin, a new C-25 metabolite from *Aspergillus variecolor*. J Chem Soc Chem Commun. 1981;1981(17):914–5.
13. Ahmed SA, Scott FE, Stenzel DJ, Simpson TJ, Moore RN, Trimble LA, et al. Studies on the biosynthesis of the mycotoxin austin, a meroterpenoid metabolite of *Aspergillus ustus*. J Chem Soc Perkin Trans 1. 1989(4):807–16.
14. McIntyre CR, Simpson TJ, Stenzel DJ, Bartlett AJ, Q'Brien E, Holker JSE. Biosynthesis of the meroterpenoid metabolites, austin and terretonin: incorporation of 3,5-dimethylorsellinate. J Chem Soc Chem Commun. 1982;(14):781–2.
15. Shiomi K, Tomoda H, Otoguro K, Omura S. Meroterpenoids with various biological activities produced by fungi. Pure Appl Chem. 1999;71:1059–64.
16. Simpson TJ. Biosynthesis of highly modified meroterpenoids in *Aspergillus variecolor*. Incorporation of 13C-labelled acetates and methionine into anditomin and andilesin C. Tetrahedron Lett. 1981;22(38):3785–8.
17. Nielsen M, Nielsen J, Rank C, Klejnstrup M, Holm D, Brogaard K, et al. A genome-wide polyketide synthase deletion library uncovers novel genetic links to polyketides and meroterpenoids in *Aspergillus nidulans*. FEMS Microbiol Lett. 2011;321(2):157–66.
18. Lo H-C, Entwistle R, Guo C-J, Ahuja M, Szewczyk E, Hung J-H, et al. Two separate gene clusters encode the biosynthetic pathway for the meroterpenoids austinol and dehydroaustinol in *Aspergillus nidulans*. J Am Chem Soc. 2012;134(10):4709–20.
19. Matsuda Y, Awakawa T, Itoh T, Wakimoto T, Kushiro T, Fujii I, et al. Terretonin biosynthesis requires methylation as essential step for cyclization. Chembiochem. 2012;13(12):1738–41.
20. Itoh T, Tokunaga K, Radhakrishnan E, Fujii I, Abe I, Ebizuka Y, et al. Identification of a key prenyltransferase involved in biosynthesis of the most abundant fungal meroterpenoids derived from 3,5-dimethylorsellinic acid. Chembiochem. 2012;13(8):1132–5.

21. Guo C-J, Knox B, Chiang Y-M, Lo H-C, Sanchez J, Lee K-H, et al. Molecular genetic characterization of a cluster in *A. terreus* for biosynthesis of the meroterpenoid terretonin. Org Lett. 2012;14(22):5684–7.
22. McDonough M, Kavanagh K, Butler D, Searls T, Oppermann U, Schofield C. Structure of human phytanoyl-CoA 2-hydroxylase identifies molecular mechanisms of refsum disease. J Biol Chem. 2005;280(49):41101–10.
23. Saikia S, Nicholson M, Young C, Parker E, Scott B. The genetic basis for indole-diterpene chemical diversity in filamentous fungi. Mycol Res. 2008;112(Pt 2):184–99.
24. Young C, McMillan L, Telfer E, Scott B. Molecular cloning and genetic analysis of an indole-diterpene gene cluster from *Penicillium paxilli*. Mol Microbiol. 2001;39(3):754–64.
25. Saikia S, Parker E, Koulman A, Scott B. Four gene products are required for the fungal synthesis of the indole-diterpene, paspaline. FEBS Lett. 2006;580(6):1625–30.
26. Tagami K, Liu C, Minami A, Noike M, Isaka T, Fueki S, et al. Reconstitution of biosynthetic machinery for indole-diterpene paxilline in *Aspergillus oryzae*. J Am Chem Soc. 2013;135(4): 1260–3.
27. Saikia S, Parker E, Koulman A, Scott B. Defining paxilline biosynthesis in *Penicillium paxilli*: functional characterization of two cytochrome P450 monooxygenases. J Biol Chem. 2007;282(23):16829–37.
28. Zhang S, Monahan B, Tkacz J, Scott B. Indole-diterpene gene cluster from *Aspergillus flavus*. Appl Environ Microbiol. 2004;70(11):6875–83.
29. Motoyama T, Hayashi T, Hirota H, Ueki M, Osada H. Terpendole E, a kinesin eg5 inhibitor, is a key biosynthetic intermediate of indole-diterpenes in the producing fungus *Chaunopycnis alba*. Chem Biol. 2012;19(12):1611–9.
30. Young C, Felitti S, Shields K, Spangenberg G, Johnson R, Bryan G, et al. A complex gene cluster for indole-diterpene biosynthesis in the grass endophyte *Neotyphodium lolii*. Fungal Genet Biol. 2006;43(10):679–93.
31. Nicholson M, Koulman A, Monahan B, Pritchard B, Payne G, Scott B. Identification of two aflatrem biosynthesis gene loci in *Aspergillus flavus* and metabolic engineering of *Penicillium paxilli* to elucidate their function. Appl Environ Microbiol. 2009;75(23):7469–81.
32. Saikia S, Takemoto D, Tapper B, Lane G, Fraser K, Scott B. Functional analysis of an indole-diterpene gene cluster for lolitrem B biosynthesis in the grass endosymbiont *Epichloë festucae*. FEBS Lett. 2012;586(16):2563–9.

Chapter 14
Ergot Alkaloids

Paul Tudzynski and Lisa Neubauer

Introduction

The History of Ergot

Ergot alkaloids are bioactive indole-derivatives produced by a wide range of fungi. Due to their structural similarity to neurotransmitters, they can have significant effects on the central nervous system of mammalia. The major producer of these important mycotoxins belong to the ascomycetous family *Clavicipitaceae* comprising mainly fungi that colonize grasses, either as endophytes or as pathogens. Best known and intensively studied is the genus *Claviceps* with more than 30 species [1] causing disease in a broad range of grasses including all economically important cereal crop plants [2]. While most *Claviceps* species have a narrow host range, *Claviceps purpurea* (Fries ex Fries) Tulasne is a broad host range pathogen infecting more than 400 plant species. The common name "ergot fungus" ("argot": French for spur) refers to the purple to dark colored sclerotia replacing seeds on infected grass ears a few weeks after infection (Fig. 14.1) [3, 4]; these surviving structures contain ergot alkaloids and have been the reason for major intoxification problems in the past.

The notorious medieval "St. Anthony's Fire" disease was caused by consumption of rye bread contaminated with alkaloid containing sclerotia. The disease symptoms described in medieval texts vary. Obviously two major groups of symptoms occurred: (1) convulsive ergotism (*Ergotismus convulsivus*) causing spasms, paranoia and

---

This chapter is dedicated to Karl Esser on the occasion of his 90th birthday.

P. Tudzynski, Dr. rer. nat. (✉) L. Neubauer
Institut für Biologie und Biotechnologie der Pflanzen, Westfälische Wilhelms Universität Münster, Münster, Germany
e-mail: tudzyns@uni-muenster.de

J.-F. Martín et al. (eds.), *Biosynthesis and Molecular Genetics of Fungal Secondary Metabolites*, Fungal Biology, DOI 10.1007/978-1-4939-1191-2_14,

Fig. 14.1 Sclerotia of *Claviceps purpurea* on rye ears. Sclerotia are formed 4–5 weeks after infection of flowering grass ears; they represent overwintering structures. In spring they germinate and differentiate stroma heads containing perithecia. The needle-like ascospores infect the stigmatic hairs of flowering grass ears and penetrate the ovaries. Around 7 days post infection (dpi) the fungus reaches the rachilla, taps the vascular tissue and starts the production of honeydew—a sugar-rich fluid containing fungal conidia, which leads to secondary infections. Around 14 dpi the formation of sclerotia is initiated [3, 4]

hallucinations; and (2) gangrenous ergotism (*Ergotismus gangraenosus*) characterized by disturbed peripheral sensation, oedema and the loss of affected limbs [5]. The reason for this variability, as we know today, is the existence of various "chemical races" of *C. purpurea* containing different sets of alkaloids [5]. It was not before the 19th century that the link between ergot contamination of rye and disease was established [6], resulting in a better control of sclerotia contamination in cereals. This led to a significant reduction in ergotism epidemics, though they were not completely banished: Ergotism outbreaks in modern times have been reported, e.g., in Germany (1879–1881), in Russia (1926–1927), and in Ethiopia (1977–1978) [5, 7, 8].

The therapeutic potential of these mycotoxins was already recognized in the Middle Ages: *Claviceps* sclerotia were used by midwives in support of childbirth or to induce abortion (see citations of medieval texts in [9]). The prevention of excessive bleeding during childbirth, one of the major reasons for maternal mortalities, has been a major application of ergot alkaloids in the last 100 years [10]. In addition, ergot alkaloids were used for blood pressure control and treatment of migraine [5] and against degenerative diseases of the central nervous system, such as Parkinson's disease. A wide range of different natural ergot alkaloids has been described, with quite different biological effects. The so-called ergopeptines mainly produced by *C. purpurea* and some related endophytes (especially *Epichloe* species) are the high-end products, derivatives of the basic ergoline ring system with attached tripeptides. Other *Claviceps* species and some members of the quite distantly related family

Trichocomaceae (e.g., *Aspergillus fumigatus*) produce intermediates/special derivatives of this pathway, the so-called clavine alkaloids. Intensive research on the chemistry and pharmacology of ergot alkaloids led to the development of highly active and more specific derivatives of the natural products [11, 12], among them the most potent hallucinogen known: lysergic acid diethylamide (LSD) [13].

Ergot alkaloids have preserved their Janus-faced role through the centuries. They still belong to the most prominent mycotoxins in cereal crops, and the risk of intoxication is high [14]. Hence the ergot disease can cause severe economic losses as often the grain might be classified as too poisonous for use though the number of seeds is not severely reduced [15].

Scientific interest in these fungi continues, because the development of new defence strategies against ergot diseases and strain improvement programs for biotechnological purposes both require a detailed understanding of the biology, physiology, and genetics. Research had focused originally mainly on *Claviceps purpurea*, but recently also the endophytic *Epichloe* species and the opportunistic human pathogen *A. fumigatus* have gained more and more interest. Several detailed and excellent reviews are available covering the biochemistry, pharmacology, and biotechnology of ergot alkaloid biosynthesis [16–19].

Pharmacological Activities and Therapeutical Applications of Ergot Alkaloids

The pharmacological activities of ergot alkaloids are mostly due to the structural similarity of the tetracyclic ergoline system to neurotransmitters: structures of noradrenalin, dopamine, and serotonin (5-hydroxytryptamine, 5-HT) fit well onto the d-lysergic acid ring structure. The substituents attached to the carboxyl group of d-lysergic acid define the mode (agonistic or antagonistic) and the intensity of the interaction with receptors for these neurotransmitters [20, 21]. The different ergot alkaloid producing fungal species and even different natural isolates of *C. purpurea* ("chemical races" or "chemotypes") can differ significantly in their alkaloid spectra: They can produce either the basic clavine alkaloids (intermediates/derivatives of the lysergic acid biosynthesis), simple lysergic acid (LA) amides such as ergonovine, or one or two of a large set of complex ergopeptines, which contain a tripeptide (with variable composition) attached to the LA carboxyl group (see Fig. 14.2 and Table 14.1 [4]). In addition, since many of these ergot alkaloids have a broad and variable specificity and therefore can probably interact with more than one subgroup of the different receptors as agonist or antagonist [22], ergot contaminated food can cause complex intoxication symptoms [20]. Even chemically pure natural alkaloids can have unfavorable side effects as pharmaceuticals. Since a de novo chemical synthesis of ergot alkaloids is hampered by the stereo-specificity of the pharmacological effects (e.g., derivatives of d-isolysergic acid, the stereoisomer of d-lysergic acid, show little or no pharmacological activity), the improvement of natural ergot alkaloids by narrowing the specificity of the compounds by chemical modification represents a major challenge in pharmaceutical research [23].

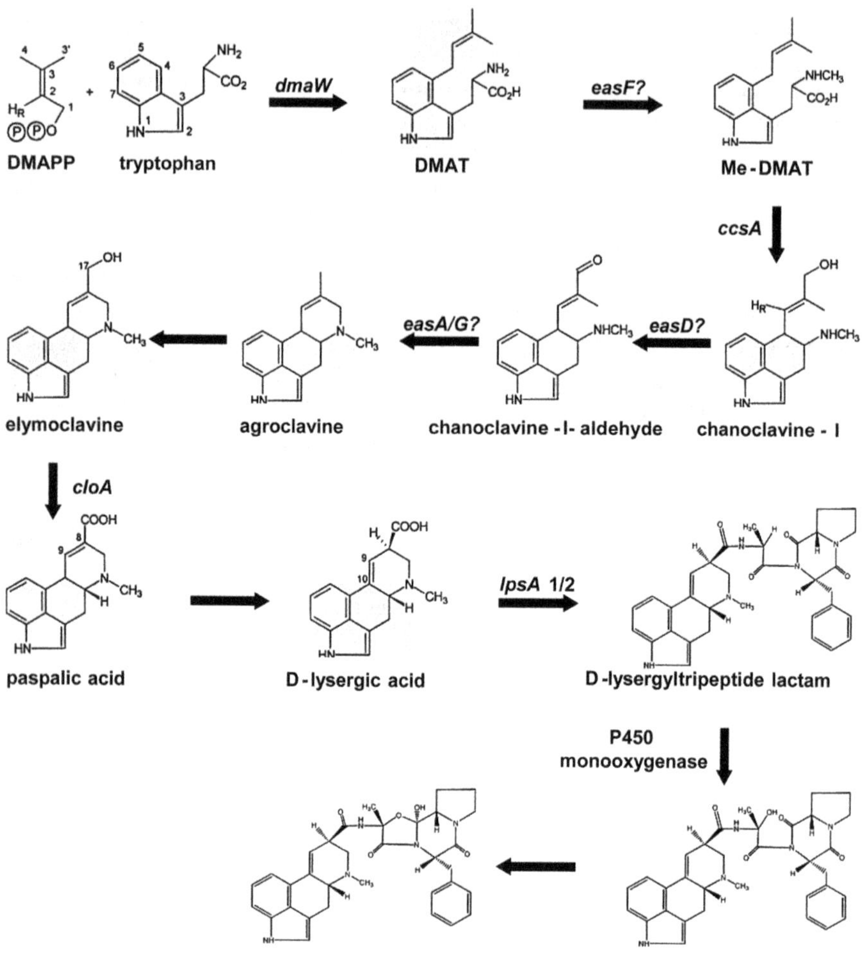

Fig. 14.2 Biosynthetic pathway of ergot alkaloids in *C. purpurea. DMAT* dimethylallyltryptophan, *Me-DMAT N*-methyl-DMAT. Modified after [58]

Table 14.1 Amino acid components of ergopeptines in *Claviceps purpurea* (the third position is always proline) (modified after [4])

	Ergotamines	Egotoxines	Erg oxines
Position I / Position II	Alanine	Valine	α(alpha)-amino butyric acid
Phenylalanine	Ergotamine	Ergocristine	Ergostine
Valine	Ergovaline	Ergocornine	Ergonine
α(alpha)-amino butyric acid	Ergobine	Ergobutyrine	Ergobutine
Leucine	α(alpha)-ergosine	α(alpha)-ergokyptone	α(alpha)-ergoptine
Isoleucine	β(beta)-ergosine	β(beta)-ergokryptine	β(beta)-ergoptine

In general, natural ergopeptines mainly have vasoconstrictive and sympatholytic-adrenolytic effects because of their high affinity for adrenergic receptors [24]. However, modification of side chains can have drastic effects; e.g., the simple derivative dihydro-ergotamine is preferentially used for the treatment of migraine since its adrenolytic effect is increased and in parallel its vasoconstrictive effect is reduced compared to the natural component ergotamine [25–27]. Dihydroergotoxin, a mixture of several ergopeptines with a saturated D ring, is used for the treatment of high blood pressure and cerebral dysfunctions in older patients [28, 29]. Natural ergotoxines, including ergocryptine, are effective inhibitors of the release of the peptide hormone prolactin. A semisynthetic derivative, 2-bromo-ergocryptine (bromocriptine) is used for efficient treatment of hyperprolactinaemia. Since this can lead to reproductive disorders such as galactorrhea and amenorrhea, acromegaly or anovulation, and even prolactin-dependent mammary carcinoma [30], bromocryptine was even used for treatment of advanced breast-cancer [31]. Because of its high affinity to dopaminergic receptors bromocriptine is also used in the treatment of Parkinson's disease [32].

The undesired side effects of the natural/semisynthetic ergot alkaloids reduced their attractiveness in pharmaceutical applications; they were partly replaced by synthetic analogs.

However, intensive activities in the last years have implicated new and promising applications of ergot derivatives, especially in the field of degenerative diseases. Recent clinical phase 3 trials show that orally inhaled dihydro-ergotamine is effective for acute migraine treatment [33]. Even clinical interest in the therapeutical potential of ergoline hallucinogens is growing, for example, for the treatment of autism [34, 35].

In recent years, the toxic side effects of ergot alkaloids also have been studied in more detail, especially with respect to biological functions beyond the receptor interactions. The long known cytotoxic effects of some ergot alkaloids had stimulated studies to test their effect as potential anticancer agents [e.g., 36]. Studies with human primary cells showed a marked cytotoxic and apoptosis-inducing effect of some ergopeptines (especially ergocristine [37]). A recent study of the same group revealed that the cytotoxicity of ergot alkaloids in human cell lines obviously depends on the type of alkaloids. Ergopeptines have a chiral center at position C-8; natural ergot alkaloids are a mixture of 8-(R) and 8(S) isomers, with the latter considered to be biologically inactive because of their weaker affinity to receptors. However, the 8(S) forms are obviously preferentially accumulated in hepatic cell lines and are mainly responsible for the apoptotic effect [38].

Biosynthesis and Molecular Genetics of Ergot Alkaloids

Ergot Alkaloid Synthesis (EAS) has been studied in detail for many years [16–18]. Based mainly on feeding experiments with labeled precursors/intermediates, cell-free extracts and purified enzymes of *C. purpurea*, *C. fusiformis*, *C. paspali*, *and A. fumigatus* (the latter species produce only ergoclavine alkaloids or simple LA derivatives, respectively) the biochemical pathway is now well established.

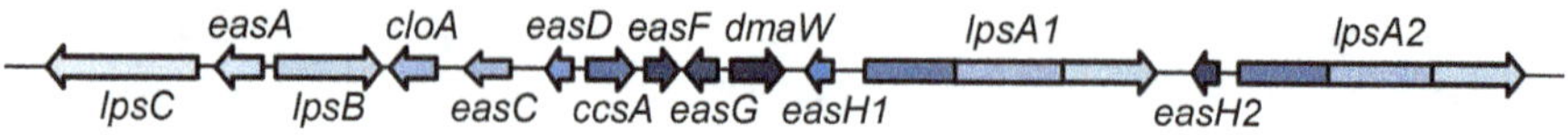

Fig. 14.3 Scheme of the alkaloid biosynthesis cluster region in *C. purpurea* strain P1. For gene designations see Table 14.2, direction of transcription is indicated by orientation of the *arrows*

The first part of the pathway generating the ergoline ring structure is highly conserved in the Ergot fungi; in *C. purpurea* it finally leads to lysergic acid (LA) (see Fig. 14.2). The first specific step of this pathway is the formation of 4-dimethylallyl-tryptophan (DMAT) by a specific prenyltransferase, dimethylallyl tryptophan synthase (DMATS), the best characterized enzyme of this pathway [39]. Since this enzyme catalyzes a determinant step of the pathway, it is strictly regulated: Tryptophan has a dual role as precursor and as inducer, whereas later steps are under feedback regulation by intermediates of the pathway (e.g., elymoclavine or agroclavine) [40].

The gene encoding DMATS (*dmaW*) was the first EAS gene identified and characterized, in a clavine-alkaloid producing *C. fusiformis* strain [41], and in *C. purpurea* strain P1, a submerse-producing mutant of the ATCC strain 20102 [42, 43]. Chromosome walking starting from the *dmaW* gene led to the detection of a cluster of 14 co-regulated genes, predicted to encode enzymes involved in EAS (Fig. 14.3) [12, 43, 44].

A recent broad comparative genomics study of alkaloid producing members of the family *Clavicipitaceae* [45] including three *Claviceps* species (*C. purpurea, C. fusiformis, C. paspali*), ten endophytes (*Epichloe* and *Neotyphodium* species), an exotic bamboo pathogen, and a morning glory symbiont (responsible for the ergot alkaloids detected in this plant) revealed conservation of a central core of genes involved in formation of the basic ergoline ring structure and high variability of peripheral genes involved in modifications of this structure. Especially the endophytes showed a high variability and complex cluster structures.

Functional analyses by gene disruption and the analysis of intermediates in *C. purpurea* and other fungi or by heterologous expression have resulted in attribution of function to most of the EAS genes. To allow comparisons among ergot alkaloid producing fungi, a systematic set of names for the genes of the ergot alkaloid pathway has been introduced [16]: EAS (*eas*) genes that have not yet been unequivocally functionally characterized are designated *easA* through *easH*. Genes with well-defined functions are named according to the enzyme activities of the encoded proteins. A list of the genes and their shown or predicted function is given in Table 14.2 [41, 46–54].

So far, seven *eas* cluster genes have been characterized by gene disruption and analysis of intermediates in *C. purpurea* P1. These include three steps of the basic pathway: apart from *dmaW*, the genes *cloA* encoding a cytochrome P450 oxidoreductase, which catalyzes the conversion of elymoclavine to paspalic acid [47], and *ccsA* (*easE*) encoding (a component of) the chanoclavine I synthase [46]. In addition, Matuschek et al. [52] could show by heterologous expression of *easG* that its product (a dehydrogenase) is involved in the biosynthesis of agroclavine. Much progress has also been made in the analysis of the terminal pathway leading to ergopeptines. The four non-ribosomal-peptide-synthetase (NRPS) genes of the EAS cluster, *lpsA*$_1$/*A*$_2$,

Table 14.2 Genes of the ergot alkaloid gene cluster in *C. purpurea* [41, 46–54]

Gene	Enzyme	Function	Reference
ccsA (*easE*)	FAD-oxidoreductase	Chanoclavine I synthase	[46]
cloA	P450 oxidoreductase	Elymoclavine oxidase	[47]
dmaW	Prenyl transferase	DMATS	[41]
easA	Oxidoreductase (old yellow enzyme)	Agroclavine synthase[a]	[48]
easC	Catalase	Chanoclavine I synthase[b]	[49]
easD	Short-chain dehydrogenase	Chanoclavine I aldehyde synthase[b]	[50]
easF	Methyl transferase	Methylation of DMAT[b]	[51]
easG	Dehydrogenase	Agroclavine synthase	[52]
lpsA1/2	Non-ribosomal-peptide synthetase	Assembly of tripeptide moiety of ergopeptines	[53]
lpsB	Non-ribosomal-peptide synthetase	Activation of lysergic acid	[53]
lpsC	Non-ribosomal-peptide synthetase	Ergonovine biosynthesis	[54]

[a]In *Neotyphodium lolii*
[b]In *Aspergillus fumigatus*

lpsB and *C* were characterized in detail. The final steps of the ergopeptine synthesis in *C. purpurea* are catalyzed by a complex of two interacting NRPSs (a new finding in fungi): D-lysergyl peptide synthetase (LPS) 1 attaching the tripeptide moiety to lysergic acid activated by LPS2 [55]. Functional analysis showed that $lpsA_1$ and $lpsA_2$ both encode LPS1 enzymes, with LPS1-1 in strain P1 being involved in the synthesis of the major alkaloid ergotamine, and LPS1-2 for the synthesis of ergocryptine [43, 56]. The monomodular NRPS enzyme encoded by *lpsC* catalyses the formation of ergonovine (=ergometrine) by attaching a single amino acid side chain to LA [54]. Thus the NRPS complex encoded by the EAS cluster in *C. purpurea* represents a unique natural combinatorial system; activated LA formed by LPS2 can be used as substrate for three different NRPS. This flexible biosynthesis scheme and the natural variability of the amino acid-binding domains of the LPS1 enzymes are the basis for the high variability of the ergot peptide alkaloid spectrum in the different natural chemical races of *C. purpurea* (see Table 14.1). This knowledge opens up interesting biotechnological perspectives: to generate *C. purpurea* strains producing single alkaloids, by knocking out *lpsA1/2* and/or *lpsC* genes (exemplified, e.g., by [56]) or even strains with new specificities by introducing "designer" *lps* genes.

Panaccione [17] pointed out that in contrast to *C. purpurea*, in other ergot fungi (especially the endophytes) chemical complexity is mainly due to "inefficient" pathways; i.e., not completely synchronized biosynthetic steps yielding accumulation of intermediates and side products, which may have specific functions.

In recent years, detailed research in other ergot fungi shed more light on the first part of the EAS pathway, i.e., from DMAT to lysergic acid. In the endophyte *Neotyphodium lolii* the product of *easA* (an oxidoreductase of the "old yellow enzyme" type) is necessary for agroclavine synthesis [48]. Since the first steps of the biosynthesis of the clavine-alkaloid fumigaclavine in the opportunistic human pathogen *A. fumigatus* are identical to the EAS pathway in *C. purpurea*, the cluster contains several genes with high homology to the *C. purpurea* cluster. Functional analyses in this system showed that *easF* encodes a methyltransferase involved in the biosynthesis of methylated DMAT [51], *easD* codes for a chanoclavine

I aldehyde synthase [50], and the product of *easC*, a catalase, is necessary for chanoclavine I synthesis [49]. Thus for most of the steps of the EAS pathway, corresponding gene functions could be identified. Recently Ryan et al. [57] could demonstrate that indeed the genes *dmaW*, *easC/E/F* are necessary and sufficient for the production of chanoclavine I by transfer of the resp. genes from *A. fumigatus* into *Aspergillus nidulans*, which lacks all EAS genes.

Despite the detailed knowledge on the structure and function of the EAS genes, their regulation is still widely not understood. Field isolates of *C. purpurea* do not produce ergot alkaloids in axenic culture, only in sclerotia. Submerse production strains have been generated by repeated mutagenesis cycles, the molecular basis of this effect has not yet been resolved. The strain P1 (see previous) needs special induction conditions for EAS: besides tryptophan as inductor and precursor, a high osmotic value and low phosphate concentration. It was shown that the "phosphate" effect acts on transcriptional level; i.e., expression of EAS genes is induced by lowering the phosphate level [58]. Lorenz et al. [59] presented evidence that epigenetic processes are involved in EAS gene regulation: Inhibitors of histone acetyl transferases induced EAS under non-inducing conditions in strain P1. This leads to the assumption that overexpression of histone deacetylases could overrule the inhibiting effect mediated by histone acetyl transferases. This could not yet be confirmed, at least not for the overexpression of two of the four classical histone deacetylases identified in the genome of *C. purpurea.* However, preliminary data show that the deletion of these genes leads to a significant reduction of alkaloid production, supporting the theory that there is a connection between chromatin remodeling and regulation of alkaloid biosynthesis.[1] Still there is yet no evidence that this epigenetic regulation is also the molecular basis of the in planta induction effect in field isolates. New insights may come from studies on signaling processes affecting the transition from the sporulating so-called *Sphacelia* stage during the honey-dew production phase to sclerotial tissue, the EA producing stage [e.g., 60]. Interestingly, most submerse production strains of *C. purpurea* are impaired in conidia production, suggesting a (negative) link to regulation of conidial differentiation processes. In *A. fumigatus* the situation seems to be inverse: Alkaloids are exclusively associated with conidia and not produced in vegetative mycelia [61]. Interestingly, the EAS gene cluster in *A. fumigatus* has been shown to be the only secondary metabolite cluster that is regulated by the conidial regulation gene *brlA* [62].

Economic Impact of Ergot Today

Today several methods have been developed to reduce the risk of ergot infection in most cereal crops with the consequence that ergotism as a human disease has almost been eliminated. Among these methods are, for example, changes in crop rotation,

[1] L. Neubauer, M. Niss, P. Tudzynski, unpublished data.

deeper ploughing, and sifting out the sclerotia [63]. Also application of fungicides, breeding for disease resistance and crossing of natural rye with hybrid rye reduced infection of rye with *C. purpurea* (summarized in [64]). In the European Union, the amount of ergot in grain used for human food is limited to 0.05 %. For animal feed 0.1 % grain samples containing ≤0.1 % of sclerotia can be tolerated. Similar standards are applied in the USA and Canada [65, 66]. Costs for cleaning of ergot-infected seed are enormous [67]. In Germany, due to changes in cultivation and crop rotation ergot did not cause any severe troubles in rye cultivation, but the increased cultivation of more susceptible hybrid rye and triticale obviously has caused more ergot infection since the 1980s [63].

Analysis of the average data of infection with *C. purpurea* between 1995 and 2004 revealed a relatively constant level of ergot contamination in the analyzed rye samples of about 0.11 % w/w. Nevertheless, depending on different climatic and weather conditions as well as differences in the employed cultivars, the contamination of rye with ergot differs considerably [68]. In 1998, massive infection of rye in some parts of Germany led to almost total loss of the harvest [64]. Also in some regions of the USA, in 2005 widespread occurrence of ergot in barley was reported [69].

In addition, ergot causes problems by poisoning grazing animals due to the consumption of ergot alkaloids. Ergot alkaloids ingested by livestock may be ultimately derived from two different sources. Ergot alkaloids produced by *Claviceps* spp. on ears of pasture grasses and feed grain crops that are ingested by livestock are particularly problematic if the animals are allowed to graze on grass that is flowering [70]. On the other hand, ergot alkaloids are also produced by endophytic fungi of the genera *Epichloë*, *Neotyphodium*, and *Balansia* belonging to the family *Clavicipitaceae* [71, 72]. The ergot alkaloids produced by these fungi play an important role in protection of their host plants against insects and grazing vertebrates [73].

Toxicosis problems suffered by animals grazing on tall fescue were first noted in the 1930s [74]. Today the estimated economic loss for the beef cattle, equine and small ruminant industry in the USA is more than $1 billion annually [75]. Affected livestock show loss of appetite and reduced weight gain, fat necrosis, loss of body temperature control (hyperthermia), convulsions, rough hair coats, reduced fertility, and lactating cows show reduced milk production [76–78]. The effects of ergot alkaloid poisoning are mainly attributed to the ergopeptine ergovaline; however, transport across ruminant gastric membranes is much higher for intermediate lysergyl compounds than for ergopeptines, suggesting intermediate ergot alkaloids may also play a significant role [79].

Biotechnology

The industrial production of ergot alkaloids (ergotamine tartrate) already had been started by Sandoz in 1921; only 30 years later companies such as Boehringer Ingelheim, Galena, Gedeon Richter, Eli Lilley, and Farmitalia joined the field (see review by [80]).

The species *C. purpurea*, *C. fusiformis*, and *C. paspali*, which were generally used as production strains, differ considerably with respect to their potential to synthesize specific alkaloids: Only *C. purpurea* produces peptide alkaloids, whereas simple lysergic acid derivatives are produced by *C. paspali*, and *C. fusiformis* is used to produce clavine alkaloids.

Today the production of simple clavine alkaloids or paspalic/lysergic acid is of major importance as a basis for semisynthetic drug development; the annual production of all ergopeptines is estimated to reach 5,000–8,000 kg, whereas about 10,000–15,000 kg of lysergic acid are produced annually [81]. In 2010, the total world production of ergopeptines and semisynthetic ergot alkaloid derivatives was about 20,000 kg [19].

Originally the field production of alkaloids on rye or triticale was the major production method, but the submersed production with specially designed strains soon prevailed [82, 83]. In 2010, the field cultivation contributed about 50 % to the total world production. An increase in the ergot yield from 400 kg per hectare in the 1940s to more than1 ton per hectare nowadays, as well as in the alkaloid content (1.5 % of the sclerotia dry mass), could be achieved by random mutagenesis and selection of highly producing strains so that the yield of field production can reach 20 kg of ergot alkaloids per hectare [19].

In wild-type strains, the alkaloid biosynthesis is induced only in planta. Only the so-called sclerotia-like cells are able to produce alkaloids [84]. So strains producing alkaloids in submersed culture have been generated by successive cycles of mutagenesis, but a degeneration process that leads to a loss of the sclerotia-like cells is a big problem. Therefore, a continuous selection to maintain good production strains is necessary [85].

Fermentation of *C. purpurea* for the production of ergopeptines requires specific conditions because the formation of sclerotia-like cells has to be induced by the cultivation media. Ergot alkaloid syntheses is positively regulated by tryptophan (as a precursor and inducer), a high concentration of a slowly metabolized carbon source (mannitol, sorbitol, or sucrose) [86] and high osmotic pressure [87]. Phosphate and ammonium inhibit alkaloid biosynthesis [88].

Conclusion

Claviceps spp., the ergot fungus, plays an important but ambivalent role in agriculture, pharmacology, and biotechnology. As a food contaminant it represents a dangerous threat for consumers of cereal products, and not all toxic substances accompanying food contamination with *Claviceps* are known yet; the availability of the genome sequence and of molecular genetic techniques will allow the identification of additional (possibly toxic) metabolites. On the other hand, *Claviceps* spp. are a historic source of valuable pharmaceuticals; modern metabolic design strategies will help to establish new strategies; e.g., for the treatment of important degenerative diseases, which has enormous economic importance.

Acknowledgment We thank M. Niss for sharing of data prior to publication, and the Deutsche Forschungsgemeinschaft (DFG) for financial support (Tu 50/18).

References

1. Taber WA. Biology of *Claviceps*. In: Demain AL, Nadine AS, editors. Biology of industrial microorganisms (Biotechnology series), vol. 6. New York: The Benjamin Cummings Publishing Co Inc.; 1985. p. 449–86.
2. Bové FJ. The story of ergot. Basel: S. Karger; 1970.
3. Tudzynski P, Scheffer J. *Claviceps purpurea*: molecular aspects of a unique pathogenic lifestyle. Mol Plant Pathol. 2004;5:377–88.
4. Schürmann J, Tudzynski P. *Claviceps*: the ergot fungus. In: Russell R, Paterson M, editors. Molecular biology of food and water borne mycotoxigenic and mycotic fungi of humans. CRC Press; 2013 (in press).
5. Eadie MJ. Convulsive ergotism: epidemics of the serotonin syndrome? Lancet Neurol. 2003;2:429–34.
6. Tulasne LR. Memoire sur l'ergot des glumacees. Ann Sci Nat (Parie Botanique). 1853; 20:5–6.
7. Barger G. Ergot and ergotism. London: Gurney; 1931.
8. Urga K, Debella A, W'Medihn YNA, Bayu A, Zewdie W. Laboratory studies on the outbreak of gangrenous ergotism associated with consumption of contaminated barley in Arsi, Ethiopian. J Health Dev. 2002;16:317–23.
9. van Dongen PWJ, de Groot ANJA. History of ergto alkaloids from ergotism to ergometrine. Eur J Obstet Gynaecol Reprod Biol. 1995;60:109–16.
10. Li XF, Fortney JA, Kotelchuck M, Glover LH. The postpartum period: the key to maternal mortality. Int J Gynaecol Obstet. 1996;54:1–10.
11. Gröger D, Floss HG. Biochemistry of ergot alkaloids—achievements and challenges. In: Cordell GA, editor. The alkaloids: chemistry and biology, vol. 50. London: Academic Press; 1998. p. 171–218.
12. Sinz A. Die Bedeutung der Mutterkorn-Alkaloide als Arzneistoffe. Pharm Unserer Zeit. 2008;4:306–9.
13. Hofmann A. Historical view on ergot alkaloids. Pharmacology. 1978;16:1–11.
14. Krska R, Crews C. Significance, chemistry and determination of ergot alkaloids: a review. Food Addit Contam. 2008;25:722–31.
15. Alderman SC, Coats DD, Crowe FJ, Butler MD. Occurrence and distribution of ergot and estimates of seed loss in Kentuckey bluegrass grown for seed in central Oregon. Plant Dis. 1998;82:89–93.
16. Schardl CL, Panaccione DG, Tudzynski P. Ergot alkaloids—biology and molecular biology. Alkaloids Chem Biol. 2006;63:45–86.
17. Panaccione DG. Ergot alkaloids. In: Hofrichter M, editor. The Mycota X. Berlin: Springer, 2010. p. 195–214.
18. Wallwey C, Li SM. Ergot alkaloids: structure diversity, biosynthetic gene clusters and functional proof of biosynthetic genes. Nat Prod Rep. 2011;28:496–510.
19. Hulvová H, Galuszka P, Frébortová J, Frébort I. Parasitic fungus *Claviceps* as a source for biotechnological production of ergot alkaloids. Biotechnol Adv. 2013;31(1):79–89.
20. Berde B, Stürmer E. Introduction to the pharmacology of ergot alkaloids and related compounds. In: Berde B, Schild HO, editors. Ergot alkaloids and related compounds. Berlin: Springer; 1978. p. 1–28.
21. Stadler PA, Giger R. Ergot alkaloids and their derivatives in medical chemistry and therapy. In: Krosgard-Larson P, Christensen CH, Kofod H, editors. Natural products and drug development. Copenhagen: Munksgaard; 1984. p. 463–85.

22. Burki HR, Asper H, Ruch W, Zuger PE. Bromocriptine, dihydroergotoxine, methysergide, D-LSD, CF 25-397, and 29-712: effects on the metabolism of the biogenic amines in the brain of the rat. Psychopharmacology (Berl). 1978;57:227–37.
23. Vendrell M, Angulo E, Casadó V, Lluis C, Franco R, Albericio F, et al. Novel ergopeptides as dual ligands for adenosine and dopamine receptors. J Med Chem. 2007;50:3062–9.
24. Görnemann T, Jähnichen S, Schurad B, Latté KP, Horowski R, Tack J, et al. Pharmacological properties of a wide array of ergolines at functional alpha$_1$-adrenoceptor subtypes. Naunyn Schmiedebergs Arch Pharmacol. 2008;376:321–30.
25. Villalon CM, de Vries P, Rabelo G, Centurion D, Sanchez-Lopez A, Saxena PR. Canine external carotid vasoconstriction to methysergide, ergotamine and dihydroergotamine: role of 5-HT$_{1B/1D}$ receptors and α_2-adrenoceptors. Br J Pharmacol. 1999;126:585–94.
26. Willems EW, Trion M, de Vries P, Heiligers JOC, Villalon CM, Saxena PR. Pharmacological evidence that α_1- and α_2-adrenoceptors mediate vasoconstriction of carotid arteriovenous anastomoses in anaesthetized pigs. Br J Pharmacol. 1999;127:1263–71.
27. Tfelt-Hansen PC, Koehler PJ. History of the use of ergotamine and dihydroergotamine in migraine from 1906 and onward. Cephalalgia. 2008;28:877–86.
28. de Groot AN, van Dongen PW, Vree TB, Hekster YA, van Roosmalen J. Ergot alkaloids. Current status and review of clinical pharmacology and therapeutic use compared with other oxytocics in obstetrics and gyneacology. Drugs. 1998;56:523–35.
29. Wadworth AN, Crisp P. Co-dergocrine mesylate. A review of its pharmacodynamic and pharmacokinetic properties and therapeutic use in age-related cognitive decline. Drugs Aging. 1992;2:153–73.
30. Crosignani PG. Current treatment issues in female hyperprolactinaemia. Eur J Obstet Gynaecol Reprod Biol. 2006;125:152–64.
31. Barrett A, Morgan L, Raggatt PR, Hobbs JR. Bromocriptine in the treatment of advanced breast cancer. Clin Oncol. 1976;2(4):373–7.
32. Thobois S. Proposed dose equivalence for rapid switch between dopamine receptor agonists in Parkinson's disease: a review of the literature. Clin Ther. 2006;28:1–12.
33. Kellerman DJ, Forst A, Combs DL, Borland S, Kori S. Assessment of the consistency of absorption of dihydroergotamine following oral inhalation: pooled results from four clinical studies. J Aerosol Med Pulm Drug Deliv. 2013;26:1–10.
34. Fantegrossi WE, Murnane AC, Reissig CJ. The behavioral pharmacology of hallucinogens. Biochem Pharmacol. 2008;75:17–33.
35. Sigafoos J, Green VA, Edrisinha C, Lancioni GE. Flashback to the 1960s: LSD in the treatment of autism. Dev Neurorehabil. 2007;10:75–81.
36. Crider AM, Lu CK, Floss HG, Cassady JM, Clemens JA. Ergot alkaloids. Synthesis of nitrosourea derivatives of ergolines as potential anticancer agents. J Med Chem. 1979;22(1):32–5.
37. Mulac D, Humpf HU. Cytotoxicity and accumulation of ergot alkaloids in human primary cells. Toxicology. 2011;282(3):112–21.
38. Mulac D, Lepski S, Ebert F, Schwerdtle T, Humpf HU. Cytotoxicity and fluorescence visualization of ergot alkaloids in human cell lines. J Agric Food Chem. 2013;61(2):462–71.
39. Gebler JC, Poulter D. Purification and characterization of dimethylallyl tryptophan synthase from *Claviceps purpurea*. Arch Biochem Biophys. 1992;296:308–13.
40. Cheng LJ, Robbers JE, Floss HG. End-product regulation of ergot alkaloid formation in intact cells and protoplasts of *Claviceps* species, strain SD58. J Nat Prod. 1980;43:329–39.
41. Tsai HF, Wang H, Gebler JC, Poulter CD, Schardl CL. The *Claviceps purpurea* gene encoding dimethylallyltryptophan synthase, the committed step for ergot alkaloid biosynthesis. Biochem Biophys Res Commun. 1995;216:119–25.
42. Keller U. Highly efficient mutagenesis of *Claviceps purpurea* by using protoplasts. Appl Environ Microbiol. 1983;46:580–4.
43. Tudzynski P, Hölter K, Correia T, Arntz C, Grammel N, Keller U. Evidence for an ergot alkaloid gene cluster in *Claviceps purpurea*. Mol Gen Genet. 1999;261:133–41.
44. Lorenz N, Wilson EV, Machado C, Schardl C, Tudzynski P. Comparison of ergot alkaloid biosynthesis gene clusters in *Claviceps* species indicate loss of late pathway steps in evolution of *C. fusiformis*. Appl Environ Microbiol. 2007;73:7185–91.

45. Schardl CL, Young CA, Hesse U, Amyotte SG, Andreeva K, Calie PJ, et al. Plant-symbiotic fungi as chemical engineers: multi-genome analysis of the clavicipitaceae reveals dynamics of alkaloid loci. PLoS Genet. 2013;9(2):e1003323.
46. Lorenz N, Olsovska J, Sulc M, Tudzynski P. The alkaloid cluster gene ccsA of the ergot fungus *Claviceps purpurea* encodes the chanoclavine-I-synthase, an FAD-containing oxidoreductase mediating the transformation of N-methyl-dimethyltryptophan to chanoclavine I. Appl Environ Microbiol. 2010;76:1822–30.
47. Haarmann T, Ortel I, Tudzynski P, Keller U. Identification of the cytochrome P450 monooxygenase that bridges the clavine and ergoline alkaloid pathways. Chembiochem. 2006;7:645–52.
48. Cheng JZ, Coyle CM, Panaccione DG, O'Connor SE. A role for old yellow enzyme in ergot alkaloid biosynthesis. J Am Chem Soc. 2010;132:1776–7.
49. Goetz KE, Coyle CM, Cheng JZ, O'Connor SE, Panaccione DG. Ergot cluster-encoded catalase is required for synthesis of chanoclavine-I in *Aspergillus fumigatus*. Curr Genet. 2011;57:201–11.
50. Wallwey C, Matuschek M, Li SM. Ergot alkaloid biosynthesis in *Aspergillus fumigatus*: conversion of chanoclavine-I to chanoclavine-I aldehyde catalyzed by a short-chain alcohol dehydrogenase FgaDH. Arch Microbiol. 2010;192:127–34.
51. Rigbers O, Li SM. Ergot alkaloid biosynthesis in *Aspergillus fumigatus*. Overproduction and biochemical characterization of A4-dimethylallyltryptophan N-methyltransferase. J Biol Chem. 2008;283:26859–68.
52. Matuschek M, Wallwey C, Xie X, Li SM. New insights into ergot alkaloid biosynthesis in *Claviceps purpurea*: an agroclavine synthase EasG catalyses, via a non-enzymatic adduct with reduced glutathione, the conversion of chanoclavine-I aldehyde to agroclavine. Org Biomol Chem. 2011;9:4328–35.
53. Correia T, Grammel N, Ortel I, Tudzynski P, Keller U. Molecular cloning and analysis of the ergopeptine assembly system in the ergot fungus Claviceps purpurea. Chem Biol. 2003;10:1281–92.
54. Ortel I, Keller U. Combinatorial assembly of simple and complex D lysergic acid alkaloid peptide classes in the ergot fungus *Claviceps purpurea*. J Biol Chem. 2009;284(11):6650–60.
55. Riederer B, Han M, Keller U. D-lysergyl peptide synthetase from the ergot fungus *Claviceps purpurea*. J Biol Chem. 1996;271:27524–30.
56. Haarmann T, Lorenz N, Tudzynski P. Use of a nonhomologous end joining deficient strain (Δku70) of the ergot fungus *Claviceps purpurea* for identification of the nonribosomal peptide synthetase gene involved in ergotamine biosynthesis. Fungal Genet Biol. 2008;45:35–44. 21.
57. Ryan KL, Moore CT, Panaccione DG. Partial reconstruction of the ergot alkaloid pathway by heterologous gene expression in *Aspergillus nidulans*. Toxins (Basel). 2013;5(2):445–55.
58. Haarmann T, Machado C, Lübbe Y, Correia T, Schardl CL, Panaccione DG, et al. The ergot alkaloid gene cluster in *Claviceps purpurea*: extension of the cluster sequence and intra species evolution. Phytochemistry. 2005;66:1312–20.
59. Lorenz N, Haarmann T, Pazoutová S, Jung M, Tudzynski P. The ergot alkaloid gene cluster: functional analyses and evolutionary aspects. Phytochemistry. 2009;70(15–16):1822–32.
60. Schürmann J, Buttermann D, Herrmann A, Giesbert S, Tudzynski P. Molecular characterization of the NADPH oxidase complex in the ergot fungus *Claviceps purpurea*: CpNox2 and CpPls1 are important for a balanced host-pathogen interaction. Mol Plant Microbe Interact. 2013;26:1151–64. Jun 18 [Epub ahead of print].
61. Coyle CM, Kenaley SC, Rittenour WR, Panaccione DG. Association of ergot alkaloids with conidiation in *Aspergillus fumigatus*. Mycologia. 2007;99(6):804–11.
62. Twumasi-Boateng K, Yu Y, Chen D, Gravelat FN, Nierman WC, Sheppard DC. Transcriptional profiling identifies a role for BrlA in the response to nitrogen depletion and for StuA in the regulation of secondary metabolite clusters in *Aspergillus fumigatus*. Eukaryot Cell. 2009;8(1):104–15.
63. Mielke H. Untersuchungen zur Bekämpfung des Mutterkorns. Nachrichtenbl Deut Pflanzenschutzd. 1993;5/6:97–102.
64. Mielke H. Studien über den Pilz *Claviceps purpurea* (Fries) Tulasne unter Berücksichtigung der Anfälligkeit verschiedener Roggensorten und der Bekämpfungsmöglichkeiten des Erregers. Mitt Biol Bundesanst Land-Forstwirtsch, vol. 375. Berlin: Parey Buchverlag; 2000.

65. Appelt M, Ellner FM. Ergot alkaloids—relevance and first own results of incidences in 2007. ALVA Mitteilungen. 2008;6:27–30.
66. Mirdita V, Dhillon BS, Geiger HH, Miedaner T. Genetic variation for resistance to ergot (*Claviceps purpurea* [Fr.] Tul.) among full-sib families of five populations of winter rye (*Secale cereale* L.). Theor Appl Genet. 2008;118:85–90.
67. Betz HG, Müller R, Wilde P, Wortmann H. Mutterkorn vermeiden. Auswertungs- und Informationsdienst für Ernährung, Landwirtschaft und Forsten. 1998;1361:3–16.
68. Lindhauer M, Münzing K, Seling S, Betsche T, Kersting HJ, Masloff S, et al. Hochwertiges Getreide durch kontinuierliche Qualitätserhebungen. Federal Research Centre for Nutrition and Food. Detmold: Forschungsreport 2; 2005.
69. Schwarz PB, Neate SM, Rottinghaus GE. Widespread occurrence of ergot in upper midwestern U.S. barley, 2005. Plant Dis. 2006;90:527.
70. Cross DL. Ergot alkaloid toxicity. In: White Jr JF, Bacon CW, Hywel-Jones NL, Spatafora JW, editors. Mycology, clavicipitacean fungi: evolutionary biology, chemistry, biocontrol and cultural impacts, vol. 19. New York: Marcel Dekker Inc.; 2003. p. 475–94.
71. Zhang HW, Song YC, Tan RX. Biology and chemistry of endophytes. Nat Prod Rep. 2006;23:753–71.
72. Schardl CL, Scott B, Florea S, Zhang D. Epichloë endophytes: clavicipitaceous symbionts of grasses. In: Deising H, editor. The mykota V: plant relationships. 2nd ed. Berlin: Springer; 2009. p. 275–306.
73. Schardl CL, Phillips TD. Protective grass endophytes where are they from and where are they going? Plant Dis. 1997;81:430–8.
74. Hoveland C. Importance and economic significance of the *Acremonium* endophytes to performance of animals and grass plants. Agric Ecosyst Environ. 1993;44:3–12.
75. Strickland JR, Looper ML, Matthews JC, Rosenkrans Jr CF, Flythe MD, Brown KR. St. Anthony's fire in livestock: causes, mechanisms, and potential solutions. J Anim Sci. 2011;89(5):1603–26.
76. Schmid SP, Osborne TG. Effects of endophyte-infected tall fescue on animal performance. Agric Ecosyst Environ. 1993;44:233–62.
77. Porter JK, Thompson FNJ. Effects of fescue toxicosis on reproduction in livestock. J Anim Sci. 1992;70:1594–603.
78. Thomson FN, Stuedemann JA. Pathophysiology of fescue toxicosis. Agric Ecosyst Environ. 1993;44:263–81.
79. Hill NS, Thompson FN, Stuedemann JA, Rottinghaus GW, Ju HJ, Dawe DL, et al. Ergot alkaloid transport across ruminant gastric tissues. J Anim Sci. 2001;79:542–9.
80. Cvak L. Industrial production of ergot alkaloids. In: Kren V, Cvak L, editors. Ergot, the genus *Claviceps*, vol. 6. Amsterdam: Harwood Academic Publishers; 1999. p. 373–409.
81. Schiff PL. Ergot and its alkaloids. Am J Pharm Educ. 2006;70:98–107.
82. Keller U, Tudzynski P. Ergot alkaloids. In: Osiewacz HD, editor. The mycota X, industrial applications. Berlin: Springer; 2002. p. 157–72.
83. Tudzynski P, Correia T, Keller U. Biotechnology and genetics of ergot alkaloids. Appl Microbiol Biotechnol. 2001;57:593–605.
84. Spalla C. Genetic problems of production of ergot alkaloids in saprophytic and parasitic conditions. In: Vaněk Z, Řeháček Z, Cudlin J, editors. Genetics of industrial microorganisms. Amsterdam: Elsevier; 1973. p. 393–403.
85. Malinka Z. Saprophytic cultivation of *Claviceps*. In: Křen V, Cvak L, editors. Ergot: the genus Claviceps. Amsterdam: Harwood Academic Publishers; 1999. p. 321–71.
86. Kren V, Pazoutová S, Rylko V, Sajdl P, Wurst M, Rehácek Z. Extracellular metabolism of sucrose in a submerged culture of *Claviceps purpurea*: formation of monosaccharides and clavine alkaloids. Appl Environ Microbiol. 1984;48(4):826–9.
87. Amici AM, Minghetti A, Scotti T, Spalla C, Tognoli L. Ergotamine production in submerged culture and physiology of *Claviceps purpurea*. Appl Microbiol. 1967;15(3):597–602.
88. Sočič H, Gaberc-Porekar V. Biosynthesis and physiology of ergot alkaloids. In: Arora DK, Elander RP, Mukerji KG, editors. Handbook of applied mycology. fungal biotechnology. New York: Dekker; 1992. p. 475–515.

Chapter 15
Fungal NRPS-Dependent Siderophores: From Function to Prediction

Jens Laurids Sørensen, Michael Knudsen, Frederik Teilfeldt Hansen, Claus Olesen, Patricia Romans Fuertes, T. Verne Lee, Teis Esben Sondergaard, Christian Nørgaard Storm Pedersen, Ditlev Egeskov Brodersen, and Henriette Giese

Iron at the Crossroads Between Life and Death

Iron is an essential component of cellular metabolism in all life and is involved in the synthesis of amino acids, DNA, and sterols. Due to its ability to transfer electrons, it acts as a catalyst in enzymatic processes and it is crucial for oxidation-reduction reactions. On one hand, organisms cannot live without iron while on the other, the level needs to be tightly controlled as excess iron can cause oxidative stress and formation of damaging reactive oxygen species [1]. In the cell, this effect is minimized by the action of catalases and peroxidases that again are dependent on iron for their function. Although present in high concentrations in the soil, iron is often a limiting growth factor in natural habitats as most of it is bound in insoluble ferric hydroxides [2]. For this reason, bacteria and fungi actively increase iron availability through the use of various iron acquisition and storage systems. In the soil, microbes and plants compete for iron in the rhizosphere and several high-affinity uptake systems have therefore evolved in this constant battle for nutrients [3]. For pathogenic organisms, the main challenge is to extract iron from the host and an obvious strategy for the host to combat the invasion is therefore to starve the pathogen.

J.L. Sørensen, Ph.D. • P.R. Fuertes, M.Sc. • T.E. Sondergaard, Ph.D. • H. Giese, Ph.D. (✉)
Department of Biotechnology, Chemistry and Environmental Engineering,
Aalborg University, Aalborg, Denmark
e-mail: hgiese@bio.aau.dk

M. Knudsen, Ph.D. • C.N.S. Pedersen, Ph.D.
Bioinformatics Research Center, Aarhus University, Aarhus, Denmark

F.T. Hansen, Ph.D. • C. Olesen, M.Sc. • D.E. Brodersen, Ph.D.
Department of Molecular Biology and Genetics, Aarhus University, Aarhus, Denmark

T.V. Lee, Ph.D.
AgResearch Structural Biology Laboratory, School of Biological Sciences,
University of Auckland, Auckland, New Zealand

J.-F. Martín et al. (eds.), *Biosynthesis and Molecular Genetics of Fungal Secondary Metabolites*, Fungal Biology, DOI 10.1007/978-1-4939-1191-2_15,

Fungi obtain iron through direct uptake of the ferrous ion, Fe^{2+}, or from heme-bound iron derived from the animal hosts. In addition, fungi have two high-affinity uptake systems, Reductive Iron Assimilation (RIA) and siderophore-mediated iron sequestration (for an excellent schematic overview, see Haas et al. [4]). RIA-mediated uptake rely on metalloreductases (ferrireductases) that carry out the reduction and solubilization of Fe(III) to Fe(II). A recent study of 29 diverse fungal genomes showed the presence of several genes encoding ferrireductases, but detailed knowledge about their function is still scarce [5]. One metalloreductase involved in RIA-mediated iron uptake in *Aspergillus fumigatus* has been described [6]. However, the redox cycling processes have been intensely studied in the model yeast, *Saccharomyces cerevisiae*, where the coordinated action of ferrireductases and ferroxidases maintain the fine balance between the two forms of iron in the cell, Fe(II) and Fe(III) [7, 8]. Like other organisms, fungi face a dilemma because iron occurs primarily in the inactive Fe(III) form, whereas the soluble Fe(II) can generate hydroxyl radicals. This conundrum has recently been solved in fungi by the discovery of a ferroxidase-permease pair that prevents the precipitation of Fe(III) [7]. Orthologues of the ferroxidase, Fet3, and the iron permease, FTR1, required for iron incorporation into the cell, have been identified in many fungi including *A. fumigatus* [9], *Ustilago maydis* [10], *Fusarium graminearum* [11], and *Cryptococcus neoformans* [12].

Fungi employ more than one strategy for iron uptake depending on the conditions, but when iron is scarce, the high-affinity siderophore-mediated uptake systems are employed. *S. cerevisiae* and some other yeasts lack the molecular machinery required for siderophore biosynthesis and rely entirely on RIA. The siderophore iron carriers, which are complex and highly modified polyketide/polypeptide compounds, are synthesized by the non-ribosomal peptide synthetase (NRPS) pathway (see section "Siderophore Biosynthesis"). The siderophores are small molecules (M_r < 1,500 Da) that chelate the ferric iron with an extremely high affinity, thereby allowing them to strip iron from other sources where it may already be tightly bound, such as ferritin. Siderophores are produced by the same molecular mechanisms in bacteria and fungi while monocot plants, notably barley, produce the phytosiderophore, mugineic acid (MA), by a different pathway involving condensation of S-adenosyl-methionine units to produce the precursor, nicotianamine [13]. Siderophores are produced in several different forms that satisfy the requirements for iron acquisition, transport, and storage and their production is triggered by iron availability. Some siderophores are excreted into the environment to sequester extracellular iron, either from inert materials or living hosts. Other compounds provide a safe mechanism for storage and transport of iron. The extracellular siderophores and their iron cargo are taken back up by the universal fungal siderophore transporter uptake mechanism (SIT), a unique yet widespread fungal pathway. Regardless of their ability to produce siderophores, all fungi exploit the SIT mechanism to take iron chelators of diverse origin from the extracellular environment to their own advantage, even those produced by other organisms, such as bacteria [4, 14].

Fungi lack a distinct iron secretion mechanism and have therefore developed specialized storage mechanisms to control intracellular iron levels. In yeast and other fungi that do not produce the high-affinity siderophores, any excess iron is most likely stored in vacuoles in the form of polyphosphates [15]. Apart from vacuolar storage, it has been speculated that fungi may possess ferritin-like molecules that maintain iron homeostasis as seen in bacteria, plants, and animals. In fact, extensive sequence analysis has revealed the presence of ferritin-like proteins in ascomycetes, zygomycetes, and chytrids, but further studies are required to determine their exact function in fungi [16]. Furthermore, in fungal hyphae and in conidia, it is known that intracellular siderophores such as ferricrocin act as iron storage compounds [4]. Finally, it has been demonstrated that ferricrocin is also involved in intra- and transcellular transport in *A. fumigatus* [17]. Iron-free siderophores have been observed in vesicles of *U. maydis* [18], which suggest that they may be secreted with the exocytic pathway. After re-entry in the fungi, extracellular siderophores can be hydrolyzed and recycled [19]. The transfer of iron from extracellular siderophores to internal siderophores, metabolism, or vacuole precedes recycling of siderophore breakdown products [20].

Iron appears to be important not only for growth in fungi, but also sexual development as observed in, for example, *Aspergillus nidulans* and *Cochliobolus heterostrophus* [21, 22]. In the heterothallic fungus, *C. heterostrophus*, ascospores fail to develop in deletion strains lacking the storage siderophore ferricrocin [22]; and in *A. fumigatus* and *Magnaporthe grisea*, asexual sporulation is affected upon deletion of the siderophore responsible for iron storage [17, 23, 24]. The effect of siderophore deletion is believed to be a reduction of the iron content in conidia, which again impacts enzyme activity, conidia size, and oxidative stress resistance [22]. High levels of iron storage siderophores have been found in fungal spores of *Neurospora crassa* and *Aspergillus ochraceus* [25]. When extracellular iron is added to ferricrocin deletion mutants, spore function is not completely restored and it is therefore speculated that siderophores may play a role in delivering iron at appropriate time-points during development to support enzyme activity and prevent oxidative damage [22].

In endophytic relationships it has been shown that extracellular siderophores are required for the maintenance of the beneficial mutualistic relationship [26]. For example, the interaction between *Epichloë festucae* and the rye grass, *Lolium perenne*, is dependent on an extracellular siderophore for mutual benefit. RIA is also a part of the relationship, but this system alone is unable to compensate for loss for the siderophore used in iron uptake from the apoplastic fluid [26]. In such mutant strains, the unbalanced mutual iron metabolism leads to uncontrolled hyphael growth and loss of symbiotic benefits [26]. Iron deficiency in *Arabidopsis thaliana* has been shown to impede infection by the pathogen, *Botrytis cinerea* [27], again stressing the importance of iron status for plant-fungal relationships.

Pathogenic fungi appear to depend on siderophores for acquisition of iron from their host environment and it has long been known that strains deficient in extracellular siderophores show reduced virulence [4]. This has been shown to be the case in *C. heterostrophus* and *Alternaria alternata* [28, 29], while in *U. maydis*, impairment of siderophore biosynthesis is even lethal [30]. *A. fumigatus* is infamous for its ability to cause life-threatening invasive diseases in humans; this fungus also depends on both extracellular and intracellular siderophores for survival in an animal host [4, 31].

Studies of pulmonary aspergillosis in mouse model systems have shown that the siderophores are functionally redundant and can to some extent compensate for each other [24].

Animals also maintain a tight control of intracellular iron levels but microbial siderophores can cannibalize their ferritin and the transport protein, transferrin, to harvest iron [32]. The counterattack against the microbial siderophores comes via the innate immune system through siderophore-binding proteins known as siderocalins [33]. The siderocalins can furthermore bind simple catechols (dihydroxybenzene), which are able to rescue iron from the microbial siderophores [34, 35]. Because microbial infections are highly influenced by iron availability, iron supplementation can worsen microbial infections in humans [36], and in plants, such as *A. thaliana*, it was shown that depletion of iron causes resistance towards a fungal pathogen [27].

Regulation of Siderophore Production in Response to Environmental Signals

Siderophores are central for survival of the majority of known fungi and as such require tight regulation in response to environmental changes. Recent reviews emphasize the complexity of the regulation as not only iron availability, but also pH and carbon availability influence siderophore production [16, 37]. Linde et al. employed a systems biology approach to attempt to unravel the regulatory interactions in iron homeostasis in fungi [38]. The regulation is based on two central transcription factors—the GATA-factor, SRE, and the bZip-factor, HapX—which interconnect via a negative feedback loop [39, 40]. The SRE transcription factor is found in a wide range of fungi with different names, including URBS1 in *U. maydis* [41], SREA in *Aspergillus* [42, 43], SREP in *Penicillium* [39], SRE in *N. crassa* [44], Sfu1 in *Candida albicans* [45], SREB in *Blastomyces dermatitidis* [46], and Fep1 in *Schizosaccharomyces pombe* [47]. SRE is produced when iron is in abundance and acts as a transcriptional repressor by binding directly to GATA motifs in the promoter region of target genes. In *A. fumigatus*, all genes involved in the high-affinity iron uptake pathway, siderophores and RIA, are controlled by SRE [43]. In some fungi, SRE is constitutively expressed while in others it is repressed by the HapX factor during iron deprivation, which in turn is downregulated by SRE when iron is abundant. There are numerous variations on this theme among fungi involving additional transcription factors as described in Haas et al. [4], Haas [37], and Canessa and Larrondo [16]. HapX generally downregulates iron-consuming pathways and up-regulates the siderophore biosynthesis during iron starvation; orthologues have been identified in several genera including *Fusarium* [48] and *Aspergillus* [49]. In *A. nidulans*, HapX is furthermore known to downregulate the hyphal siderophore and up-regulate the extracellular siderophore [40]. HapX functions by binding to the heterotrimeric CCAAT-binding complex (CBC), which is a conserved multimeric transcriptional activator in eukaryotes consisting of the subunits, Hap2p, Hap3p, Hap4p, and Hap5p. CBC is a Hap2p/Hap3p/Hap5p heterotrimer that binds DNA, whereas Hap4p is in charge of transcriptional activation [50].

The global fungal repressor, Tup1, is required for iron-dependant regulation of RIA-mediated iron uptake pathway in *Candida albicans* [51]. Interaction of Tup1 with SRE has been demonstrated in *Schizosaccharomyces pombe* [52] and also with HapX in *Candida albicans* [53]. A further regulatory link between the RIA and siderophore high-affinity uptake pathways was revealed by the identification of the zinc cluster transcription factor, AcuM, in *A. fumigatus* [54]. AcuM is presumed to influence iron acquisition by repression of SRE and activation of HapX, thereby increasing expression of genes involved in RIA and siderophore-mediated iron uptake at the same time. The exact mechanism by which this takes place is not known as deletion of *acuM* does not fully prevent siderophores biosynthesis [54].

The distribution of SRE and HapX orthologues throughout the fungal genera suggests a conserved regulatory mechanism for fungal siderophore biosynthesis. This basic system has then been adapted to suit the needs of individual species as demonstrated by the increasing number of control layers that have been uncovered. The complexity observed in regulation of iron homeostasis in fungi probably results from the need to balance several iron uptake and storage strategies. In the end, the systems biology approach described by Linde et al. [38] may provide the tools required to unravel the complex regulatory networks in individual fungal species.

Siderophore Structure and Functional Specialization

Several hundred unique microbial siderophores have been identified to date, and although they are structurally very different, they can be classified into three groups based on the type of iron chelating functional group: hydroxamate, hydroxycarboxylate, or catecholate [55, 56]. All fungal siderophores isolated so far have been hydroxamates, except for the polycarboxylate rhizoferrin [57]. The hydroxamate-type siderophores, which contain N^5-acetyl-N^5-hydroxyornithine residues as the iron-binding ligands, can be further subdivided into three structural classes: ferrichromes, fusarinines, and coprogens [58]. The N^5-acetyl-N^5-hydroxyornithines originate primarily from L-ornithine (see section "Siderophore Biosynthesis" and Fig. 15.1a [59]), except for neurosporin, which contains residues derived from D-ornithine [60]. Ferrichrome (Fig. 15.1b) was the first identified siderophore isolated from *Ustilago sphaerogena* [61]. Subsequently, a range of derivatives—e.g., ferricrocin, ferrichrysin, ferrirhodin, ferrirubin, and ferrichrome A—have been identified from a wide variety of both ascomycetes and basidiomycetes [59]. Ferrichromes are usually cyclic hexapeptides consisting of three L-ornithine residues coupled to three proteogenic amino acids, which usually are alanine, serine, and/or glycine. Although a ferrichrome transporter has been identified in *F. graminearum* [62], these compounds are primarily found to be involved in intercellular iron storage, whereas fusarinines and coprogens assimilate iron extracellularly. More than one ferrichrome synthetase are sometimes present in the same species [63, 64], suggesting that the compounds can have different roles despite similar structures.

Fig. 15.1 Stucture of L-ornithine (**a**) and ferrichromes (**b**), which has been derived from Renshaw et al. [59]

a

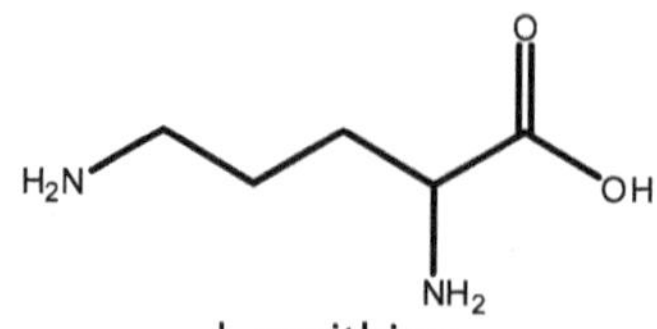

L-ornithine

b

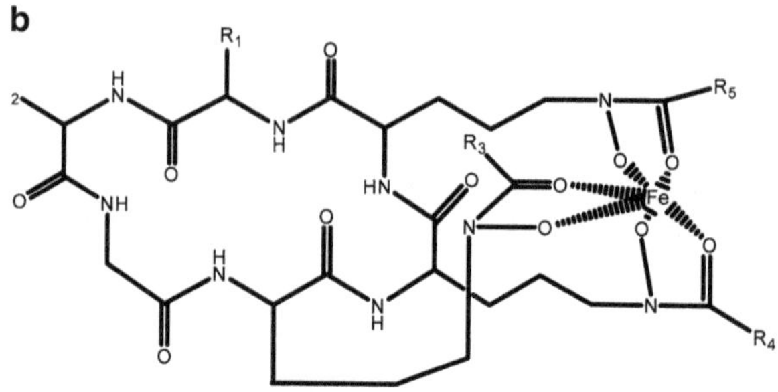

	R1	R2	R3	R4	R5
Ferrichrome	H	H	CH_3	CH_3	CH_3
Ferrichrome A	CH_2OH	CH_2OH			
Ferrichrome C	H	CH_3	CH_3	CH_3	CH_3
Ferrichrysin	CH_2OH	CH_2OH	CH_3	CH_3	CH_3
Ferricrocin	H	CH_2OH	CH_3	CH_3	CH_3
Ferrirubin	CH_2OH	CH_2OH	A	A	A
Ferrirhodin	CH_2OH	CH_2OH	D	D	D
Malonichrome	H	CH_3	F	F	F
Sake colorant A	CH_2OH	CH_3	CH_3	CH_3	CH_3
Asperchrome A	CH_2OH	CH_3	A	A	A
Asperchrome B1	CH_2OH	CH_2OH	CH_3	A	A
Asperchrome B2	CH_2OH	CH_2OH	A	CH_3	A
Asperchrome B3	CH_2OH	CH_2OH	A	A	CH_3
Asperchrome C	CH_2OH	CH_2OH	C	A	A
Asperchrome D1	CH_2OH	CH_2OH	A	CH_3	CH_3
Asperchrome D2	CH_2OH	CH_2OH	CH_3	A	CH_3
Asperchrome D3	CH_2OH	CH_2OH	CH_3	CH_3	A
Asperchrome E	CH_2OH	CH_2OH	D	A	A
Asperchrome F1	CH_2OH	CH_2OH	G	A	A
Asperchrome F2	CH_2OH	CH_2OH	A	G	A
Asperchrome F3	CH_2OH	CH_2OH	A	A	G

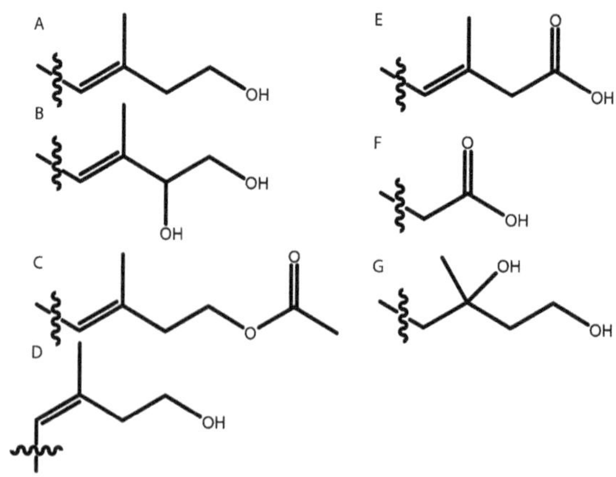

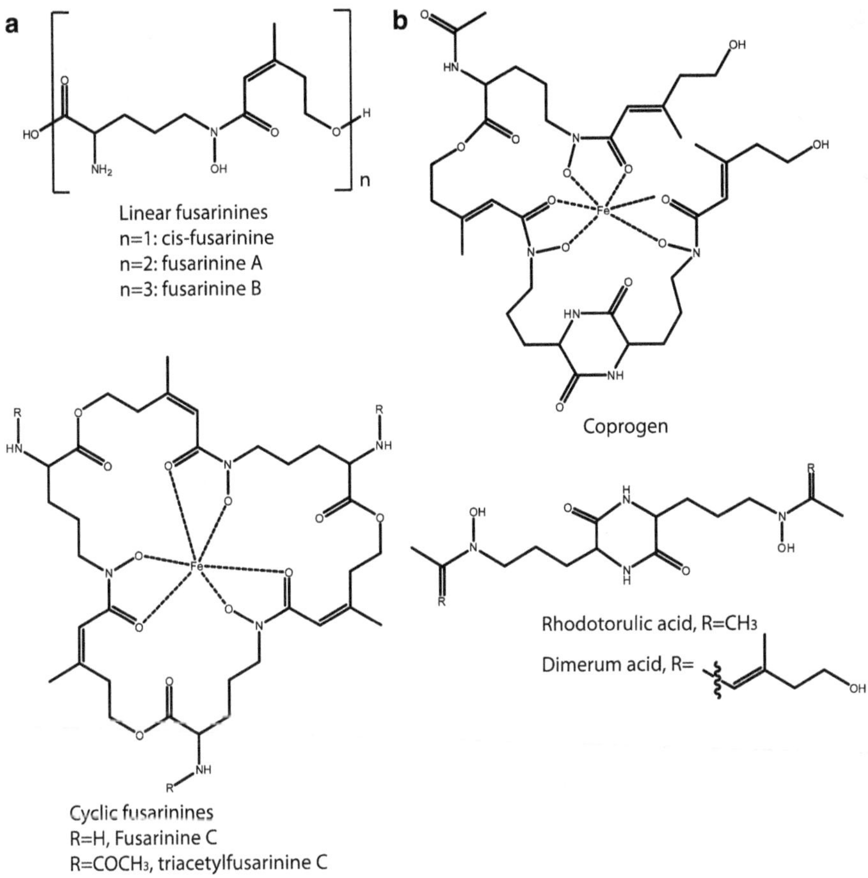

Fig. 15.2 Structures of fusarinines (**a**) and of coprogens (**b**)

Fusarinines (Fig. 15.2a) are monomers, linear dimers, or linear or cyclic trimers of N^5-acyl-N^5-hydroxyornithine, where the acyl group is an anhydromevalonyl moiety [59]. The fusarinines are furthermore connected head-to-tail via ester bonds between the acyl group and the ornithine carboxyl group and the hydroxyl group of the acyl residue forming linear dimers (fusarinine A), trimers (fusarinine B) and cyclic trimers (fusarinine C). The ester bonds are highly susceptible to hydrolysis [19], and metabolite extracts from fungi producing the cyclic trimers often contain various forms of both dimers and monomers [65]. It is not clear whether this is a result of diverse biosynthesis or progressive hydrolysis of the compounds.

Coprogens are linear dimers or trimers of N^5-acetyl-N^5-hydroxyornithine, two of which are joined head-to-head to form a six-membered diketopiperazine ring and the optional third unit is linked by an ester bond [59] (Fig. 15.2b). Several coprogens have been identified and their diversity is mainly due to differences in the attached acyl moieties, which are derived from trans-anhydromevalonyl CoA or acetyl CoA. Dimerum acid and rhodotorulic acid are the only known coprogens dimers

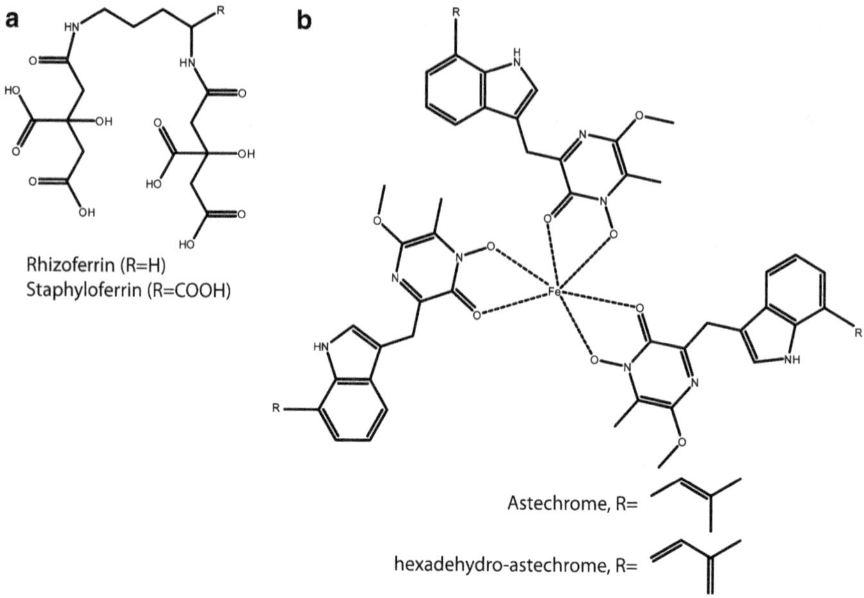

Fig. 15.3 Structure of rhizoferrin and staphyloferrin (**a**) and of astechrome and hexadehydro-astechrome (**b**)

differing in the acyl groups, which for dimerum acid are *trans*-anhydromevalonyl moieties [66], while they are acetyl groups in rhodotorulic acid [67]. Several trimeric coprogens have been identified in fungi including *M. grisea* [68], *Fusarium dimerum* [69], and *N. crassa* [70]. Coprogens are involved in extracellular iron assimilation and specific transporters have been identified in *N. crassa* [71].

Rhizoferrin (Fig. 15.3a) consists of two citric acid units linked to 1,4-diaminobutane through two amide bonds; it was originally isolated from *Rhizopus microspores* [72], but is also produced by several families of the Mucorales [73]. The molecular basis for production of rhizoferrin has not been identified, but the structurally related staphyloferrin A [74, 75] is synthesized by a NRPS-independent siderophore (NIS) pathway in the gram-positive bacterium, *Staphylococcus aureus* [76]. Staphyloferrin A is synthesized by two NIS synthetases, where the first synthetase catalyzes amide bond formation of D-ornithine and citric acid to form an intermediate compound to which another citric acid unit is condensed by the second synthetase [76]. The known biosynthetic pathways for bacterial polycarboxylate siderophores utilize the NIS pathway [77], but whether rhizoferrin is produced in a similar manner in fungi is still unknown.

An iron-containing trimeric compound, hexadehydro-astechrome (HAS, Fig. 15.3b) was recently identified in *A. fumigatus* by overexpression of the associated transcription factor [78]. The HAS monomer is structurally related to the *A. terrreus* non-ribosomal peptide astechrome [79] and the responsible gene cluster has only been found in pathogenic *Aspergillus* species. The HAS gene cluster contains one NRPS, two transcription factors, a transporter, an *O*-methyltransferase, a 7-dimethylallyltryptophan synthase, a putative FAD-binding protein, and a cytochrome P450 [78]. The HAS monomer is synthesized by a two-domain NRPS, which condenses tryptophan and alanine and the structural difference between HAS and astechrome is due to the conversion of the prenyl to a methylbutadienyl side chain by the FAD-binding domain protein, which is present in the *A. fumigatus* gene cluster but not in *A. terreus* [78]. Although the compounds contain iron and are potentially excreted, they were not considered likely to compete with the more potent, iron-binding siderophores [78].

Siderophore Biosynthesis

As described previously, siderophores are synthesized from primary metabolites, including both proteogenic and non-proteogenic amino acids as well as a range of relatively hydrophobic acyl groups [4]. A common component of fungal siderophores is the hydroxamate group, which arises from modification of the non-proteogenic amino acid, L-ornithine. In the first step of the biosynthetic pathway, L-ornithine is hydroxylated by a N^5-L-ornithine monooxygenase to form N^5-hydroxy-*L*-ornithine. These enzymes are class B flavoprotein monooxygenases that utilize NADPH and FAD as cofactors [80] and the genes encoding them have been identified in a wide range of fungal species including *U. maydis* (*sid1*) [81], *A. nidulans* (*sidA*) [82], *A. oryzae* (*dffA*) [83], *A. fumigatus* (*sidA*) [9], *F. graminearum* (*Gibberella zeae*) (*SID1*) [84], and *C. heterostrophus* (*SIDA*) [85].

In the second step of hydroxamate synthesis, an acyl group is transferred from acyl-CoA by N^5-*L*-ornithine transacylase to form hydroxymate or N^5-acyl-N^5-hydroxy-L-ornithine. This modified amino acid is then incorporated along with the proteogenic amino acids, serine and glycine, into the siderophore core structure (Fig. 15.4). Fungi that employ multiple acyl chains in their various siderophores, encode several transacylases, each specific for a particular acyl group. In *A. fumigatus*, for example, *sidL* encodes an acetyl-specific transacylase [6] while *sidF* encodes an anhydromevalonyl-specific enzyme [24]. Other identified fungal N^5-*L*-ornithine transacylases include Fer5 from *U. maydis* [86] and the putative N^5-*L*-ornithine transacylase from *O. olearius*, Ato1 [87]. Yet other enzymes are involved in producing the acyl chains incorporated by these transacylases such as Fer4, which synthesizes methylglutaconyl in *U. maydis* [86] and SidI and SidH, which are responsible for anhydromevalonyl production in *A. fumigatus* [88].

In siderophores, the amino acid building blocks are linked by either peptide (amide) or ester bonds. The enzymes responsible for this process are the non-ribosomal peptide synthetases (NRPSs), which is a group of large, multifunctional enzymes capable of synthesizing peptides in the absence of mRNA and the ribosome. The only known exceptions are the rhizoferrin siderophores produced by

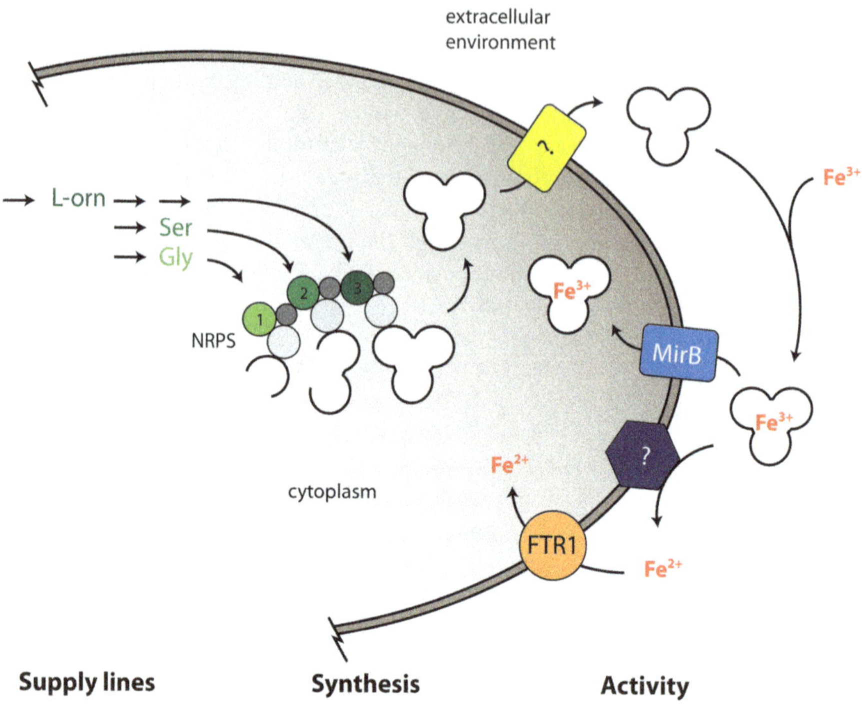

Fig. 15.4 Extracellular iron can be assimilated by Fe^{2+} permeases (FTR1) or by NRPS produced siderophores, which are imported back into the cells through specific transporters (MirB)

Zygomycetes, which are carboxylate siderophores and synthesized in a NRPS-independent manner [73]. Thus, both the information about the correct order of the amino acids and the enzymatic machinery to put them together are contained within the NRPSs. The enzymes have a modular architecture in which each unit (consisting of multiple domains) recognizes and incorporates a single amino acid into the resulting peptide product. For some siderophores, such as triacetylfusarinine and hydroxyferricrocin, the molecule is further modified after release from the NRPS enzyme. For triacetylfusarinine, SidG has been identified as the enzyme responsible for this step in *A. fumigatus* [24].

Once the siderophore synthesis is complete, the molecule is excreted to the extracellular environment by an unknown mechanism (Fig. 15.4). Genes showing homology to ABC transporters are often found clustered with siderophore biosynthesis genes and it is expected that these transporters are involved in siderophore secretion, although this has not been experimentally validated [26, 89]. Once the siderophores have bound iron, they are taken up again through secondary transporters of the UMF/SIT subfamily, which most likely function as proton symporters utilizing the cell membrane potential for iron uptake. A number of these siderophore transporters have been identified, including those in *S. cerevisiae* [90–92], *C. albicans* [90], *A. nidulans* [57], and *A. fumigatus* [93].

Domain Architecture of Siderophore Producing NRPS Enzymes

Generally, filamentous fungi have the ability to synthesize siderophores from more than one family; i.e., the fusarinines, the coprogens, and the ferrichromes (Figs. 15.1b and 15.2) [59, 94]. For example, *Trichoderma longibrachiatum* simultaneously produces siderophores from all three structural families at the same time [95] and the biosynthetic pathway for each of these families is dependent on NRPSs [96–98]. The substrates of the enzyme are amino acid residues, which are specifically recognized by the adenylation domain (A) and adenylated to aminoacyl-AMP in an ATP dependant reaction [99]. The activated amino acid is transferred by a peptidyl carrier protein domain (PCP or T) through a thioester linkage to a condensation domain (C), where peptide bond formation occurs [100]. The initiation module consists of an A and T domains, while the following elongation modules consists of and A, T, and C domains. The NRPSs involved in biosynthesis of most siderophores are nonlinear and characterized by having additional PCP and C domains in the final module, these are referred to as didomains [101]. However, the NRPSs responsible for synthesis of fusarinines and coprogens are of the single-module type, consisting of only one module. In contrast, most ferrichromes are synthesized by NRPSs with usually three or more modules (Fig. 15.5).

Siderophore-synthesizing NRPSs do not follow the collinear sequence of modules where the number and order of modules determines the length and structure of the final product. Rather, iterative use of modules, where a single module can be used multiple times to incorporate the same substrate several times is common. The characteristic T-C didomain seen in each family is predictive of such iterative nature. An example of iterative use of modules is the production of the hexapeptide ferrichrome, ferricrocin. The NRPS responsible for the synthesis consists of only three modules; hence iterative use of one or more modules is needed for synthesis of the final hexapeptide product [102]. Ferrichrome NRPSs exhibit a great variation in their domain architecture and six different modular types of domain architecture have been found [63, 97]. It has been hypothesized that the characteristic domain architecture of the ferrichrome NRPSs is derived from a hexamodular ancestral

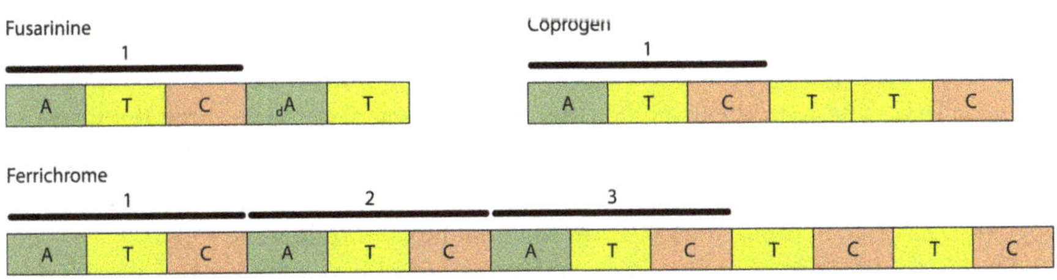

Fig. 15.5 Organization of siderophore nonribosomal peptide synthetases (NRPSs) containing adenylation domains (A) (or degenerated A domains, dA), Peptidyl carrier protein domain (T) and condensation domains (C). Fusarinine and coprogen NRPSs are monomodular, whereas the ferricrocine NRPS contains three modules

gene, and that loss of either individual A domains or entire modules have led to the current diversity [63]. NRPSs producing coprogens are of the single-module type, consisting of only one module followed by the T-C didomain. This domain architecture is similar to what is found in the fusarinines.

The Evolution of the Siderophore Non-ribosomal Peptide Synthetases

The NRPSs responsible for synthesizing fungal siderophores are generally highly divergent, but some conservation is seen among the ascomycetes and basidiomycetes. As mentioned previously, single-cell yeasts appear to be devoid of NRPS genes altogether, except for a few examples such as *S. pompe*, which contains a single gene encoding a siderophore synthetase [103]. In contrast, all ascomycetes contain one to three genes encoding siderophore NRPSs [63]. Phylogenomic analysis and structural conservation of domain architecture has suggested that the single-modular NRPSs are older in origin than the multimodular NRPSs, such as the ferrichrome NRPSs [103]. The ferrichrome synthetases can be sorted into six different families depending on their modular makeup, where five of these consist of three to four complete modules often followed by T-C repeats. The majority of the ferrichrome synthetases only contain three modules, except NRPS2 of *C. heterostrophus*, which has six modules and a T-C repeat, and the ferrichrome synthetase of *C. cinerea*, which only consists of a single module with T-C repeats [63]. An example of a very widespread as well as structurally and functionally conserved siderophore synthetase is NRPS6, encountered in both *C. heterostrophus* and *F. graminearum* among many other species [96, 103]. NRPS6 is present as a single copy in most species and has the same modular makeup consisting of an ATC module followed by a degenerated adenylation domain and a T domain ($_{d}$A, Fig. 15.5).

The evolution of the fungal siderophore synthetases appears to have happened independently from their bacterial counterparts [103]. This conclusion is based on the observation that both the extra and intracellular synthases group together separately from the bacterial synthetases when their sequences are analyzed, and basically suggest that bacteria and fungi independently evolved separate systems for acquiring iron from their surroundings [103]. Phylogenetic analysis further supports the hypothesis that the domain organization within the ferrichrome synthetase family took place as gene duplication or deletion events from a common hexamodular ancestor [63]. The hypothesis is that this ancestor was formed in three duplication steps, starting with the duplication of the first module and followed by duplication of this bimodular gene into a tetramodular gene. Finally, the two terminal domains appear to have been duplicated to form the final hexamodule gene [63].

Studies by Bushley et al. have shown that the ferrichrome synthetases likely consist of two distinct lineages: those homologous to *C. heterostrophus* NRPS2 and those related to *A. nidulans* SidC [63]. The duplication event leading to the two separate lineages appears to have happened after the divergence of ascomycetes and basidiomycetes, as the NRPS2 lineage is only found in ascomycetes while the SidC lineage is found in both. The evolutionary history of the T-C repeats among the ferrichrome synthetases paints a complex picture, involving a series of tandem duplication events of either T-C repeats or entire modules followed by a loss of the A domains. In contrast, there appears little evidence for alterations within the C domains, suggesting that entire domains were duplicated or deleted instead. This together with the hexameric ancestor gene hypothesis also provides a good explanation for the diverse domain population of the synthetases. When looking closer at the modules of the ferrichrome synthetases, especially *C. heterostrophus* NRPS2 and *F. graminearum* NRPS2, from a structural point of view, little can be guessed about their specificity. However, studies of models of the A domains based on existing bacterial structures as scaffolds have indicated that the terminal A domains are specific for the hydroxylated ornithine residues (AHO) while the others are specific for glycine and serine [63].

Structural Studies of Fungal Siderophore Biosynthesis Enzymes

To our knowledge, only two structures of fungal NRPSs have been determined to date, namely the N^5-*L*-ornithine monooxygenase, SidA, from the human pathogen *A. fumigatus* [104] and the third adenylation domain of the epichloënin synthesis NRPS, SidNA3, from the grass endophyte, *Neotyphodium lolii* [105]. Both these structures aid in solving the mechanism behind substrate selectivity by the fungal NRPSs.

SidA binds *L* ornithine and hydroxylates it to form N^5-hydroxy-*L*-ornithine. It is a member of the class B flavoprotein monooxygenases; the structure shows strong structural homology to the ornithine hydroxylase from *Pseudomonas aeruginosa*, PvdA [106]. SidA crystallizes as a homotetramer (Fig. 15.6a), the same oligomeric state as seen in solution [104]. Unlike many other members of the class B flavoprotein monooxygenases, SidA shows strong selectivity for ornithine as a substrate [107–109], a property it shares with PvdA [106]. SidA was crystallized both bound to ornithine and lysine in the active site, which showed that ornithine bound in an extended conformation with the terminal amino group positioned in the ideal location to be hydroxylated by the bound hydroperoxyflavin intermediate [104]. Lysine binds to the enzyme in a similar manner to ornithine, but the terminal amino group is positioned much closer to the hydroperoxyflavin intermediate, leading to much more inefficient hydroxylation [104].

SidN is the NRPS responsible for synthesizing the unusual ferrichrome-type siderophore, epichloënin [26, 110]. The structure of the third adenylation domain of SidN, SidNA3, is so far the only known structure of a eukaryotic NRPS domain (Fig. 15.6b). The structure revealed that fungal NRPS adenylation domains show strong structural similarity to the previously determined structures of bacterial domains [111–113]. NRPS adenylation domains are members of the ANL superfamily of

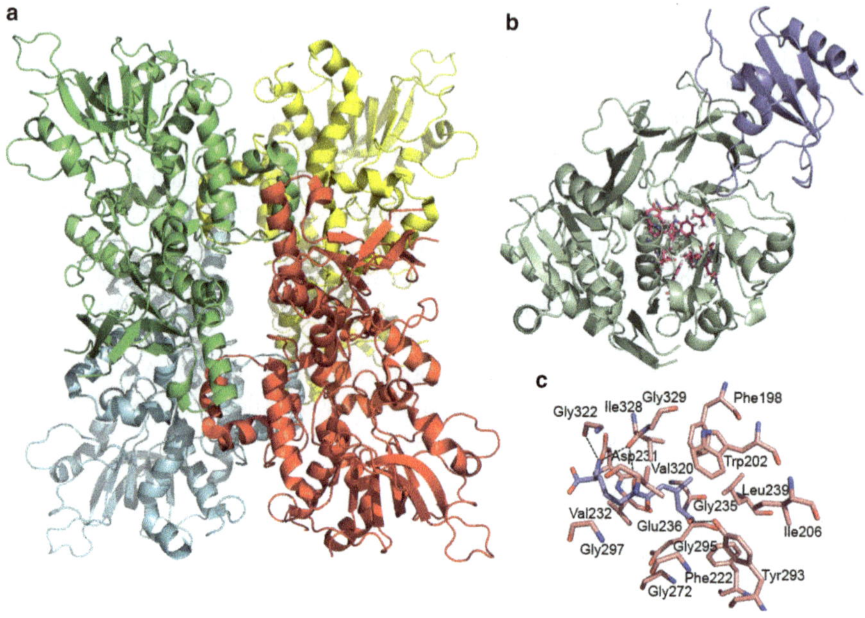

Fig. 15.6 (**a**) Crystal structure of the homotetrameric N^5-*L*-ornithine monooxygenase, SidA, from the human pathogen *A. fumigatus* [104]. (**b**) Structure of the third adenylation domain of the epichloënin synthesis NRPS, SidNA3, from the grass endophyte, *Neotyphodium lolii* [105]. (**c**): Close-up of the active site of SidNA3 showing a modeled substrate molecule, $N^{\delta(\text{delta})}$-*cis*-anhydromevalonyl-$N^{\delta(\text{delta})}$-hydroxy-L-ornithine (*blue sticks*). Residues involved in substrate binding are indicated

adenylating enzymes, which are composed of a core N-terminal domain and a mobile C-terminal lid domain [114]. The mobility of the lid domain allows the enzymes to reconfigure the active site to perform the two reactions that they catalyze [114].

SidNA3 binds and activates N^5-acyl-N^5-hydroxy-*L*-ornithine [105], which in this case is N^5-*trans*-anhydromevalonyl-N^5-hydroxy-*L*-ornithine [110] and there is a large binding pocket in the core domain of SidNA3 to accommodate this amino acid [105]. The structure of SidNA3 was determined without any substrate bound in the binding site, but modeling of the ligand in the binding site suggests that this large substrate binds in an extended conformation in the large binding pocket (Fig. 15.6c) [105]. Interestingly, SidNA3 shows some substrate promiscuity, being able to bind and activate N^5-*trans*-anhydromevalonyl-N^5-hydroxy-*L*-ornithine and N^5-*cis*-anhydromevalonyl-N^5-hydroxy-*L*-ornithine as well, but not N^5-*trans*-anhydromevalonyl-*L*-ornithine (Lee et al., unpublished results).

The large binding pocket of SidNA3 highlights some of the difficulties in an area that is very challenging for fungal NRPS research, prediction of the amino acid substrates from the adenylation domains primary sequences. Following the determination of the initial structure of a bacterial NRPS adenylation domain, PheA [111], methods were developed to predict the specificities of these domains based on the nine residues observed to be lining the substrate binding pocket of PheA [115–119].

These methods tend to work well for bacterial domains but not fungal domains [105]. The SidNA3 structure revealed that the residues lining its binding pocket overlap with the nine residues from PheA, but with the larger pocket extending deeper into the protein being lined by an additional eight residues (Fig. 15.6c) [105]. This indicated that the failure of substrate prediction for fungal domains is most likely the result of the lack of prior specificity data for fungal domains rather than that the fungal domains bind their amino acids in a different manner compared to bacterial domains.

Prediction of Fungal NRPS Products

Natural products synthesized by NRPSs comprise, apart from the siderophores, a wide range of compounds of pharmaceutical interest, such as the antibiotics penicillin and vancomycin, the immunosuppressant cyclosporine, and the anticancer agent bleomycin. One approach to uncover potential new pharmaceuticals involves analysis of NRPS primary sequence data to predict their products using *in silico* methods. The first step in discovering the function of unknown NRPSs is the localization of the genes coding for intact NRPSs or individual NRPS domains within the genomic data. The first implementation of a method specifically designed to locate NRPS genes and domains was NRPS–PKS [120], which is based on a manually curated database of NRPS domains from 22 experimentally characterized biosynthetic gene clusters. The web server takes an amino acid sequence as input and performs pairwise local alignments using BLAST with all domains in the database [121]. Software for gene and domain localization has improved significantly in the decade since NRPS–PKS was launched. Besides benefitting from larger and more diverse datasets, current state-of-the-art prediction servers such as antiSMASH [122] and SMURF [123] also incorporate more advanced machine learning methods such as hidden Markov models [124] to guide predictions. antiSMASH covers both bacterial and fungal genomes and furthermore performs specificity predictions for all a domains located. SMURF predicts genes, and must thus be used in connection with other software for domain localization, but it is tailored specifically to fungal genomes.

The next step following domain localization is to predict the specificities of the individual adenylation (A) domains. Predicting the specificity of an A domain by simple BLAST comparisons of entire domains does not perform well, likely because subtle differences are masked by the overall evolutionary relationship. The problem was first overcome by focusing solely on ten residues in the active site of the A domains, which became known as the specificity-conferring, or Stachelhaus, code [119]. Using this approach, a prediction performance above 80 % was possible, albeit only for bacterial A domains, and is used to this day by web servers such as NRPS–PKS [120]. Similar to the NRPS gene and domain prediction servers mentioned previously, more recent prediction tools such as NRPSpredictor2 [118] and NRPSsp [116] employ machine learning methods such as Support Vector Machines (SVMs) [125] and profile hidden Markov models (pHMMs) [126] to infer the specificities of

A domains from their amino acid sequences. Common to these machine learning based prediction tools is that they use a dataset of A domains with known specificity (called the training set) to infer the parameters of a mathematical model, which can subsequently be used for prediction of the specificities of unknown A domains. Basing prediction tools on well-established machine learning methods such as SVMs and pHMMs has the advantage of having access to theoretically well-founded and computationally efficient training and prediction algorithms.

NRPSpredictor2 extends the specificity-conferring code by considering the identity of all amino acids within an 8 Å radius of the binding site in PheA [118]. This results in 34 positions, each of which is subsequently annotated with 15 numerical attributes describing physiochemical and biochemical properties of the amino acids. An entire A domain is thus represented by $34 \times 15 = 510$ features and is therefore conceptually a point in a 510-dimensional space. NRPSpredictor2 is based on a training set of about 500 A domains with known specificities. The training phase then computes SVMs that separate the corresponding points in the 510-dimensional feature space using hyperplanes. The specificity of an unknown A domain is inferred based on to which sides of the hyperplanes its corresponding point lies in the 510-dimensional feature space.

NRPSsp uses a somewhat different approach, in which the specificity-conferring code and the structure of the active site are ignored. Instead, it considers the amino acid sequences of entire A domains and is based on a training set of almost 1,600 A domains with known specificity. It groups the A domain sequences in the training set according to their specificity and builds a pHMM for each group capturing its sequence signature. The specificity of an unknown A domain is then inferred based on which pHMM resembles it the most. A recent prediction tool [127] also uses pHMMs for specificity prediction and achieves an overall good performance comparable to NRPSsp. It uses a manually curated training set of 1,044 A domains, all related to NRPSs obtained from various sources. The training set of NRPSsp is larger, but it also contains A domains from non-NRPS gene clusters that might affect its predictions.

NRPSpredictor2 and NRPSsp perform well on NRPSs of bacterial origin, but significantly worse on fungal NRPSs. First and foremost, this is due to the lack of experimentally characterized training data. For example, of the almost 1,600 training sequences used in NRPSsp, only 120 are of fungal origin and they bind only 14 different substrates. In order for machine learning methods to make progress on fungal NRPSs, let alone smaller subsets thereof such as siderophores, it is of crucial importance to obtain more data. In general, the prediction of products synthesized by NRPSs is made difficult by many factors. As mentioned previously, many NRPSs do not operate in a simple, colinear fashion where one A domain in the NRPS corresponds to an amino acid in the final product. Many NRPSs are iterative with groups of A domains used multiple times, and this ability cannot currently be predicted based on their amino acid sequences alone [128]. Furthermore, products may undergo posttranslational modifications (e.g., cyclization), rendering it difficult to determine the order in which the building blocks of the product were incorporated by the NRPS. There are numerous examples of cyclic, polycyclic, and branched peptides in the comprehensive NORINE database of non-ribosomal peptides [129]. While the specificity predictors described previously perform well on individual A domains, putting the building blocks correctly together still remains very challenging.

Hybrid peptide synthetases comprising two modules put together from different organisms have been created in an attempt to understand the molecular mechanisms underlying synthesis. Measurements of substrate uptake by such engineered synthetases indicate that the resulting specificities and dynamics are likely more complicated than what is predicted by considering the individual domains independently in isolation [130]. For example, it has been shown that the specificity of one A domain affects the specificity of the downstream module, and in addition that the C domains may play a role in dictating the specificity of their cognate A domains. Perhaps the C domains participate in some sort of quality-control mechanism preventing incorporation of near-cognate amino acids that may be activated by the A domains due to their chemical similarity to the correct substrate.

References

1. Halliwell B, Gutteridge JMC. Oxygen toxicity, oxygen radicals, transition metals and disease. Biochem J. 1984;219:1–14.
2. Lindsay WL, Schwab AP. The chemistry of iron in soils and its availability to plants. J Plant Nutr. 1982;5:821–40.
3. Miethke M. Molecular strategies of microbial iron assimilation: from high-affinity complexes to cofactor assembly systems. Metallomics. 2013;5:15–28.
4. Haas H, Eisendle M, Turgeon BG. Siderophores in fungal physiology and virulence. Annu Rev Phytopathol. 2008;46:149–87.
5. Grissa I, Bidard F, Grognet P, Grossetete S, Silar P. The Nox/Ferric reductase/Ferric reductase-like families of Eumycetes. Fungal Biol. 2010;114:766–77.
6. Blatzer M, Schrettl M, Sarg B, Lindner HH, Pfaller K, Haas H. SidL, an Aspergillus fumigatus transacetylase involved in biosynthesis of the siderophores ferricrocin and hydroxyferricrocin. Appl Environ Microbiol. 2011;77:4959–66.
7. Kosman DJ. Redox cycling in iron uptake, efflux, and trafficking. J Biol Chem. 2010;285:26729–35.
8. Kosman DJ. Iron metabolism in aerobes: managing ferric iron hydrolysis and ferrous iron autoxidation. Coord Chem Rev. 2013;257:210–7.
9. Schrettl M, Bignell E, Kragl C, Joechl C, Rogers T, Arst HN, et al. Siderophore biosynthesis but not reductive iron assimilation is essential for Aspergillus fumigates virulence. J Exp Med. 2004;200:1213–9.
10. Eichhorn H, Lessing F, Winterberg B, Schirawski J, Kaemper J, Mueller P, Kahmann R. A ferroxidation/permeation iron uptake system is required for virulence in Ustilago maydis. Plant Cell. 2006;18:3332–45.
11. Park Y-S, Kim J-H, Cho J-H, Chang H-I, Kim S-W, Paik H-D, et al. Physical and functional interaction of FgFtr1-FgFet1 and FgFtr2-FgFet2 is required for iron uptake in Fusarium graminearum. Biochem J. 2007;408:97–104.
12. Jung WH, Hu G, Kuo W, Kronstad JW. Role of ferroxidases in iron uptake and virulence of Cryptococcus neoformans. Eukaryot Cell. 2009;8:1511–20.
13. Kobayashi T, Nishizawa NK. Iron uptake, translocation, and regulation in higher plants. Annu Rev Plant Biol. 2012;63:131–52.
14. Saha R, Saha N, Donofrio RS, Bestervelt LL. Microbial siderophores: a mini review. J Basic Microbiol. 2013;53:303–17.

15. Van Ho A, Ward DM, Kaplan J. Transition metal transport in yeast. Annu Rev Microbiol. 2002;56:237–61.
16. Canessa P, Larrondo LF. Environmental responses and the control of iron homeostasis in fungal systems. Appl Microbiol Biotechnol. 2013;97:939–55.
17. Wallner A, Blatzer M, Schrettl M, Sarg B, Lindner H, Haas H. Ferricrocin, a siderophore involved in intra- and transcellular iron distribution in Aspergillus fumigatus. Appl Environ Microbiol. 2009;75:4194–6.
18. Ardon O, Nudelman R, Caris C, Libman J, Shanzer A, Chen YN, Hadar Y. Iron uptake in Ustilago maydis: tracking the iron path. J Bacteriol. 1998;180:2021–6.
19. Kragl C, Schrettl M, Abt B, Sarg B, Lindner HH, Haas H. EstB-mediated hydrolysis of the siderophore triacetylfusarinine C optimizes iron uptake of Aspergillus fumigatus. Eukaryot Cell. 2007;6:1278–85.
20. Gsaller F, Eisendle M, Lechner BE, Schrettl M, Lindner H, Mueller D, et al. The interplay between vacuolar and siderophore-mediated iron storage in Aspergillus fumigatus. Metallomics. 2012;4:1262–70.
21. Eisendle M, Schrettl M, Kragl C, Mueller D, Illmer P, Haas H. The intracellular siderophore ferricrocin is involved in iron storage, oxidative-stress resistance, germination, and sexual development in Aspergillus nidulans. Eukaryot Cell. 2006;5:1596–603.
22. Oide S, Krasnoff SB, Gibson DM, Turgeon BG. Intracellular siderophores are essential for ascomycete sexual development in heterothallic Cochliobolus heterostrophus and homothallic Gibberella zeae. Eukaryot Cell. 2007;6:1339–53.
23. Hof C, Eisfeld K, Antelo L, Foster AJ, Anke H. Siderophore synthesis in Magnaporthe grisea is essential for vegetative growth, conidiation and resistance to oxidative stress. Fungal Genet Biol. 2009;46:321–32.
24. Schrettl M, Bignell E, Kragl C, Sabiha Y, Loss O, Eisendle M, et al. Distinct roles for intra- and extracellular siderophores during Aspergillus fumigatus infection. PLoS Pathog. 2007;3:1195–207.
25. Matzanke BF, Bill E, Trautwein AX, Winkelmann G. Role of siderophores in iron storage in spores of Neurospora crassa and Aspergillus ochraceus. J Bacteriol. 1987;169:5873–6.
26. Johnson LJ, Koulman A, Christensen M, Lane GA, Fraser K, Forester N, et al. An extracellular siderophore is required to maintain the mutualistic interaction of Epichloë festucae with Lolium perenne. PLoS Pathog. 2013;9:e1003332.
27. Kieu NP, Aznar A, Segond D, Rigault M, Simond-Cote E, Kunz C, et al. Iron deficiency affects plant defence responses and confers resistance to Dickeya dadantii and Botrytis cinerea. Mol Plant Pathol. 2012;13:816–27.
28. Chen LH, Lin CH, Chung KR. A nonribosomal peptide synthetase mediates siderophore production and virulence in the citrus fungal pathogen Alternaria alternata. Mol Plant Pathol. 2013;14:497–505.
29. Lee BN, Kroken S, Chou DYT, Robbertse B, Yoder OC, Turgeon BG. Functional analysis of all nonribosomal peptide synthetases in Cochliobolus heterostrophus reveals a factor, NPS6, involved in virulence and resistance to oxidative stress. Eukaryot Cell. 2005;4:545–55.
30. Leong SA, Winkelmann G. Molecular biology of iron transport in fungi. Met Ions Biol Syst. 1998;35(35):147–86.
31. Schrettl M, Haas H. Iron homeostasis-Achilles' heel of Aspergillus fumigatus? Curr Opin Microbiol. 2011;14:400–5.
32. Correnti C, Strong RK. Mammalian siderophores, siderophore-binding lipocalins, and the labile iron pool. J Biol Chem. 2012;287:13524–31.
33. Clifton MC, Corrent C, Strong RK. Siderocalins: siderophore-binding proteins of the innate immune system. Biometals. 2009;22:557–64.
34. Bao G, Clifton M, Hoette TM, Mori K, Deng S-X, Qiu A, et al. Iron traffics in circulation bound to a siderocalin (Ngal)-catechol complex. Nat Chem Biol. 2010;6:602–9.
35. Correnti C, Richardson V, Sia AK, Bandaranayake AD, Ruiz M, Rahmanto YS, et al. Siderocalin/Lcn2/NGAL/24p3 does not drive apoptosis through gentisic acid mediated iron withdrawal in hematopoietic cell lines. PLoS One. 2012;7(8):e43696.

36. Cherayil BJ. The role of iron in the immune response to bacterial infection. Immunol Res. 2011;50:1–9.
37. Haas H. Iron—a key nexus in the virulence of Aspergillus fumigatus. Front Microbiol. 2012;3:28.
38. Linde J, Hortschansky P, Fazius E, Brakhage AA, Guthke R, Haas H. Regulatory interactions for iron homeostasis in Aspergillus fumigatus inferred by a systems biology approach. BMC Syst Biol. 2012;6:6.
39. Haas H, Angermayr K, Stoffler G. Molecular analysis of a Penicillium chrysogenum GATA factor encoding gene (sreP) exhibiting significant homology to the Ustilago maydis urbs1 gene. Gene. 1997;184:33–7.
40. Hortschansky P, Eisendle M, Al-Abdallah Q, Schmidt AD, Bergmann S, Thoen M, et al. Interaction of HapX with the CCAAT-binding complex—a novel mechanism of gene regulation by iron. EMBO J. 2007;26:3157–68.
41. Voisard C, Wang J, McEvoy JL, Xu PL, Leong SA. urbs1, a gene regulating siderophore biosynthesis in Ustilago maydis, encodes a protein similar to the erythroid transcription factor GATA-1. Mol Cell Biol. 1993;13:7091–100.
42. Haas H, Zadra I, Stoffler G, Angermayr K. The Aspergillus nidulans GATA factor SREA is involved in regulation of siderophore biosynthesis and control of iron uptake. J Biol Chem. 1999;274:4613–9.
43. Schrettl M, Kim HS, Eisendle M, Kragl C, Nierman WC, Heinekamp T, et al. SreA-mediated iron regulation in Aspergillus fumigatus. Mol Microbiol. 2008;70:27–43.
44. Zhou LW, Haas H, Marzluf GA. Isolation and characterization of a new gene, sre, which encodes a GATA-type regulatory protein that controls iron transport in Neurospora crassa. Mol Gen Genet. 1998;259:532–40.
45. Lan CY, Rodarte G, Murillo LA, Jones T, Davis RW, Dungan J, et al. Regulatory networks affected by iron availability in Candida albicans. Mol Microbiol. 2004;53:1451–69.
46. Gauthier GM, Sullivan TD, Gallardo SS, Brandhorst TT, Vanden Wymelenberg AJ, Cuomo CA, et al. SREB, a GATA transcription factor that directs disparate fates in blastomyces dermatitidis including morphogenesis and siderophore biosynthesis. PLoS Pathog. 2010;6(4):e1000846.
47. Pelletier B, Beaudoin J, Mukai Y, Labbe S. Fep1, an iron sensor regulating iron transporter gene expression in Schizosaccharomyces pombe. J Biol Chem. 2002;277:22950–8.
48. Lopez-Berges MS, Capilla J, Turra D, Schafferer L, Matthijs S, Joechl C, et al. HapX-mediated iron homeostasis is essential for rhizosphere competence and virulence of the soil-borne pathogen Fusarium oxysporum. Plant Cell. 2012;24:3805–22.
49. Tanaka A, Kato M, Nagase T, Kobayashi T, Tsukagoshi N. Isolation of genes encoding novel transcription factors which interact with the Hap complex from Aspergillus species. Biochim Biophys Acta. 2002;1576:176–82.
50. Forsburg SL, Guarente L. Identification and characterization of HAP4: a third component of the CCAAT-bound HAP2/HAP3 heteromer. Genes Dev. 1989;3:1166–78.
51. Knight SAB, Lesuisse E, Stearman R, Klausner RD, Dancis A. Reductive iron uptake by Candida albicans: role of copper, iron and the TUP1 regulator. Microbiology. 2002;148:29–40.
52. Znaidi S, Pelletier B, Mukai Y, Labbe S. The Schizosaccharomyces pombe corepressor Tup11 interacts with the iron-responsive transcription factor Fep1. J Biol Chem. 2004;279:9462–74.
53. Hsu PC, Yang CY, Lan CY. Candida albicans Hap43 is a repressor induced under low-iron conditions and is essential for iron-responsive transcriptional regulation and virulence. Eukaryot Cell. 2011;10:207–25.
54. Liu H, Gravelat FN, Chiang LY, Chen D, Vanier G, Ejzykowicz DE, et al. Aspergillus fumigatus AcuM regulates both iron acquisition and gluconeogenesis. Mol Microbiol. 2010;78:1038–54.
55. Boukhalfa H, Crumbliss AL. Chemical aspects of siderophore mediated iron transport. Biometals. 2002;15:325–39.

56. Oves-Costales D, Kadi N, Challis GL. The long-overlooked enzymology of a nonribosomal peptide synthetase-independent pathway for virulence-conferring siderophore biosynthesis. Chem Commun (Camb). 2009;43:6530–41.
57. Haas H. Molecular genetics of fungal siderophore biosynthesis and uptake: the role of siderophores in iron uptake and storage. Appl Microbiol Biotechnol. 2003;62:316–30.
58. Winkelmann G. Structures and functions of fungal siderophores containing hydroxamate and complex one type iron binding ligands. Mycol Res. 1992;96:529–34.
59. Renshaw JC, Robson GD, Trinci APJ, Wiebe MG, Livens FR, Collison D, Taylor RJ. Fungal siderophores: structures, functions and applications. Mycol Res. 2002;106:1123–42.
60. Engwilmot DL, Rahman A, Mendenhall JV, Grayson SL, Vanderhelm D. Molecular structure of ferric neurosporin, a minor siderophore-like compound containing N.delta.-hydroxy-D-ornithine. J Am Chem Soc. 1984;106:1285–90.
61. Neilands JB. A crystalline organo-iron pigment from a rust fungus (Ustilago sphaerogena). J Am Chem Soc. 1952;74:4846–7.
62. Kim J-H, Kim H-W, Heo D-H, Chang M, Baek I-J, Yun C-W. FgEnd1 is a putative component of the endocytic machinery and mediates ferrichrome uptake in F. graminearum. Curr Genet. 2009;55:593–600.
63. Bushley KE, Ripoll DR, Turgeon BG. Module evolution and substrate specificity of fungal nonribosomal peptide synthetases involved in siderophore biosynthesis. BMC Evol Biol. 2008;8:328.
64. Hansen FT, Sørensen JL, Giese H, Sondergaard TE, Frandsen RJ. Quick guide to polyketide synthase and nonribosomal synthetase genes in Fusarium. Int J Food Microbiol. 2012;155: 128–36.
65. Lehner SM, Atanasova L, Neumann NKN, Krska R, Lemmens M, Druzhinina IS, Schuhmacher R. Isotope-assisted screening for iron-containing metabolites reveals a high degree of diversity among known and unknown siderophores produced by Trichoderma spp. Appl Environ Microbiol. 2013;79:18–31.
66. Diekmann H. Metabolic products of microorganisms. 81. Occurrence and structures of coprogen B and dimerum acid. Arch Mikrobiol. 1970;73:65–76.
67. Atkin CL, Neilands JB. Rhodotorulic acid, a diketopiperazine dihydroxamic acid with growth-factor activity. I. Isolation and characterization. Biochemistry. 1968;7:3734–6.
68. Antelo L, Hof C, Welzel K, Eisfeld K, Sterner O, Anke H. Siderophores produced by Magnaporthe grisea in the presence and absence of iron. Z Naturforsch C. 2006;61:461–4.
69. Jalal MA, Love SK, van der Helm D. N alpha-dimethylcoprogens. Three novel trihydroxamate siderophores from pathogenic fungi. Biol Met. 1988;1:4–8.
70. Winkelmann G, Zahner H. Metabolic products of microorganisms. 115. Uptake of iron by Neurospora crassa. 1. To the specificity of iron transport. Arch Mikrobiol. 1973;88:49–60.
71. Ernst JF, Winkelmann G. Enzymatic release of iron from sideramines in fungi NADH:sideramine oxidoreductase in Neurospora crassa. Biochim Biophys Acta. 1977;500:27–41.
72. Drechsel H, Metzger J, Freund S, Jung G, Boelaert JR, Winkelmann G. Rhizoferrin: a novel siderophore from the fungus Rhizopus microsporus var. rhizopodiformis. Biol Met. 1991;4:238–43.
73. Thieken A, Winkelmann G. Rhizoferrin: a complex one type siderophore of the mocorales and entomophthorales (Zygomycetes). FEMS Microbiol Lett. 1992;94:37–42.
74. Konetschnyrapp S, Jung G, Meiwes J, Zahner H. Staphyloferrin A: a structurally new siderophore from staphylococci. Eur J Biochem. 1990;191:65–74.
75. Meiwes J, Fiedler HP, Haag H, Zahner H, Konetschnyrapp S, Jung G. Isolation and characterization of staphyloferrin A, a compound with siderophore activity from Staphylococcus hyicus DSM 20459. FEMS Microbiol Lett. 1990;67:201–5.
76. Cotton JL, Tao J, Balibar CJ. Identification and characterization of the Staphylococcus aureus gene cluster coding for staphyloferrin A. Biochemistry. 2009;48:1025–35.
77. Challis GL. A widely distributed bacterial pathway for siderophore biosynthesis independent of nonribosomal peptide synthetases. Chembiochem. 2005;6:601–11.

78. Yin W-B, Baccile JA, Bok JW, Chen Y, Keller NP, Schroeder FC. A nonribosomal peptide synthetase-derived iron(III) complex from the pathogenic fungus Aspergillus fumigatus. J Am Chem Soc. 2013;135:2064–7.
79. Arai K, Sato S, Shimizu S, Nitta K, Yamamoto Y. Metabolic products of Aspergillus terreus: 7. Astechrome, an iron-containing metabolite of the strain IFO 6123. Chem Pharm Bull. 1981;29:1510–7.
80. van Berkel WJH, Kamerbeek NM, Fraaije MW. Flavoprotein monooxygenases, a diverse class of oxidative biocatalysts. J Biotechnol. 2006;124:670–89.
81. Mei BG, Budde AD, Leong SA. sid1, a gene initiating siderophore biosynthesis in Ustilago maydis: molecular characterization, regulation by iron, and role in phytopathogenicity. Proc Natl Acad Sci U S A. 1993;90:903–7.
82. Eisendle M, Oberegger H, Zadra I, Haas H. The siderophore system is essential for viability of Aspergillus nidulans: functional analysis of two genes encoding L-ornithine N-5-monooxygenase (sidA) and a non-ribosomal peptide synthetase (sidC). Mol Microbiol. 2003;49:359–75.
83. Yamada O, Nan SN, Akao T, Tominaga M, Watanabe H, Satoh T, et al. dffA gene from Aspergillus oryzae encodes L-ornithine N-5-oxygenase and is indispensable for deferriferrichrysin biosynthesis. J Biosci Bioeng. 2003;95:82–8.
84. Greenshields DL, Liu GS, Feng J, Selvaraj G, Wei YD. The siderophore biosynthetic gene SID1, but not the ferroxidase gene FET3, is required for full Fusarium graminearum virulence. Mol Plant Pathol. 2007;8:411–21.
85. Turgeon BG, Oide S, Bushley K. Creating and screening Cochliobolus heterostrophus non-ribosomal peptide synthetase mutants. Mycol Res. 2008;112:200–6.
86. Winterberg B, Uhlmann S, Linne U, Lessing F, Marahiel MA, Eichhorn H, et al. Elucidation of the complete ferrichrome A biosynthetic pathway in Ustilago maydis. Mol Microbiol. 2010;75:1260–71.
87. Welzel K, Eisfeld K, Antelo L, Anke T, Anke H. Characterization of the ferrichrome A biosynthetic gene cluster in the homobasidiomycete Omphalotus olearius. FEMS Microbiol Lett. 2005;249:157–63.
88. Yasmin S, Alcazar-Fuoli L, Gruendlinger M, Puempel T, Cairns T, Blatzer M, et al. Mevalonate governs interdependency of ergosterol and siderophore biosyntheses in the fungal pathogen Aspergillus fumigatus. Proc Natl Acad Sci U S A. 2012;109:E497–504.
89. Munawar A, Marshall JW, Cox RJ, Bailey AM, Lazarus CM. Isolation and characterisation of a ferrirhodin synthetase gene from the sugarcane pathogen Fusarium sacchari. Chembiochem. 2013;14:388–94.
90. Heymann P, Gerads M, Schaller M, Dromer F, Winkelmann G, Ernst JF. The siderophore iron transporter of Candida albicans (Sit1p/Arn1p) mediates uptake of ferrichrome-type siderophores and is required for epithelial invasion. Infect Immun. 2002;70:5246–55.
91. Lesuisse E, Simon-Casteras M, Labbe P. Siderophore-mediated iron uptake in Saccharomyces cerevisiae: the SIT1 gene encodes a ferrioxamine B permease that belongs to the major facilitator superfamily. Microbiology. 1998;144:3455–62.
92. Yun CW, Tiedeman JS, Moore RE, Philpott CC. Siderophore-iron uptake in Saccharomyces cerevisiae - Identification of ferrichrome and fusarinine transporters. J Biol Chem. 2000;275:16354–9.
93. Raymond-Bouchard I, Carroll CS, Nesbitt JR, Henry KA, Pinto LJ, Moinzadeh M, et al. Structural requirements for the activity of the MirB Ferrisiderophore transporter of Aspergillus fumigatus. Eukaryot Cell. 2012;11:1333–44.
94. Tobiasen C, Aahman J, Ravnholt KS, Bjerrum MJ, Grell MN, Giese H. Nonribosomal peptide synthetase (NPS) genes in Fusarium graminearum, F. culmorum and F. pseudograminearium and identification of NPS2 as the producer of ferricrocin. Curr Genet. 2007;51:43–58.
95. Anke H, Kinn J, Bergquist KE, Sterner O. Production of siderophores by strains of the genus Trichoderma. Isolation and characterization of the new lipophilic coprogen derivative, palmitoylcoprogen. Biol Met. 1991;4:176–80.

96. Oide S, Moeder W, Krasnoff S, Gibson D, Haas H, Yoshioka K, Turgeon BG. NPS6, encoding a nonribosomal peptide synthetase involved in siderophore-mediated iron metabolism, is a conserved virulence determinant of plant pathogenic ascomycetes. Plant Cell. 2006;18:2836–53.
97. Reiber K, Reeves EP, Neville CM, Winkler R, Gebhardt P, Kavanagh K, Doyle S. The expression of selected non-ribosomal peptide synthetases in Aspergillus fumigatus is controlled by the availability of free iron. FEMS Microbiol Lett. 2005;248:83–91.
98. Varga J, Kocsube S, Toth B, Mesterhazy A. Nonribosomal peptide synthetase genes in the genome of Fusarium graminearum, causative agent of wheat head blight. Acta Biol Hung. 2005;56:375–88.
99. Strieker M, Tanovic A, Marahiel MA. Nonribosomal peptide synthetases: structures and dynamics. Curr Opin Struct Biol. 2010;20:234–40.
100. Finking R, Marahiel MA. Biosynthesis of nonribosomal peptides. Annu Rev Microbiol. 2004;58:453–88.
101. Schwecke T, Gottling K, Durek P, Duenas I, Kaufer NF, Zock-Emmenthal S, et al. Nonribosomal peptide synthesis in Schizosaccharomyces pombe and the architectures of ferrichrome-type siderophore synthetases in fungi. Chembiochem. 2006;7:612–22.
102. Johnson L. Iron and siderophores in fungal-host interactions. Mycol Res. 2008;112:170–83.
103. Bushley KE, Turgeon BG. Phylogenomics reveals subfamilies of fungal nonribosomal peptide synthetases and their evolutionary relationships. BMC Evol Biol. 2010;10:26.
104. Franceschini S, Fedkenheuer M, Vogelaar NJ, Robinson HH, Sobrado P, Mattevi A. Structural insight into the mechanism of oxygen activation and substrate selectivity of flavin-dependent N-hydroxylating monooxygenases. Biochemistry. 2012;51:7043–5.
105. Lee TV, Johnson LJ, Johnson RD, Koulman A, Lane GA, Lott JS, Arcus VL. Structure of a eukaryotic nonribosomal peptide synthetase adenylation domain that activates a large hydroxamate amino acid in siderophore biosynthesis. J Biol Chem. 2010;285:2415–27.
106. Olucha J, Meneely KM, Chilton AS, Lamb AL. Two structures of an N-hydroxylating flavoprotein monooxygenase: ornithine hydroxylase from Pseudomonas aeruginosa. J Biol Chem. 2011;286:31789–98.
107. Chocklett SW, Sobrado P. Aspergillus fumigatus SidA is a highly specific ornithine hydroxylase with bound flavin cofactor. Biochemistry. 2010;49:6777–83.
108. Mayfield JA, Frederick RE, Streit BR, Wencewicz TA, Ballou DP, DuBois JL. Comprehensive spectroscopic, steady state, and transient kinetic studies of a representative siderophore-associated flavin monooxygenase. J Biol Chem. 2010;285:30375–88.
109. Romero E, Fedkenheuer M, Chocklett SW, Qi J, Oppenheimer M, Sobrado P. Dual role of NADP(H) in the reaction of a flavin dependent N-hydroxylating monooxygenase. Biochim Biophys Acta. 2012;1824:850–7.
110. Koulman A, Lee TV, Fraser K, Johnson L, Arcus V, Lott JS, et al. Identification of extracellular siderophores and a related peptide from the endophytic fungus Epichlöe festucae in culture and endophyte-infected Lolium perenne. Phytochemistry. 2012;75:128–39.
111. Conti E, Stachelhaus T, Marahiel MA, Brick P. Structural basis for the activation of phenylalanine in the non-ribosomal biosynthesis of gramicidin S. EMBO J. 1997;16:4174–83.
112. May JJ, Kessler N, Marahiel MA, Stubbs MT. Crystal structure of DhbE, an archetype for aryl acid activating domains of modular nonribosomal peptide synthetases. Proc Natl Acad Sci U S A. 2002;99:12120–5.
113. Tanovic A, Samel SA, Essen LO, Marahiel MA. Crystal structure of the termination module of a nonribosomal peptide synthetase. Science. 2008;321:659–63.
114. Gulick AM. Conformational dynamics in the acyl-CoA synthetases, adenylation domains of non-ribosomal peptide synthetases, and firefly luciferase. ACS Chem Biol. 2009;4:811–27.
115. Challis GL, Ravel J. Coelichelin, a new peptide siderophore encoded by the Streptomyces coelicolor genome: structure prediction from the sequence of its non-ribosomal peptide synthetase. FEMS Microbiol Lett. 2000;187:111–4.
116. Prieto C, Garcia-Estrada C, Lorenzana D, Martin JF. NRPSsp: non-ribosomal peptide synthase substrate predictor. Bioinformatics. 2012;28:426–7.

117. Rausch C, Weber T, Kohlbacher O, Wohlleben W, Huson DH. Specificity prediction of adenylation domains in nonribosomal peptide synthetases (NRPS) using transductive support vector machines (TSVMs). Nucleic Acids Res. 2005;33:5799–808.
118. Röttig M, Medema MH, Blin K, Weber T, Rausch C, Kohlbacher O. NRPSpredictor2-a web server for predicting NRPS adenylation domain specificity. Nucleic Acids Res. 2011;39: W362–7.
119. Stachelhaus T, Mootz HD, Marahiel MA. The specificity-conferring code of adenylation domains in nonribosomal peptide synthetases. Chem Biol. 1999;6:493–505.
120. Ansari MZ, Yadav G, Gokhale RS, Mohanty D. NRPS-PKS: a knowledge-based resource for analysis of NRPS/PKS megasynthases. Nucleic Acids Res. 2004;32:W405–13.
121. Altschul SF, Gish W, Miller W, Myers EW, Lipman DJ. Basic local alignment search tool. J Mol Biol. 1990;215:403–10.
122. Blin K, Medema MH, Kazempour D, Fischbach MA, Breitling R, Takano E, Weber T. antiSMASH 2.0—a versatile platform for genome mining of secondary metabolite producers. Nucleic Acids Res. 2013;41:W204–12.
123. Khaldi N, Seifuddin FT, Turner G, Haft D, Nierman WC, Wolfe KH, Fedorova ND. SMURF: genomic mapping of fungal secondary metabolite clusters. Fungal Genet Biol. 2010;47:736–41.
124. Rabiner LR. A tutorial on hidden Markov models and selected applications in speech recognition. Proc IEEE. 1989;77:257–86.
125. Cortes C, Vapnik V. Support-vector networks. Mach Learn. 1995;20:273–97.
126. Eddy SR. Profile hidden Markov models. Bioinformatics. 1998;14:755–63.
127. Khayatt BI, Overmars L, Siezen RJ, Francke C. Classification of the adenylation and acyltransferase activity of NRPS and PKS systems using ensembles of substrate specific hidden Markov models. PLoS One. 2013;8(4):e62136.
128. Mootz HD, Schwarzer D, Marahiel MA. Ways of assembling complex natural products on modular nonribosomal peptide synthetases. Chembiochem. 2002;3:490–504.
129. Caboche S, Pupin M, Leclère V, Fontaine A, Jacques P, Kucherov G. NORINE: a database of nonribosomal peptides. Nucleic Acids Res. 2008;36:D326–31.
130. Doekel S, Marahiel MA. Dipeptide formation on engineered hybrid peptide synthetases. Chem Biol. 2000;7:373–84.

Index

J.-F. Martín et al. (eds.), *Biosynthesis and Molecular Genetics of Fungal Secondary Metabolites*, Fungal Biology, DOI 10.1007/978-1-4939-1191-2,

B

G

R

S

T

V

X

GPSR Compliance
The European Union's (EU) General Product Safety Regulation (GPSR) is a set of rules that requires consumer products to be safe and our obligations to ensure this.

If you have any concerns about our products, you can contact us on

ProductSafety@springernature.com

In case Publisher is established outside the EU, the EU authorized representative is:

Springer Nature Customer Service Center GmbH
Europaplatz 3
69115 Heidelberg, Germany

www.ingramcontent.com/pod product compliance
Lightning Source LLC
Chambersburg PA
CBHW060743150926
51501CB00003B/81
* 9 7 8 1 4 9 3 9 1 1 9 0 5 *